E. Roy John
Editor

Machinery of the Mind

Data, Theory, and Speculations About Higher Brain Function

With Co-Editors
Thalia Harmony
Leslie S. Prichep
Mitchell Valdés-Sosa
Pedro A. Valdés-Sosa

Based on the First International Conference on Machinery of the Mind, February 25–March 3, 1989, Havana City, Cuba

With 250 Illustrations

1990

Birkhäuser
Boston • Basel • Berlin

E. Roy John, M.D.
Brain Research Laboratories, New York University Medical Center, New York,
New York 10020, USA

Thalia Harmony
National University of Mexico, Ciudad de Los Deportes, Mexico

Leslie S. Prichep
Brain Research Laboratories, New York University Medical Center, New York, New York
10020, USA

Mitchell Valdés-Sosa
Centro Nacional de Investigaciones Científicas Ciudad de la Habana, Cuba

Pedro A. Valdés-Sosa
Centro Nacional de Investigaciones Científicas Ciudad de la Habana, Cuba

Library of Congress Cataloging-in-Publication Data
Machinery of the mind: data, theory, and speculations about higher
   brain function / E. Roy John ... [et al.], editors.
      p. cm.
   Includes bibliographical references and index.
   ISBN 0-8176-3461-4
   1. Brain—Congresses. 2. Cognition—Congresses.
3. Electrophysiology—Congresses. 4. Higher nervous activity—Congresses.
5. Electroencephalography—Congresses. I. John, E. Roy (Erwin Roy)
QP395.M33 1991
612.8'2—dc20   90-38098

QP
395
.M33
1989

Typeset by Ampersand Publisher Services, Inc., Owings Mills, Maryland.
Printed and bound by Braun-Brumfield, Inc., Ann Arbor, Michigan.
Printed in the United States of America.

9 8 7 6 5 4 3 2 1

ISBN 0-8176-3461-4
ISBN 3-7643-3461-4

Preface

In the spring of 1987, I was in Havana, Cuba, where I was participating in planning a large-scale longitudinal study of the neurophysiological, neurochemical, and behavioral characteristics of cohorts of patients with cerebrovascular disease, depression, senile dementia, schizophrenia, or learning disabilities; and also part of this study were their first-degree blood relatives. This study was the outgrowth of a long-term project on the practical application of computer methods for the evaluation of brain electrical activity related to anatomical integrity, maturational development, and sensory, perceptual, and cognitive processes, especially in children. For many years, that project had been supported by the United Nations Development Program (UNDP), the National Scientific Research Center of Cuba (CNIC), and the Ministries of Public Health and of Education of Cuba. Since its inception, I had served as a technical advisor to the UNDP project.

When the project began, I became acquainted with Dr. Jose M. Miyar Barrueco, who was at that time the Rector of the Medical School of the University of Havana. Because of his keen interest in the new computer technology and its potential utility in developing countries, we met from time to time during my visits. These occasional meetings continued after he became Secretary of the Cuban Council of State, so that he could remain apprised of progress and problems with which he might help.

Thus it happened that during my visit in the spring of 1987, I found myself at lunch with Dr. Miyar and Dr. Selman, a physician who is an advisor to Premier Fidel Castro on medical scientific matters. In the course of the conversation, Dr. Miyar asked whether I was satisfied with the rate of progress on the medical computer project and the longitudinal study. I said I felt that the groundwork was being laid for what might well become an important contribution to our understanding of the major cognitive and psychiatric disorders and that in view of the large scale of this undertaking, the progress was quite impressive.

I went on to confess that although world neuroscience was making great strides toward better understanding of basic brain mechanisms on the one hand and of the solution of practical medical problems on the other hand, I felt that the problem that was of greatest human relevance and was indeed the central, ultimate question to which the modern tools of neuroscience should be applied—the physiological basis of subjective experience, the material origin of human self-awareness—had been almost completely neglected. We knew little more about this topic than had Aristotle or Leibnitz or Descartes, in spite of the efflorescence of information on the unitary or molecular level. The problems of integrative processes and of the transformation of ionic currents and molecular syntheses to personal subjective experience were of little concern for contemporary research.

Was this because the proscription of such questions by behaviorism had effectively legislated them out of the legitimate domain of experimental neuroscience, relegating them to the speculations of philosophers? Or was it because the numerous powerful new methods that led to the remarkable explosion of factual information about events on the microscopic and molecular levels in the brain have focused us on a reductionistic level, yielding little better understanding about integrative processes?

To my surprise, Dr. Miyar and Dr. Selman not only became intrigued by the scientific problem but were quite disturbed by the perspective that I had presented of a widespread lack of concern with such questions in contemporary neuroscience. The discussion gradually focused on whether anything useful might be done to change this situation. One of the ideas that seemed attractive was that of convening an international meeting of eminent researchers in neuroscience to focus upon these issues and thereby generate some worthwhile insights. Further, a book containing the presentations of these workers might serve to rekindle student interest in more molar neuroscientific pursuits. In order to facilitate such an endeavor, Dr. Miyar proposed that, if I were willing to help organize it, the meeting would be sponsored by the National Scientific Research Center and the Ministry of Public Health, which would defray all related expenses.

Thus was born the First Havana International Workshop on the Machinery of the Mind. The format decided upon by myself and Dr. Mitchell Valdes, director of the Laboratory of Neurophysiology at CNIC who joined me in the organizational tasks, was that there should be an International Conference on Advances in Neuroscience, to be called Neurosciences '89. In addition to the papers delivered by participants in Neurosciences '89, about 50 experts spanning many of the major areas of rapid research progress would be invited from all over the world to give plenary lectures. The purpose of these didactic lectures was to provide an overview of some of the current research frontiers.

A subset of these international experts was selected because of their expressed interest and potential relevance to the topic of the physiological basis of conscious experience. The lectures that they delivered to the plenary sessions of Neurosciences '89 represented that aspect of their work and current thinking that seemed to them most relevant to this problem. This volume contains those lectures. The abstracts of all of the papers presented in the platform and poster sessions of Neurosciences '89 are published in an issue of the *International Journal of Neurosciences*.

After the conclusion of Neurosciences '89, most of the contributors to this volume, joined by a few Cuban and international scientists whose works are not included here, spent 3 days in the workshop on the Machinery of the Mind. This closed workshop, held in a small and comfortable conference center, had an unusual format. On the first day, those participants who chose to do so presented a proposal for an experiment that if it were to be carried out, would in their opinion lead to improved understanding of the origin of conscious subjective experience. Presentations were brief and conceptual. Intensive, open-ended discussions followed each presentation, as group members explored the experimental ideas.

After these initial proposals had been aired, a unique aspect of the workshop began. A number of "common interest" groups were formed to focus upon a few issues that had surfaced and that seemed of critical importance to sufficient persons to warrant formation of a discussion group. The charge to each discussion group was first to design an experiment that the group members agreed would cast light upon the selected issue, were it to be actually performed. Second, the group was to explore concrete ways in which the experiment designed by the group could actually be carried out as a multilaboratory collaboration. After 2 days of these intensive experimental design sessions, the groups reassembled and presented the products of their deliberations to the full workshop. Further suggestions and modifications were made and a "final" design was agreed upon.

For many of us, this workshop was an extraordinary experience. As one might have predicted, there was an initial period in which some of us felt quite uncomfortable. People argued about nonessential points, and there was the amount of pontificating that a group of individualistic prima donnas might be expected to produce. One participant burst out in rage about how stressful he found the lack of structure. Then, suddenly, when the detailed experimental planning began, the atmosphere abruptly changed. Competitive, self-assertive, declamatory behavior disappeared as people became intrigued by the challenge of designing experiments that might effectively address the theoretical issues. Constructive contributions to complex, multidisciplinary problems began to be made by experienced workers who had been invited precisely because they were so knowledgeable in the relevant experimental techniques.

As a result, experiments were constructed that were far different from anything that the individual collaborators would have conceived by themselves before these interactions. Scientists from different disciplines and different countries, with different experimental strengths and different scientific traditions, found that there was merit in unfamiliar viewpoints and experimental benefit to be derived from complementary methodologies.

Several multidisciplinary, multilaboratory, multinational collaborative studies were agreed upon. Some of them may actually be performed. At least in one case, involving collaborators from Canada, Cuba, the Netherlands, the United States, and the Soviet Union, several meetings have taken place since the workshop and some pilot studies are in progress.

Whether or not these collaborative studies are actually completed, two major benefits have already emerged from the workshop. One was the realization that this was a new kind of scientific meeting, producing interactions such that many participants departed with ideas quite different from those with which they came. The second was a widespread consensus that two problems appear to be of central importance for the endeavor to advance our understanding of the nature of the integrative mechanisms that must constitute the physiological basis of subjective experience, consciousness, and self-awareness. One of these problems is the relationship between single-unit activity, the spontaneous EEG, the event-related potential, and the processing of information by the brain. The second problem is how to explain the observation, shared by several of the participants, that under certain circumstances apparently simultaneous events occur in anatomically separated regions so synchronously that they cannot be accounted for by synaptic transactions. Such observations were of particular interest to those participants interested in chaos theory and strange attractors.

Plans are already in progress for the Second Workshop on the Machinery of the Mind, which will also be sponsored by the National Scientific Research Center and the Ministry of Public Health of Cuba. On behalf of the participants in the first workshop, I want to express our appreciation to Premier Fidel Castro and the Government of Cuba for their willingness to defray the very substantial expenses of this intellectual adventure and our appreciation to the philosophy of science that inspired Dr. Miyar and Dr. Selman to become the protagonists of this workshop. Finally, I want to express our hope that this volume contains ideas that may inspire future generations of neuroscientists to ask themselves how they can contribute to the development of better understanding of the physiological basis of the human experience. Is there a more important question for science to address?

E. Roy John

Contents

Contents

Part VII   Brain Imaging

21   The Statistical Analysis of Brain Images
     Pedro A. Valdés-Sosa and Rolando Biscay Lirio 405

22   The Physical Basis of Electrophysiological Brain Imaging: Exploratory
     Techniques for Source Localization and Waveshape Analysis of Functional
     Components of Electrical Brain Activity
     *Roberto D. Pascual-Marqui, Rolando Biscay Lirio,
     and Pedro A. Valdés-Sosa* 435

Contributors

Alfredo Alvarez Amador
Centro Nacional de Investigaciones
  Cientifícas
Ciudad de la Habana, Cuba

Paul Bach-y-Rita
Department of Rehabilitation Medicine
University of Wisconsin Medical School
Madison, Wisconsin, USA

Erol Başar
Institute of Physiology
Medical University of Lübeck
Lübeck, Federal Republic of Germany

C. Başar-Eroglu
Institute of Physiology
Medical University of Lübeck
Lübeck, Federal Republic of Germany

Jacqueline Becker
National University of México
Ciudad de Los Deportes, Mexico

A. Berghold
Institute of Biomedical Engineering
Graz University of Technology
Graz, Austria

Rolando Biscay Lirio
Centro Nacional de Investigaciones
  Cientifícas
Ciudad de la Habana, Cuba

Maria A. Bobes
Centro Nacional de Investigaciones
  Cientifícas
Ciudad de la Habana, Cuba

Christian Bohm
Institute of Physics
University of Stockholm
Stockholm, Sweden

Konstantin P. Budko
Institute of Higher Nervous Activity and
  Neurophysiology
USSR Academy of Sciences
Moscow, USSR

Jan Bures
Czechoslovak Academy of Sciences
Institute of Physiology
Prague, Czechoslovakia

Olga Buresova
Czechoslovak Academy of Science
Institute of Physiology
Prague, Czechoslovokia

R. Chabot
Brain Research Laboratories
New York University Medical Center
New York, New York, USA

D. Crisp
Brain Behavior Laboratory
Department of Psychology
Simon Fraser University
Burnaby, British Columbia, Canada

Ana E. Díaz DeLeón
National University of México
Ciudad de Los Deportes, Mexico

Robert W. Doty
Department of Physiology
University of Rochester Medical Center
Rochester, New York, USA

P. Easton
Brain Research Laboratories
New York University Medical Center
New York, New York, USA

Thalia Fernandez-Harmony
National University of México
Ciudad de Los Deportes, Mexico

Walter J. Freeman
Department of Physiology-Anatomy
University of California
Berkeley, California, USA

Joaquin M. Fuster
Department of Psychiatry and Brain
    Research Institute
UCLA Medical Center
Los Angeles, California, USA

Yu. L. Gogolitsin
Department of Neurophysiology
Institute for Experimental Medicine
USSR Academy of Medical Sciences
Leningrad, USSR

Alexander M. Gorbach
Institute of Higher Nervous Activity and
    Neurophysiology
USSR Academy of Sciences
Moscow, USSR

Torgny Greitz
Department of Neuroradiology
Karolinska Institutet
Stockholm, Sweden

H. Haken
Institut für Theoretische Physik und
    Synergetik
Universität Stuttgart
Stuttgart, Federal Republic of Germany

H. Hantke
Institute of Pathobiochemistry
Academy of Medicine Magdeburg
Magdeburg, German Democratic Republic

Thalia Harmony
National University of México
Ciudad de Los Deportes, Mexico

U. Haselhorst
Institute of Pathobiochemistry
Academy of Medicine Magdeburg
Magdeburg, German Democratic Republic

Hans-Jochen Heinze
Department of Neurology
Medical School of Hannover
Hanover, Federal Republic of Germany

Steven A. Hillyard
Department of Neuroscience
University of California, San Diego
La Jolla, California, USA

E. Roy John
Brain Research Laboratories
New York University Medical Center
New York, New York, USA

B. Johnson
Brain Behavior Laboratory
Department of Psychology
Simon Fraser University
Burnaby, British Columbia, Canada

Willem Kamphuis
Department of Experimental Zoology
University of Amsterdam
Amsterdam, The Netherlands

R. E. Kiryanova
Department of Neurophysiology
Institute for Experimental Medicine
USSR Academy of Science
Leningrad, USSR

W. Klimesch
Department of Physiological Psychology
Institute of Psychology
University of Salzburg
Salzburg, Austria

A. Krusche
Institute of Pathobiochemistry
Academy of Medicine Magdeburg
Magdeburg, German Democratic Republic

Dietrich Lehmann
Neurology Department
University Hospital
Zurich, Switzerland

Fernando H. Lopes da Silva
Workgroup Neural Plasticity
Department of Experimental Zoology
University of Amsterdam
Amsterdam, The Netherlands

Steven J. Luck
Department of Neurosciences
University of California, San Diego
La Jolla, California, USA

George R. Mangun
Department of Neurosciences
University of California, San Diego
La Jolla, California, USA

Erzsébet Marosi
National University of México
Ciudad de Los Deportes, Mexico

F. Mas
Brain Research Laboratories
New York University Medical Center
New York, New York, USA

W. Mohl
Institute of Biomedical Engineering
Graz University of Technology
Graz, Austria

V. B. Nechaev
Department of Neurophysiology
Institute for Experimental Medicine
USSR Academy of Science
Leningrad, USSR

D. D. Orlovskaja
Institute of Mental Health
Academy of Medical Science
Moscow, USSR

Roberto D. Pascual-Marqui
Centro Nacional de Investigaciones
   Científicas
Ciudad de la Habana, Cuba

Gerd Pfurtscheller
Institute of Biomedical Engineering
Graz University of Technology
Graz, Austria

Jan Pieter M. Pijn
Department of Experimental Zoology
University of Amsterdam
Amsterdam, The Netherlands

Karl H. Pribram
Center for Brain Research and
   Informational Sciences
Radford University
Radford, Virginia, USA

Leslie S. Prichep
Brain Research Laboratories
New York University Medical Center
New York, New York, USA

A. W. Robertson
Brain Behavior Laboratory
Department of Psychology
Simon Fraser University
Burnaby, British Columbia, Canada

J. Röschke
Institute of Physiology
Medical University of Lübeck
Lübeck, Federal Republic of Germany

H. Schenk
Institute of Pathobiochemistry
Academy of Medicine Magdeburg
Magdeburg, German Democratic Republic

H. Schimke
Department of Physiological Psychology
Institute of Psychology
University of Salzburg
Salzburg, Austria

J. Schult
Institut fur Physiologie
Medizinische Hochschule Lübeck
Lübeck, Federal Republic of Germany

George A. Sharaev
Institute of Higher Nervous Activity and
   Neurophysiology
USSR Academy of Sciences
Moscow, USSR

Igor Shevelev
Institute of Higher Nervous Activity and
    Neurophysiology
USSR Academy of Sciences
Moscow, USSR

V. B. Shvyrkov
Institute of Psychology
USSR Academy of Sciences
Moscow, USSR

Robert W. Thatcher
School of Medicine
University of Maryland
Baltimore, Maryland, USA

Eugeny N. Tsicalov
Institute of Higher Nervous Activity and
    Neurophysiology
USSR Academy of Sciences
Moscow, USSR

N. A. Uranova
Institute of Mental Health
Academy of Medical Science
Moscow, USSR

Mitchell Valdés-Sosa
Centro Nacional de Investigaciones
    Cientifícas
Ciudad de la Habana, Cuba

Pedro A. Valdés-Sosa
Centro Nacional de Investigaciones
    Cientifícas
Ciudad de la Habana, Cuba

L. Henk van der Tweel
Laboratory Medical Physics
University of Amsterdam
Amsterdam, The Netherlands

Jan M.A.M. van Neerven
Department of Experimental Zoology
University of Amsterdam
Amsterdam, The Netherlands

Harold Weinberg
Brain Behavior Laboratory
Department of Psychology
Simon Fraser University
Burnaby, British Columbia, Canada

Introduction: Brain and Consciousness
A Wealth of Data

Karl H. Pribram

Introduction

The explicit purpose of this conference was, in the words of E. Roy John, to discuss the evidence available for understanding "the physiological basis of subjective experience, the material origin of human self-awareness." I attended the presentations and discussions and have devoted many hours to studying the written contributions that evolved from the conference. The presentations and contributions were and are impressive and I am honored to be invited to write an introduction to the resulting publication. What I hope to do in this introduction is examine some of the assembled contributions most relevant to the explicit purpose of the conference within the framework of systematizations I have achieved of my own observations and experimental results.

Quantum Theory

Walter Freeman poses the challenge in chapter 2: "On the fallacy of assigning an origin to consciousness." According to this challenge the title of our symposium is misleading. There is no *the* mind and therefore there can be no "machinery" of mind in the sense of Newtonian mechanistic explanation. According to Freeman, mind (derived from minding as pointed out by Gilbert Ryle, 1949) is process, actually a set of processes coordinating the functions of our material body with the panoply of environmental events. Of course, the brain has a special relation to organizing these processes and Freeman presents evidence on how some of these organizations develop and become replaced with others. He makes the further point that certain activities (usually cyclic) initiate the organizing process.

A great beginning. I can clearly hear some of my philosopher friends, however, shaking their heads, grumbling: "But where does this leave us with regard to our conscious experience of a more or less unitary self, the progenitor and executive of the minding process?" I counsel patience. In chapter 1, Robert Doty, in fact, takes the next step in addressing the issue of "the unity of mind." Doty furthers the idea that Newtonian mechanics fails to describe the role of brain in organizing minding. Doty is forthright in proposing that brain processes are nonlocal cooperatives such as those that describe interactions in quantum physics.

In chapter 3, E. Roy John comprehensively reviews his own considerable amount of evidence that ensembles of distributed neural events cooperate to produce recognizable brain patterns that correlate with recognizable behavioral patterns. John points out that ensemble cooperation is best described in the spectral domain, though spatiotemporal factors operate to constrain the patterns of cooperation. John, however, does not clearly distinguish temporal from spectral organization, a distinction necessary for understanding the possibility enunciated by Doty that in the brain cortex ensemble cooperativity is akin to processes described in quantum microphysics.

The distinction between spectral and temporal organization (as well as between spectral and spatial organization) is mathematical. Spectra are measured in terms of frequencies, which in mathematics are Fourier transforms of patterns of time (and space). Spectral analyses of the EEG determine power in various bandwidths of frequency. The time of occurrence of any particular frequency becomes enfolded into the totality of the bandwidth measurement.

In 1946, Dennis Gabor devised a scheme to measure the efficiency with which telephone communication (via Atlantic cable) could proceed. Following Hartley (1928), Gabor pointed out that there is a tradeoff between frequency and time (and therefore space on the cable) taken for transmission. He suggested that the frequency and duration of a signal be plotted simultaneously. The resultant plot within which the signal was described is a phase space—and Gabor used Hilbert's mathematics to develop the plot. In such a phase space the duration of a signal must cover at least a half wavelength or else the frequency of the signal is indeterminate. This minimum is mathematically identical with that described by Heisenberg to describe a quantum in microphysics. Gabor therefore called his minimum a "quantum of information."

During the 1970s it became clear that this Gabor elementary function, as the quantum of information is now called, is the best descriptor of a receptive field of a neuron in the visual cortex submitted to harmonic analysis (see review by DeValois and DeValois, 1988; Pribram and Carlton, 1986). Whether this means that dendritic processing as reflected in receptive field organization, is a microphysical quantum process remains an open question, van der Tweel in chapter 8 remarks that though such an hypothesis can not be disproved, he remains unconvinced. Unfortunately, he provides no explanation for this stance. In view of the overwhelming mass of data in favor of Gabor-like processing in both the visual and auditory systems, it remains tenable that the laws that were formulated to describe quantum physics apply as well to sensory psychophysics (Licklider, 1951) and to the neurophysiology of sensory processing. (For detailed review see Pribram, 1990a and 1990b, Lectures 1, 2 and 4.)

To summarize: The contributions of Doty, Freeman, and John, that make up Part 1 of this book, point us toward an explanation of the role of brain organization in mental processing as obeying laws akin to those developed to describe quantum rather than mechanistic (Newtonian) physics. Evidence is available that, indeed, processing is distributed among ensembles of neural (dendritic) events but constrained by the temporal and spatial "initial conditions" describing the anatomic connectivity and functional properties of the neural network.

The fact that these considerations are reached by use of harmonic analysis based on spread functions such as the Fourier or Fourier-like transformations indicated that, by means of convolutions and inverse transforms, correlations among patterns can be readily achieved. Such procedures are the essence of image processing in computerized tomography. In processing sensory inputs, such correlations can be accomplished through movements that allow the extraction of invariant properties, constancies, to produce symmetry groups, which when inverse transformed are, (at least computationally) coordinate with the perception of objects. (For detailed reviews, see Pribram, 1990b, Lecture 5 and Appendix B.)

According to such a computational scheme, over any wide range of movements the only "object" that remains invariant is the mover, the corporeal self. (For the precise neural processes involved in the perception of a corporeal self, see Pribram 1990, Lecture 6). The challenge initiated by Freeman and posed by philosophers is, at least as an initial possibility, met by following through the proposals (based on their data) made by Doty and John.

Chaos Theory

Harmonic analysis undertaken along the lines indicated by Gabor yield linear, invertible, computational transformations. Much more popular at the moment are closely related nonlinear procedures such as those used in Synergetics (see Haken, chapter 7, and Haken and Stadler, 1990) and in the applications of so-called chaos theory to the analysis of brain electrical activity. Freeman introduces this type of analysis in chapter 2 and Parts II and III of this book are filled with carefully executed studies using these techniques.

The quantum of information is an elementary function that changes (e.g., its frequency and therefore minimum duration) as conditions change. Further, neurologically, these quanta describe receptive field nodes in ensembles of cooperative dendritic events. Due to spontaneous neural activity and inputs from sensory and chemical receptors and from other neural networks, the activity in these ensembles is continuously changing. When the trajectories of these changes lead to temporary, i.e., quasistable, patterns within the ever-changing "enchanted loom" as Sherrington (1911/1947) once described brain electrical activity, these quasistabilities are called "attractors." When such patterns cannot be identified, the system is said to be chaotic. I believe this terminology is misleading: a holographic encoding of the nodes of interference patterns—Fourier coefficients—looks to be "random" or "chaotic" but in fact is a distributed domain that contains all of the information (by Gabor's definition) necessary to display, when inverse transformed, images of objects.

The reason for applying these nonlinear techniques is that simple Fourier transformations do not supply sufficiently rich computational power to describe either neuroelectric or psychophysical phenomena per se, much less the relationships between them. To the extent that one can stay linear, as in quantum physical description, to that extent computation is simplified. When, however, nonlinearities must be introduced, as when irreversible choices or trajectories need to be described, they can be conceived either as basic—or as critical variables that affect a basically linear process. Başar, in chapter 5, and Haken, chapter 7, take the latter route and their analyses and models are readily compatible with what I have described so far. For instance, in Başar's contribution, the "strange attractor" is a fractal (i.e., a noninteger) and fractals derive from the Fourier transformation, which allows not only translational but dilational invariance, the essense of fractal geometry (see Pribram, 1990b, Lecture 5 for details). Could it be that the trajectory leading from what appears to be chaos to what is described as the "strange attractor" is simply one manifestation of a Fourier-like process? In such a process, constraints reflect nonlinear rather than linear couplings of neural resonators, as indicated by the contribution of Alvarez Amador, Pascual-Margui, and Valdés-Sosa (chapter 4). Evidence produced with microelectrodes by Singer (1989) indicates that such resonances among dendritic processes within a cortical column in the visual cortex do, in fact, occur and can account for the conjoining of features in visual processing.

Chaos theory is distinguished from other binding techniques using non-linear dynamics, such as that of Thatcher in chapter 20 of this volume, in specifying attractors, tendencies toward which ensembles of momentary configurations (maps) of neural microprocesses converge. The principles of chaos theory take a step further the optimization principles of: 1) paths of least action in macrophysics (Hamiltonians that describe the tendency of systems to converge on least energy expenditure); 2) the action integral path in quantum mechanics (Feynman, 1985) in which Hamiltonians become vectors in phase space and therefore make possible stabilities above energy minima; to 3) attractors which describe (Prigogine and Stengers, 1984) temporary stabilities far from equilibrium (i.e., far from points of minimum energy). The binding problem, the fact that imaging is experienced as a unitary flow while the brain functions in terms of discrete microstates, is addressed by Lehmann in chapter 10.

Lehmann notes that the apparently continuous "flow" that characterizes conscious experience is found neurologically to be composed of discrete "moments." Lehmann describes momentary microstates displayed as EEG maps, each map lasting about 10 msec. The metaphors of cinematography and television bring this issue into everyday experience: discrete frames or scans are experienced as continuous flows of images. Philosophers have discussed this issue as a part of the "grain problem" that deals with the fact that the grain or scale of our descriptions (and measurements) of physical events is often different from the scale of description of consciously perceived events. We perceive tables and chairs, not molecules, atoms, or quarks. In the case of discrete neural events such as those describing momentary microstates, the issue is posed as to how such moments become bound into larger and larger units until continuity is achieved. Lehmann's data identify the initial steps as reflected in the neuroelectric record. His data suggest that "even though the individual functional microstates are very short, their reoccurrence over a certain period of time would permit the brain to reenter into a given global condition." Lehmann's description is reminiscent of that used by Gerald Edelman (1987, 1989). Edelman tackles the binding problem by describing putative reentrant processes that accomplish correlations and integrations between the activities of "multiple reentrantly connected" cortical systems. In older engineering terminology, feedbacks and feedforwards were described to accomplish integration (see, e.g., Pribram, 1971) and currently, "backwards processing" is the backbone of neural network (PDP) approaches to the problem (Rumelhart and McClelland, 1986, Vols 1 and 2). E.R. John and various collaborators in this volume (see Prichep et al., chapter 23) utilize cross-spectral coherence as well as factor analytic methods to determine groupings of processes. Each of these techniques addresses the issue of binding from a slightly different perspective. What the techniques have in common with each other and with chaos theory is their expression in mathematical language that can be exploited computationally.

Conscious Sensation

To grasp fully the importance of the processes described by these mathematical techniques it is necessary to first explore the relation between experience and the input that triggers the experience. The question can be framed as to what it is that is "added" to conscious experience by the transformational processes that lead from sense organ to brain. The issue is clearly illustrated at the sensory receptor and primary sensory projection system for the experience of color.

Radiant energy is diffracted by reflection and the optical structure into a continuous spectrum. This spectrum is convolved with three photochemical receptor systems (as suggested by Helmholtz 1909/1924 and identified by Wald, 1964) to form three output functions from the retinal cone system. The tuning curves of these functions center on maxima that do not correspond to what we know to be primary colors from ordinary mixing experiments. Rather, the cone receptor outputs are in turn convolved with lateral inhibitory networks of the horizontal and especially the amacrine layers of the retina to produce, by subtraction, outputs from the ganglion cells that correspond to opponens pairs (Hering, 1964; Hurvitch and Jameson, 1957). DeValois (1988) has shown that the receptive fields of units at the lateral geniculate nucleus demonstrate such opponens processing: e.g., red will excite, while green will inhibit one cell's response, and the opposite may be true of another.

Two opponens pairs have been identified: red-green and blue-yellow. These interact with an achromatic dimension to form color "images" in a three-dimensional Cartesian color space.

An additional dimension comes into play when the excitatory region of the receptive field and its inhibitory flank respond in an opposite manner to a particular color and still another opposition with its opponent (Zeki, 1980; DeValois and DeValois, 1988). To achieve color constancy depends on utilizing the double opponens processing capabilities for the three sets of color pairs. Hurlbert and Poggio (1988) used computational techniques to simulate color constancy in color spaces composed of Mondrians of different reflectances. Their implementation utilized parallel processing in simple analog networks composed of the linear units devised by Poggio (see Pribram, 1990b, Lecture 1). One of these implementations uses a "gradient descent" method that, over iterations, minimizes the least mean square of the error between actual and desired output (as determined by psychophysical experiment). This method is similar to that employed in Occam and the thermodynamic models (for review, see Pribram, 1990b, Lecture 2). As Hurlbert and Poggio pointed out, this procedure is closely related to optimal Bayesian estimation. In their computations, they used vectors to represent sample input sequences of Gaussian stochastic processes with zero mean (similar to the difference of Gaussians, reviewed in Pribram, 1990b, Lecture 4). The property (e.g., Blue) is then fully specified by a computation similar to a regression equation.

It is necessary that the operator in this computation be space invariant, that is, it must not change with a change in location. This is accomplished by computing in the spectral domain (Fourier transforming). The computation thus becomes "equivalent to the formation of an optimal [matched] filter" (p. 239). The computation utilizes the power spectrum (amplitude-modulated frequencies) of the inputs. From an ensemble of such inputs, the cross power spectrum is computed. These computations carried out by processes in the primary visual systems (Pribram, 1991) are all critical to our conscious *experience* of color. Are other sensory experiences equally dependent on transformations of sensory input by neural operations?

Epicritic and Protocritic Sensory Processes

An excellent candidate for attempting to answer this question is provided by the chapter on recovery from brain damage by Bach-y-Rita (chapter 18). Central to his theme is the conscious experiencing of pain. Pain is the foundation of one category of hedonic experience. As such it is critical to the issue of defining consciousness. There is a deep implicit relation between hedonic valuation and conscious experience: in fact, in the French language "conscience" (to know together) means both consciousness and conscience. The deep meaning of this relationship cannot be deciphered from observations of behavior alone: An operational behavioral definition such as "pain is that which produces withdrawal or agonistic responding" fails to describe the full spectrum of the phenomenon: aches and suffering cannot always be handled by simple responding; and, masochistic behavior actually seeks what in other contexts would be described as pain-inducing stimulation.

The neural side of the hoped for equation brings its own puzzles. Any free nerve ending, which under ordinary circumstances mediates sensations of deformation, can, when stimulated excessively, serve as a pain receptor. Two separate paths to the central nervous system carry signals that become interpreted as "pain": A system of "A" delta fibers that transmits relatively rapidly and a system of "C" fibers that transmits signals slowly. In the spinal cord, the pain-transmitting fibers are inexorably intertwined with those that determine the sensation of temperature as well as those entering the cord via the B fiber visceroautonomic afferent system (Prechtl and Powley, 1990). At the brain stem level the paths become peculiarly complex: endpoints are reached in the periaqueductal grey and in the thalamic grey matter surrounding the third ventrical but other tracts carry signals back downward into the spinal cord! Finally, at the cerebral level somatosensory thalamocortical pathways such as those involved in the transmission of pain are traced to the parietal lobe but parietal lobectomy is useless in ameliorating intractable pain, whereas frontal lobectomy or leukotomy serve well. Add to all this the observation that, on the whole, brain tissue is insensitive to manipulation and that no single brain neuron or groups of neurons can be identified to respond solely to signals that ordinarily result in pain and we have on our hands what appears to be an insurmountable task of explanation.

But the matter is far from hopeless. The presence of pathways descending from the brain stem to spinal cord has suggested that the sensation of pain may be subject to a gating procedure (Melzack and Wall, 1965). Tracing these fibers to layer V of the substantia gelatinosa of the dorsal horn of the cord, the origin of the spinothalamic pain and temperature tract gives anatomical support to the gate theory.

The discovery that electrical stimulation of the periaqueductal grey (the terminus of a large number of the "C" fiber system of the spinal pain and temperature system) can, depending on the frequency of the stimulus, not only lower but also *raise* the threshold of withdrawal from ordinarily noxious stimulation has provided specific evidence for gating (Liebeskind, Guilbaud, Besson, and Oliveras, 1973; Liebeskind, Mayer, and Akil, 1974).

When it was discovered that the periaqueductal grey is the site of preference for the absorption of opiates another important lead to explanation was opened. Soon it became apparent that a set of chemicals secreted by the pituitary and within the nervous system were also absorbed selectively by the periaqueductal grey system and showed the pain-protecting properties of opiates. These enkephalins, endorphins (endogenous "morphines"), and dynorphins are all derived from a protein molecule that is also the origin of the adrenocorticotrophic (ACTH) hormone secreted by the adrenal cortex in situations producing chronic stress, i.e., discomfort.

At the brain stem and cerebral level electrical stimulation of these same systems of neurons has produced both deterrence and reinforcement of behavior. These systems of neurons extend from the periaqueductal grey to the limbic forebrain (Olds and Milner, 1954; Olds, 1955). Here, as in the case of pain and temperature tracts in the spinal cord, the deterrence and reinforcement systems are to a large extent indistinguishably intertwined.

Pain and temperature, though an apparently odd couple, continue their close relationship into the amygdala of the limbic forebrain and the closely related cortex of the temporal pole and orbital surface of the frontal lobe. As in the case of pain, and, in contrast to other somatosensory submodalities, the discrimination of temperature is hardly affected by parietal lobectomy or stimulation. Temperature discrimination, as is avoidance of pain, is, on the other hand, severely disrupted by lesions or stimulations of the amygdala and related cortical systems (Pribram, 1977).

There is more evidence that the pain and temperature senses are in some way intimately related. During frontal leukotomy, marked warming of extremities is experienced and observed when the final quadrant of fibers is transected. Electrical stimulation in humans of sites that produce endorphin secretion are often accompanied by feelings of chilliness (Richardson and Akil, 1974). And the production of endorphins has been directly implicated in the raising of the threshold of the production of chills and thrills.

With one assumption, based on physiological experiment and evidence obtained in arctic research, a neural model for the pain (and temperature) process can be constructed. The assumption is that in mammals metabolism is anchored in maintaining a stable basal temperature. The evidence for this assumption has been reviewed in detail by Brobeck (1963). In addition, it is the maintenance of a stable basal temperature that is perceived as comfort.

Pain and temperature sensations, therefore, serve as initiators of experiences of suffering and comfort. It is well known that maintaining basal temperature and metabolic stability entails homeostatic (or better, homeorhetic, because their set point is resettable) processes. What can now be added is that pain and suffering are also regulated homeorhetically.

The homeorheteic nature of pain and suffering explains phenomena such as the insensitivity to pain during competitive sports and other strenuous activity. Administration of Naloxin, as endorphin antagonist, during a "runners high" collapses the euphoric feeling.

Masochism is also explained. The appetitive phase of discomfort is ordinarily experienced as itch, the consummatory phase as pain. During skillful sadomasochistic interaction, the masochist's threshold for pain (endorphin level?) is raised more or less gradually so that what would ordinarily be experienced as pain is a phenomenon akin to an itch.

Essentially, the homeorhetic processing of pain and discomfort is the result of a dual "gating" mechanism: (1) top-down connections from the frontolimbic forebrain to the thalamic and periaqueductal grey (Fulton, et al. 1949), which, in turn, connect with the dorsal horn cells of the substantia gelatinosa of the spinal cord; (2) the threshold-raising endorphins, which are controlled by these top-down neural systems.

The fact is, therefore, that together with other homeorhetic processes, such as those regulating eating, drinking, and sexual behaviors, the pain and temperature senses are processed to a large extent by a frontolimbic rather than by a parietal forebrain system. This fact can be conceptualized in terms of experiments in which peripheral nerves are sectioned. During the initial stages only fine fibers of the C type compose the regenerating nerve. During this stage only diffuse relatively uncomfortable and difficult to localize (in time and space) sensations are experienced. Once the normal fiber size spectrum has been attained, sensations are once again experienced as normal, i.e., they demonstrate what neurologists call local sign. Henry Head (1920) distinguished these normal sensation as *epicritic* and the system of spinal cord and brain stem fibers involved has been traced to parietal lobe cortical terminations.

By contrast, as reviewed here, the fiber systems involved in the sensations that Head called protopathic, and which are initially mediated by the small nerve fiber system, terminate frontolimbically. Because this system operates not only in pathological circumstances but also normally in the intact organisms, the term *protocritic* is more appropriate.

In summary: a protocritic set of systems based on sensations perceived as pain and temperature has been identified. These protocritic systems are homeorhetic and join other homeorhetic brain stem and frontolimbic systems that regulate the well-being of the organism. A major characteristic of these systems is their top-down control over receptor processing.

Forms of Conscious and Unconscious Processing

Mitchell Valdés-Sosa and Maria Bobes in their chapter 13 review the issues concerning implicit (unconscious) and explicit (conscious) processing. These issues are ordinarily discussed in terms of mechanisms of memory. These mechanisms concern primarily the coding and retrieval operations of remembering rather than those of memory storage. The question asked by Valdés-Sosa and Bobes is whether coding and retrieval are unitary or modular. Their evidence added to those of others clearly support modularity, i.e., separate systems can be identified to process different remembrances.

Their analysis raises additional questions. Are implicit, unconscious, processes all of a kind? Furthermore, if explicit conscious processes are modular, to what does modularity apply: states, contents or only the processes that relate the states and the contents of consciousness?

Shvyrkov in chapter 17 on animal experience addresses the organization of what in humans would be called implicit processing. Human subjectivity is in many respects a meta-consciousness, i.e., a subjective awareness of awareness. For example, bilateral resections of the medial portions of the temporal lobe produce a profound disturbance of subjective memory in humans without loss of skill memory. In monkeys with such resections I have observed retention of a visual discrimination at the 90–98% level after 2 years (with controls performing in the low 80% range). The lesioned monkeys were sufficiently conscious to perform the task, and there was no adequate test of meta-consciousness for monkeys to determine whether a loss similar to that sustained by humans had occurred.

Inspired by P.K. Anokhin, what Shvyrkov did find out was that the skills he was testing were organized according to the behavior *acts*, their objective environmental consequences and not the "functions" of the body parts implementing these behaviors. Shvyrkov's findings support those of Bernstein (1967) and my own which have demonstrated that the classical precentral motor cortex encodes the environmental consequences, the environmental loads placed on the motor system, and not the metric contractions or movements of muscles (Malis, Pribram, and Kruger, 1953; Pribram, 1971; Pribram, Sharafat, and Beekman, 1984; Pribram, 1984; Pribram, 1991).

The conclusion can be reached that objective awareness of the contents of consciousness is shared with animals. This leaves open the question of shared subjectivity and of the nature of unconscious processing. These questions are related: Sherrington (1911/1947) noted that to the extent that behavior is "reflex" to that extent "mind" does not enter the process. As noted, the term "mind" derives from "minding", i.e., "paying" attention.

The contributions of Steven Hillyard et al., (chapter 9), Joaquin Fuster (chapter 15), Thalia Harmony et al., (chapter 19), Robert Thatcher (chapter 20), Gerd Pfurtscheller (chapter 12), et al., in this volume address the extensive set of data obtained from studying the relation of attention and para-attentional processing by means of recorded electrical brain activity. Two overlapping classes of electrical brain activity have been identified (John, Herrington, and Sutton, 1967): those (extrinsic) which closely reflect sensory processing and those which depend more intimately on the contributions of brain processing (intrinsic). The extrinsic components have been shown to reflect more or less objective, content oriented automatic para-attentional processes while the intrinsic components are identified with more subjectively controlled, conscious attention (see, for example, Hillyard and Picton, 1979; Näätänen, 1990; and Pribram and McGuinness, 1991 for comprehensive reviews).

William James once stated that he was just about ready to jettison the concept "consciousness" in favor of thinking only in terms of "attention". If we accept the definition of attention (and para-attentional processes) derived from brain electrical recording as determined by a set of control operations on sensory processing, we can next ask: What brain processes can be shown to provide these controls?

The Orienting Reaction: Key to Conscious Experience

Control over sensory input constitutes attention. Control is exercised not only over protocritic sensory processing but over epicritic as well. The pathways by which such control is exercised have been documented by studies utilizing the orienting reaction. Sokolov (1963) in a classical series of experiments demonstrated that orienting occurs not only in response to extraordinary sensory stimulation but to any stimulus that mismatches a neuronal model representing the prior experience of the organism. This demonstration was accomplished by presenting tones or light flashes in a regular series and then omitting an "expected" stimulus; or by presenting tones or light flashes in a regular series and then omitting an "unexpected" stimulus; or by presenting tones or flashes of a certain intensity and then presenting one of lower intensity. In each instance a strong orienting reaction was obtained on the unfamiliar occasion.

Sokolov used visceroautonomic indicators such as heart and respiratory rate and galvanic skin conduction as well as behavioral (turning of head and eyes) indicators of orienting in his experiments. In my laboratory, an additional series of studies showed that in both monkeys and humans the visceroautonomic responses could be dissociated from the behavioral indicators by resections of the amygdala and temporal pole and by lesions of the orbital portions of the frontal lobe (Bagshaw and Benzies, 1968; Bagshaw and Coppock, 1968; Bagshaw et al., 1965; Bagshaw et al., 1965, 1970a, b, 1972; Bagshaw and Pribram, 1968; and Luria et al., 1964).

Ordinarily the orienting reaction habituates in a few (three or five to ten) trials indicating familiarity with the situation. After amygdala/temporal pole or orbitofrontal damage, however, the visceroautonomic components of the orienting reaction do not occur and the behavioral components fail to habituate. Familiarization apparently entails viceroautonomic arousal.

The fact that behavioral orienting continues after amygdalectomy indicates that another processing system is involved in the total orienting reaction. This other system centers on the striatum of the basal ganglia (caudate and putamen) and a lateral strip of frontoparietal cortex surrounding the Rolandic somatic sensory-motor projection systems. Behavioral orienting ceases when this strip of cortex and/or the related basal ganglia are damaged; in fact, total neglect of the stimulating event is produced when damage is severe (Heilman and Valenstein, 1972). In a sense, this system, when activated, readies, i.e., prepares, the organism's orienting process. Readiness entails maintaining attention.

As noted, the process whereby attention is maintained has been studied using the late, intrinsic, components of event-related brain electrical potential changes (ERPs). Beginning at around 300 msec, after a relevant but unexpected stimulus, a process is initiated that "updates" the neuronal model (Donchin, 1981). Updating is signalled by a late (ca 400 msec) negatively.

The updating process involves a circuit in which the thalamic reticular nucleus plays a critical part. Activation of this nucleus inhibits activity in the thalamocortical sensory projection pathways. In turn, activity in the reticular nucleus is controlled by an extralemniscal tecto-tegmental brain stem system that itself is influenced by input from the brain stem reticular formation and collaterals from the lemniscal sensory projection pathways. An additional input to the reticular nucleus of the thalamus originates in the orbitofrontal-amygdala system. This input is antagonistic to the input from the tecto-tegmental system: whereas the tecto-tegmental input activates the thalamic reticular nucleus, and therefore inhibits—gates-sensory projection activity, orbitofrontal-amygdala input inhibits the activity of the thalamic reticular nucleus and thus enhances activation of the lemniscal sensory projection systems, allowing behavioral orienting to occur. In turn, this orbitofrontal-amygdala system is controlled by the basal ganglia via the intralamnar and central nuclei of the thalamus.

The question arises as to the order in which activation and gating occur. Utilizing depth recordings in humans, Velasco and Velasco (1979; Velasco et al., 1973) showed that the late components of the ERP recorded from thalamic electrodes *precede* those recorded from brain stem sites: Updating is a top-down process. The orienting stimulus acts as a trigger, releasing attentive readiness to alter the neuronal model on the basis of just experienced consequences.

According to the model of attentional controls proposed by Pribram and McGuinness (1976; 1990) a third system is involved in mediating, when necessary, between familiarization and the maintenance of readiness. This mediating system is centered on the hippocampal formation. Excitation of the hippocampal system makes innovation possible (see Pribram, 1988, for review). The system, when idling, reflects comfortable, unstressed exploratory behavior. When engaged, the systems makes it possible for attention to be *paid* and effort to be expended in the updating process. The hippocampal system exerts its influence on arousal by way of frontocorticothalamic connections and on the maintenance of readiness posteriorly by way of brain stem connectivities (see Pribram, 1991) for review.

Subjective, Objective, and Intentional Consciousness

These controls on attention add a dimension to consciousness that goes beyond awareness of sensory-driven perceptions. The additional dimension is referred to as reflective self-awareness or intentionality (Brentano, 1960; von Uexkll, 1926). Intentionality combines subjective and objective awareness. Subjective awareness, as noted, devolves on the arousal systems centered on the amygdala to provide familiarity. Dsyfunction of these systems produce the clinical syndromes of déjà and jamis vu. These syndromes entail a feeling of familiarity in strange situations (déjà) and a feeling of unfamiliarity in situations that have been encountered repeatedly (jamais) in which case habituation has failed to take place. In the extreme, during a psychomotor seizure which most likely spreads to the hippocampal formation, the events experienced fail to become part of "the remembered present" (as Edelman has aptly called the coding process by which current experience remains accessible, 1989). .

Disturbances of the hippocampally centered systems do not interfere with experiencing objective consciousness. A patient with such lesions responds and experiences himself and the contents of his perceptions normally until a distracting event occurs that does not allow the updating process to come to completion. Experienced events are apparently stored haphazardly and can be retrieved only with special probing techniques (see Weiskrantz, 1986, for review). Each experienced episode is self contained, never to become integrated into the stream of consciousness. Thus, it is the hippocampal system that completes the "stream of consciousness" that constructs subjectivity.

Disturbances of the systems entailing the maintenance of readiness can lead to neglect syndromes, as has already been described. These systems make possible the awareness of excitations of the corporeal self as distinct from the primary sensory projection systems that make possible responding per se to sensory contents. Cases such as those of blind sight occur when the primary sensory thalamocortical projection pathways are damaged (Weiskrantz, 1986). Cases of neglect demonstrate disturbances of self-reference. Acting together, normal functioning of these systems makes possible the intentional distinction between the conscious experience of being a perceiver and the sensed contents of that experience. It is these systems, therefore, that are critically concerned in the production of intentional consciousness.

The Brain Consciousness Connection

According to the data reviewed here, a great deal is known about the relations, in general, between brain processes and conscious experience. What remains to be explored in most instances are the specifics of the brain processes entailed in the systems relations that have been uncovered. It is to these explorations that the contributions in the current volume are addressed. These contributions have as their basic assumption some form of identity between brain processing and conscious experience. I have suggested elsewhere (Pribram, 1986) that this identity can best be understood in terms of the metaphor of computer programming: I use English in addressing my word processor. The word processing system, the operating system, the ASCII assembler, and hexadecimal codes are all stages in the transformation of English to binary, which is the language used by the computer. In a similar fashion, subjective experience is transformed in steps to the language(s), codes, used by the brain. Many of the transformational steps are performed by the wetware of the sensory and neural processes reviewed in this volume; some others by the relations among changes in the environment, as for instance those produced by movement (see Gibson, 1977, 1979, for review).

What becomes clear from these considerations is that some sort of order remains invariant across all of the processing steps involved in the procedure that constitutes the transformations from English to binary and those from conscious experience to dendritic processing. We usually call this order "in-formation"—the form within. Measures of the amount of information are mathematically related to the amount of ordering of energy, that is the amount of entropy. Entropy reflects the efficiency with which a system operates. Relations between measures of entropy and measures of synergy; and the amount of organization of ensembles of Gabor quanta of information; and of "chaotic" attractors have been developed (see for example Pribram 1990, 1991).

These contributions may be only a first step toward understanding, but one that promises to break down the brain/consciousness barrier once and for all.

References

Bagshaw, M. H., Benzies, S. (1968). Multiple measures of the orienting reaction and their dissociation after amygdalectomized monkeys. *Exp. Neurol.* 20, 175–187

Bagshaw, M. H., Coppock, H. W. (1968). Galvanic skin response conditioning deficit in amygdalectomized monkeys. *Exp. Neurol.* 20, 188–196

Bagshaw, M. H., Kimble, D. P., Pribram, K. H. (1965). The GSR of monkeys during orienting and habituation and after ablation of the amygdala, hippocampus and inferotemporal cortex. *Neuropsychologia* 3, 111–119

Bagshaw, M. H., Mackworth, N. H., Pribram, K. H. (1970a). Method for recording and analyzing visual fixations in the unrestrained monkey. *Percept. Mot. Skills* 31, 219–222

Bagshaw, M. H., Mackworth, N. H., Pribram, K. H. (1970b). The effect of inferotemporal cortex ablations on eye movements of monkeys during discrimination training. *Int. J. Neurosci.* 1, 153–158

Bagshaw, M. H., Mackworth, N. H., Pribram, K. H. (1972). The effect of resections of the inferotemporal cortex or the amygdala on visual orienting and habituation. *Neuropsychologia* 10, 153–162

Bagshaw, M. H., Pribram, K. K. (1968). Effect of amygdalectomy on stimulus threshold of the monkey. *Exp. Neurol.* 20, 197–202

Bernstein, N. (1967). The co-ordination and regulation of movements. New York: Pergamon Press

Brentano, F. (1960/1967). The distinction between mental and physical phenomena. In: *Realism and the background of phenomenology*, Chisholm, R. M. (ed.). New York: Free Press, 39–61

Brobeck, J. R. (1963). Review and synthesis. In: *Brain and behavior* vol. 2. Brazier, M.A.B. (ed.). Washington, D.C.: American Institute of Biological Sciences, 389–409

DeValois, R. L. DeValois, K. K. (1988). *Spatial vision* (Oxford psychology series, No. 14). New York: Oxford University Press

Donchin, E. (1981). Surprise—Surprise? *Psychophysiology* 18, 493–513

Edelman, G. M. (1987). *NeuroDarwinism: The theory of neuronanal group selections*. New York: Basic Books

Edelman, G. M. (1989). *The Remembered present: A biological theory of consciousness*. New York: Basic Books

Feynman, R. P. (1985). *QED*. Princeton, NJ:Princeton University Press

Fulton, J. F., Pribram, K. H., Stevenson, J. A. F., Wall, P. (1949). Interrelations between orbital gyrus, insula, temporal tip and anterior cingulate gyrus. *Trans. Am. Neurol. Assoc.* 175–179

Gabor, D. (1946). Theory of communication. *J. Inst. Electrical Engineers* 93, 429–441

Gibson, J. J. (1977). On the analysis of change in the optic array in contemporary research in visual space and motion perception. *Scand. J. Psychol.* 18, 161–163

Gibson, J. J. (1979). *The ecological approach to visual perception*. Boston: Houghton Mifflin

Haken, H, Stadler, M. (1990). *Synergetics of Cognition*. Berlin: Springer-Verlag

Hartley, R. V. L. (1928). Transmission of information. *Bell. System. Tech. J.* 7, 535

Head, H. (1920). *Studies in neurology* (Vol. 1 and 2). London: Oxford University Press

Heilman, K. M., Valenstein, E. (1972). Frontal lobe neglect. *Neurology* 28, 229–232

Helmholtz, H. von (1909/1924). *Handbook of physiological optics*, (3rd ed.). Rochester, NJ: Optical Society of America

Hering, E. (1964). *Outlines of a theory of the light sense*. Cambridge, MA: Harvard University Press (original work published in 1878)

Hillyard, S. A., Picton, T. W. (1979). Event-related brain potentials and selective information processing in man. In: *Cognitive Components in Cerebral Event-Related Potentials and Selective Attention* Desmedt, J. E. (ed.)

Hurlbert, A., Poggio, T. (1988). Synthesizing a color algorithm from examples. *Science* 239, 482–485

Hurvich, L., Jameson, D. (1957). An opponent-process theory of color vision. *Psychol. Rev.* 64, 384–404

John, E., Herrington, R., Sutton, S. (1967). Effects of visual form on the evoked response. *Science* 155, 1439–1442

Licklider, J. C. R. (1951). Basic correlates of the auditory stimulus. In: *Handbook of experimental psychology*. Stevens, S. S. (ed.). New York: John Wiley and Sons, 985–1039

Liebeskind, J. C., Guilbaud, G., Besson, J. M., Oliveras, J. L. (1973). Analgesia from electrical stimulation of the periaqueductal gray matter in the cat: Behavioral observations and inhibitory effects on spinal cord interneurons. *Brain Res.* 50, 441–446

Liebeskind, J. C., Mayer, D. J., Akil, H. (1974). Central mechanisms of pain inhibition: Studies of analgesia from focal brain stimulation. In: *Advances in neurology, Vol. 4, Pain* J. J. Bonica (ed.). New York: Raven Press

Luria, A. R., Pribram, K. H., Homskaya, E. D. (1964). An experimental analysis of the behavioral disturbance produced by a left frontal arachnoidal endothelloma (meningioma). *Neuropsychologia* 2, 257–280

Malis, L. I., Pribram, K. H., Kruger, L. (1953). Action potential in "motor" cortex evoked by peripheral nerve stimulation. *J. Neurophysiol.* 16, 161–167

Melzack, R., Wall, P. D. (1965). Pain mechanisms: A new theory. *Science* 150, 971–979

Näätänen, R. (1990). The role of attention in auditory information processing as revealed by event-related potentials and other brain measures of cognitive function. *Behavioral Brain Sci.* 13, 201–288

Olds. J. (1955). Physiological mechanisms of reward. In: *Nebraska Symposium on Motivation*. Jones, R. R. (ed.). Lincoln, NB: University of Nebraska Press, 73–138

Olds, J., Milner, P. (1954). Positive reinforcement produced by electrical stimulation of septal area and other regions of rat brain. *J. Comp. Physiol. Psychol.* 47, 419–427

Prechtl, J. C., Powley, T. L. (1990). B-Afferents: A fundamental division of the nervous system mediating homeostasis? *Behavioral Brain Sci.* 13, 289–331

Pribram, K. H. (1971). *Languages of the brain: Experimental paradoxes and principles in neuropsychology*. Englewood Cliffs, NJ: Prentice-Hall

Pribram, K. H. (1986). The cognitive revolution and the mind/brain issues. *Am. Psychol.* 41, 507–520

Pribram, K. H. (1988). Brain systems involved in attention and para-attentional processing. In: *Attention, cognitive and brain processes and clinical applications*, Sheer, D., Pribram, K. H. (eds.). New York: Academic Press

Pribram, K. H. (1990). Frontal cortex—Luria/Pribram rapproachment. In: *Contemporary neuropsychology and the legacy of Luria*, Goldberg, E. (ed.). Hillsdale, NJ: Lawrence Erlbaum Associates

Pribram, K. H. (1991). *Brain and perception: holonomy and structure in figural processing*. Hillsdale, NJ: Lawrence Erlbaum Associates

Pribram, K. H., and Carlton, E. H. (1986). Holonomic brain theory in imaging and object perception. *Acta. Psychol.* 63, 175–210

Pribram, K. H., McGuinness, D. (1976). Arousal, activation and effort in the control of attention. *Psychol. Rev.* 82, 116–149

Pribram, K. H., McGuinness, D. (1991). Attention and para-attentional processing: Event-related brain potentials as tests of a model. *New York Academy of Science*

Pribram, K. H., Sharafat, A., Beekman, G. J. (1984). Frequency encoding in motor systems. In: *Human motor actions: Bernstein reassessed*, pp. 121–156. Whiting, H. T. A. (ed.). North-Holland: Elsevier

Prigogine, I., Stengers (1984). *Order out of chaos.* New York: Bantam Books

Richardson, D. E., Akil, H. (1974). Chronic self-administration of brain stimulation for pain relief in human patients. *Proceedings American Association of Neurological Surgeons*, St. Louis, MO.

Rumelhart, D. E., McClelland, J. L., (PDP Research Group) (1986). *Parallel distributed processing, Vol. I and II.* Cambridge, MA: MIT Press

Ryle, G. (1949). *The concept of mind.* London: Hutchinson. (Republished by University of Chicago Press, 1984)

Sherrington, C. (1911/1947). *The integrative action of the nervous system.* New Haven, CT: Yale University Press

Singer, W. (1989). Search for coherence: A basic principle of cortical self-organization. *Concepts in Neuroscience* 1(1), 1–25.

Sokolov, E. N. (1963). *Perception and the conditioned reflex.* New York: MacMillan Publishing

Uexkull, J. von (1926). *Theoretical biology.* San Diego, CA: Harcourt Brace Jovanovich

Velasco, F., Velasco, N. (1979). A reticulo-thalamic system mediating propriceptive attention and tremor in man. *Neurosurgery* 4, 30–36

Velasco, N., Velasco, F., Machado, J., Olvera, A. (1973). Effects of novelty, habituation, attention, and distraction on the amplitudes of the various components of the somatic evoked responses. *International Journal Neuroscience* 5, 3–13

Wald, G. (1964). The receptors of human color vision. *Science*, 145, 1007–1017

Weiskrantz, L. (1986). *Blindsight: A case study and implications.* Oxford: Clarendon Press

Zeki, S. M. (1980). The representation of colours in the cerebral cortex. *Nature*, 284, 412–418

Part I

Integrative Processes

1

Forebrain Commissures and the Unity of Mind

ROBERT W. DOTY

"It is often said that physicists invented mechanistic-reductionistic philosophy, taught it to biologists, and then abandoned it themselves."

Paul Davies (1988)

1.1 The Physical World

There is something profoundly amiss between the nature of reality and the ability of the human mind to comprehend it. The incongruity does not arise from mere ignorance, but from paradoxical observations. For instance, single electrons fired one at a time toward a pair of narrow slits can go through both slits simultaneously, as evidenced by the ensuing interference pattern. Yet, this pattern is absent if only a single slit is present or if the passage of the electron through the slits is recorded (Feynman, 1985)! This familiar peculiarity of the wave particle, and the fatal intrusiveness of observing it, is now further compounded by the demonstration that quantum mechanical effects can violate physical "locality" (Aspect et al., 1982; Bell, 1986); that is, the influence of observation can be conveyed at a speed exceeding the velocity of light.

The background of these latter experiments derives from Einstein's distrust of the probabilistic character of quantum mechanics (Bohr, 1969). He and his colleagues (Einstein et al., 1935) devised an argument that the experiments of Aspect et al. (1982) have directly addressed and contradicted. Consider this imaginary and exaggeratedly simplistic illustration of this principle (Davies and Brown, 1986). A single identifiable particle is confined to a pair of communicating boxes. The communication between the boxes is then closed and the boxes separated, leaving the particle in one or the other box. According to the uncertainty principle, however, its position can be affirmed only when it is observed, and until that time it remains as a probability function in both boxes. The boxes are put on two space ships and transported until they are light minutes, hours, or years apart. One of the boxes is then opened to examine it for the presence or absence of the particle. This act of observation instantaneously affects both boxes, so that upon subsequent comparison the observer discovers that if the particle is found in one box it is absent from the other, and vice versa. Note, of course, that the act of observation at one location cannot communicate any information to an observer of the second box. Yet, the challenge nevertheless remains that the act of observation simultaneously alters the condition of the particle vis-à-vis both boxes, regardless of their separation.

Now the point of this exercise in the counterintuitive features of quantum mechanics is that these paradoxes compound the already slippery task of stating something sensible about the physical nature of mental activity. It would appear that there is a heretofore unexpected interplay between the mind and the world; merely observing the world, however passive one's intent, alters it. Of course, that one's muscles may alter the physical world is not surprising, but that capturing photons in one's photoreceptors (if it is at this point that the paradox really occurs!?) may instantaneously transform the situation elsewhere in the universe does make seeing even a bit more miraculous than it already is. Is there not somewhere in the incomprehension of this interaction between mind and matter some hidden clue as to how the former is devised from the latter (or vice versa as Bishop Berkeley, but scarcely anyone else, would have us believe)? Many of the greatest minds of this century have wrestled with this problem, but no credible suggestions have yet been offered.

1.2 The Neuronal World

If it exists, the hidden clue most likely is to be found in the brain, with neurons being unequivocally the elements from which mentation is devolved. Yet, how do the digital discharges of widely dispersed neurons create the grainless panorama of three-dimensional, colored space that the normal, seeing eye conveys to the brain? Is there some "nonlocality" here, not in the sense of the quantum mechanical observations of Aspect et al. (1982), for information is obviously conveyed across the visualized field and the times involve the conduction velocity of nerve fibers rather than the speed of light, but rather in the instantaneous apprehension of all parts and characteristics of the imparted scene?

Consider the details. The mosaic of retinal photoreceptors (see Williams, 1988) and their attendant neural apparatus shade off in numbers with distance from the fovea. Each receptor's capture of about 500 photons/sec (on an overcast day; Wald, 1961) is converted into a stream of nerve impulses, about 0.3 to perhaps 100 Hz, by the retinal ganglion cells and thence into the optic nerve. In other words, each eye sends about 10 million asynchronous impulses/sec into the optic nerves, and it is from these impulses that the visual scene is flawlessly reconstructed. Indeed, it was from the discovery of this connection of the eyes into the brain that Alkmaion almost 2500 years ago first discerned the linkage between brain and mind, introspection otherwise leaving mentation wholly undefined as to bodily locus (Doty, 1965). However, the difficulty of relating visual experience to neuronal discharge only begins with this incoming flood of digital events from the optic nerves. Massive analytical processing ensues in the striate cortex; much of it is probably involved with unconscious operations, such as maintaining a constancy of perception in the face of continual shifting in position of the eyes and hence of the pattern of input from the multitudinous elements of the retinal mosaic. The striate cortex also serves as a great switchboard for channeling different streams of analysis into different loci in the neocortex so that "color" and "movement" are projected into widely dispersed areas (DeYoe and Van Essen, 1988; Zeki and Shipp, 1988). Furthermore, of course, the dispersion is enormously complicated in that each half of the visual field and all its attendant processing are precisely duplicated in each cerebral hemisphere.

The "nonlocality" of the neuronal flux from which the visual world is created is thus apparent. Bits and pieces, color here, analysis of orientation there, appraisal of movement elsewhere, all proceed throughout the brain, yet fuse in a gapless continuum in which all parts are apprehended simultaneously.

Several features of this transformation from neural processes to perception should be noted. First, perhaps, is the fact that the conversion derives not solely from the present input but necessarily from the early history of the system, from its imprinted memory. Visual systems not properly schooled during infancy thereafter fail in various ways to achieve the normal interpretation of their retinal input (Boothe et al., 1985).

Another item of note is the total absence of sensation where neuronal processes are lacking, i.e., at the optic disc, an invisible "blind spot" in the monocular visual field. That there is no corresponding "hole" in the visual scene (unless specifically revealed by quite simple testing) emphasizes both the necessity of neuronal activity for sensation or recognition of its absence and the elegant ease with which extrapolation enters construction of the visual scene. My colleague, James Ringo, astutely comments that the absence of any "granularity" in perception of the visual world, despite manifold variation in the density of receptor and processing elements and a corresponding reduction in acuity across the scene, can arise from this same process of extrapolation.

A particularly significant clue as to how neurons relate to mind is the fact that only certain types of neuronal activity (or activity in certain classes of neurons?) are associated with conscious perception (Doty, 1976a). There seem to be great systems of neurons, such as those controlling secretion of hypophyseal hormones, whose activity forever lies outside consciousness. Libet and his colleagues (Libet, 1973), recording from human cortex exposed at the time of surgery, noted that prominent electrical components could be evoked by direct or thalamic stimulation that were entirely lacking in subjective awareness until repeated at relatively high frequency. Thus, neuronal discharge per se is accompanied by sensation only when conforming to particular patterns. An even more striking example is that of "blindsight," in which an individual lacking the neuronal system of the striate cortex is incapable of subjectively experiencing even the existence of visual stimuli at some loci while remaining uncannily capable of accurately pointing to them when asked to guess (Weiskrantz, 1986).

In contrast, even those neurons, the activity of which can form the basis of conscious experience, clearly do so only under specific conditions. The presence or absence of "attention" (whatever that imparts to their pattern of activity) smoothly switches whole fields of neuronal discharge, great or small, in or out of perception; the lurking tiger may be unseen even though transmitted in the retinal pattern. Simpler examples are the total loss of sensation in a monocular Ganzfeld or under conditions of binocular rivalry while the input from retina to striate cortex continues unaltered (Bolanowski and Doty, 1987; Riggs and Wooten, 1972).

A reasonable hypothesis here would be that the brainstem somehow plays a permissive but essential role in whether or not activity in neurons capable of producing conscious experience actually does so. Certainly, in the absence of medial mesencephalic pathways, the human forebrain is doomed to unconsciousness (Ingvar and Sourander, 1970), and the brainstem is the major focus of processes affecting transitions between sleep and wakefulness (Hobson et al., 1986). An observation by Trevarthen (1987) on a "split-brain" patient strikingly illustrates the power of this brainstem switching of visual consciousness. The patient was to use her left hand to mark a white object on a black cloth in the right visual field. However, she reported that the instant she began to move her left hand toward this object, it disappeared! In other words, as her attention shifted to her right hemisphere for movement of the left hand, visual attention was likewise transferred from the right to the left visual half-field; without the corpus callosum and anterior commissure the visual world was neatly partitioned between the right and left hemispheres and which hemisphere "saw" depended upon which was queried.

The sum of all these observations suggests that a conscious experience arises only when there is a conjunction of brainstem input with, or producing appropriate patterns in, activity of neocortical neurons specialized for this purpose. Since the conduction velocity of neurons projecting from the mesencephalon to neocortex is exceptionally slow, their participation might be expected to add a certain sluggishness to neural processes contributing to conscious perception. Whether or not this is the source, Libet (1973, 1987) has demonstrated a remarkably lengthy period for elicitation of sensation by direct excitation of neurons in the somatosensory cortex or for erasure there of an incipient sensation elicited by cutaneous stimuli. In the case of direct cortical excitation, liminal stimuli must be applied for about 500 msec to elicit a subjective experience; shorter pulse trains remain ineffective, although they do elicit neuronal responses with each pulse. A liminal stimulus applied to the skin elicits a neuronal response at the somatosensory cortex within about 20 msec and will invariably yield a conscious experience. However, if a liminal train of electrical pulses to the corresponding point in the somatosensory cortex is begun up to 125 to 200 msec or even 500 msec after the cutaneous stimulus is applied, the cutaneous stimulus is not perceived (Libet, 1973). There is thus a protracted delay between the onset of discharge in cortical neurons and the realization of their effect in conscious experience.

1.3 The Two Cerebral Hemispheres

A still largely unexploited means of analyzing the relation of neurons to conscious experience is examination of the timing of the neuronal discharge that unifies the two cerebral hemispheres. This is particularly relevant in vision in which the field is split along the vertical meridian, the two halves of the scene projecting separately to each hemisphere. There is reason to believe that particularly large neurons in the circumstriate cortex (Shoumura et al., 1975) subserve this purpose, providing a rapidly conducting pathway to interchange visual meridional information between the hemispheres. However, these neurons and presumably their axons are uniquely large in the callosal system, and the vast majority of fibers in the human corpus callosum are likely to be very small and slow (Innocenti, 1986; LaMantia and Rakic, 1984; Swadlow, 1985), with interhemispheric conduction times ranging as slow as 300 msec!

Such long delays and the general paucity of callosal as opposed to intrahemispheric fibers suggest that action in the two cerebral hemispheres has a large degree of independence (Bogen, 1990). This characteristic, as well as the possibility of a specific modus operandi for the callosal system to conserve mnemonic space (Cook, 1986; Doty and Negrão, 1973; Doty et al., 1973), no doubt encourages the development of the well-known and markedly different intellectual propensities displayed by the two hemispheres (Bryden, 1982)

Given that the two hemispheres, when isolated by transection of the forebrain commissures, each maintain a wholly human but divided consciousness (Sperry, 1984), the profound conclusion is inescapable: the commissures allow one mind to communicate with another and produce thereby a single conscious entity. There are occasional quibbles; that with intact human subjects one hemisphere might be so dominant that the contribution of the other is negligible. This view, however, is not supported by the electrophysiological (Petsche et al., 1986) or metabolic (Roland and Frieberg, 1985) data nor the disturbance of mental life by minor lesions of the nondominant hemisphere (see Doty et al., 1986; Doty, 1989a, for review). Indeed, it must be inferred that a continual interaction between the two independently competent hemispheres is essential to achieve the normal unity of mental life and action. It is plausible also to suppose that pathology of that interchange would severely disrupt mentation, even to the point of psychosis (Doty, 1989a). On the other hand, there is probably a very wide range in the normal human capacity for interhemispheric interchange, since there is a twofold variation in cross-sectional area of the corpus callosum and a sevenfold variation in size of the anterior commissure, independently of sex or brain weight (Demeter et al., 1988).

Among the peculiarities of having two fully formed and effective hemispheres is that they maintain a dynamic equilibrium based on competition. This equilibrium is revealed with particular clarity, either anatomically or functionally, when the forebrain commissures are transected (see Cook, 1986; Doty, 1989a; Innocenti,

1986; Pearson et al., 1985; Trevarthen, 1987). In such circumstances the behavioral congruence of action must be maintained via the brainstem. In some human patients this congruence remains unsatisfactory and inefficient, and conflicting actions and predilections of the two hemispheres seriously complicate everyday behavior (Ferguson et al., 1985). Fortunately, this working of the two hemispheres at cross-purposes is uncommon in the chronic condition of split-brain patients. To some degree this is surprising, for the evidence is rather good that the two hemispheres normally have differing emotional outlooks (Gainotti, 1987) probably because of the unilaterality of the first-order connections from the amygdala (Doty, 1989b). How this difference in emotional tone of the two hemispheres is accommodated in normal individuals is as obscure as any other interchange between them. A well-documented case exists, however, in which the subject, initially troubled by abrupt, spontaneous emotional transitions, learned to control these moods; laboratory findings much later revealed that her control was achieved by switching dominance from one to the other hemisphere (Gott et al., 1984).

In view of the fact that brainstem influences are essential to the processes subserving consciousness, it is relevant to examine the interhemispheric exchange at this level. Nothing presently useful is known physiologically, but the anatomical arrangement is one of extraordinary complexity. In the serotonergic raphé system one side of the mesencephalon and pons projects to both cerebral hemispheres. These neurons in turn are under the control of the habenulointerpeduncular system, which itself is perhaps the most contorted system of the brain in regard to laterality (Doty, 1989b; Groenewegen et al., 1986). A multifarious loop thus exists, interconnecting the two hemispheres at the forebrain and brainstem levels: the forebrain commissures (corpus callosum and anterior commissure) passing bidirectionally between the hemispheres, the output of each hemisphere projecting bilaterally into the brainstem and into systems controlling the criss-crossing habenulointerpeduncular system, which passes to the raphé from which each side projects upon both sides of the forebrain!

Highly efficient and rapid mnemonic interchange occurs via forebrain commissures or components thereof as evidenced by experiments on macaques whose visual mnemonic capacities so closely resemble those of man (Doty et al., 1988a,b; Lewine et al., 1987). More surprising is the finding of a bilateral effect on mnemonic processing even in the absence of the forebrain commissures (Lewine and Doty, 1988). A hemisphere in split-brain macaques completely lacks access to specific visual memories held by the other hemisphere, but its performance is nevertheless burdened by the presence of memories in the other hemisphere. Thus, the two hemispheres must share some mechanism, presumably in the brainstem, that limits their individual capabilities.

There is strong evidence that even in split-brain human subjects some subtle degree of interhemispheric informational communication is maintained (Cronin-Golomb, 1986; Sergent, 1986, 1987) yet, as presumably is also true in the foregoing case with macaques, this communication can scarcely be designated as a conscious process. The patients are able to infer from some "hunch," reminiscent of the situation with blindsight, what information has been presented to the hemisphere contralateral to that being queried, but they have no specific details of what the contralateral hemisphere has seen or experienced. In other instances the two hemispheres can cooperate to sum information provided separately to each, with neither being aware of the total picture on which the decision of one or the other is based (Sergent, 1987). Thus, although highly organized neuronal processing may occur without overt reflection in consciousness, it is only via the privileged intercommuniation across the forebrain commissures that the consciousness of the two hemispheres, as distinct from the overt behavior of the individual, can be unequivocally unified.

1.4 Downward Causation

Up to this point consideration has been given only to what manner of neuronal activity is linked to conscious experience and how this experience comes to surmount the spatiotemporal dispersion of the underlying processes. There is a glimmer of understanding about the link between neuronal activity and consciousness, at least in the prospect of identifying the necessary elements, yet the latter remains enigmatic. For explanation one tends to take refuge in the fecund vastness of the numbers of impulses and of neurons interconnected in unimaginable intricacy, the equivalent of the entire population of the earth at service to the effect. Yet, it is to no avail, for the transformation from momentary influxes of sodium ions to the intangible experience of sight leaves too great a gap to set aside the mystery. Yet endeavoring to think of force fields or hyperspace to effect the unifying transition seems silly, for how are such explanations to be tied to the creation of the unity accomplished by the ion-interchanging axons of the neocommissures?

This conundrum is challenge enough, but in addition the feedback of consciousness into the physical world—the fact that acts are guided by the contents of consciousness (Sperry, 1980)—remains to be understood. It is one thing to think of conscious experience as arising, passively, consequent to action in the neuronal network, but quite another to concoct a credible means by which this impalpable state now works to control the firing patterns of the constituent cell assemblies. That consciousness does guide or initiate behavior is simply too common an experience to deny, as so lucidly argued by Ryle (1949). True, as evidenced by the accuracy of blindsight-pointing (Weiskrantz, 1986), the unconscious concatenation of information for decisions by split-brain subjects (Sergent, 1987), and multiple instances in normal subjects, very complex and effective behavior can ensue without benefit of consciousness. However, it is equally clear that conscious action is a highly efficient mechanism for survival, creating the spear and cooperation of one's fellows against the imagined possibility of the tiger. This, in essence, is Sperry's argument, that consciousness is too useful to have evolved merely for the delectation of the organism; it is too efficient not to have access to the neuronal engine.

How is this access to be achieved? Lacking any convincing understanding of how consciousness is induced by neuronal action, it is doubly incomprehensible how it exerts control upon the flux or lack thereof for the ions driving the attendant cell discharge. Is it here that the strange interaction of the quantum particle with the far-flung world comes into play, the point at which the observer enters and alters the course of physical events? Both physics and physiology are too primitive even to suggest a likely mechanism. This seems more than just another example of Braitenberg's (1984) "law of uphill analysis and downhill invention." Although there is an astonishing difficulty in discerning the principles of operation of his ingenious vehicles despite the bald simplicity of their construction, the problems of consciousness are of another category. It is not behavior that is to be understood, but awareness, a property fundamentally detectable only by the participant even though it may be paramount to the behavior.

The nervous system has at its disposal a superlative capacity for controlled amplification of the action of single neurons, as in the design of the retinal rods for normal operation as detectors of single photons (Wald, 1961). There are also demonstrable arrangements, such as my "centers" (Doty, 1976b) or Minsky's (1986) "agents," which form interlocking hierarchies that, once triggered, yield smoothly integrated, automatic outcomes. It is thus not too far-fetched to suggest that rather minor intrusions into thresholds at the afferent portal of selected hierarchies could control at least major elements of behavior. Yet, quantum mechanics notwithstanding, just what would be the nature of an "intrusion"; how do molecules or ions dance to the tune of the consciousness that they produce a tune somehow conveyed by visible and vulnerable fibers to unite the mind of one hemisphere with that of the other? In this riddle lies the nature of man.

References

Aspect, A., Grangier, P., Roger, G. (1982): Experimental realization of Einstein-Podolsky-Rosen-Bohm Gedankenexperiment: A new violation of Bell's inequalities. *Phys. Rev. Let.* 49, 91–94

Bell, J. (1986): BBC radio interview. In: *The ghost in the atom.* Davies, P.C.W., Brown, J.R. (eds.). Cambridge: Cambridge Univ. Press, pp. 45–57

Bogen, J.F. (1990): Partial hemispheric independence with the neocommissures intact. In: *Brain circuits and functions of the Mind.* Trevarthen, C. (ed.). Cambridge: Cambridge Univ. Press, pp. 215–230

Bohr, N. (1969): Discussion with Einstein on epistemological problems in atomic physics. In: *Albert Einstein: Philosopher-scientist.* Schilpp, P.A. (ed.). La Salle, IL: Open Court Publishing Co (Republished in *The world of physics:* Vol. 3. Weaver, J.H., (ed.). New York: Simon and Schuster, 1987, pp. 801–834)

Bolanowski, S.J. Jr., Doty, R.W. (1987): Perceptual "blankout" of monocular homogeneous fields (Ganzfelder) is prevented with binocular viewing. *Vision Res.* 27, 967–982

Boothe, R.G., Dobson, V., Teller, D.Y. (1985): Postnatal development of vision in human and nonhuman primates. *Ann. Rev. Neurosic.* 8, 495–545

Braitenberg, V. (1984): *Vehicles—Experiments in synthetic psychology.* Cambridge, MA: MIT Press

Bryden, M.P. (1982): *Laterally—Functional asymmetry in the intact brain.* New York: Academic Press

Cook, N.D. (1986): *The brain code—Mechanisms of information transfer and the role of the corpus callosum.* London: Methuen

Cronin-Golomb, A. (1986): Subcortical transfer of cognitive information in subjects with complete forebrain commissurotomy. *Cortex* 22, 499–519

Davies, P. (1988): *The cosmic blueprint.* New York: Simon and Schuster

Davies, P.C.W., Brown, J.R. (1986): *The ghost in the atom.* Cambridge: Cambridge Univ. Press

Demeter, S., Ringo, J.L., Doty, R.W. (1988): Morphometric analysis of the human corpus callosum and anterior commissure. *Hum. Neurobiol.* 6, 219–226

DeYoe, E.A., Van Essen, D.C. (1988): Concurrent processing streams in monkey visual cortex. *Trends Neurosci.* 11, 219–226

Doty, R.W. (1965): Philosophy and the brain. *Persp. Biol. Med* 9, 23–34

Doty, R.W. (1976a): Consciousness from neurons. *Acta Neurobiol. Exp.* (Warsaw) 35, 791–804

Doty, R.W. (1976b): The concept of neural "centers." In: *Simple networks. An approach to patterned behavior and its foundations.* Fentress, J. (ed.). Sunderland, MA: Sinauer, pp. 251–265

Doty, R.W. (1989a): Schizophrenia: A disease of interhemispheric processes at forebrain and brainstem levels? *Behav Brain. Res.* 34, 1–33

Doty, R.W. (1989b): Some anatomical substrates of emotion and their bihemispheric coordination. In: *Emotions and the dual brain.* Gainotti, G., Caltagirone, C. (eds.). Berlin: Springer-Verlag, pp. 56–82.

Doty, R.W., Negrão, N. (1973): Forebrain commissures and vision. In: *Handbook of sensory physiology:* vol. VII/3B. *Central processing of visual information,* Part B. Jung, R. (ed.). Berlin: Springer, pp. 543–582

Doty, R.W., Negrão, N. Yamaga, K. (1973): The unilateral engram. *Acta Neurobiol. Exp.* (Warsaw) 3, 711–728

Doty, R.W., Ringo, J.L., Lewine, J.D. (1986): Interhemispheric mnemonic transfer in macaques. In: *Two hemispheres—one brain: Functions of the corpus callosum.* Leporé, F., Ptito, M., Jasper, H.H. (eds.). New York: Liss, pp. 269–279

Doty, R.W., Ringo, J.L., Lewine, J.D. (1988a): Human-like characteristics of visual mnemonic system in macaques. In: *Cellular mechanisms of conditioning and behavioral plasticity.* Alkon, D., Woody, C.D. (eds.). New York: Plenum, pp. 303–312

Doty, R.W., Ringo, J.L., Lewine, J.D. (1988b): Forebrain commissures and visual memory: A new approach. *Behav. Brain Res.* 29, 267–280

Einstein, A., Podolsky, B., Rosen, N. (1935): Can quantum-mechanical description of physical reality be considered complete? *Phys. Rev.* 47, 777–780

Ferguson, S.M., Rayport, M., Corrie, W.S. (1985): Neuropsychiatric observations on behavioral consequences of corpus callosum section for seizure control. In: *Epilepsy and the corpus callosum.* Reeves, A.G. (ed.). New York: Plenum, pp. 501–514.

Feynman, R.P. (1985): *QED, the strange theory of light and matter.* Princeton: Princeton Univ. Press

Gainotti, G. (1987): Disorders of emotional behaviour and of autonomic arousal resulting from unilateral brain damage. In: *Duality and unity of the brain—Unified functioning and specialisation of the hemispheres.* Ottoson, D. (ed.). New York: Plenum, pp. 161–179

Gott, P.S., Hughes, E.C., Whipple, K. (1984): Voluntary control of two lateralized conscious states: Validation by electrical and behavioral studies. *Neuropsychologia* 22, 65–72

Groenewegen, H.J., Ahlenius, S., Haber, S.N., Kowall, N.W., Nauta, W.J.H. (1986): Cytoarchitecture, fiber connections, and some histochemical aspects of the interpeduncular nucleus in the rat. *J. Comp. Neurol.* 249, 65–102

Hobson, J.A., Lydic, R., Baghdoyan, H.A. (1986): Evolving concepts of sleep cycle generation: From brain centers to neuronal populations. *Behav. Brain Sci.* 9, 371–448

Ingvar, D.H., Sourander, P. (1970): Destruction of the reticular core of the brainstem. A patho-anatomical followup of a case of coma of three years' duration. *Arch. Neurol.* 33, 1–8

Innocenti, G.M. (1986): General organization of callosal connections in the cerebral cortex. In: *Cerebral cortex:* vol 5. *Sensory-motor areas and aspects of cortical connectivity,* Jones, E.G., Peters, A. (eds.). New York: Plenum, pp. 291–353

LaMantia, A-S, Rakic, P. (1984): The number, size, myelination, and regional variation of axons in the corpus callosum and anterior commissure of the developing rhesus monkey. *Abstr. Soc. Neurosci.* 10, 1081

Lewine, J.D., Doty, R.W. (1988): Mnemonic capacity in split-brain macaques. *Abstr. Soc. Neurosci.* 14, 2

Lewine, J.D., Doty, R.W., Provencal, S., Astur, R. (1987): Monkey beats man in efficiency of mnemonic retrieval. *Abstr. Soc. Neurosci.* 13, 206

Libet, B. (1973): Electrical stimulation of cortex in human subjects, and conscious sensory aspects. In: *Handbook of sensory physiology:* vol 2. *Somatosensory system.* Iggo, A. (ed.). Berlin: Springer, pp. 743–790

Libet, B. (1987): Consciousness: Conscious, subjective experience. In: *Encyclopedia of neuroscience:* vol 1. Adelman, G. (ed.). Boston: Birkhäuser, pp. 271–275

Minsky, M. (1986): *The society of mind.* New York: Simon and Schuster

Pearson, R.C.A., Sofroniew, M.V., Powell, T.P.S. (1985): Hypertrophy of cholinergic neurones of the rat basal nucleus following section of the corpus callosum. *Brain Res.* 338, 337–340

Petsche, H., Pockberger, H., Rappelsberger, P. (1986): EEG topography and mental performance. In: *Topographic mapping of brain electrical activity.* Duffy, F.H. (ed.). Boston: Butterworths, pp. 63–98

Riggs, L.A., Wooten, B.R. (1972): Electrical measures and psychophysical data on human vision. In: *Handbook of sensory physiology:* vol VII/4. *Visual psychophysics.* Jameson, D., Hurvich, L.M. (eds.). Berlin: Springer, pp. 690–731

Roland, P.E., Friberg, L. (1985): Localization of cortical areas activated by thinking. *J. Neurophysiol.* 53, 1219–1243

Ryle, G. (1949): *The concept of mind.* London: Hutchinson (republished by Univ. of Chicago Press, 1984)

Sergent, J. (1986): Subcortical coordination of hemispheric activity in commissurotomized patients. *Brain* 109, 357–369

Sergent, J. (1987): A new look at the human split brain. *Brain* 110, 1375–1392

Shoumura, K., Ando, T., Kato, K. (1975): Structural organization of 'callosal' OBg in human corpus callosum agenesis. *Brain Res.* 93, 241–252

Sperry, R.W. (1980): Mind-brain interaction: Mentalism, yes; dualism, no. *Neuroscience* 5, 195–206

Sperry, R.W. (1984): Consciousness, personal identity and the divided brain. *Neuropsychologia* 22, 661–673

Swadlow, H.A. (1985): The corpus callosum as a model system in the study of mammalian cerebral axons. A comparison of results from primate and rabbit. In: *Epilepsy and the corpus callosum.* Reeves, A.G. (ed.). New York: Plenum, pp. 55–71

Trevarthen, C. (1987): Subcortical influences on cortical processing in 'split' brains. In: *Duality and unity of the brain—Unified functioning and specialisation of the hemispheres,* Ottoson, D. (ed.). New York: Plenum, pp. 382–415

Wald, G. (1961): General discussion of retinal structure in relation to the visual process. In: *The structure of the eye.* Smelser, G.K. (ed.). New York: Academic Press, pp. 101–115

Weiskrantz, L. (1986): *Blindsight—A case study and implications.* Oxford: Clarendon Press

Williams, D.R. (1988): Topography of the foveal cone mosaic in the living human eye. *Vision Res.* 28, 433–454

Zeki, S., Shipp, S. (1988): The functional logic of cortical connections. *Nature* 335, 311–317

2

On the Fallacy of Assigning an Origin to Consciousness

WALTER J. FREEMAN

2.1 Introduction

Recent developments in the theory of nonlinear dynamics that have been applied to brain theory have substantially expanded our understanding of the neural mechanisms by which large-scale patterns of brain activity are self-organized. In particular, these new concepts give us fresh insight into the neurodyanmics of goal-seeking behavior, how it emerges within the brain, and how it regulates the influx of sensory information into the cerebrum. It has become clear that the stimulus-response paradigm fails to address the most basic properties of biological intelligence, which are its autonomy and its creative powers. Chaotic dynamic systems not only destroy information (in the Shannon-Weaver sense) but they also create it. Our experimental studies of the electroencephalogram (EEG) have shown us that brains are chaotic systems that do not merely "filter" and "process" sensory input; they use sensory stimuli as "instructions" to create perceptual patterns that replace the stimuli.

A key concept in our hypothesizing and modeling is that of reafference. When a neural activity pattern emerges by chaotic dynamics that expresses a drive toward a goal, it has two facets. One is in the form of a motor command that activates the descending motor systems. The other is a set of messages to the central sensory systems that prepare those systems for the consequences of motor actions that are about to take place. Studies in anatomy and experimental surgery indicate that the neural activity patterns that express goal-directed tendencies emerge in the limbic system, which disseminates reafference information, and that the sensory consequences of actions are fed back into the limbic system through the entorhinal cortex. Philosophical and psychological considerations suggest that the cyclical processes of emergent goal-seeking, reafference, and sensory feedback constitute the basis for what we perceive as subjective consciousness.

This cycle suggests a further inference: the physiological basis for our human conception of cause and effect lies in the mechanism of reafference; namely, that each intended action is accompanied by motor command ("cause") and expected consequence ("effect") so that the notion of causality lies at the most fundamental level of our capacity for acting and knowing. This trait results in the replacement of sensory stimuli by self-organized activity patterns that are contingent on past experience, present motivational state, and expectancy of future action. We cannot truly gain access to what is "out there," the *Ding an Sich* of Kant (1787), yet we experience the perceptual feedback as the consequence of our action. I conclude that the intuition of causality is essential for human understanding and action but that it cannot validly be applied to the process by which the intuition emerges. Hence, consciousness cannot be said to have a cause, an origin, a seat, or beginning in time and place within the brain. Therefore, it is fallacious to seek for its cause, location, or time of onset in phylogeny and ontogeny.

Cognitive research necessarily shuttles among several disciplines, including computer science, linguistics, psychology, neuroscience, and philosophy. These last two are the fields of most concern to this discussion, for investigation of the physiological basis of cognition and consciousness ultimately leads to the consideration not only of the origin of consciousness but also of the notion of origin itself.

It should not be surprising, however, that cognitive scientists seldom investigate consciousness itself, for they have no need to introduce it into their models. As Laplace remarked when asked about the place of God in his deterministic model of the universe, "I do not have need of that hypothesis." On the contrary, most biologists are less wary of pronouncing on something so evanescent as consciousness—they have sufficient reason, biological as well as philosophical, to do so. They can accept consciousness as a self-evident fact of life in humans and animals because it is subject to known biological forces: gradation between species over the evolutionary scale, elaboration and decay over individual life spans, and the vicissitudes of biochemical accidents and manipulations. The important questions then are not whether it exists or what is its cause, but how does it arise in the brain and what is its biological role in behavior.

Consciousness is bidirectional: it simultaneously reaches out into the world as an expression of the brain's goal-directed dynamics and folds in reflexively upon its own operations. As John Dewey (1896) observed:

[C]onsciousness is always in rapid change, for it marks the place where the formed disposition and the immediate situation touch and interact. It is the continuous readjustment of self and the world in experience. "Consciousness" is the more acute and intense in the degree of the readjustments that are demanded, approaching the nil as the contact is frictionless and interaction fluid. It is turbid when meanings are undergoing reconstruction in an undetermined direction, and becomes clear as a decisive meaning emerges.

According to Dewey's philosophical perspective, consciousness is a continuous give-and-take between self and world from which meaning emerges. This point of view had been espoused and developed by other brain scientists and philosophers in the century following its expression as, for example, by Sherrington (1937), Merleau-Ponty (1942), Herrick (1924), and Maturana and Varela (1980). Beyond this insight into the experiential nature of consciousness, there extends a more fundamental line of questions, which are at once physiological and philosophical, about the arising and working of consciousness as an intrinsic process of the brain.

The concept of emergence is central to any understanding of consciousness and the brain. As a technical term, it describes a process by which order appears "spontaneously" within a system (Prigogine, 1980). When many elements are allowed to mingle, they form patterns among themselves as they interact. This emergence of order from a network of interacting parts may appear sudden to a casual observer, but it is in fact the result of a sequence of small steps, each based on a preceding step and leading to the next—a continuum, in other words. What seems fast to us may to the process itself be slow and vice versa.

Consciousness arises from the interactions of immense numbers of nerve cells that individually act in a thousandth of a second, but that, taken as a whole, produce an evolving sequence of patterns much more slowly, as we know well from introspection. Within each pattern lies the seed of a bodily action that at once grows from a preceding pattern-and-action pair and prepares for the one to follow.

Consider the origin of a string quartet. The composer plays with a group of elements—a melody, an isolated chord, an emotion—aiming somehow to make a piece of music with them. At some point, some of these elements gel, apparently of their own accord, and the collection of ideas becomes a sketch for a scherzo. Eventually, this process culminates in the emergence of a polished quartet.

In contrast, when I turn on the radio to listen to a performance of the same quartet, although sound comes out of the speaker, the process is not emergent in the technical sense. Both the sound that I hear and the act of music making that preceded it are externally driven (my hand on the volume knob controls the sound I hear, the printed score structures the musicians' playing), not internally determined and self-organized as is the composition process. It is possible, however, that the four musicians may be so attuned to one another as they play that they are collectively transported to an intense emotional experience. In this rare and wonderful case, the performance process *is* emergent, by virtue of which the recording may become a classic.

This problematic relation of a complicated system (the quartet) and the interactions of elements within it (a diminished seventh chord in the score or the feeling of anticipation it may produce in a listener) has long plagued cognitive and neuroscience. The aim of this chapter is to enable the reader to understand consciousness globally by examining its emergence from a network of neurological interactions. Considering consciousness as an emergent, self-organizing phenomenon complicates a simple cause-effect (or stimulus-response) model of cognition and consequently has much to teach about the notion of origin itself.

2.2 Functional Neurology of Consciousness

Among physiologists and psychologists, the prevailing model of brain function is that of a physical mechanism controlled by input. According to this view, the brain responds to external stimuli in much the same way as man-made machines respond to human commands. This model rests upon the assumption that the brain works according to a simple, determinist pattern of stimulus and response (Gibson, 1979;

Kandel and Schwartz, 1981; Pavlov, 1927). It is as if the brain were a sort of keyboard; to produce a specific response, all an engineer would have to do would be to press the fingers on the appropriate keys.

This model, relying as it does on external control to explain brain function, does not satisfactorily reflect either the way brains are built or they way they work. Brains are self-organizing. Behavior is determined not by the motor and sensory cortices found in the outer brain shell but also by systems inside the shell. The reticular system, for example, functions as the nervous system's on-off switch and the brain's activating core (Klopf, 1982); the limbic lobe is its organizing system (Freeman, 1984). The name "limbic lobe" comes not from theological Limbo, as might be supposed, but from the Latin *limbus,* meaning border or belt; it was first applied to this part of the brain by Paul Broca in 1873. "La grande lobe limbique," as Broca called it, is indeed a circular structure: parts of it can be seen in Figure 2.1 as the septum, the amygdala, the fornix, the hippocampus, and the entorhinal cortex (O'Keefe and Nadel, 1978).

Sensory input converging onto the entorhinal area is channeled into the hippocampus, whence it is returned to the entorhinal and motor cortices and finally passed on to the motor system (Ramón y Cajal, 1904). It follows, then that the entorhinal area plays a crucial role in the organization of sensory input and the staging of motor action. It is also, I believe, uniquely situated for the emergent properties relating to cognition and consciousness in both animals and humans.

The most common physiological model for the way the limbic system processes sensory information traces an almost Pavlovian reflex arc (Pavlov, 1927): sensory input enters the thalamus from below, goes up to the cortex, passes via the limbic system into the motor system, and then goes out again. This conventional stimulus-bound view of the nervous system, like the keyboard metaphor for brain control, misses the most distinctive feature of brain functioning: the self-organized, dynamic production of goal-oriented behavior. In so doing, it leaves no room for autonomy, self-organization, goal-directedness, or even consciousness (Ramón y Cajal, 1904).

My own work has, in contrast, led me to believe that perception instead begins with an internally generated neural process that prepares the organism to seek out future stimuli in the outside world. What actually occurs is the emergence of a motor action pattern—a command, if you like—in the nervous system, which is then fed out to the muscles and glands through the brainstem. It is important to note, however, that most of this outflow remains within the brain itself where, redirected to the receiving areas of the cortex, it alerts the sensory systems to expect the sensory consequences of motor action (Helmholtz, 1879). Through this self-reflexive update, now known as corollary discharge (von Holst and Mittelstaedt, 1950), by the process of reafference (Sperry, 1950), the brain actively grasps for the consequences of action into the environment. Corollary discharge is what distinguishes looking from seeing and listening from hearing. It is also an essential part of what we experience as consciousness (Sacks, 1985).

In summary, the chain of events in the limbic system occurs as follows: A motor command—to reach for a slice of pie, for example—leads to the expectation of

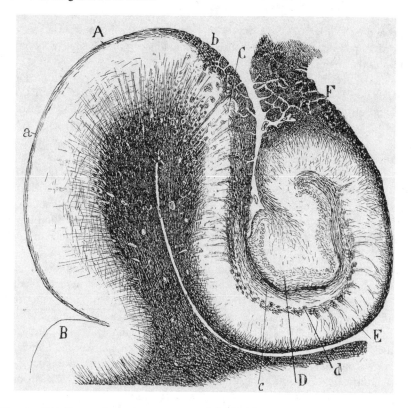

FIGURE 2.1. Some principal components of the limbic system are shown in this section through the adult human brain at the hand of the master anatomist Ramon y Cajal (1904). The entorhinal cortex (A) receives its input through stages from all parts of the sensory neocortex (B), and it sends its output to the hippocampus (D,E) by way of a connecting axonal tract (c) that is known as the "perforant path." The hippocampus returns its output by another tract (d) first to the subiculum (C), which directs its output into motor systems of the forebrain, and second to the deeper layers of the entorhinal cortex (A), from whence its output goes back to all of the sensory neocortex (B). Reprinted from Ramon Y Cajal, S (1904) *Textura del sistema nerviosa del hombre y los vertebrados*, p. 1028. Madrid: Nicolas Moya.

This anatomical organization and the attendant flow of neural activity are found in vertebrates at all levels of the phylogenetic tree. Here I speculate that the convergence of all forms of sensory input provides a basis for the unity of consciousness; that the transmission to and assembly in the hippocampus provide a short-term memory for integration across the few seconds of the psychological "here and now," that the subicular path provides for the issuance of motor commands, and that the entorhinal return paths provide the basis for reafference updates of the sensory systems.

All parts of the brain participate in the elaboration of teleological behavior through and by this fundamental circuit identified by Paul Broca (1873) and elaborated by C. Judson Herrick (1924) and James Papez (1937). All vertebrates have some degrees of ability to predict, plan, and control, and the differences in degree of ability are reflected in the complexities of these other parts. For example, the superiority of humans over cats and dogs is based on the greater development of the frontal lobes (Goldman-Rakic, 1986), as well as other parts of the cerebral hemispheres. What has not heretofore been recognized widely is the role of the entorhinal-hippocampal system in reafference.

sensory input, which is mediated by corollary discharge, and then to a burst of sensory input, which is shaped by the action's sensory consequences. The combination of corollary discharge with the burst of sensory input, both proprioceptive and exteroceptive, produces a transformation of neural activity patterns that have been established by and within the cortex. It is this transformation then that we call perception (Helmholtz, 1879). Perception is also molded by the innumerable influences during past experiences with the same and similar stimuli. That is, the basis for remembering a particular experience lies embedded in the modified strengths of neural connections (Hebb, 1949; Viana di Prisco, 1984).

Once the activity patterns from the several sensory systems have been sent by stages into the entorhinal cortex for assembly, concept formation follows: the creation of an organic whole from the combined modalities of all sensory input. Transmitted to and temporarily stored in the hippocampus, the integrated input serves to update the brain's central world model or cognitive map (O'Keefe and Nadel, 1978). The output from the hippocampus returns to the entorhinal cortex where it will serve to shape the next motor command.

In this continuous give-and-take between action and concept formation, experiential data from the cortex are united with that of the limbic system over time and modality, thus laying the physiological foundation for the unity of consciousness (Kant 1787).

2.3 A View of Consciousness at Work

To understand how this process actually works, let us examine electroencephalograms ("brain waves") recorded (Skarda and Freeman, 1987; Viana di Prisco and Freeman, 1986) from a rabbit's olfactory system (Figure 2.2). The animal's breathing appears as an alternation between two different graphic patterns: when the rabbit inhales, the machine registers a burst of activity; and when it exhales, the pattern collapses.

Each inhalation is the consequence of a motor command. As the rabbit sniffs, exploring its environment, its olfactory bulb is under direct control of the limbic system, which drives the olfactory analyzer to expect input from odors. What is seen in the EEG is the bulb's transition from an elemental, chaotic, basal state (exhalation) to an ordered one (inhalation). Order thus emerges from chaos on limbic command. Each inhalation contributes afresh to the world image evolving in the animal's brain. The world image is built on ordered patterns that emerge from the chaos in which the brain is immersed by its own hand, so to speak (Viana di Prisco and Freeman, 1986).

Similarly erratic, unpredictable, and apparently "noisy" brain waves have been recorded from other sensory cortices, and some instances have displayed the same pattern that are observed here in the sense of smell. I postulate that order may arise from chaos in the visual system, for example, each time the subject moves its eyes, the visual equivalent of taking a sniff. Everywhere throughout the brain is disorder—chaos—from which order emerges (Prigogine, 1980).

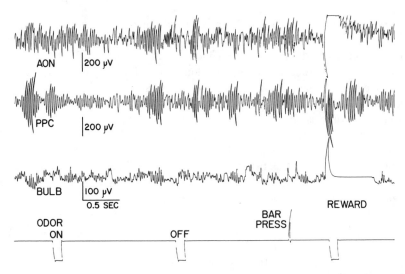

FIGURE 2.2 The activity of the brain gives rise to electrical signs that we use to decipher brain operations. Here are shown traces of those signs that were recorded as electroencephalographic (EEG) waves from the olfactory system of a cat as it performed a conditioned response—pressing a bar to receive water upon receiving a conditioned stimulus, an odor of amyl acetate (banana oil). The electromyogram (EMG) was used for detection of the motor act. Each "burst" in the EEG from the olfactory bulb and the prepyriform cortex (PPC) is the sign of a limbic "command" to sample the olfactory environment with the nose by sniffing.

There has been considerable interest lately in the role played by chaotic generators in the origin of neural activity. This kind of activity, although it looks erratic, random, and unpredictable, does have structure, as can be shown by plotting one recording against others in a graph on paper or an oscilloscope. If we take our anatomical diagrams, express them as networks of equations, and then solve the equations with a digital computer (Freeman, 1987), the solutions are remarkably similar to the rabbit's brain wave patterns. This shows that we should not think about the rabbit's brain waves as "noise." Instead, we should consider them as manifesting forms of nerve cell activity deliberately generated and shaped to meet the brain's requirements.

We can say that the chaos then is an intrinsic property of large masses of neurons interacting with others of a similar kind and thus is a characteristic of the neural system as an interactive whole (Skarda and Freeman, 1987). One property of a chaotic generator that will be of the utmost importance in future philosophical inquiries is its capacity to create information, as well as to destroy it. We do not yet know how this occurs in perception or how to replicate it in the laboratory. Yet, it follows that we should look to these patterns of activity for information about the neurological origins of consciousness and the creative powers of the brain.

In experiments, if we disconnect two particular elements of the olfactory neurons, what results is not the chaos-order alternation that we saw above, but a rigid three/sec spike-and-wave pattern (Freeman, 1986). This pattern (Figure 2.3) resembles that of a well-known form of epilepsy in humans, an illness in which the victim temporarily loses memory, awareness, and contact with the outside world, but maintains postural control and reflexive protection of the body. If you put your hand up to the face of someone suffering from this kind of seizure, he or she will back away from you but not engage you. The activity is no longer goal-directed; the epileptic patient is turned off, and the flow of consciousness is interrupted. The patient will not remember anything that happens during the seizure, nor does he or she retain during that time any residue of experiential learning. Children who suffer these attacks in classrooms may be accused by their teachers of daydreaming and not paying attention; they cannot defend themselves from contumely. This example shows that consciousness depends on the proper functioning of the physiological reflexive systems, not merely on their anatomical integrity, and that imbalances and malfunctions may occur abruptly, as if an electric light had been extinguished.

We may also suppose from the relative rarity of this form of epilepsy that the brain state that maintains normal consciousness is enormously robust and capable of withstanding many sorts of shocks and impacts as long as the neural connections remain in reasonably active condition. What happens when this proviso is not met is truly tragic. One of the best known instances results from the brain deterioration of Alzheimer's disease, in which the failure of brain connections is especially marked along the pathway connecting the entorhinal cortex to the hippocampus. The decay of this connection, known as the perforant path and labeled with particular genetic and biochemical markers, is accompanied by profound memory loss and deterioration of consciousness. The disease has even

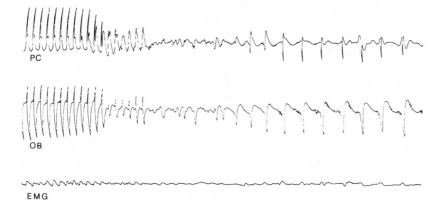

FIGURE 2.3 When a functional overload is placed on a brain pathway by electrical stimulation, causing it to fail temporarily to transmit from the bulb to the prepyriform cortex, a stereotypic epileptic attack supervenes. Along with the seizure discharge as shown, there is loss of normal behavior and of consciousness. Behavior and the EEGs both return simultaneously and abruptly to normal after the seizure has run its course in less than a minute.

been characterized as a "disconnect" syndrome, owing to the virulence of this degeneration. Alzheimer's disease is not unique in this regard, as loss of the optic nerves results in blindness, loss of motor tracts results in paralysis, and so forth; however, unlike these other degenerative conditions, Alzheimer's disease may leave the reflex neural machinery of sensation and motion untouched while destroying the goal-directed, self-organized behavior characteristic of consciousness. Although the literature containing further evidence is vast, we need look no further than this fact for resounding proof of the unity of brain function and consciousness.

2.4 What Is the Origin of Consciousness?

Let us review briefly the physiological medium in which we find consciousness as an intrinsic emergent property. We can say that a conscious "willed" action begins as a self-organized pattern of neural activity in the limbic system. The two principal threads in this pattern are, on the one hand, motor discharge descending through the brainstem and spinal cord to the motor neurons and, on the other, corollary discharge arriving at all the sensory systems, preparing them for the modified sensory input that will accompany the commanded action. Corollary discharge also instructs the somatosensory system to preserve a copy of the command so that the brain has the necessary information to determine whether or not the command has been carried out. In addition, it prepares our visual system to interpret apparent motion of the world as self-induced when we move our eyes (Helmholtz, 1879), prepares the auditory system to distinguish the sound of our own voices when we speak, prepares the gustatory system to expect the taste of coffee when we order one, and so on.

As any behavior takes place, the integrative action of the nervous system combines corollary discharge with the body sense coming from the muscles and joints (proprioception) and from the viscera (interoception) with the activity of the outward-looking senses (exteroception). In each sensory cortex, activity is conditioned by past experience incorporated into the strengths of the synaptic connections between nerve cells (Hebb, 1949) and by those cells' chemical milieu (hormones, neurotransmitters, and neuromodulators) that expresses their excitability, sensitivity, and readiness to respond.

This integrand, which in accordance with Kant (Kant, 1787) we refer to as a concept, must be combined with those preceding it over the previous few seconds in order to guide and shape properly the next command arising out of this evolving pattern of neural activity and in order to produce the proper action. In answer to the introductory question of what is consciousness, it is this flickering process that combines corollary discharge with the messages on all the sensory lines. These sensory lines at once carry fresh input and are shaped by previous experience. Like them, consciousness bears the imprint of both the recent past and the expectation of future action, real or imagined, that will shortly involve the most intimate reaches of the brain and body. The purpose of consciousness is to report in the brain on the consequences of that organ's actions, thereby preparing the next step in the brain's evolution. We experience consciousness just as we do time, force, space, and the extension of our own bodies. These are the unarguable givens that form the basis of being human (Kant, 1787). All of these are gifts from our brains. Since there are cases of brain damage in which each of these has been lost selectively (Freeman, 1986; Sacks, 1985), we are led to conclude that consciousness is an intrinsic and essential attribute of a functioning brain operating in a responsive body.

In these terms we can define what consciousness is, how it arises, what it does, and how, within certain limits, it may be controlled. Yet when we ask what is the origin of consciousness, the problem becomes much more difficult. An origin, of course, is a source or a cause. To use the term implies that there is some precise point at which a thing begins; the word begs impossibly frustrating questions. When on the phylogenetic scale does consciousness emerge? Which species have it and which do not? At what age does it start? When and how do we regain it as each day begins anew? When does it leave us under anesthesia or in old age?

The word too implies that there is some exact place within the brain where consciousness is located; we thus are tempted to ask is this place in the hippocampus, the entorhinal cortex, the brainstem reticular formation, the thalamus, or the neocortex? If we continue with this line of origin-induced questioning, we are ultimately led to ask what causes consciousness and how does it cause us to act.

None of these questions has a natural answer. The necessarily imposed answers that we do give are carefully selected for their social, scientific, legal, or statistical utility and not for their intrinsic relevance to the process itself. Consciousness occurs in a continuum accompanying the flow of matter and energy in and through brains. We can identify brain states that cover the extent of the cooperative interactions holding during a period of time, and we can assign degrees or kinds of consciousness to those brain states based on our subjective experiences and observations. Yet, we cannot pinpoint times or locations of onset and termination. These are the wrong questions to ask. Nor can we assign any single cause to consciousness. In the beginning is the chaos of neural activity, which is already present in the womb when the brain is no bigger than a grain of rice. Is the brain then conscious? Is consciousness caused by conception, by birth, by feeding, or exercise, or education? These questions seem to mock us.

Why do we ask them? We obviously derive some deep satisfaction from characterizing sequences of events in terms of cause and effect. This human tendency, which shapes our way of looking at everything from car accidents to world history, including the workings of consciousness itself, is a direct expression of the neurodynamic relationships that constitute that very consciousness. Each of the half-million motor commands that daily shoot through the brain triggers a change in proprioceptive feedback that informs the brain that the body's position has changed while exteroceptive and interoceptive feedback inform it that the environment has changed.

I suggest that this sequence of command, corollary discharge, and confirmation of change is the most elementary experience we have and that it holds the essence of our relationship to our world. More than a metaphor for our actions into the world with our brains and our bodies, it is the ultimate basis for our understanding anything at all. "I act" means "I cause," and "this then happens" means "this is the effect." We cannot escape these understandings of our actions—nor should we try—and we are equally entitled to express our understandings of our fellow human beings in these terms. By metaphorical extension, we can apply the same framework of understanding to the direct actions of animate and inanimate entities in our world. To do this, we must reduce them first to objects existing and acting forward on the scales of time and distance corresponding to our brains and bodies and second to discrete, unitary actions that correspond to the play of "I" inside onto the "it"outside. These reductions are akin to our intuitions that our floors are flat and that we move through time and three spatial dimensions.

A human being has a birth and a death; a river has a spring and a mouth. The finger pulls a trigger and makes a person die; a flood erodes a bridge foundation and makes the span collapse. We conceive these events in our terms immediately and without difficulty. Yet, when we address the dynamics of systems, each with many elements and many reflexive pathways for the exchange of energy, matter, and information and when these putative actions and reactions reach far over time, distance, and sequential stages, we can only comprehend relationships within the whole, and the cause-effect metaphor is overwhelmed. Any circular relation that we can characterize as a reflexive or feedback loop will ultimately surpass our metaphor. Does the chicken cause the egg or vice versa? When the furnace turns itself off with its own thermostat, does it heat the room or cool it?

It is not that we are forbidden to speculate or understand, but we must be sure that the terms we use are adapted to the needs of the phenomenon observed and not the observer. In brief, the search for an "origin" for consciousness is an anthropomophic fallacy, presupposing that the phenomenon to be understood can be squeezed between the forceps of a human syllogism. The process is much too complex, contradictory, and (to our good fortune) chaotic to be so treated.

There are then many things to ask and learn about consciousness, this physiological reality of human experience, but its origin and cause are not among them. Physiologists have some interesting and fruitful information, techniques, and ideas with which to address this reality. The visionary poet William Blake (1793) declared that "if the doors of perception were cleansed, every thing would appear to man as it is, infinite." The infinite, however, can paralyze, as Blake knew well from his visions. We too know that we could be stunned by the infinite and that the doors are a necessary condition for our existence. It is my hope that together we can begin to remove the accumulated philosophical grime from these doors in order to examine them and better understand what lies both within and beyond them.

Acknowledgments. The author wishes to express his thanks to Catherine Brown, Department of Comparative Literature, University of California at Berkeley, for her assistance in preparing this report. The research was supported by a grant MH06686 from the National Institute of Mental Health, United States Public Health Service.

References

Blake, William (1793): *The marriage of heaven and hell.* New York: Random House (published in 1944)

Broca, P. (1873): *Memoires and d'anthropologie.* Paris: C. Reinwald

Dewey, J. (1896): Mind and consciousness. In: *Intelligence in the modern world.* Ratner, J. (ed.). New York: Random House (published in 1939)

Freeman, W.J. (1984): La fisiologia de las imagenes mentales. *Salud Mentale* 7, 3–8

Freeman, W.J.: (1986): Petit mal seizure spikes in olfactory bulb and cortex caused by runaway inhibition after exhaustion of excitation. *Brain Res. Rev.* 11, 259–284

Freeman, W.J. (1987): Simulation of chaotic EEG patterns with a dynamic model of the olfactory system. *Biol. Cybernetics* 56, 139–150

Gibson, J.J. (1979): *The ecological approach to visual perception.* New York: Houghton Mifflin

Goldman-Rakic, P. (1986): Circuitry of primate prefrontal cortex and regulation of behavior by representational memory. In: *Handbook of physiology—The nervous system V.* Plum, F. (ed.). Baltimore, MD: William & Wilkins, pp. 373–417

Hebb, Donald O. (1949): *The organization of behavior.* New York: Wiley

Helmholtz, H. von (1879): *Physiological optics: vol 3: The perceptions of vision.* Southall, J.P.C. (trans. and ed.). Rochester, NY: Optical Society of America (published in 1925).

Herrick, C. Judson (1924): *Neurological foundations of animal behavior.* New York: Hafner Publishing Company

Kandel, E., Schwartz, J.H. (1981): *Principles of neural science.* New York: North Holland/Elsevier

Kant, Immanuel. (1787): *Kritik der reinen Vernunft.* Frankfurt: Suhrkamp Taschenbücher Wissenschaft (reprinted in 1968)

Klopf, A. H. (1982): *The hedonistic neuron.* Washington, DC: Hemisphere Publishing Corporation

Maturana, H. R., Varela, F. (1980): *Boston studies in the philosophy of science: vol 42. Autopoiesis and cognition.* Cohen, R.S., Wartovsky, M.W. (eds.). New York: Reidel

Merleau-Ponty, M. (1942): *La structure du comportement.* Paris: Presses Universitaires de France

O'Keefe, J., Nadel, L. (1978): *The hippocampus as a cognitive map.* Oxford: Clarendon Press

Papez, J. W. (1937): A proposed mechanism of emotion. *Arch. Neurol Psychiatr.* 38, 725–743

Pavlov, I.P. (1927): Conditioned reflexes. Oxford: Oxford Univ. Press

Prigogine, I. (1980): *From being to becoming: Time and complexity in the physical sciences.* San Francisco: W. H. Freeman

Ramon Y Cajal, S. (1904): *Textura del sistema nerviosa del hombre y los vertebrados.* Madrid: Nicolas Moya

Sacks, O. (1985): *The man who mistook his wife for a hat.* New York: Summit Books

Skarda, C. A., Freeman, W.J. (1987): How brains make chaos in order to make sense of the world. *Behav. Brain Sci.* 10, 161–195

Sherrington, C. S. (1937): *Man on his nature.* Cambridge: Cambridge Univ. Press

Sperry, R. W. ((1950): Neural basis of the spontaneous optokinetic response produced by visual inversion. *J. Comp. Physiol.* 43, 482–289

Viana di Prisco, G. (1984): Hebb synaptic plasticity. *Prog. Neurobiol.* 22, 89–102

Viana di Prisco, G., Freeman, W. J. (1986): Representation espcacio-temporal dinamica de la informacion sensorial en el bulbo olfatorio. *Acta Cientifica Venezolana* 37, 526–531

von Holst, E., Mittelstaedt, H. (1950): Das Reafferenzprinzip (Wechselwirckung zwischen Zentralnervensystem und Peripherie). *Naturwissenschaften* 37, 464–476

3

Representation of Information in the Brain

E. Roy John

Neurons in the brain have a resting potential across their membrane, which is usually about 90 mV with the outside positive with respect to the inside. When this membrane potential is sufficiently decreased, or depolarized, by an input stimulus, activation results and the cell fires. Firing results in a powerful after-hyperpolarization, which often lasts 80 to 100 msec and is inhibitory (Llinas and Yarom, 1986). During this inhibitory period, the cell is refractory, and the probability of response to otherwise effective input is absent or much reduced.

Many cells, especially in the thalamus, have a tendency to autorhythmicity. That is, a depolarizing membrane potential rebound occurs after this hyperpolarization so that the cell tends to fire again (Yarom and Llinas, 1987), leading to an oscillatory discharge that can persist for some period. Thus, isolated or repetitive "spontaneous" firing can occur in the absence of input from outside the cell.

The *probability* that a neuron will discharge is proportional to the local extracellular potential as measured by an extracellular microelectrode (Elul, 1972; Fox and O'Brien, 1965; John and Morgades, 1969; Laufer and Verzeano, 1967; Ramos et al., 1976; Schwartz et al., 1976) (Figure 3.1).

Furthermore, the *frequency* at which a neuron will oscillate depends on the resting membrane potential at the time of discharge (Jahnsen and Llinas, 1984). Coherent ensemble activity can bias local extracellular potentials, which can modulate local oscillatory rhythms as a function of shared activity in the neural population.

Adjacent cells in an ensemble appear to affect each other in a synergistic way, which indicates the existence of effective electrical coupling among them (Llinas and Yarom, 1981a and b; Sotelo et al., 1974) possibly due to effects of firing on ionic concentrations in local electrolyte pools.

As the number of neurons within a local ensemble firing close together in time increases, the change in extracellular potential will increase. This will not only result in an increased probability that more cells in the ensemble will fire but also increases the probability that they will enter a similar oscillatory pattern. Interaction between temporal patterns of afferent input and these shared oscillatory patterns generates complex interference patterns that can produce syncopated rather than periodic patterns of coherent firing.

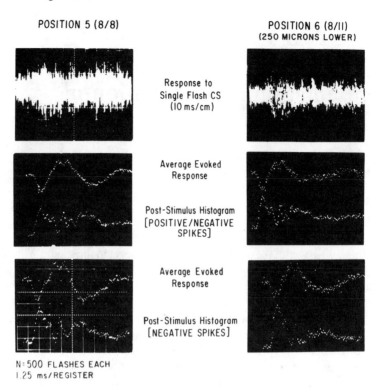

POSITION 5 (8/8)

POSITION 6 (8/11)
(250 MICRONS LOWER)

Response to
Single Flash CS
(10 ms/cm)

Average Evoked
Response

Post-Stimulus Histogram
[POSITIVE/NEGATIVE
SPIKES]

Average Evoked
Response

Post-Stimulus Histogram
[NEGATIVE SPIKES]

N=500 FLASHES EACH
1.25 ms/REGISTER

FIGURE 3.1. The relationship between neuronal firing probability and local extracellular potential can be seen in these recordings from a chronically implanted, movable microelectrode located in the lateral geniculate nucleus of an unrestrained conditioned cat. *Left Column: Top panel,* Multiple unit responses to a single flash conditioned stimulus. *Middle panel, upper trace,* average evoked response (N = 500 flashes) recorded from microelectrode versus a frontal sinus reference, using a low-pass filter (0.2-100 Hz). *Lower trace,* PSH of positive/negative spikes (N=500 flashes) from unit selected out of multiple unit ensemble using a Schmidt trigger after a high-pass filter (500-5000 Hz). *Lower panel, upper trace,* average evoked response recorded as in middle panel. *Lower trace,* PSH from unit producing only negative spikes, recorded as in middle panel. *Right Column,* Analogous data as in left column, but recorded from a different neuronal ensemble 3 days later after lowering the movable microelectrode 250 mV further into the nucleus. Reprinted with permission from *Brain Research Bulletin*, vol 1, Ramos A, Schwartz E, John ER, Evoked potential-unit relationship in behaving cats, copyright 1976, Pergammon Press

These syncopated oscillatory discharges can spread from a few cells to an ensemble, and syncopated rhythms in ensembles can project to other anatomical regions where synchronous firing can be induced in large ensembles of cells (Andersen and Andersen, 1968). Integrated excitatory and inhibitory postsynaptic potentials, which correspond well to these coherent syncopated ensemble discharges, can be recorded from the surface of the human scalp as EEG and event-related potential (ERP) waveshapes.

The frequency composition of these macropotentials seems to reflect a functional neuroanatomical organization that is species-specific and dependent on maturation. Normative age-regression equations have been derived that describe the statistical composition of the resting EEG spectrum in different brain regions of healthy individuals with great accuracy, independent of ethnic background (John et al., 1980, 1987, 1989a). ERP waveshapes in different brain regions of normal individuals can be described as a suitably weighted combination of standardized factor waveshapes obtained by principal component analysis (John et al., 1989b). Sherg (1989) has shown by three-dimensional dipole spatial localization that these factors can reasonably be attributed to coherent activity of generators, in particular subcortical and cortical neuronal ensembles. The basic and clinical implications of these orderly features of brain electrical activity are discussed in this volume by Drs. Prichep and Harmony (Chapters 19 and 23) and elsewhere by John et al. (1989b).

These features of brain electrical activity suggest that some form of frequency encoding may play a significant role in informational transactions within and between brain structures. At the same time, the orderly isomorphic mapping of receptor surfaces upon central structures and of one central structure upon another makes it seem likely that the spatial loci at which activity arises must also be of informational utility, at least in the early stages of encoding. For these reasons, our attention is directed to the role of spatiotemporal patterns of brain electrical activity in the representation of information.

The difference between the syncopated coherent rhythms, which emerge in a neuronal ensemble processing information, and the normative resting local EEG spectrum, which reflects the inherent cortico-cortico and thalamocortical transactions generating baseline oscillatory rhythms, defines the locally encoded information. This can be conceptualized as an informational modulation of a nonstationary carrier wave. Part of this carrier wave is assumed to reflect a process for rhythmic scanning of these representational ensembles (Pitts and McCulloch, 1947).

If we observe a single neuron in a thalamic relay nucleus, such as the lateral geniculate body, using a chronically implanted microelectrode in an unanesthetized and freely moving cat in a dimly lit room (such as in Figure 3.1), we will see that it discharges spontaneously from time to time, as the local extracellular potential fluctuates. If we now present a repetitive flash of light that brightly illuminates the room around the cat, we observe highly variable responses in the lateral geniculate single cell (Figure 3.2). Sometimes it fails to discharge at all in response to the flash because it is refractory. Sometimes it responds to the flash by a repetitive burst of discharges. Yet, on those presentations when it does fire, it fires at different instants of time relative to the time of the flash and at variable rates, presumably depending upon local factors in its environment.

If we use spike height discrimination techniques, we can fractionate a small ensemble and observe the firing patterns of different single cells to the same stimulus. In Figure 3.3, we see that four closely adjacent cells in the lateral geniculate body of an unanesthetized cat display patterns of discharge to a single flash that are not at all synchronized.

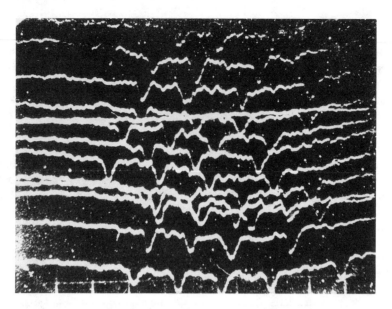

FIGURE 3.2. Variable responses of a single unit to successive flashes of light recorded from a movable microelectrode chronically implanted in the lateral geniculate body, as in Figure 3.1. Each flash triggered the oscilloscope. Time base, 1 msec per division.

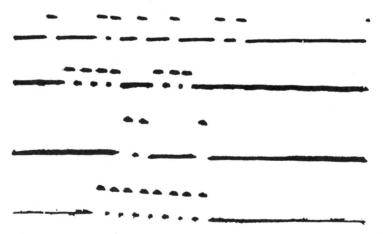

FIGURE 3.3. Simultaneous recording of the discharges of four different closely adjacent cells in the lateral geniculate body of an unrestrained trained cat in response to a single flash of light, recorded as in Figure 3.1 and electronically separated by a spike height discriminator. The oscilloscope sweep was triggered by the flash. Note the absence of synchronization between these cells. Reprinted with permission of Lawrence Erlbaum Associates, Inc. from Thatcher RW, John ER (1977): *Functional Neuroscience, Vol. I. Foundations of Cognitive Processes*, p 255.

Thus, the single cell in a thalamic relay nucleus is an unreliable reporter. It may fire when no change has occurred in that aspect of the environment that it monitors, it may fail to fire when a profound change takes place in the environment, and if it does respond "appropriately" to occurrence of a stimulus event in the modality primarily monitored by this thalamic nucleus, it does so in a highly variable fashion. How then can a reliable report be extracted from a noisy, unreliable, and inconsistent reporter?

If we construct poststimulus histograms (PSHs) showing neural responses of each of the four cells seen in the Figure 3.3 as a function of time, summated to 256 stimuli (Figure 3.4), it can be seen that the probability of firing as a function of time after a light flash is closely similar for the four cells. In some fashion, a cooperative process within the ensemble has influenced the firing probability so that individual cells, although unsynchronized, display a highly similar syncopated average firing pattern.

Using pulse height discrimination, the responses of a single cell to a stimulus can be separated out of the activity of a neural ensemble and averaged to construct a "poststimulus time histogram." The PSH contour that emerges after 5000 repetitions of the stimulus reveals the relative probability of discharge of that cell as a function of time. Recording from the same microelectrode but setting the electronic threshold so that responses of multiple units in the neuronal ensemble are averaged, the average firing pattern of the ensemble converges to the same temporal function more rapidly, yielding almost the identical PSH contour after only 500 repetitions of the stimulus (Figure 3.5).

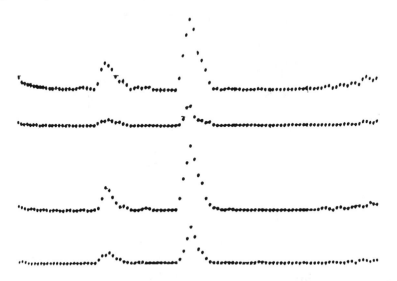

FIGURE 3.4. PSHs showing the average temporal distribution of discharge in the same four cells seen in Figure 3.3 in response to 256 repetitions of the flash stimulus. Note that the four asynchronous cells converge to a very similar *average* firing pattern. Reprinted with permission of Lawrence Erlbaum Associates, Inc. from Thatcher RW, John ER (1977): *Functional Neuroscience, Vol. I. Foundations of Cognitive Processes*, p 255.

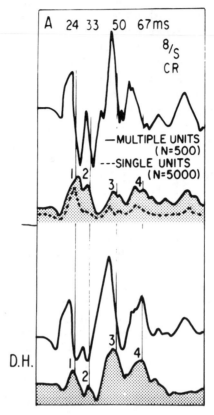

FIGURE 3.5. When stimuli are presented repetitively, the average firing pattern of a single cell within a neural ensemble in the lateral geniculate body converges slowly (*dotted line,* N=5000) to the contour characteristically displayed by the whole ensemble to the same stimulus after a much smaller number of repetitions (*solid line,* N=500). Note the close correspondence between the locally recorded average evoked response (*upper trace*) and the ensemble firing probability shown by the PSH (*shaded area*). Reprinted with permission of Academic Press from John ER, Morgades P (1969): Patterns and anatomical distribution of evoked potentials and multiple unit activity by conditioned stimuli in trained cats. *Comm. Behav. Biol.* 3:204.

Thus, although different elements of an ensemble are subject to influences that produce similarities in their long-term statistical behavior, the temporal pattern of departure from random discharge emerges much sooner from the average behavior of a multineuronal ensemble than from the behavior of a single cell. This indicates an *ergodic* aspect of neural activity. The behavior of a single element of an ensemble converges over a long time period to the short-term behavior of the ensemble as a whole. The "information content" in a neuronal ensemble is proposed to be the statistically significant departure from randomness (i.e., coherence) in the ensemble firing pattern across some period of time, which we will term the "perceptual frame" (Efron, 1970).

The basic ensemble in the brain is probably the single cortical column, which contains a large number of neurons. Cooperative ensembles can be envisaged, ranging from groups of columns to cortical architectonic areas to entire thalamic nuclei. Presumably, columnar information cascades in a convergent manner to yield multicolumnar patterns that are "sensations."

If we compute the PSH within a local ensemble in response to a particular conditioned stimulus (CS1), it has a characteristic spatial and temporal pattern of deviation from randomness, a "coherence signature." The PSH elicited within the same ensemble in response to a different conditioned stimulus, CS2, has a quite different coherence signature.

Using moving microelectrodes to explore the difference between PSHs and evoked potentials elicited by two different conditioned stimuli, CS1 and CS2, within neuronal ensembles at different locations within an anatomical structure, it was found that the variance between coherence signatures elicited within the same ensemble by different stimuli was greater than the variance between coherence signatures elicited across different ensembles by the same stimulus (John and Morgades, 1969). This finding indicates that information about the stimuli is encoded, not as local discharge in "labeled lines" consisting of different cells dedicated to representation of specific restricted information, but rather as different coherence signatures in sets of cells that can represent many different items of information.

During learning, EEG and ERP waveshapes that are initially disparate in different anatomical regions become similar, forming a "representational system." Novel stimuli elicit ERPs with only short latency components. After learning, long latency components augment the ERP waveshapes, which become highly similar in different brain regions. Furthermore, these waveshapes can become highly synchronized and even apparently *simultaneous* in regions distant from one another. This closely coupled activity is difficult to explain by conventional concepts of neural transmission and suggests a shift to some kind of resonant oscillation (Figure 3.6) (John, 1967). Some of these temporal patterns appear to be exogenous, but others are endogenous (John, 1967; John and Killam, 1959 and 1960).

Neuronal ensembles in anatomical regions with grossly different connectivity and morphological structure, such as the lateral geniculate body (LG) and the dorsal hippocampus (DH), display very similar PSHs and ERPs during correct behavioral responses to conditioned stimuli (Figure 3.7A) (John and Morgades, 1969). The LG and DH show similar responses but a different coherence signature when a different behavior is correctly performed to another conditioned stimulus (Figure 3.7B). However, when conditioned response failure occurs (Figure 3.7C) or novel stimuli appear (Figure 3.7D), this similarity of the coherence signature in different parts of the "representational system" deteriorates.

In generalization to novel stimuli, certain regions display such activity as an interference pattern compatible with the absent conditioned stimulus, as well as with the actual physical stimulus, as seen in Figures 3.8 and 3.9 (John, 1967).

If the ERP waveshape elicited during failure to respond to a novel stimulus (NR "no response") is subtracted from the waveshape elicited during generalization,

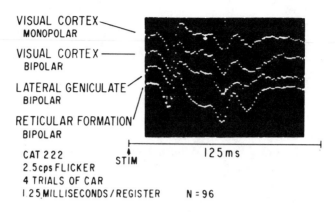

VISUAL CORTEX
MONOPOLAR

VISUAL CORTEX
BIPOLAR

LATERAL GENICULATE
BIPOLAR

RETICULAR FORMATION
BIPOLAR

CAT 222
2.5 cps FLICKER
4 TRIALS OF CAR
I 25 MILLISECONDS / REGISTER

STIM

I25ms

N = 96

FIGURE 3.6. Average evoked responses recorded simultaneously from four different electrode derivations chronically implanted in the brain of a cat trained to perform differentiated avoidance (CAR) and approach (CR) responses to two different repetition rates of a flicker-conditioned stimulus. From top to bottom, the tracings are from (1) an epidural electrode over the visual cortex, recorded *monopolar* versus a frontal sinus reference electrode; (2) the same visual cortex electrode recorded *bipolar* relative to a second epidural electrode located 2 mm anterior to the first; (3) a *bipolar* electrode pair located in the lateral geniculate body and separated by 1 mm; (4) a *bipolar* electrode pair located in the mesencephalic reticular formation and separated by 1 mm. Number of stimuli = 96, resolution = 1.25 msec/bin. Note the high similarity and virtual simultaneity of these locally generated voltage oscillations, which modulated the response probability of neuronal populations in these distant brain regions. Reprinted with permission from John, E. (1967): *Mechanisms of memory.* New York: Academic Press.

when a previously learned behavioral response is performed to the novel stimulus, a "difference waveshape" remains. This difference wave or "readout potential" arises with apparent simultaneity in an extensive anatomical system, including the sensory-specific cortex (visual cortex in this example) and the midbrain reticular formation, and thence propagates centrifugally to more "distal" regions. (Figure 3.10) (John, 1967). This seems to indicate that neuronal ensembles capable of representing the same past experience are distributed in different brain regions and somehow enter a highly synchronized or *resonant* mode.

In order to control for pseudoconditioning and other nonspecific origins of these endogenous facsimiles of responses to absent stimuli, the method of "differential generalization" has been utilized (John, 1972). In this method, stimuli intermediate (CS3) between stimuli CS1 and CS2 are randomly interspersed in a random sequence of CS1 and CS2. When CS3 is treated as CS1, certain brain regions display ERPs as if CS1 had actually been presented. When CS3 is treated behaviorally as CS2, these regions display ERP waveshapes as if CS2 had been presented (Figure 3.11) (John, 1972). Thus, the same physical stimulus, V3, can evoke different ERP waveshapes or coherence signatures depending upon which memory is most activated by that input.

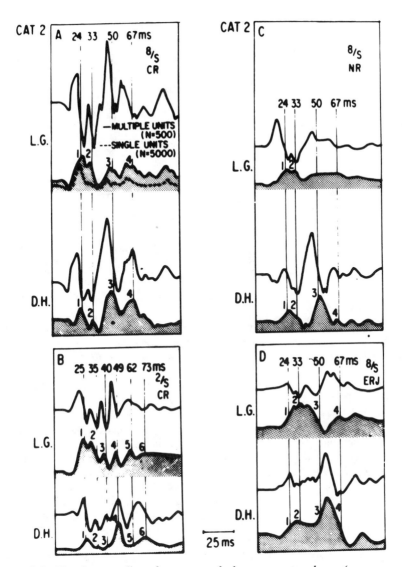

FIGURE 3.7. The close coupling of average evoked response waveshapes (*upper curves*) and the latencies of peaks in the PSHs (*shaded areas*) between different brain regions is a reflection of dynamic system organization. *Panel A* shows high similiarity between the left lateral geniculate (LG) and the right dorsal hippocampus (DH) when a cat makes a correct conditioned response to the *left* lever on a work panel in response to an 8/sec flicker-conditioned stimulis (8/s CR). *Panel B* shows comparable high similarity between LG and DH, while a different coherence signature is displayed, when the cat makes a correct conditioned response to the *right* lever on the work panel in response to a 2/sec flicker SC (2/sec CR). *Panel C* shows uncoupling of LG and DH, especially after 33 msec, when an error of omission occurs a few minutes later and the cat fails to perform (8/sec NR). *Panel D* shows disparate response patterns in the LG and DH when the flicker usually constituting the 8/sec CS illuminates the experimenter's face (8/sec ERJ). Reprinted with permission of Academic Press from John ER. Morgades P (1969): Patterns and anatomical distribution of evoked potentials and multiple unit activity by conditioned stimuli in trained cats. *Comm. Behav. Biol.* 3:181–207.

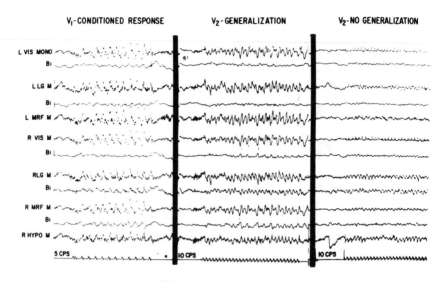

FIGURE 3.8. EEG recordings obtained in the same session from a cat trained to obtain food by pressing a lever in response to a 5 cps flicker (V_1). MONO=monopolar recording relative to a frontal sinus reference; B1 = bipolar recording between two closely spaced electrodes 1-2 mm apart; VIS = visual cortex; LG = lateral geniculate, MRF = mesencephalic reticular formation; HYPO = dorsal hippocampus. *Left panel*, During correct performance to V_1; *middle panel*, during generalization to a 10 cps flicker (V_2); *right panel*, during a presentation of V_2 when no generalization occurred. During correct response to V_1, a clear 5/sec rhythm corresponding to the actual stimulus can be seen in the monopolar recording from every structure. The bipolar records display a similar rhythm, at lower amplitude, indicating that it is being generated locally in each of these regions between the closely spaced electrodes. During generalization to V_2, a complex pattern that resembles an interference pattern between 5 and 10 Hz waves appears in the visual cortex and reticular formation, especially on the right side, in both monopolar and bipolar derivations. At the same time, the lateral geniculate and hippocampus are dominated by 10/sec activity. This suggests that the 10/sec input from the actual stimulus V_2, via the lateral geniculate, is activating a representation of the familiar 5/sec V_1 in both the cortex and the reticular formation. When V_2 fails to elicit generalization, no 5/sec rhythms appear, suggesting that the memory model was not activated.

If the same differential generalization maneuver is carried out while recording from single cells, two kinds of neurons can be identified (Ramos et al., 1976; Schwartz et al., 1976). "Stable" neurons display firing patterns and PSHs characteristic for each stimulus independent of the subsequent behavioral response. These cells are statistically consistent reporters of environmental events. "Plastic" neurons display firing patterns and PSHs independent of the physical stimulus, but predictively characteristic of the subsequent behavioral response. When activated by an appropriate syncopated oscillatory pattern reflecting input to nearby stable cells or by input from remote memory systems, these cells release a stored pattern of similarly syncopated oscillatory activity representing previous experience. Stable and plastic cells can be encountered in a variety of anatomical structures, reflecting the diffuse representation of both present events and previous memories (Figure 3.12A and B) (Ramos et al., 1976; Schwartz, et al., 1976).

I have tried to build a model that reconciles the phenomena described here with various other known features of brain anatomy and physiology. In this model, sequential steps in information processing are related to temporal features of the cortical ERP. Although the intermittent stimulation characteristic of ERP studies is very different from the continuous input that occurs in natural life, the ERP methodology permits us to track the events that take place as a discrete packet of information, delivered at a known instant, is processed by the brain. In each of the figures that illustrate this model, the same set of structural elements is displayed. Solid lines represent anatomical pathways newly active at each stage, and dotted lines represent pathways already activated, which may continue to transmit.

Step I (Figure 3.13): Stimulation of the receptor surface in a sensory modality has two major consequences. First, large numbers of stable cells in the principal thalamic relay nucleus and primary sensory cortex of that modality are activated selectively. Afferent input to axosomatic synapses of stable cells causes depolarization of cell body membranes, especially in cortical layer IV, and the discharge of some of these cells, accompanied by a surface-positive voltage shift that corresponds to component P1 of the ERP. We know from the work of Libet that direct electrical stimulation of the somatosensory cortex does not produce a subjective sensation for about 200 msec, so this discharge is not sufficient for the information to be perceived (Libet, 1973).

At the same time, activity in collateral pathways enters the midbrain reticular formation (MRF). The spatiotemporal pattern of this input activates neuronal ensembles of plastic cells, which are those memory models with the lowest threshold because of their congruence with recent environmental events, internal state, and the momentary input. Temporally patterned output from MRF enters the intralaminar nuclei of the thalamus and structures in the limbic system where ensembles of plastic cells that resonate to the "coherence signature" of this input begin to fire.

Step II (Figure 3.14): Non-sensory-specific axodendritic input from the intralaminar nuclei facilitates the discharge of many of the excited stable cells in the cortex. This causes an axosomatic input to ensembles of plastic cells in the sensory and association cortex. The axodendritic depolarization causes a surface negative shift that corresponds to the beginning of component N1 of the ERP. At the same time, the plastic cell ensembles in the MRF and the intralaminar nuclei responsible for that cortical input enter a hyperpolarized refractory state and are inhibited.

Step III (Figure 3.15): Convergent distributions of stable cell axons on plastic cell soma impress the coherence signature that describes momentary reality. This is one-half of the information required for the plastic cells to serve as comparators (John, 1967; Sokolov, 1963). The other half is provided by the coherence signature of the axodendritic input reflecting the most relevant memory model that was activated in the nonspecific system. The synchronized modulation of somatic and

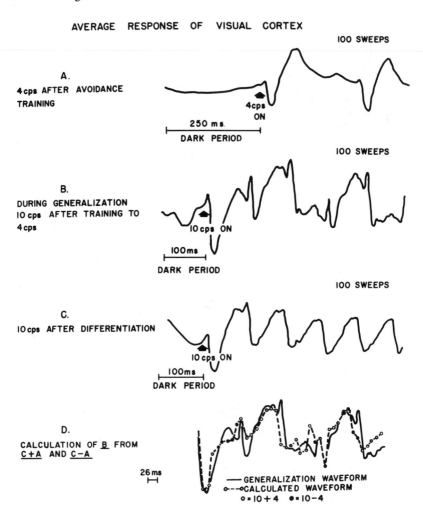

FIGURE 3.9. Average evoked responses recorded from bipolar electrodes chronically implanted in the visual cortex (*Left*) and the mesencephalic reticular formation (*Right*) of a trained cat. Each EP is the average of 100 stimuli. A, During trials resulting in correct performance of a conditioned avoidance response to a 4 cps flicker that was the conditioned stimulus used during training. B, During trials resulting in generalization to a novel 10 cps flicker after reaching a high performance criterion to the 4 cps CS. C, During trials resulting

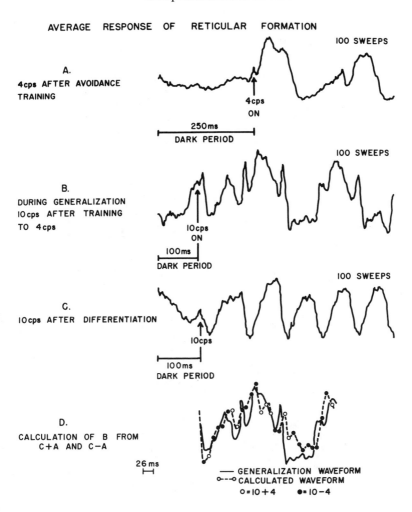

AVERAGE   RESPONSE   OF   RETICULAR   FORMATION

100 SWEEPS

A.
4cps AFTER AVOIDANCE
TRAINING

4cps
ON

250ms
DARK PERIOD

100 SWEEPS

B.
DURING GENERALIZATION
10cps AFTER TRAINING
TO 4cps

10cps
ON

100ms
DARK PERIOD

100 SWEEPS

C.
10cps AFTER DIFFERENTIATION

10cps

100ms
DARK PERIOD

D.
CALCULATION OF B FROM
C+A AND C−A

26 ms

——— GENERALIZATION WAVEFORM
○---○ CALCULATED WAVEFORM
○=10+4   ●=10−4

in no generalization after differential training to respond to 4 cps but not to 10 cps. D, Superposition of waveshape B, actually recorded during generalizations (*solid waveform*) when 10 cps was responded to as if it were 4 cps, and the interference pattern (*dotted waveform*) generated from the waveshapes elicited by the 4 cps CS after initial avoidance training and by the 10 cps inhibitory stimulus after differentiation. Reprinted with permission from John, E. (1967): *Mechanisms of memory.* New York: Academic Press.

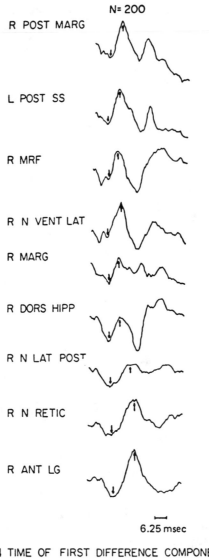

N= 200

R POST MARG

L POST SS

R MRF

R N VENT LAT

R MARG

R DORS HIPP

R N LAT POST

R N RETIC

R ANT LG

6.25 msec

ꟾ TIME OF FIRST DIFFERENCE COMPONENT
ꟾ TIME OF SECOND DIFFERENCE COMPONENT

FIGURE 3.10. Difference waves constructed by subtracting the ERP waveshape elicited during failure to respond to a novel stimulus from the ERP elicited by the same stimulus when behavioral generalization occurs. Note the similarity of the difference waveshape and the closely synchronized onset of the earliest difference components, marked by the arrows, in some cortical, reticular, and thalamic regions. Reprinted with permission from John, E. (1967): *Mechanisms of memory*. New York: Academic Press.

CAT 6 (LLG$_b$)

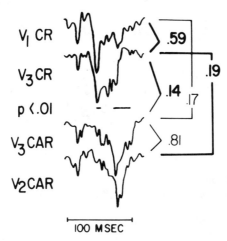

FIGURE 3.11. ERPs elicited in lateral geniculate body (bipolar derivation) in a cat differentially trained to perform a conditioned approach response (CR) to one flicker frequency (V_1) and a conditioned avoidance response (CAR) to a second flicker frequency (V_2), during differential generalization to an intermediate flicker frequency (V_3). When V_3 elicited the CR, the resulting ERP waveshapes (V_3CR) closely resembled the waveshape (V_1CR) usually elicited by the conditioned stimulus V_1, which was the appropriate cue for that behavior. Conversely, when V_3 elicited the CAR, the ERP waveshape (V_3CAR) closely resembled the waveshape (V_2CAR) usually elicited by the cue V_2 for that behavior. Numbers to the right represent the correlation coefficients between the waveshapes indicated by the brackets. The solid line segments between V_3CR and V_3CAR indicate the latency intervals during which the significance of the difference then exceeded the 0.01 level. Thus, the same physical stimulus, V_3, can elicit significantly different ERP waveshapes depending upon the *meaning* attributed to that event, i.e., which memory is most activated.

dendritic excitability selects that subset of plastic cells that resonate to the coincidence. Their discharge reports a "match" between present reality and past experience, a prerequisite for *identification* of the momentary sensation. The concomitant deep depolarization causes the surface potential to begin to swing positive.

This cortical output volley is still characterized by the coherence signature for which the responsive plastic cell ensemble had been tuned. Arriving in the MRF, this input resonates with the plastic cell ensembles that initiated the representation of the model, now rebounding from their hyperpolarization. This multisensory

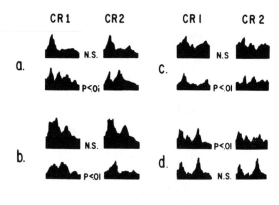

FIGURE 3.12. A. PSHs constructed from stable and plastic cells isolated electronically from multiple unit recordings obtained from chronically implanted microelectrodes in trained cats during differential generalization to a novel visual stimulus, V_3, resulting in pressing of the left pedal (CR_1) or the right pedal (CR_2) on a work panel to obtain food. The cats had initially been trained to perform CR_1 in response to a flicker CS at frequency V_1 and CR_2 to a CS at frequency V_2. The two columns of PSHs on the left come from cells in the lateral geniculate nucleus (LGN) and on the right from cells in area 18 of the visual cortex (cortex). Each pair of cells—a, b, c, and d—come from a different cat. In each pair, the stable cell shows no significant difference (N.S.) between the PSHs elicited by V_3 during CR_1 and CR_2, whereas the closely nearby, simultaneously *recorded plastic cell* shows two firing patterns that are significantly different ($P < 0.01$). Reprinted with permission from *Brain Research Bulletin*, vol 1, Ramos A, Schwartz E, John ER, *An examination of the participation of neurons in readout from memory*, copyright 1976, Pergammon Press. B. PSHs recorded from a plastic cell in LGN of one cat and in the visual cortex of a second cat during the same differential generalization test described in Figure 3.12A. V_1CR_1 = PSH during correct left pedal responses to V_1; V_3CR_1 = PSH during left pedal generalization responses to V_3; V_2CR_1 = PSH during erroneous left pedal responses to V_2, for which right pedal responses would be correct; V_2CR_2 = PSH during correct right pedal responses to V_2; V_3CR_2 = PSH during right pedal generalization responses to V_3. All PSHs in Figures 3.12A and B were recorded in response to about 200 individual flashes of light distributed across 15 to 20 trials resulting in the indicated behavior. In each trial, data collection stopped at the first overt movement of the cat. (*Figure continued.*)

FIGURE 3.13. Step I: Stimulation of the receptor surface in a particular sensory modality activates that set of stable cells in the principal thalamic nucleus of that modality that are responsive at the moment when the afferent input reaches the thalamus. Cortical projections from these cells activate axosomatic synapses of stable cells, especially in layer IV of the corresponding primary sensory cortex. Discharge of some of these cells causes the surface-positive shift that is component P_1 of the ERP. This is the central representation best corresponding to unimodal present reality, which is not perceived (see text).

At the same time, activity in collateral pathways enters the midbrain reticular formation (MRF), where it activates neuronal ensembles of plastic cells that are those memory models most congruent with internal state, recent events, and "present reality." Spatiotemporally patterned output from MRF is projected to the intralaminar nuclei of the thalamus and to limbic structures. Ensembles of plastic cells in those structures that resonate to the coherence signature of the afferent input begin to fire.

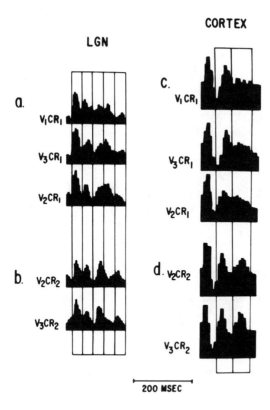

LGN

CORTEX

a.

V₁CR₁

V₃CR₁

V₂CR₁

b.

V₂CR₂

V₃CR₂

c.

V₁CR₁

V₃CR₁

V₂CR₁

d.

V₂CR₂

V₃CR₂

200 MSEC

FIGURE 3.12B. *continued.*

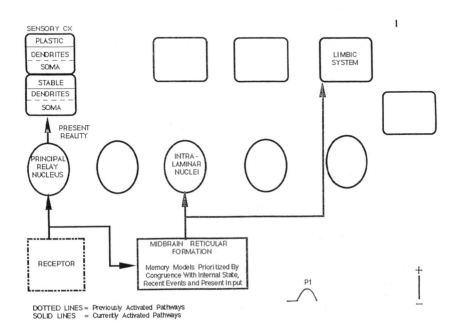

SENSORY CX

| PLASTIC |
| DENDRITES |
| SOMA |
| STABLE |
| DENDRITES |
| SOMA |

LIMBIC SYSTEM

PRESENT REALITY

PRINCIPAL RELAY NUCLEUS

INTRA-LAMINAR NUCLEI

RECEPTOR

MIDBRAIN RETICULAR FORMATION

Memory Models Prioritized By Congruence With Internal State, Recent Events and Present Input

P1

DOTTED LINES = Previously Activated Pathways
SOLID LINES = Currently Activated Pathways

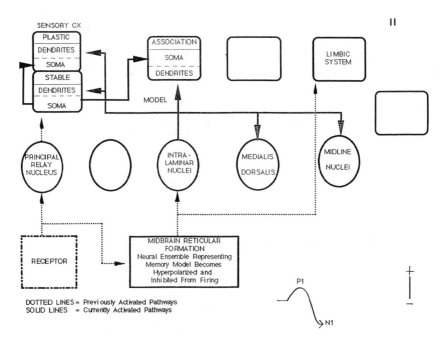

FIGURE 3.14. Step II: Axodendritic input from the intralaminar nuclei facilitates the discharge of stable cells in the sensory signature reflecting the concordance between the axosomatic sensory-specific and axodendritic non-sensory-specific coincident activation. The axonal output from these stable cells impinges upon the soma of the adjacent plastic cells in sensory cortex and in more distant association cortex, also receiving axodendritic input from the intralaminar nuclei. Axodendritic depolarization causes onset of a surface negative shift that will become component N_1 of the ERP. The coherence signature also propagates to nucleus medialis dorsalis and the midline nuclei of the thalamus. At the same time, plastic cell ensembles in MRF and intralaminar nuclei previously engaged by the coherence signature of the activated model become hyperpolarized and are inhibited.

feedback confirms that the afferent stimulus configuration has been recognized. Arriving in nucleus reticularis, this feedback selects a resonating set of neurons that receive the template for the *focus of attention*. In addition, this template is installed in medialis dorsalis. The cortical cells responsible for the feedback become hyperpolarized.

The coherence signature of these ensembles, representing an aspect of perceptual reality, is reflected by a corresponding macropotential waveshape composed of the integrated excitatory and inhibitory postsynaptic potential fluctuations during this interval. John et al. (1967) showed that squares and circles produced discriminable ERP waveshapes independent of stimulus size. Hudspeth (personal communication, 1969) has shown that color (red, green), form (circle, diamond), or conceptual (equivalent pictures and words) attributes contribute identifiable components to ERP waveshapes in this latency region.

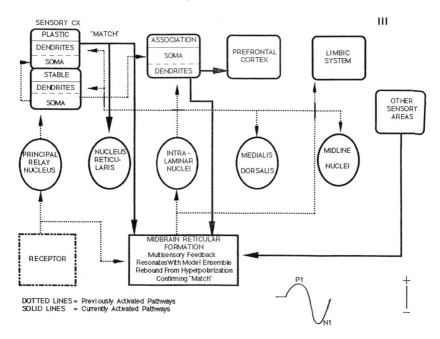

FIGURE 3.15. Step III: Synchronous modulation of excitability of plastic cells in the cortex by convergeance of the axosomatic coherence signature from the stable cortical cells, representing momentary reality, and the axodendritic coherence signature from the non-specific system, representing the most relevant memory model, selects the subset of plastic cells that resonate to the coincidence. Their discharge confirms a "match" between present reality and past experience, which permits identification of the momentary sensation and causes a deep depolarization. The surface potential begins to swing positive after the peak of component N_1. The cortical comparator cells become hyperpolarized.

The efferent cortical volley is characterized by the coherence signature that achieved congruence. Arriving in MRF, this descending input from many cortical areas resonates with the plastic cell ensembles that represented the activated model, now rebounding from hyperpolarization. This feedback confirms multisensory recognition. Arriving in nucleus reticularis, it specifies a template for the *focus of attention.*

Step IV (Figure 3.16): Nucleus reticularis imposes this template upon the primary relay nucleus, intralaminar nuclei, and midline nuclei. This corresponds to the "searchlight hypothesis" of Crick (1984). Attention is focused upon the familiar environmental stimuli. The activated memory ensemble in MRF and intralaminar nuclei become more dominant as additional cells are recruited by the resonance between multisensory feedback, afferent input, and the model. The cortical ensemble, synchronized by rebound from the hyperpolarization, responds more vigorously to the coincidence between sensory-specific and non-sensory-specific input. Cortical activity becomes more coherent and resonates with activity in the thalamus and MRF, causing a surface positively corresponding to P2 of the ERP.

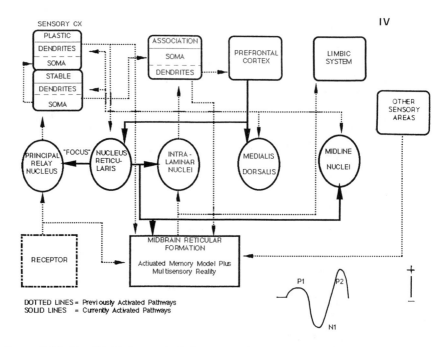

FIGURE 3.16. Step IV: Nucleus reticularis imposes the template upon the relay nuclei, the intralaminar nuclei, the MRF, and the midline nuclei. The searchlight of attention acts as a filter. The interaction between the multisensory feedback and the activated plastic ensembles in cortex and MRF, synchronized by rebound from hyperpolarization, recruits additional cells into a more vigorous resonance between cortex, thalamus, and MRF. These processes coincide with surface-positive component P_2 of the ERP. Conscious *perception*—that is, the content of present reality filtered through the focus of attention and transmuted from sensations in the context of past experience—arises during this interval.

As we saw earlier, when novel stimuli or unattended stimuli are presented, ERPs display only short latency exogenous components. After further experience and learning, similar waveshapes emerge in many regions, and longer latency endogenous components appear. Psychophysical studies indicate that there is a minimum duration of a perception, a perceptual frame that lies between 120 and about 200 msec (Efron, 1970; von Bekesy, 1971). Retroactive inhibition, or backward masking, can be produced by a blank flash delivered about 100 msec after a flash containing information. Using direct electrical stimulation of the brain in cats, I found that disruption of long latency components completely blocked conditioned response performance, although orienting responses to the conditioned stimuli were still displayed and unconditioned responses were performed. Disruption of early components had no behavioral effects (John, 1967). In humans, Penfield (1958) and Libet (1973) have shown that the conscious experience of a peripheral sensory stimulus can be abolished by electrical stimulation of the corresponding sensory cortex for 100 msec, beginning up to 200 msec after the peripheral event. All of these various phenomena were anticipated long ago in the formulation of "cortical scansion" postulated by Pitts and McCulloch (1947).

For all these reasons, I believe that the time interval corresponding to component P2 of the ERP is when *conscious perception* of the figure-ground structure of momentary reality arises, filtered through the focus of attention and made interpretable in the context of past experience. This conception corresponds with that of Hassler (1979) based upon the results of direct brain stimulation in neurosurgical patients. Hassler concluded that, if the specific and nonspecific responses overlapped, conscious perception occurred. Exploring this idea, he was able to restore conscious experience to comatose patients by intermittent stimulation of the intralaminar nuclei. Conversely, inadvertent coagulation of intralaminar nuclei in stereotaxic procedures for Parkinsonism resulted in the loss of all reactions to stimuli, or "coma vigil." (Figure 3.17).

Step V (Figure 3.18): A second hyperpolarization of cortical and thalamic ensembles occurs after the resonance that generates perceptual awareness, causing the surface negative potential corresponding to component N2 of the ERP. Nucleus reticularis gates on the midline nuclei to permit the pattern representing multisensory reality and the matching memory model to reach the limbic system, where *affective evaluation* of the percept occurs. Elsewhere in non-sensory-specific structures, the activated ensembles of plastic cells continue syncopated oscillation, recruiting more cells into the representational system.

Step VI (Figure 3.19): The rebound from the second hyperpolarization releases ensembles of plastic cells in the cortex and thalamus to permit syncopated coherent oscillations that link the posterior cortex, association cortex, prefrontal cortex, limbic system, thalamic nuclei, and MRF into a resonant common mode. A *conscious cognitive experience* ensues, accompanied by a surface positive wave corresponding to component P3 in the ERP.

This common mode of syncopated oscillation or resonance has the result that, during mental activity, EEG and ERP waveshapes in different brain regions become similar and highly coherent (Adey et al., 1961; Bressler, 1987; Gevins et al., 1989; John, 1967; Livanov, 1977). This similarity is superimposed upon the orderly spectral composition of the EEG, which can be described by age-regression equations (John et al., 1987), presumably arising as a physiological manifestation of the functional neuroanatomical organization of the human brain.

The resting normal brain displays a high coherence, reflecting this common mode of syncopated activity. The average resonance level for 109 normal subjects is shown for the 10/20 system in the top half of Figure 3.20. The mean amount of local processing, defined as 100% minus the resonant common mode, is shown below. Note that 40–64% of cortical EEG activity is coherent across the whole cortex.

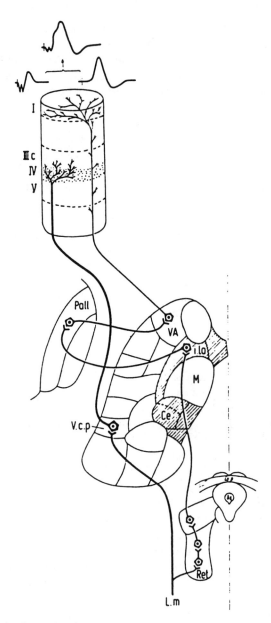

FIGURE 3.17. Anatomical scheme showing sensory specific (*heavy lines*) and nonspecific (*light lines*) pathways enabling somatosensory input to reach axosomatic (layer IV) and axodendritic (layer I) synapses. Axosomatic input alone from direct stimulation of nucleus VCP of the thalamus results in positive-negative ERP waveshape shown to left above diagram (positive down). Axodendritic input alone from intralaminar stimulation causes negative-positive ERP waveshape shown on right. Dual stimulation, or natural somatosensory stimuli, causes triphasic ERP seen at top. Data from brain stimulation of human patients by Hassler (1979), who concluded that conscious perception required overlap of specific and nonspecific activation.

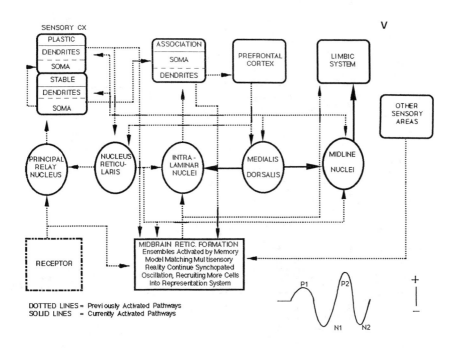

FIGURE 3.18. Step V: A second hyperpolarization of cortical and thalamic ensembles occurs after the resonance that generates awareness, causing surface negative ERP component N_2. Nucleus reticularis gates on the midline nuclei, permitting the output of the system matching reality with memory to reach the limbic system, where *affective evaluation* of the percept occurs. At the same time, ensembles of plastic cells in the previously activated brain regions continue synchronous oscillations, recruiting more cells into resonance.

Figures 3.13 to 3.19 reflect the temporal sequence of events when large regions of cortex are synchronized by a strong intermittent or phasic stimulus, such as is characteristically used in ERP studies. I propose that a similar temporal sequence occurs when a vertical thalamo-cortical-thalamic loop activates a single cortical column (Andersen and Andersen, 1968). Adjacent loop systems can be recruited by strong stimuli or focus of attention (Verzeano, 1963; Verzeano and Negishi, 1960). Thus, the discrete serial sequence of (1) afferent input, (2) matching input from reality with models from memory, (3) focusing of attention, (4) perceptual identification, (5) affective evaluation, and (6) cognitive interpretation, unmasked by the application of averaging methodology to neuronal systems synchronized by the use of unnatural phasic stimuli, actually proceeds asynchromously in parallel but phase-shifted subsystems of thalamo-cortico-thalamic loops under more natural circumstances. The continuously fluctuating waveshapes of the spontaneous EEG reflect summation of postsynaptic potentials from cells in these circulating asynchronous loops, each involving the syncopated coherent behavior of ensembles containing huge numbers of cells.

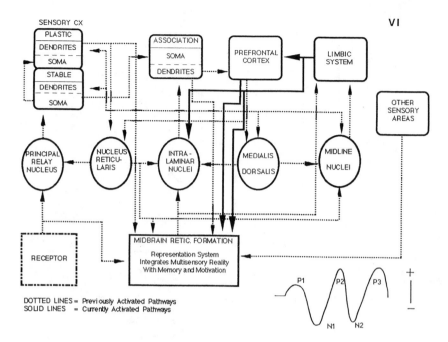

FIGURE 3.19. Step VI: Rebound from the second hyperpolarization releases ensembles of plastic cells in cortex and thalamus. Coherent syncopated oscillations link posterior, association and prefrontal cortex; limbic system; thalamic-specific and nonspecific nuclei; and MRF into a resonant common mode. A *conscious cognitive* experience ensues, accompanied by surface-positive component P_3 of the ERP.

Perceptual frame and backward masking experiments and similar work provide evidence that information processing may occur in discrete steps with a limited time resolution. These experimental situations may reveal the processing limitations when information is constrained to a single loop system. In any recurrent loop circuit of the sort just described, alternations occur between intervals in which neurons in particular levels of the system are excitable and can participate in informational transactions and intervals in which they are nonresponsive, either because of refractoriness caused by their own activity or inhibitory influences from other neurons. However, under ordinary circumstances subjective experience is continuous and permits the discrimination of temporal sequences with far higher resolution than the 50 to 100 msec occupied by each of the processes seen in our ERP analysis. Thus, the continuity of successive moments of consciousness must be mediated by different but parallel loop circuits of cells, which act as multiplexers mediating substantially similar representations of information in sequential instants of time.

How can the temporal continuity of subjective awareness be explained? Each sensory modality is encoding a continuous stream of information from external

(COH)²		MEAN RESONANCE		109 NORMALS
	54.07		54.48	
49.36	61.74	63.17	61.78	50.34
44.67	61.79	64.08	61.79	45.82
40.73	56.01	58.00	53.67	41.12
	44.19		43.42	

1- (COH)²		MEAN LOCAL ORIGIN (%)		
	45.93		45.52	
50.64	38.26	36.83	38.22	49.66
55.33	38.21	35.92	38.21	54.18
59.27	43.99	42.00	46.33	58.88
	55.81		56.58	

FIGURE 3.20. The average cross-spectral coherence of every lead versus every other lead in the International 10/20 Electrode Placement System was computed across a 2-minute resting EEG sample for the delta, theta, alpha, and beta frequency bands in 109 normal subjects. *Top,* The average amount of power in each brain region that was coherent with the rest of the brain was calculated and expressed as "common mode resonance." *Bottom,* "Local processing" was calculated as 100% minus the common node resonance. Array represents the 10/20 system viewed from above, face upward.

reality into discontinuous packets transmitted as the departure from randomness in coherent syncopated discharges of large ensembles, each serving as a channel in a multiplexer. When a series of stimuli is delivered in an ERP paradigm, each stimulus encounters some subpopulation of cells in a responsive state while many others are refractory or already engaged. The ensemble that is responsive encodes the stimulus so as to reflect the coherence signature arriving from the receptors, and the ensuing afferent volley undergoes the sequence of steps in information processing that has just been described. The next stimulus in the series encounters a different responsive ensemble, a different channel in the multiplexer, and the ensuing afferent volley undergoes the same sequence of steps, as shown in Figure 3.21. The method of averaging, which yields the ERP, computationally phase-locks these similar sequences mediated by different neural ensembles so that the similarity of process is revealed as a characteristic ERP waveshape. This obscures the fact that the same temporal sequence of events is being summated from different asynchronous ensembles, as seen in Figures 3.3 and 3.4.

In natural conditions, stimuli do not impinge on receptors at discrete instants of time, with no stimulation between those time points. Instead, stimuli constantly bombard the organism. Refractory periods and recovery cycles select a set of neural elements capable of responding, which changes continuously. Some subset will always be responsive, and the change in the statistical behavior of that

E R P

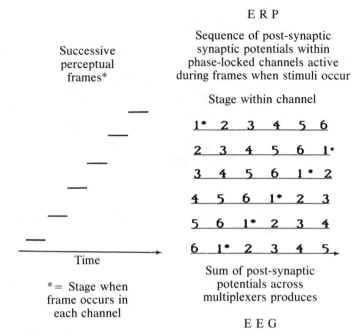

Successive
perceptual
frames*

Sequence of post-synaptic
synaptic potentials within
phase-locked channels active
during frames when stimuli occur

Stage within channel

Time

* = Stage when
frame occurs in
each channel

Sum of post-synaptic
potentials across
multiplexers produces

E E G

FIGURE 3.21. Consider an ERP as reflecting six steps in information processing (1) afferent input, (2) matching reality with a memory model, (3) focusing attention, (4) perceptual identification, (5) affective evaluation, and (6) cognitive interpretation. Each step in this process engages ensembles of neurons in multiple anatomical regions and is reflected by excitatory and inhibitory postsynaptic potentials that can be recorded at the scalp. When successive stimuli are delivered in an ERP paradigm, each stimulus engages some anatomically distributed subensemble of cells that happen to be in a responsive state, and the consequent voltage fluctuations are recorded relative to the time of stimulus delivery. Although each stimulus engages a different subensemble, the ERP waveshape of the averaged voltage fluctuations is reproducible because every subensemble activated by stimulus delivery processes the information through the same sequence of steps and the potentials. Therefore, those steps are time-locked to stimulus delivery by the averaging computer.

Each such subensemble can be envisioned as a separate channel in a multiplexed information processing system. Within a single channel, a "perceptual frame" is represented by the time interval from afferent input (Stage 1) to perceptual identification (Stage 4), as shown in this figure, and lasts about 100 msec (P_{100} to P_{200}). However, because the multiplexed channels are asynchronous, the temporal resolution of the system is much greater than the duration of a perceptual frame. The spontaneous EEG is the sum of the postsynaptic potentials integrated across all of the neural ensembles comprising the full set of multiplexed channels.

randomly selected ensemble brought about by the input is exactly the same as for any other ensemble that might have been responsive.

The spontaneous EEG is the sum of the postsynaptic potentials integrated across all of the asynchronous channels, which are active simultaneously. If we accept the evidence that information processing in any channel constituted by such an ensemble is discontinuous, but that information in each multiplexer channel is processed the same as in any other, it becomes clear that the apparent continuity of subjective awareness, the "stream of consciousness," requires that consciousness be mediated by a system that senses the invariant nonrandom temporal features of a sequential set of processes in some region of space, independent of which neurons contributed to those features at any moment, as shown in Figure 3.22.

The content of consciousness arises from two systems, which can be identified and separated conceptually even though they may be composed of neural elements that are physically interchangeable and indistinguishable, and the interaction between those systems. The first system consists of ensembles of cells in the different major subdivisions of the cortex, the thalamic nuclei, and other subcortical structures, which constitute a species-characteristic functional neuroanatomical organization common to all human beings. This functional neuroanatomical organization is reflected in the orderly composition and evolution of the EEG spectrum in different brain regions, which has been precisely described by our neurometric equations (John et al., 1987). The high volume of transactions between these anatomically distributed ensembles is reflected by the remarkably high coherence of the resting EEG between different brain regions, shown in

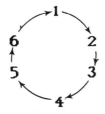

"Hyperneuron"

Constantly updated and refreshed pattern with invariant statistical features independent of cells contributing each feature

Continuous "stream of consciousness"

FIGURE 3.22. Information processing in any subensemble (channel) is discontinuous, involving a number of discrete steps, and all neurons have intervals in which they cannot be responsive. The "stream of consciousness" is subjectively continuous. The overall information content (departure from randomness or coherence signatures in different neural ensembles) of the multiplexed system shown in Figure 3.21 is the nonrandom coherence in an anatomically extensive neural ensemble, with statistical features that are invariant and do not depend on the particular neurons that contribute to those statistics. Subjective experience is an emergent property of this statistical behavior of the ensemble.

Figure 3.20. The resting but conscious brain displays an extremely high level of dynamic organization, which is maintained continuously although necessarily mediated by different neurons from moment to moment. This persisting species-characteristic nonrandom organization of the statistical behavior of widespread neural ensembles provides a stable background of consciousness, responsible for the fundamental *humanness* of our shared subjective experience, as well as for the stability of our body image and our self-perception from instant to instant.

The second system consists of the ensembles of cells at different levels of the sensory-specific and nonsensory systems that serve as the multiplexed channels that transmit incoming information and activated memory models—describing, recognizing, and interpreting our external and internal environments. These cells, which may be recruited from the first system by the fortuitous coincidence between demand and availability, generate activity the quality of which is dominated though not determined by our individual experience. This activity becomes more and more individualized as our lives reflect unique experiences and superimposes an equally nonrandom but personal organization upon the statistical behavior of neural ensembles embedded among and interacting with the ongoing activity in the first, species-characteristic system.

The modulation of the nonrandom neural activity generated in a way inherent in our human nervous systems by the nonrandom neural activity, which is generated in a way uniquely determined by our personal histories and our interaction with our present environment, constitutes the content of consciousness, a continuous and internally consistent subjective experience and self-awareness. The anatomical distribution of the participating ensembles, together with the discontinuous characteristics of neuronal activity, makes it clear that consciousness, subjective experience, and self-awareness cannot be attributed to any process localizable to any discrete set of neurons, connected to all other neurons so as to assess nonrandomness in all remote neural ensembles and responsive continuously.

For these reasons, it seems necessary to postulate that consciousness is an emergent property arising as a physical consequence of nonrandomness in the electrochemical behavior of a very large number of very small elements enclosed within a relatively small volume. This nonrandomness causes very large local voltage gradients across very small distances, establishes electrical resonances, changes ionic gradients, and alters intercellular impedances. Clearly, as the amount of coherent neural activity increases, many characteristics of the physicochemical system containing the neurons, glia, interstitial fluids, and other elements change in dramatic ways. It is to some process generated by these effects on the system as a whole to which we should turn in our search for the origin of consciousness, not to a subset of the elements of which the system is composed.

I call this process a "hyperneuron" to emphasize my belief that it is an emergent property of ensembles of cells that is not derivable from a knowledge of the individual cellular elements. There is presently no basis to decide whether consciousness arises directly from the nonrandom distribution of charge in space or from some action of the charge upon the neurons responsible for the distribution.

References

Adey, W. R., Walter, D. O., Hendrix, C. E. (1961): Computer techniques in correlation and spectral analysis of cerebral slow waves during discriminative behavior. *Exp. Neurol.* 3, 501–524

Andersen, P., Andersen, S. (1968): *Physiological basis of the alpha rhythm.* New York: Appleton-Century-Crofts

Bressler, S. L. (1987): Relation of olfactory bulb and cortex. I, Spatial variation of bulbo-cortical interdependence. *Brain Res.* 409, 285–293

Crick, F. (1984): Function of the thalamic reticular complex: The Searchlight hypothesis. *Proc. Nat. Acad. Sci. USA* 81, 4586–93

Efron, R. (1970): The relation between the duration of a stimulus and the duration of a perception. *Neuropsychologia,* 8:37–35

Elul, R. (1972): The genesis of the EEG. *Int. Rev. Neurobiol.* 15, 228–272

Fox, S., O'Brien, J. (1965): Duplication of evoked potential waveform by curve of probability of firing of a single cell. *Science* 147, 888–890

Gevins, A. S., Bressler, S. L., Morgan, N. H., Cutillo, B. A., White, R. M., Greer, D. S., Illes, J. (1989): Event related covariants during a bimanual visumotor task, I.Methods and analysis of stimulus-and time-locked data. *EEG Clin. Neurol.* 74, 58–75

Hassler, R. (1979): Striatal regulation of adverting and attention directing induced by pallid stimulation. *Appl. Neurophysiol.* 42, 98–102.

Jahnsen, H., Llinas, R. (1984): Ionic bases for the electroresponsiveness and oscillatory properties of guinea-pig thalamic neurones in vitro. *J. Physiol.* 349, 227–247

John, E. R. (1976): *Mechanisms of memory.* New York: Academic Press

John, E. R. (1972): Switchboard versus statistical theories of learning and memory. *Science* 177, 850–864

John, E. R., Killam, K. F. (1959): Electrophysiological correlates of avoidance conditioning in the cat. *J. Pharm. Exp. Ther.* 125, 252–274

John, E. R., Killam, K. F. (1960): Electrophysiological correlates of differential approach-avoidance conditioning in cats. *J. Nerv. Ment. Dis.* 131, 183–201

John, E.R., Morgades, P. (1969): Patterns and anatomical distribution of evoked potentials and multiple unit activity by conditioned stimuli in trained cats. *Comm. Behav. Biol.* 3, 181–207

John, E.R., Herrington, R., Sutton, S. (1967): Effects of visual form on the evoked response. *Science* 155, 1439–1442

John, E. R., Ahn, H., Prichep, L., Trepetin, M., Brown, D., Kaye, H. (1980): Developmental EEG equations for the electronencepalogram. *Science* 210, 1255–1258

John, E. R., Prichep, L., Easton, P. (1987): Normative data banks and neurometrics: Basic concepts, methods and results of norms constructions. In: *Handbook of electroencephalography and clinical neurophysiology; vol III. Computer analysis of the EEG and other neurophysiological signals.* Remond, A. (ed.). Amsterdam: Elsevier, pp. 449–495

John, E. R., Prichep, L. S., Ahn, H., Kaye, H., Brown, D., Easton, P., Karmel, B. Z., Toro, A., Thatcher, R. (1989a): *Neurometric evaluation of brain function in normal and learning disabled children.* Ann Arbor, MI: Univ. of Michigan Press

John, E. R., Prichep, L. S., Friedman, J., Easton, P. (1989b): Neurometric topographic mapping of EEG and evoked potential features: Application to clinical diagnosis and cognitive evaluation. In: *Topographic brain mapping of EEG and evoked potentials.* Maurer, K. (ed.). New York: Springer-Verlag, pp. 90–111

Laufer, M., Verzeano, M. (1967): Periodic activity in the visual system of the cat. *Vision*

Res. 1, 215–229

Libet, B. (1973): Electrical stimulation of cortex in human subjects, and conscious sensory aspects. In: *Handbook of sensory physiology: vol. II.* Iggo, A. (ed.). New York: Springer-Verlag, pp. 743–790

Livanov, M. N. (1977): *Spatial organization of cerebral processes.* New York: John Wiley & Sons.

Llinas, R., Yarom, Y. (1981a): Electrophysiology of mammalian inferior olivary neurones in vitro. Different types of voltage-dependent ionic conductances. *J. Physiol.* 315, 549–567

Llinas, R., Yarom, Y. (1981b): Properties and distribution of ionic conductances generating electroresponsiveness of mammalian inferior olivary neurones in vitro. *J. Physiol.* 315, 569–584

Llinas, R., Yarom, Y. (1986): Oscillatory properties of guinea-pig inferior olivary neurones and their pharmacological modulation: An *in vitro* study. *J. Physiol.* 376, 163

Penfield,W. (1958): *The excitable cortex in conscious man.* Liverpool: Liverpool Press

Pitts, W., McCulloch, W. (1947): How we know universals. *Bull. Math. Biophysics* 9, 127–147

Ramos, A., Schwartz, E. L., John, E. R. (1976): Stable and plastic unit discharge patterns during behavioral generalization. *Science* 192, 393–396

Schwartz, E. L., Ramos, A., John, E. R. (1976): Single cell activity in chronic unit recording: A quantitative study of the unit amplitude spectrum. *Brain Res. Bull.* 1, 57–68

Scherg, M. (1990): Fundamentals of dipole source potential analysis. In: *Auditory evoked magnetic fields and electrical potentials:* Grandori, F., Hoxe, M., Romani, G. L. (eds.). Basel: S. Karger, pp. 40–69

Sokolov, E. (1963): *Perception and the conditioned reflex.* New York: MacMillan

Sotelo, C., Llinas, R., Baker, R. (1974): Structural study of inferior olivary nucleus of the cat: Morphorlogical correlates of electronic coupling. *J. Neurophysiol.* 37, 541–559

Verzeano, M. (1963): Las funciones del sistema nervioso. *Acta Neurol. Latinoam.* 9, 296–307

Verzeano, M., Negishi, K. (1960): Neuronal activity in cortical and thalamic networks. *J. Gen. Physiol.* 43(suppl. 177)

von Bekesy, G. (1971): Auditory backward inhibition in concert halls. *Science* 171, 529–536

Yarom, Y., Llinas, R. (1987): Long-term modifiability of anomalous and delayed rectification in guinea pig inferior olivary neurons. *J. Neurosci.* 7, 1166–1177

Part II

Strange Attractors and Synchronization

4

Spatiotemporal Properties of the α Rhythm

ALFREDO ALVAREZ AMADOR, ROBERTO D. PASCUAL-MARQUI, AND
PEDRO A. VALDÉS-SOSA

4.1 Introduction

Ontogenetic development of the brain is reflected in the EEG, the most prominent changes consisting of variations in its frequency composition. These variations lead to the establishment of the α rhythm in late childhood (Niedermeyer, 1987; Petersén et al., 1975). This phenomenon has been the subject of many studies that have been carried out first by manual measurement of the dominant frequency of the EEG (Lindsley, 1939) and later using more accurate quantitative techniques (Gibbs and Knott, 1949; John et al., 1980; Matoušek and Petersén, 1973a, and b; Petersén et al., 1975) based on the use of EEG spectral analysis.

The α rhythm itself has been the subject of many experimental studies (Andersen and Andersson, 1968; Ilmoniemi et al., 1988; Lopes da Silva and van Leeuwen, 1977, 1978; Lopes da Silva et al., 1973; Vvedensky et al., 1988a and b, 1989; Williamson et al., 1989) and, more recently, of biophysical modeling (Lagerlund and Sharbrough, 1988, 1989; Lopes da Silva et al., 1974, 1976; Nuñez, 1981; Wright and Kydd, 1984a, b, and c; Wright et al., 1984).

Such sustained interest in the origin and development of the α rhythm might seem academic if this type of activity, appearing in relaxed and mentally unoccupied subjects, is just considered to be an epiphenomenon of an "idling" brain. That this is not the case is suggested by several lines of evidence following the establishment of "EEG Developmental Equations" by John et al. (1980, 1987).

The transcultural validity of these quantitative descriptions of brain maturation (Alvarez et al., 1987; John et al., 1980) and their sensitivity to different types of cognitive dysfunction (Gasser et al., 1983a and b; Chapter 19) suggest that they are a reflection of basic mechanisms that are generic to the species. It has been hypothesized that such generality is a consequence of the anatomical and functional determinants of nervous systems associated with the fact of being a human (Chapter 3).

This chapter summarizes research (Alvarez et al., 1985, 1987, 1989; Pascual et al., 1985, 1988; Valdés et al., 1985, 1987) carried out on the quantitative modeling of EEG maturation. Attention is given not only to those aspects in which the

methodology presented has been successful but also to the exceptions and difficulties that, it is hoped, will point out the direction of future refinement.

4.2 Aspects of EEG Studied

The aspects of the EEG that are scrutinized in quantitative EEG studies are
determined by the type of descriptive model used to summarize the full set of
recorded tracings. To date, most quantitative studies of EEG maturation have been
based on the use of the EEG power spectrum (EPS). This is not surprising since
the EPS is easy to compute via the fast Fourier transform (FFT), and under very
general conditions, it provides a complete description of the EEG (Valdés, 1984).
The changes associated with the emergence of the α rhythm are reflected in this
statistic and consist of modifications of both *spectral size* and *spectral shape*.

However, there are many reasons to consider the EEG spectrum itself as a
blurred image of the real processes. Many workers (Barlow, 1985; Jansen et al.,
1981; Lehmann, 1986; Praetorius et al., 1977; Chapter 10) have pointed out an
apparent microstructure of the EEG. Usual spectral analysis is a mathematical

Log (S)

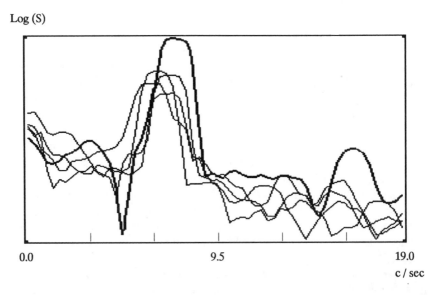

0.0   9.5   19.0

c / sec

FIGURE 4.1. EEG power log spectra, each computed from an EEG segment of 256 points
sampled every 10 msec. Note the differences in the positions of the α peaks. The darker
spectrum shows a well-defined β peak appearing at approximately two times the α frequency.

operation made up of averaging different "segments." This is illustrated in Figure 4.1 in which very short-term spectral analyses are shown (Jimenez et al., unpublished results). There is an appreciable frequency "jitter." The effect of this jitter on usual spectra, which consist of the average of many such short-term spectra, is well known from analogous situations in the time domain with time jitter of evoked response peaks.

Nevertheless, and for the reasons given in the introduction to this chapter, long-term (1 min) frequency spectra may be considered a first approximation to the characterization of system properties. Their study is a first step toward establishing macroscopic neurodynamic relationships that may guide neural theory.

Among the procedures most often used in EEG quantitative analysis, broad band spectral parameter analysis (Matoušek and Petersén, 1973a and b) and normalized slope descriptors (Hjorth, 1970) seem to be of use only as condensations of spectral information since they are not capable of approximating the shape of the spectrum to a satisfactory degree.

In addition, the autoregressive (AR) model, even though it offers the possibility of reconstructing the spectrum very closely, is difficult to interpret in the sense that there is no direct relationship associating changes in its parameters with changes in spectral shape. To overcome the lack of interpretability of AR parameters, an alternative approach was developed by Zetterberg (1969): spectral parameter analysis (SPA). The derived parameters related to spectral peaks are obtained from the AR parameters. SPA has been one of the most successful attempts to describe the power spectrum. For the first time, it enabled the EEG spectrum to be decomposed into different components. Wennberg and Zetterberg (1971) distinguish components of Type I and II. However, this approach shares with AR models the difficulty of the order estimation.

Parametric model fitting in the frequency domain (Brillinger, 1981, 1983; Davies, 1983; Kingma et al., 1976) seems to hold the greatest promise, though serious statistical difficulties have yet to be overcome. First, the estimation procedures are statistically nonoptimal (least squares is generally used). Second, there are no criteria for model building or for model comparison. Different methods are not even compared empirically on the same data set. Even worse, there is no theoretical approach to judging the efficiency of different models. An additional problem is that no practical extensions from univariate to multivariate models are generally available.

In this chapter, a new family of models for the description of the EEG spectrum (Pascual et al., 1988) is discussed. It allows the formulation of a methodology for EEG analysis that appears to have significant advantages over previously reported procedures. This methodology is used to decompose the EEG into two independent processes with possibly different anatomical and physiological correlates. This chapter focuses on the decomposition, using EEG modeling, of frequency components that usually are analyzed together. This family of models represents a numerical dissection of the basic phenomena related to the origin of the EEG.

4.3 EEG Univariate Analysis

4.3.1 Spectra Description

The results of visual inspection of EEG spectra of 3200 individual spectra from 400 children (5 to 12 years) showed two fundamental types of EEG sample power spectra depending on whether or not a well-defined α rhythm was present. In the first case, a peak at about 10 c/sec (α) is present. These peaks are present in most children from 7 years of age. The α peak is superimposed on a background activity that attains a maximum at the low frequency (below 0.4 c/sec) and slowly decays for higher frequencies (Figure 4.2A). The absence of the α rhythm peak is typical of children younger than 5 years of age and of older children with an under-developed or pathological EEG. In such cases, only the background activity is present (Figure 4.2B).

A small percentage of normal children had EEGs that did not conform to the above description. These presented multiple α peaks or peaks in the β band. Until recently, attention has been centered on the typical pattern described in Figure 4.2, and the extension of the model to describe normal variants has not been carried out.

In order to compare different models of EEG spectral changes with age, a description as nonparametric as possible is needed. This may be provided by carrying out a regression with age of each spectral value $S(\omega)$ in a group of normal subjects, thus producing a set of narrow band spectral parameter (NBSP) equations for each derivation. The number of equations in the set depends on the number of frequencies in the FFT analysis (49 discrete frequencies here). The NBSP equa-

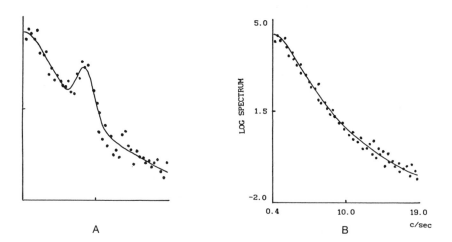

FIGURE 4.2. Single-channel sample EEG power spectra (logarithm) for derivation 01. A, With α rhythm; B, Without α rhythm.

tions provide no data reduction and were calculated only for comparative purposes. Substitution of age t into the regression equations for each frequency provides an estimate of the average spectrum $\bar{S}_t(\omega)$ for any age t (Figure 4.3).

The methodology for producing the regression surfaces which are distinct from regression curves is provided by a particular case of the model for growth curves (Khatri, 1966):

$$f(\omega,t) = \sum_{q}^{Q}\sum_{p}^{P} \beta_{qp}G_p(t)H_q(\omega) + E(\omega,t) \qquad [1]$$

where $f(\omega,t)$ is the mean value of the spectrum at frequency ω, for age t; $G_p(t)$ and $H_q(\omega)$ are known functions of age (t) and frequency (ω), respectively; β_{qp} are unknown parameters, and $E(\omega,1)..E(\omega,t)$ are Gaussian errors having zero mean and dispersion matrix ε. Preliminary results revealed that polynomial functions of degree 3 for G, and of degree 4 for H, provide an adequate description of the full EEG spectrum between 5 and 97 years of age (Galan et al., 1989).

In relation to the two statistical difficulties pointed out previously—estimation procedures are not optimal and no comparison method is available—Pascual et al. (1988) used the likelihood function as a unified approach. It is assumed that the discrete Fourier transform (DFT) of the EEG has an asymptotic complex Gaussian distribution (Brillinger, 1983), from which maximum likelihood (ML) estimators can then be derived. An extensive literature documents the fact that ML estimators possess optimal properties under very general conditions (Cox and Hinkley, 1974, Serfling, 1980). An additional advantage of the use of ML estimation is that a methodology for testing hypotheses can be developed on this basis. Pascual et al. (1988, page 91, eq. 3) described a statistic for measuring the distance between the sample spectrum $\psi(\omega)$ and a theoretical model $\phi(\omega)$. This statistc is the yardstick

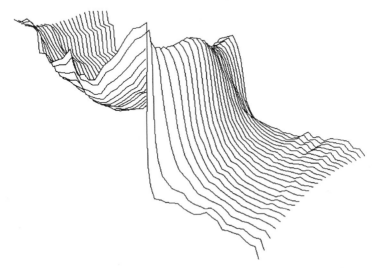

FIGURE 4.3. Regression surface for 01 spectra. Left to right corresponds to frequency (0.5 to 19.0 c/sec). Front to back (perspective) corresponds to age (5 to 97 years).

against which prospective models are to be compared. The likelihood ratio test statistic (λ) for the comparison is:

$$\lambda(\phi,\psi) = 2\ N_s \sum_{\omega}\ [\ (\psi(\omega)/\phi(\omega)) - 1n(\psi(\omega)/\phi(\omega))-1\ ] \qquad [2]$$

where N_s is the number of EEG segments used in estimating the sample spectrum $\psi(\omega)$. This criterion is a metric for the comparison of different spectral models. Higher values of λ correspond to greater discrepancy.

4.3.2 Univariate ξα Model

The fact that there are mainly two different types of spectra led Pascual et al. (1988) to the tentative identification of two different processes—the ξ and the α process—with the following properties. The ξ process is characterized by a spectral peak with its maximum located at low frequency and with a slow decrease in power at increasing frequencies. It includes activity that has traditionally been described as belonging to the δ, θ, α, and β bands. It always appears at all leads. The term "ξ process" was preferred by the authors in order to avoid confusion with traditional slow-wave activity δ. The α process is characterized as a spectral peak superimposed on the ξ process, which is usually centered in the classical frequency range of 7-13 c/sec. It may be absent, especially at lower ages and in more frontal derivations. Its appearance and growth in height and in peak frequency are characteristic of EEG maturation.

The formulation and improvement of the $\xi\alpha$ model were the result of three successive phases.

4.3.2.1 Selection of the Functional Form for Spectral Components

Pascual et al. (1988) assumed that the EEG power spectra may be described as:

$$f(\omega)=\xi(\omega) + \alpha(\omega) \qquad [3]$$

where $f(\omega)$ denotes the EEG power spectra, $\alpha(\omega)$ and $\xi(\omega)$ denote the two component processes described above, and ω denotes frequency. It must be noted that eq. 3 expresses the additional assumption that the two processes are independent and additive. Although other forms of interaction are conceivable, linear interactions are the most tractable mathematically.

There are some conceptual arguments in favor of these assumptions. If the α process is more or less functionally decoupled from the ξ process, as a first approximation they may be considered statistically independent. Since the measured quantities are electrical in nature, additivity follows from the superposition principle for electromagnetic fields. The acceptance of these assumptions will, of course, depend on the fit between their consequences and the experiment.

The functional form of these components must allow the representation of spectral peaks. The general form of a component is:

$$g(\omega;A,B,U)=A.h(\ (\omega - U)/B\ ) \qquad [4]$$

where the function h(.) is positive and unimodal and integrates to unity over the real line. The parameters A, B, and U have the following meaning: A = the maximum height of the peak (peak amplitude), B = a parameter related to the width of the peak at half the maximum amplitude, and U = the peak position (the frequency at which the maximum occurs).

The functional forms g(.) explored were:

·Student t type density:

$$g(\omega,A,B,C,U) = A/(1 + ((\omega - U)/B)^2)^C \qquad [5]$$

where C is a parameter related to the kurtosis of the peak;
·Cauchy functions, which are a particular case of eq. 5 for C = 1;
·Gaussian functions (Kingma et al., 1976), which are also a particular asymptotic case of eq. 5 in which C tends to infinity; and
·Functions derived from the AR model

Each of the four types of functional forms for g(.) was fitted to 30 individual EEG spectra randomly sampled from a data base. Estimation procedures for model parameters were developed, based on the maximum likelihood principle (Cox and Hinkley, 1974). Since the maximum number of parameters for the first three models is 7 (Student's t test), the maximum order for the AR model was taken as this number (for the ξ process, U_ξ is set to zero).

Goodness of fit measures based on the likelihood function were studied by testing the hypothesis that the spectrum can be described by the model against the alternative hypothesis that the spectrum is not restricted to any particular model (test statistic given by eq. 2).

The model with Student's t type functional form was the only one for which the hypothesis of model adequacy was accepted for the 30 cases tested. All other models were rejected consistently. It was concluded that the most adequate functional form for component description is the Student's t type curve. The model specified by the combinations of eq. 3 and 5 shall be henceforth called the $\xi\alpha$ model.

In Figure 4.4A a sample power spectrum is shown with the estimated $\xi\alpha$ model superimposed. The sample power spectrum is similar to Figure 4.4B where a comparison with the BBSP model could be made. The advantage of the $\xi\alpha$ model is obvious: with only seven parameters an almost exact point-to-point representation of the data is obtained.

4.3.2.2 Exploratory Study of the Model: Parameter Reduction

An empirical study of 3200 EEG spectra revealed that some of the $\xi\alpha$ parameters were essentially constant over the sample and that it was possible to assign fixed values to them without affecting $\lambda(S,\xi\alpha)$ significantly. These parameters and their assigned values were $U\xi = 0.0$; $C\xi = 1.5$, and $C\alpha = 60$ (Alvarez et al., 1989).

This result seems to suggest that both the ξ and α processes do not vary in "shape." Interindividual differences are mostly due to scale and location param-

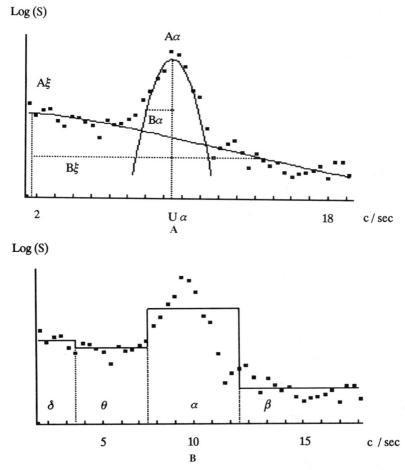

FIGURE 4.4. Sample EEG power spectra (*dots*) for derivation 01. A, Student *t* type curves (continuous lines) representing the ξ and α processes. The ξα model made a satisfactory fit to the power spectrum. Each process is completely specified by its height (A), width (B), and position (U). B, Estimated broad band spectral parameter (BBSP) model (*horizontal lines*). Vertical lines are the EEG band limits for the model (δ:1.5–3.5, θ:3.5–7.5, α:7.5–12.5, β:12.5–19.0 c/sec). This model consists of a step function approximation to the data.

eters. It was thus possible to fix the two C parameters, thereby reducing the number of parameters to be estimated to five for each derivation.

As discussed above, there are EEG power spectra in which no α process is observable. An initial estimation criterion was to consider that the α process was missing if fitting both a ξ and an α process produced no significant improvement in the test criteria, as compared to only fitting a ξ process. A more adequate solution to this problem is based on the method described below, which gives as a by-product age regression norms. In the visual inspection of the 3200 EEG spectra, it was seen that the probability of missing α increases in younger children and in more frontal derivations. Summary statistics for the parameters of the α process were obtained by regression analysis using a variable sample size, with observations for derivations with a missing α being excluded. This allowed a modified estimation procedure.

4.3.2.3 Empirical Bayes Estimation Procedure

The information gained in the exploratory study served to modify the estimation procedure for the ξα model. The modifications were based on the assumption that the variability of EEG power spectra stems from two different sources: (1) intraindividual variations around the ideal EEG spectra described by the ξα model parameters, which are assumed to be fixed for a given subject and (2) interindividual variations caused by random fluctuation of the ξα parameters around fixed population parameters. In other words, the chance experiment of sampling an EEG spectra may be conceptualized as proceeding in two stages. First, a subject is observed. This is equivalent to sampling a set of ξα parameters, θ, from the parameter distribution $p(\theta; \psi)$, where ψ denotes the population parameters. In the second stage, given θ, the actual sample spectrum f is obtained from the spectral distribution $p(f; \theta)$.

Under these assumptions, the log-likelihood of a sample spectrum f and its parameter vector θ is:

$$\ln p(f;\ \theta,\ \psi) = \ln p(f;\ \theta) + \ln p(\theta;\ \psi) \qquad [6]$$

Eq. 6 may be used as the basis for an empirical Bayes estimation procedure. The initial values for the iterative scheme are now the values of θ, these being the a priori most probable values. The likelihood is now "weighted" by the distribution of the parameters ψ, incorporating a priori knowledge about these parameters into the estimation. Since ψ is known to vary with age, it was estimated during the exploratory study using regression analysis methods. The description and properties of this nonstandard Bayes procedure were developed by Biscay et al. (1986).

The modified estimation procedure for the five ξα model parameters (fixed C parameters) was evaluated in the whole data set. Z values reflecting goodness of fit were very similar to those obtained in the exploratory model in spite of the smaller number of parameters.

Due to the new assumptions embodied in eq. 6, there were now no missing α. Instead, some subjects presented very small Aα values. A unified description for all children was thus possible. This circumvented the need for statistical methods that deal with missing parameters (Elashoff and Afifi, 1966).

4.3.3 ξα Model Evaluation: Univariate Data

One important approach to EEG spectra description has been broad band spectral parameters (BBSPs) analysis. The resulting approximation to the EEG spectrum is a step function in which any fine detail is lost (Alvarez et al., 1989). (Fig.

4.4B). For this reason, it could be stated that the BBSPs have more practical than theoretical value. In this section, comparisons between BBSP and $\xi\alpha$ models are made.

4.3.3.1 Comparison Among Different Spectral Models

Recently, Alvarez et al. (1989) compared $120\xi\alpha$, BBSP, and AR models concerning their ability to fit the EEG spectrum. They evaluated 30 randomly selected EEG spectra from the total sample using the goodness of fit statistic λ. The results showed that the $\xi\alpha$ model is a much better description of the power EEG spectra than the other two. They found that the best fits were $\lambda(S,BBSP) = 9.2$ and $\lambda(S,AR) = 7.2$ for a fifth-order AR model, whereas the worst case for $\lambda(S,\xi\alpha)$ was 2.3.

Moreover, they monotonically increased the number of bands of the BBSP model to prove that the results were not due to the fact that the $\xi\alpha$ model has more parameters than the BBSP model (five versus four, respectively). They found in all cases that approximately 16 bands were needed to provide a similar value of λ as the $\xi\alpha$ model. Figure 4.5 illustrates this result for one spectrum. The straight line at 1.72 is the value of $\lambda(S,\xi\alpha)$. The points correspond to the values of $\lambda(S,BBSP)$ for modified BBSP models with equal band sizes plotted for 1, 2, 3, 4, 6, 8, 12, 16, and 24 bands. The inset shows the raw spectrum that represents the NBSP model (corresponding to 48 bands).

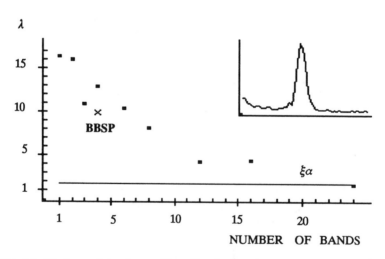

FIGURE 4.5. The $\xi\alpha$ model and several broad band spectral models (consisting of different number of bands) are compared and tested for goodness of fit (λ) with a sample spectrum (*upper right corner insert*). Points correspond to BBSP models of 1, 2, 3, 4, 6, 8, 12, 16, and 24 frequency bands. The cross (x) corresponds to the BBSP model based on the four clinical frequency bands—δ, θ, α, and β ($\lambda = 9.87$ normal deviates). The horizontal line corresponds to the $\xi\alpha$ model ($\lambda = 1.72$ normal deviates). The BBSP model achieves a descriptive power similar to the $\xi\alpha$ model only for 24 bands ($\lambda = 1.67$).

4.3.3.2 ξα Development Equations

The variations with age of the ξα parameters give an exact characterization of the maturation process of the EEG. The ξα developmental equations of the EEG have been presented elsewhere (Alvarez et al., 1989). Using the five ξα parameters these authors "reconstructed" 70 normative spectra (70 different values of age were used in the age range 5 to 12; the increment between ages was 0.1 year) defined as:

$$\xi\alpha_t(\omega) = f[\omega;\ A\xi(t),B\xi(t),0] + f[\omega;\ A\alpha(t),B\alpha(t),U\alpha(t)] \qquad [7]$$

where f[°] is given by eq. 5.

They found that the changes in spectral shape had the following common features: the position of the α process shifted steadily toward increasing frequencies, augmenting in height, albeit with a practically constant width, and the ξ process shifted energy toward higher frequencies. These changes are illustrated in Figure 4.6, which shows $\xi\alpha_t(\omega)$ for 01 at four different ages.

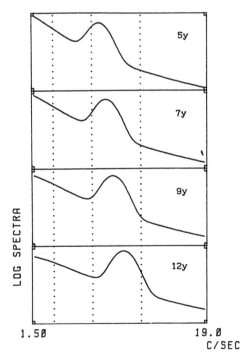

FIGURE 4.6. "Normative theoretical spectra" reconstructed from the evaluation of age regression equations of ξα parameters for 01 derivation. From top to bottom are the spectra corresponding to 5, 7, 9, and 12 years of age. The vertical lines are at 3.5, 7.5, and 12.5 c/sec.

4.3.3.3 Comparison of BBSP and $\xi\alpha$ Norms

Alvarez et al. (1989) used the λ statistic to compare $\overline{S}_t(\omega)$ with both $\xi\alpha_t(\omega)$ and $BBSP_t(\omega)$ for each age t. They found that the $\xi\alpha$ model provides a constant and close fit to the average spectrum for all ages as revealed by low values of $\lambda(\overline{S}_t,\xi\alpha_t)$. On the other hand, the BBSP model fits rather badly at lower ages, and $\lambda(\overline{S}_t,BBSP_t)$ is greater than $\omega(\overline{S}_t,\lambda\alpha_t)$ at all ages. These results are presented in Figure 4.7.

As has been stated previously, the adequacy of any proposed model for the EEG spectrum can be estimated by the goodness of fit statistic λ, introduced in Pascual et al. (1988). Figure 4.5 shows that the descriptive power of BBSP models increases with the number of bands, reaching a perfect description when the number of bands is equal to the number of FFT coefficients (NBSP model). Also evident from this figure is that a very large number of bands (greater than 12) is needed for a good description by the BBSP model.

The $\xi\alpha$ model is capable of providing new information on spectral changes with age. This is shown in Figure 4.7 in which both sets of developmental equations are compared with the average spectrum. This can be best described in terms of what happens to the component processes. Figure 4.6 illustrates that the ξ process changes its shape, transferring energy to higher frequencies. The α process

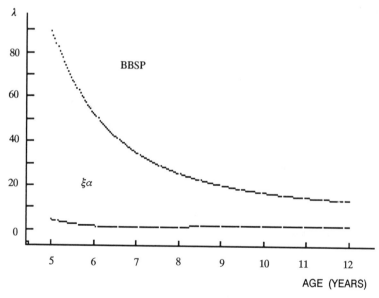

FIGURE 4.7. The $\xi\alpha$ model and the BBSP model are compared and tested for goodness of fit (λ) with the narrow band spectral model for the average sample spectra at different ages (from 5 to 12 years, increments of 0.027). The upper curve corresponds to the BBSP model and the lower one to the $\xi\alpha$ model. The $\xi\alpha$ model provides a very close fit to the average spectrum for all ages. In contrast, the BBSP model fits rather poorly, especially at lower ages.

"slides" toward higher frequencies with a constant shape. The height of the α process increases only in frontal derivations, whereas that of the ξ process decreases with maturation.

On the basis of all these results, Alvarez et al. (1989) concluded that the ξα equations constitute an additional level of refinement in the description of EEG maturation. The results previously discussed show that

The ξα model offers a more parsimonious and better description of spectral shape (as defined by the goodness of fit measure, λ).

The ξα equations essentially "predict" previously described EEG developmental equations (John et al., 1980, 1987) based on BBSPs (Alvarez et al., 1989).

The ξα equations provide a closer fit to narrow band developmental equations.

4.4 EEG Multivariate Analysis

In this section, the properties of the single-channel EEG spectra are generalized to the multichannel case. The appropriate probability model is a multivariate stochastic process. The derived statistical methodology falls within the framework of multiple time series analysis (Brillinger, 1981). The basic assumption here is that the statistical distribution of the vector formed by the DFTs of each EEG channel is asymptotically multivariate complex normal, with a zero mean. As in the univariate case, different frequencies are statistically independent.

It is thus possible to eliminate intrachannel correlations by moving to the frequency domain, reducing the study of a multiple time series to a frequency-by-frequency analysis of interchannel dependencies. The price paid is the introduction of complex valued multivariate statistics (Krishnaiah et al., 1983).

It is to be remembered that, in the single channel case, the power spectrum was a real valued function of frequency proportional to the variance of the DFT at each frequency. The multivariate extension of the power spectrum is the cross-spectral matrix. This is a covariance matrix in which each element is a function of frequency. The diagonal elements correspond to the univariate power spectra of each single channel. The off diagonal elements represent the covariance between the DFTs of each pair of channels at each frequency. These covariances are complex numbers, in order to reflect the introduction of the concept of phase, which is unknown in usual multivariate statistics. The cross-spectral matrices are hermitian; that is, the covariance of channel i with j is the complex conjugate of the covariance of channel j with i. This corresponds to a 180° phase shift.

The complex covariance is best interpreted when it is normalized to produce a complex analog of the correlation coefficient, which is known as the coherence. The magnitude of the coherence is limited to the range (0,1) and is a measure of the amplitude coupling between two DFTs at a certain frequency. The phase of the coherence is the expected phase shift between the two DFTs. As in the univariate case, the cross-spectrum is a complete description of Gaussian stochastic processes.

Unfortunately, a simple graph representation does not exist for the cross-spectrum since a great number of functions are needed. If there are r channels, then the cross-spectrum contains r univariate power spectra (as the diagonal elements) and r(r − 1)/2 complex valued covariance spectra, one for each pair of channels. It is therefore impossible to study graphically the properties of the cross-spectrum, without the aid of auxiliary methods for dealing with multivariate data. These methods were developed and applied by Pascual et al. (1988) for the construction of a multivariate extension of the $\xi\alpha$ model.

4.4.1 Generalized Spectrum, Power, and Coherence

In this section, some summary statistics for the cross-spectral matrices are presented, which are of aid in the graphic representation of multivariate information. These new methods for exploratory data analysis have revealed multivariate properties and structure of the multichannel EEG that are essential for model development (Pascual et al., 1988).

In classical parametric multivariate statistics, a summary of the covariance matrix is the so-called generalized variance, defined as the determinant of the covariance matrix. This statistic is used for condensing the information contained in the covariance matrix. This definition may be generalized to the case of an hermitian covariance matrix, which is also a real number. This suggests the possibility of using the determinant of the cross-spectral matrix as a summary statistic.

The determinant of the cross-spectral matrix is related to that of its corresponding coherence matrix in a simple way:

$$\ln |F(\omega)| = \ln |R(\omega)| + \ln |\text{diag }(F(\omega))| \qquad [8]$$

where $F(\omega)$ = the cross-spectral matrix at frequency ω, $R(\omega)$ = its associated coherence matrix, diag(A) = the diagonal matrix corresponding to A, obtained from A by setting all off-diagonal elements to zero, and $|A|$ = the determinant of matrix A. $\ln |F(\omega)|$ is the natural logarithm.

Based on these properties, Pascual et al. (1988) made the following definitions:

The *generalized power* is the determinant of the diagonalized cross-spectra, interpretable as the energy at each frequency produced by the system if all derivations were functionally decoupled. Setting the covariances to zero is a "numerical cortical isolation."

The *generalized spectrum* is the determinant of the cross-spectral matrix, in analogy with the generalized variance in multivariate statistics. It may be interpreted as the amount of energy produced at each frequency by the system. From eq. 8, it may be deduced that a correction is introduced for energy sources shared by two or more derivations.

The *generalized coherence* is the determinant of the coherence matrix. It decomposes a measure of total brain correlation into contributions by frequency. This is due to the fact that $\{- \ln |R(\omega)|\}$ is proportional to the likelihood ratio test statistic for complete independence between all leads. The larger its value, the more likely that the electrical activity is correlated significantly. In order to illustrate this point further, two extreme cases are exemplified:

The activity at each electrode is uncorrelated with every other, which means that the coherence matrix has the value one on the diagonal and zero elsewhere. The determinant of this matrix is 1, and its logarithm is zero, which gives the minimum value attainable by the generalized coherence.

The activity at one electrode can be expressed as an exact linear combination of the activities at some of the other electrodes. This corresponds to the case of exact colinearity, for which the coherence matrix has determinant zero and the test statistic has an infinitely large value.

The three related summary statistics above have the advantage of being real valued scalar functions of frequency, and their graphic representations may be used to study the global pattern of interelectrode relationships.

These summary statistics were obtained for the 400 normal children. The graphs fall into two general categories, which correspond to the absence and presence of α activity. In Figure 4.8B, no α rhythm is present. The generalized spectra do not differ in form from the simple univariate power spectrum in Figure 4.2B. The most important feature in Figure 4.8B is the nearly constant generalized coherence, which is independent of frequency. These results might tempt one to conclude that the coherence matrices for the ξ process have a structure that is constant over all frequencies. This is further explored by frequency domain principal component analysis. Another important finding is that the hypothesis of channel independence is rejected at very high significance levels for the ξ process.

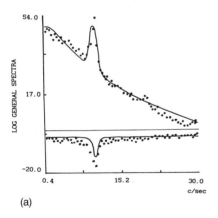
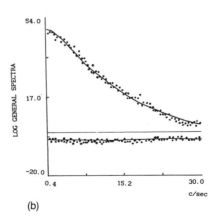

(a)   (b)

FIGURE 4.8. Log generalized spectrum (*upper curve*) and coherence spectrum (*lower curve*) for a multichannel EEG with (a) and without (b) α rhythm. The points represent the sample spectra, and the continuous curves correspond to the theoretical ξα model computed from the estimated parameters. The ξα model is able to reproduce the fundamental characteristics of the summarizing statistics.

In Figure 4.8A, a case with the presence of α rhythm is presented. Again, the generalized spectra do not differ in form from one of the corresponding univariate power spectra. The feature to note here is the peak of stronger correlation in the α rhythm frequency band that is superimposed on the nearly constant generalized coherence of the ξ process. This peak corresponds to the peaks in the generalized spectra and power. This result suggests that the α process is characterized by higher coherences than the ξ process. The physiological interpretation is that α rhythm is not only the appearance of quasi-periodic activity at a certain frequency. The emergence of the α process also involves a functional reorganization that is reflected in a greater degree of dependency between the activity recorded at different electrode sites (Pascual et al., 1989).

These results are typical of the whole sample studied. There was an exact correspondence between the appearance of the α peak in the graphs of generalized spectra and power and the peak in the graph of generalized coherence. In cases when the individual pairwise coherences were too noisy to permit observation of the α peak, it could be seen in the generalized coherence. The generalized coherence thus produces an enhancement of the signal-to-noise ratio of system property estimates.

An interesting feature of these measures is that any linear transform only affects the measure by producing a DC shift over the whole frequency range. A typical linear transform found in EEG practice is a change of recording montage (Nuñez, 1981). It is therefore not surprising that these curves exhibited the same qualitative behavior for 96 bipolar EEG recordings (Damiani et al., 1987).

An important question is how much of the dependency measured by the generalized coherence and spectrum is due to volume conduction or the reference electrode used. Both introduce linear dependencies that can be considered linear transforms of uncontaminated data. Therefore, the qualitative results mentioned above would seem to be valid even if instrumental redundancy could be eliminated. Work is in progress with source derivations to confirm this conclusion.

4.4.2 Test for Zero Phase Shifts

Further insight into the multivariate structure of the ξ and α processes may be obtained by testing specified statistical hypotheses. One example of such a hypothesis was explained in the last section: is there a general independence of the electrical activity over the scalp at frequency ω? The construction of the test statistic for this hypothesis led to the definition of generalized coherence.

Pascual et al. (1989) considered the following hypothesis: *a cross-spectral matrix at frequency ω has real valued elements.* Stated in other words, this means that the coherences between each electrode pair have zero phase shifts.

The formulation of such a test was motivated by the empirical observation that the cross-spectral matrices had negligible imaginary components for the ξ process, whereas in the α frequency band these were not so evident. If the phases are zero, then one possible interpretation is that the measured activity is being transmitted by volume conduction (practically instantaneous conduction). However, this is not the only possible explanation for zero phases. A possible example of a system with zero phase coherences would be one in which fiber conduction of activity is homogeneous and isotropic.

Figure 4.9 illustrates the asymptotic χ^2 likelihood ratio test statistic plotted for each frequency, together with the $p = 0.01$ critical value for rejection of the hypothesis that the cross-spectral matrix has zero imaginary parts. In Figure 4.9A, the α process is present. Figure 4.9B corresponds to a multichannel EEG without any α rhythm. These results again confirm the existence and distinct nature of the two processes α and ξ.

The ξ process, when prevailing, has zero phase coherences at each frequency. This result seems to be consistent with the hypothesis that the ξ process has its origin in some generalized "diffuse" system, in which projection to the cortex and cortico-cortical transmission do not follow any preferred pathway (Pascual et al., 1988).

The α process is characterized by significant nonzero phase coherences. This can be interpreted as evidence refuting the idea that such a process is transmitted from one lead to another exclusively by volume conduction, although volume conduction cannot be discarded entirely. This result does not contradict the hypothesis that the generating mechanism and the propagation of the α process are due to systems with preferred location and direction. The confirmation of this assertion is more difficult to obtain (see below).

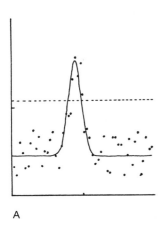
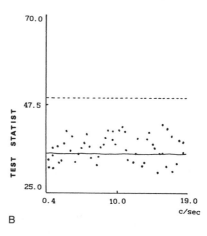

FIGURE 4.9. Zero phase test statistic for cross-spectral matrices. Horizontal line is critical value for rejection ($p < 0.01$). The points represent the test values for the sample cross-spectra, and the continuous curves correspond to the values for the theoretical ξα model computed from the estimated parameters. A, Cross-spectrum with α rhythm present. B, Cross-spectrum without the presence of α rhythm.

4.4.3 Principal Components in the Frequency Domain

The methods presented up to now in this chapter to study cross-spectral matrices may be considered as "summarizing" since the results are scalar functions of frequency that are useful for measuring or testing some aspect of the system.

An alternative technique for the analysis of the covariance structure in usual multivariate statistics is to represent the data as the linear combination of hypothetical "components" that are assumed to be statistically independent. This is the technique of principal component analysis (PCA). The number of "components" necessary to describe the data reveals the inherent dimensionality of the data. It permits reduction of covariance matrices to a standard form.

The extension of these ideas to the case of multivariate time series has been described fully by Brillinger (1981). PCA is carried out in the frequency domain for the cross-spectral matrices at each frequency. This analysis involves the computation of the eigenvalues and eigenvectors of the cross-spectral matrices. The new aspect is that the cross-spectral matrices are hermitian, which reflects the fact that interchannel phase relationships are also to be analyzed.

A useful form of presentation of results is as eigenspectra, a plot of the eigenvalues against frequency. They may be viewed as the energy of each principal component as a function of frequency. These are usually plotted on a logarithmic scale. It can be proved that the sum of the log eigenspectra is equal to the log of the generalized spectrum (Pascual et al., 1985). Eigenspectra can thus be viewed as a decomposition of the generalized spectrum into spectra produced by ideal statistically independent components.

Figures 4.10 and 4.11 show the typical eigenspectra produced by a frequency domain PCA of the cross-spectra. In Figures 4.11A and 4.11B the results correspond to a multichannel EEG without any α rhythm. The eigenspectra mimic the generalized spectrum of this type of EEG. The eigenspectra are also approximately parallel to each other. This would be the result to be expected in accordance with the model that has been evolving up to now: here, only the ξ process is present, and its coherence structure is constant over all frequencies (Figure 4.11B).

Figures 4.10A and 4.10B correspond to an EEG with an α process. These eigenspectra are also similar to the generalized spectra of EEG with an α process. These graphs are all approximately flat, except in the region of the α peak. Once again, this is consistent with the following view: in those portions of the frequency domain where the ξ process dominates, the covariance structure of the EEG corresponds to that of a process with constant coherences. In the band where the α process dominates, it imposes its coherence structure, which seems to have a higher degree of organization, as reflected in higher interchannel dependencies.

These conclusions are reinforced by PCA of the coherence matrices, which is a decomposition of the generalized coherence. It may also be viewed heuristically as a PCA of a multichannel EEG that has been transformed to white noise. Thus, the eigenspectra are not affected by the presence of spectral peaks at each lead. Rather, they only reflect interchannel dependency structures.

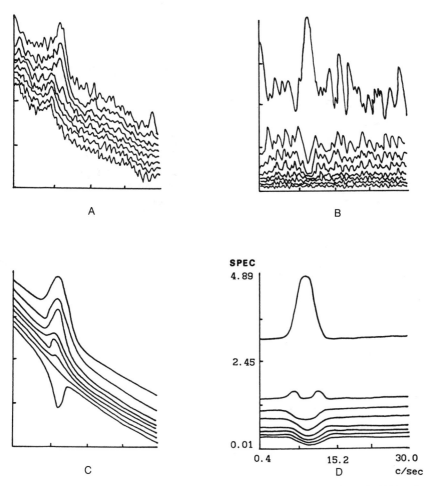

FIGURE 4.10. Principal component analysis was made on each spectral matrix (PS) and on each coherence matrix (CS) of an EEG with α rhythm. *Left,* Log eigenspectra for the PS. *Right,* Eigenspectra for the CS. *Top,* Based on experimental data. *Bottom,* Based on the estimated α model.

The eigenspectra of coherence matrices of EEGs without an α process are flat and parallel (Figure 4.11B). Those of EEGs with an α process (Figure 4.10B) show the same structure as in Figure 4.11B, except in the region where the α process predominates. This is additional evidence of a more highly correlated structure in this band.

Little attention has been given to the study of the eigenspectra of the EEG. Nuñez (1981) made PCA of isolated frequencies of the EEG cross-spectrum. Complete eigenspectra of EEG have been published by Gersch et al. (1977),

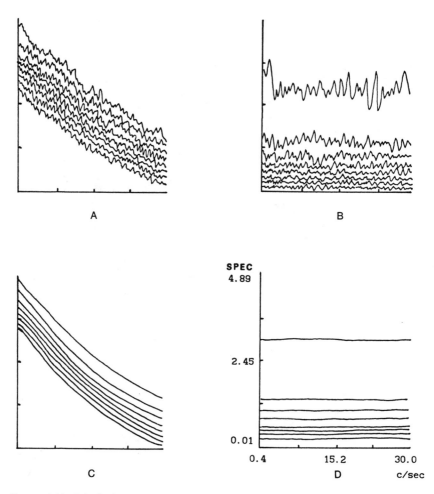

FIGURE 4.11. Principal component analysis performed on each spectral matrix (PS) and on each coherence matrix (CS) of an EEG without α rhythm. *Left,* Log eigenspectra for the PS. *Right,* Eigenspectra for the CS. *Top,* Based on experimental data. *Bottom,* Based on the estimated ξα model.

obtained from cross-spectral matrices estimated by a multivariate AR model. Visual inspection of their figures reveals striking similarities to those presented here.

4.4.4 Multivariate ξα Model

Most multiple time series models published to date are based on multivariate generalizations of the ARMA (autoregressive moving average) model. Applications to the EEG have been limited to the multivariate AR model (Gersch et al., 1977). Multivariate AR models have provided a certain amount of data reduction.

However, as in the univariate EEG analysis, the estimation of the order of the model is fraught with difficulties. The interpretation of the estimated parameters is impossible; their rapidly increasing number becomes unmanageable. Additionally, the AR parameters of a single channel considered as a univariate time series are not the same as when the estimation is based on all channels. Adding a new derivation may change all estimates.

A step toward interpretability might be multivariate extensions of spectral parameter analysis. However, the rather indirect estimation of parameters in the univariate case becomes a formidable difficulty in the multivariate case. This difficulty may be avoided by extending the approach outlined previously for model building in the univariate case, i.e., direct parametric model fitting in the frequency domain, based on the statistical distribution of the cross-spectral matrix. This approach leads to a multivariate generalization of the $\xi\alpha$ model.

The formulation of the multivariate $\xi\alpha$ model was based on the following assumptions:

The EEG recorded at each channel is composed of two additive and independent processes—ξ and α—that may be described as in the univariate case.

The ξ coherence corresponding to an electrode pair is a frequency independent real valued parameter. If there are r electrodes, then there are $r(r - 1)/2$ such parameters.

The α coherence corresponding to an electrode pair is a frequency independent complex valued parameter (r electrodes produce $r(r - 1)/2$ parameters).

Let $A(\omega)$ and $\Xi(\omega)$ denote, respectively, the diagonal matrices formed by the square roots of the spectra of the α and ξ for each channel. $P_{\alpha\alpha}$ and $P_{\xi\xi}$ designate the frequency independent coherence matrices for each process. The multivariate $\xi\alpha$ model for the cross-variance matrix of the EEG may then be expressed as:

$$\Sigma_{xx}(\omega) = A(\omega) \cdot P_{\alpha\alpha} \cdot A(\omega) + \Xi(\omega) \cdot P_{\xi\xi}(\omega) \cdot \Xi(\omega) \qquad [9]$$

4.4.4.1 The Multivariate $\xi\alpha$ Model: Estimation

The problems of model estimation in the univariate $\xi\alpha$ model are practically the same in the multivariate extension. The amount of computation for the multivariate $\xi\alpha$ model (even for eight electrodes only) is almost prohibitive for most small computer systems.

A variant of the ML estimation method that simplifies computations was therefore used, which consists of two main steps. First, the MLEs for the parameters of each univariate $\xi\alpha$ model are computed. Second, MLEs for $P_{\xi\xi}$ and $P_{\alpha\alpha}$ are obtained by a weighted least-squares algorithm that is based on asymptotic distribution properties.

Since the algorithm was designed to decrease the goodness of fit statistic, this measure was used to compare the performance of this model with other possible representations of the cross-spectral matrix. In particular, a structured multivariate AR model was considered. The matrices of AR coefficients were assumed to be diagonal, a case in which the univariate EEGs are correlated with one another due to the covariance structure of the multivariate noise process. In this example, the improvement of the goodness of fit statistic was a decrease from values as large as 300 down to values as small as 20.

4.4.5 ξα Model Evaluation: Multivariate Data

The adequacy of the model was shown by its success in modeling the behavior of the exploratory statistics described in previous sections. For this purpose, the estimated model parameters are used to reconstruct a series of "theoretical" cross-spectral matrices, one for each frequency. These are a very smooth approximation of the original data.

Figure 4.8A illustrates the generalized spectrum and coherence of a EEG record with the α process (dots). The continuous line illustrates the same statistics computed from the cross-spectral matrix reconstructed from the estimated model parameters. The ξα model is able to reproduce the salient characteristics of the summarizing statistics. In Figure 4.8B is a similar illustration for an EEG without α rhythm, and again the model reproduces the results obtained with the sample data.

In Figures 4.9A and 4.9B, the points represent the test values for the sample cross-spectra (with and without α rhythm, respectively), and the continuous lines are based on the fitted multivariate ξα model. Note the satisfactory agreement of the theoretical statistic with the sample statistic.

This close agreement of the model was also observed for principal component analysis in the frequency domain. Eigenspectra of an EEG record with α rhythm are plotted in Figures 4.10A and 4.10B, respectively, corresponding to the raw cross-spectra and the coherence spectra. Figures 4.10C and 4.10D are the eigenspectra of the reconstructed cross-spectral matrix. In Figure 4.11, a similar comparison can be made for an EEG without α rhythm.

4.5 Properties of ξ and α Processes

4.5.1 ξ Coherence Matrix

The ξ coherence matrix consists of real numbers, as proven by the test for zero phase shifts. It may be concluded that the ξ process has no significant phase shifts between derivations.

In Figure 4.12, the ξ coherences are plotted as a function of interelectrode distance, which is computed as the arc distance on a spherical head with a 10-cm

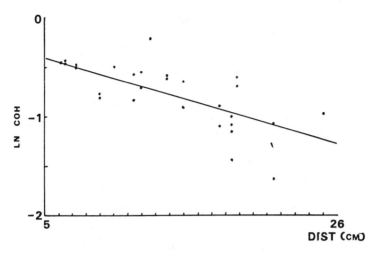

FIGURE 4.12. The log coherences for the ξ process were found to be linearly related to interelectrode distances with a significant correlation coefficient of -0.71 ($p > 0.001$) for $r_{jk}{}^{\xi} = 0.85 \exp(-0.044\, d_{jk})$, which corresponds to an homogeneous isotropic process.

radius. This illustration shows that the coherences are a smooth function (exponential relation) of interelectrode distance.

The hypothesis may be put forward that the ξ is an homogeneous and isotropic process over the scalp. Such processes are characterized by the properties mentioned above: the covariance structure for each frequency is a real symmetric nonnegative definite matrix, and the elements are some given function of distance.

Moreover, PCA was performed on the estimated ξ coherence matrices. In all cases, the first eigenvector corresponded to equal weighting of all variables and can therefore be considered a scale factor. The result of plotting the second versus the third eigenvectors was a nearly exact reconstruction of electrode positions. These results could be seen as a further confirmation of the ξ homogeneous isotropic process hypothesis.

One explanation of how a process of this type may be generated is based on the assumption that the ξ is caused by a distributed layer of stochastically independent cortical dipoles. The laws of electrodynamics applied to the inhomogeneous volume conductor with realistic head geometry will produce spatially decreasing correlations (Valdès, unpublished results), which is completely consistent with the results described above. On the other hand, the additional operation of isotropic and homogeneous neuronal connections would conserve this factor structure.

A generalization of the Box and Cox nonlinear transforms for studying complex regressions was developed by one of the authors (Pascual, unpublished results). Analysis of the regression of the coherences of the ξ process revealed no significant age dependence.

These results identify the ξ process as reflecting fundamentally anatomical constraints in a rather homogeneous fashion. The above-mentioned properties of

the ξ coherence matrix are completely consistent with the properties deduced for
the ξ process in Section 4.3.

4.5.2 α Coherence Matrix

The α coherences were generally complex valued numbers. This was proved
statistically by the test described in Section 4.3.2. The magnitudes of the individual coherences tended to decrease with increasing interelectrode distance.
Conversely, the absolute values of the phases increased with the interelectrode
distance.

PCA analysis of the α coherence matrix revealed a more highly interrelated
dependence structure than for the ξ process. This is illustrated by the fact that only
four components of the α coherence explain 99.5% of the total variance, whereas
for the ξ process seven (of eight) components explain 98%. The α coherence
matrix was always nearly singular. This finding is consistent with and explains the
behavior of the generalized coherence and the eigenspectra of EEGs with α
processes. The near singularity of the α coherence matrix and its appreciable phase
shifts might be explained by a small number of generators that originate activity
that flows sequentially to other sites.

The results of this analysis of the α coherence matrix are in close agreement
with the results published by Chapman et al. (1984) in a combined study of EEG
and MEG (magnetoencephalography), suggesting that there are a limited number
of generators for the alpha activity of the brain.

4.6 Neurophysiological Discussion

4.6.1 Simulated Factor Analysis

The BBSP equations (John et al., 1980, 1987) have provided a useful description
of what is happening to the EEG during maturation. However, it must be borne in
mind that BBSPs are only a gross summary of the power spectrum.

The BBSP model is a simple step function approximation of the power spectrum. Rather than reflecting basic neurophysiological mechanisms, BBSPs were
designed to simulate the performance of previous analog measuring systems based
on the idea of fixed bands.

The difference in viewpoint of interpreting spectral changes with age according
to the "fixed band" model of BBSP analysis versus the "frequency shifting
component" description given by the $\xi\alpha$ model is underscored in Figure 4.6. Note
that the α process peak at age 5 overlaps the traditional θ band, whereas at age 12
it slightly overlaps the traditional β band.

The functional interpretation implicitly encouraged by BBSP analysis is that the

EEG is generated as the sum of a set of resonators, each with a fixed bandwidth. According to this view, the changes observed with maturation are due to a shift in the relative importance of the generators. Support for the actual existence of these "EEG bands" has been based on factor analysis of EEG spectra (Defayolle and Dinand, 1974; Elmgren and Lowenhard, 1973; Herrmann, 1982; Herrmann et al., 1983).

In order to test this hypothesis, Alvarez et al. (1989) made a simulation experiment. They obtained a "sample" of 70 EEG spectra reconstructed from the five $\xi\alpha$ parameter trajectories for the 01 derivation. Later, at each age point t they computed a sample value Z for each parameter from the formula:

$$Z(t) = Y(t) + S \circ N(0,1) \qquad [10]$$

where $Y(t)$ is an ordinary linear regression equation of the transformed $\xi\alpha$ parameter against age and S is standard deviation of the regression equation for parameter Y. $N(0,1)$ is a standard normal deviate obtained with a pseudorandom number generator.

The values thus generated were inverse transformed and substituted into eq. 7 for values of ω from 0.8 to 30 c/sec, every 0.8 c/sec. They fed these "pseudo" sample spectra into a standard principal component analysis program. Four components were retained for Varimax rotation, which accounted for 90.57% of the total variance. Figure 4.13 shows the squared loadings for the four rotated factors. Superimposed as vertical lines are the classical band limits. Alvarez et al. (1989) proposed that Factor I may be considered as a combined δ and θ band, Factor II an α_1 band, Factor III the α_2 band, and Factor IV the β band.

Careful consideration suggests that these results are not inconsistent with the $\xi\alpha$ model of the EEG; rather these results support the view that a band structure for the EEG may arise as a by-product of an inadequate statistical analysis. Alvarez et al. (1989) attributed this result to interindividual "frequency jitter" of spectral peaks, which may produce an overestimation of the number of factors needed to describe a data set. Because of the imposed orthogonality of the resultant factors (principal components, with or without rotation), the obligatory result of such analyses is the estimation of a greater number of components than actually exist (Valdés, 1984).

Careful reconsideration of the traditional concept of bands is therefore warranted. The band structure might just be a by-product of an analytical mechanism that views the spectrum through fixed windows.

In the $\xi\alpha$ description of the EEG, the emergence of the α rhythm with maturation is not viewed as caused by changes in the frequency composition of certain fixed bands, but rather as the increased reflection in the EEG of a distinct neural subpopulation with a steadily increasing resonating frequency. The activity of this putative neural subpopulation is what is called α process, and its changes with age are reflected in the growth of the α peak central frequency.

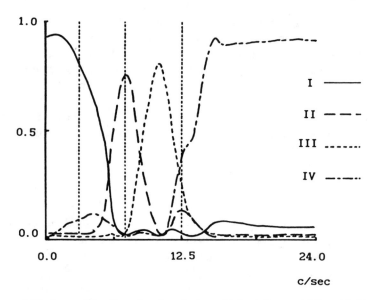

FIGURE 4.13. Squared loadings for four factors obtained from a PCA with Varimax rotation made on the normative theoretical spectra derived from the five $\xi\alpha$ parameters. These factors explain 90.57% of the total variance. Vertical lines are at 3.5, 7.5, and 12.5 c/sec. There appears to exist an approximate correspondence between factors and EEG frequency bands: Factor I—combined δ and θ band, Factor II—mostly α_1, Factor III—$\alpha_2\gamma$ band, and Factor IV—β band.

4.6.2 Projection Fiber Systems

The prime objective in EEG modeling is to achieve an accurate description that is both parsimonious and physiologically meaningful. The results presented in this chapter allow a discussion about the extent to which the $\xi\alpha$ model fulfills this objective.

The $\xi\alpha$ model provides an accurate description of EEG with a minimum of parameters. It represents a reduction by a factor of more than 30 compared with the original cross-spectrum, which is a much more reduced feature set than any other yet proposed in the literature for multichannel EEG. Yet, there is an almost exact point-to-point reconstruction of spectra and multivariate summarizing statistics. Moreover, it is possible to reconstruct satisfactorily other feature sets based on the $\xi\alpha$ parameters.

In spite of these encouraging results in data reduction, there is still much redundancy in the $\xi\alpha$ parameter set (Pascual et al., 1985, Chapter 21). It remains to be established how much of this redundancy is due to measurement problems, such as volume conduction, and how much is inherent in anatomical and physiological system constraints, a point discussed further in Valdés et al. (1985).

In particular, there is a remarkable concordance in the results of analyses of the variation of the univariate $\xi\alpha$ parameters. A pattern consisting of an increase in Uα, relative decrease in Aξ, and increase in Bξ characterizes maturation. Subjects with pathological EEGs tend to present changes in the opposite direction. Finally, PCA results show that the individual variability around mean tendencies falls along the same unidimensional continuum. It is quite possible that the shape of the EEG spectrum is a function of a single parameter.

The experimental results presented support the notion that the multichannel EEG is composed of two component processes: the ξ process and the α process. Evidence has also been presented for the differential development of the two processes during maturation. The multivariate analyses gave further support to this hypothesis and help specify a concrete model for EEG cross-spectra that depends logically on two assumptions: (1) at a given cortical site, the two processes are generated independently; and (2) the influence of a given process at a cortical site is only exerted on processes of the same type at other cortical areas.

It is customary to differentiate the generalized thalamocortical system (GTCS) from the specific projection system (SPS). These two systems might be related to the two processes discussed in this chapter.

The ξ process is characterized by the absence of phase shifts between derivations. This absence seems to point to a "diffuse" projection system. It should also be remembered that the ξ process is remarkably well modeled by an homogeneous and isotropic stochastic process over the head, the degree of coupling decreasing exponentially with interelectrode distance. These results are consistent with the GTCS as the generator of the ξ process and subsequent cortical propagation via the short-range cortical connections. Nonetheless, as noted above, volume conduction may also produce zero phase relationships. The relative importance of volume conduction and neural connectivity in the emergence of the properties of the ξ process remains to be determined.

From the analysis of structure of the α process coherence matrix, it is evident that the classical "α rhythm" is more than just the appearance of a spectral peak in certain leads. It actually corresponds to an increased functional coupling of brain regions, as reflected in a higher generalized coherence. The reduced dimensionality of the α coherence matrix is consistent with the existence of a small number of generators. This, together with significant phase shifts between different derivations, suggests transmission to the cortex via the SPS pathways and subsequent propagation by long-range association fibers. The nonhomogeneous spatial distribution of the local α process properties and the orderly phase shifts between derivations seem to preclude an important role for volume conduction in the production of the observed couplings. Such a view of the propagation of the α rhythm has been emerging from the use of coherence measures (Lopes da Silva et al., 1974) and is the basis of most recent models (Nuñez, 1981).

The decomposition of the EEG into component processes is a point of view that has been reached by other authors by analyzing different data sets with different methods. Thatcher et al. (1986) has proposed a two-compartment model for the EEG, (see Chapter 20). The degree of convergence with the formulation discussed here is striking.

When analyzing the possible physiological relevance of the $\xi\alpha$ model, it is necessary to take into account the following considerations that may affect the interpretation of the results presented in the previous sections. First, this chapter limits its attention to second-order properties of the EEG as summarized in the cross-spectrum. It is quite possible that nonlinear interactions, such as harmonic coupling of α and β peaks (Gasser, 1977), might occur that are of importance in formulating an EEG model. In fact, in Figure 4.1 such a β peak (darker spectrum) seems to be present for one EEG segment. In spite of this, it would seem that as a first approximation a complete description of the simpler linear properties should be a prerequisite for theoretical work.

The work described here could be compared to the establishment of a macroscopic "neurodynamics,"similar to nineteenth-century thermodynamics. The detailed microscopic "statistical neurdynamics" have yet to be worked out. Yet, it is necessary to point out that the $\xi\alpha$ model is essentially nonlinear in the time domain. Linear models, such as ARMA, lead to rational spectra. That is not the case in $\xi\alpha$. Future work will be directed to explain the ξ and α processes from more fundamental concepts. Nonlinearity will surely enter this formulation, and the inclusion of higher-order spectral moments will probably be mandatory.

Another difficulty with this model is its implicit assumption of stationarity. The experimental conditions from which the $\xi\alpha$ model was derived allow one to exert sufficient control so that a steady-state EEG was obtained. For that reason, the model is limited to EEG segments that fulfill the stationarity conditions.

A further limitation is that there was no explicit spatial modeling. This could allow not only a more complete parameter reduction (there are still redundancies in the $\xi\alpha$ parameter set) but perhaps also increase the insight into the two processes of the model (see Chapter 21).

Finally, the criteria of normality were very strict. This limited the model to fit EEG spectra with or without α peak, but double peaks in the α frequency range or peaks in the θ or β bands generally produce unsatisfactory descriptions.

From the logical point of view, the most important assumption in the model is the statistical independence of the ξ and α processes. Additive combination of the two processes to produce the observed EEG would follow because of the superposition principle for electromagnetic fields. In that sense, the model is a linear one. This raises the question about the anatomical substrata of these two hypothetical generator systems. The relation between the generators and the observed EEG could be nonlinear. Recently, there has been great interest in chaotic systems. Nevertheless, those results, even though very exciting, need more experimental confirmation for future practical usefulness. This is an issue that may be verified experimentally.

References

Alvarez, A., Ricardo, J., Valdés, P., Pascual, R. (1985): Clinical evaluation of the Xi-Alpha model for the EEG. Comparison with broad band spectral features. Neuroscience Branch, Technical Report No. NH-001, Havana, National Center for Scientific Research

Alvarez, A., Valdés, P., Pascual, R. (1987): EEG developmental equations confirmed for Cuban schoolchildren. *Electroenceph. Clin. Neurophysiol.* 67, 330–332

Alvarez, A., Valdés, P., Pascual, R., Galan, L., Bosch, J., Biscay, L. (1989): On the structure of EEG. *Electroenceph. Clin. Neurophysiol.* 73, 10–19

Andersen, P., Andersson, S.A. (1968): *Physiological basis of the alpha rhythm.* New York: Meredith Corporation

Barlow, J.S. (1985): Methods of analysis of nonstationary EEGs, with emphasis on segmentation techniques: A comparative review. *J. Clin. Neurophysiol.* 2, 267–304

Biscay, R., Pascual, R., Valdés, P. (1986): Estimació Bayesiana empírica sin el uso de la densidad marginal. V Conferencia Cientíca de Ciencas Naturales. Universidad de la Habana, Ciudad de la Habana. Cuba

Brillinger, R.D. (1981): *Time series: Data analysis and theory* (expanded edition). New York: Holt, Rinehart and Winston

Brillinger, R. D. (1983): The finite Fourier transform of a stationary process. In: *Handbook of statistics: vol 3.* Brillinger, D.R., Krishnaian, P.R. (eds.). Amsterdam: Elsevier, pp. 21–37

Chapman, R.M., Ilmoniemi, R., Barbanera, S., Romani, G.L. (1984): Selective localization of alpha from brain activity with neuromagnetic measurements. *Electroenceph. Clin. Neurophysiol.* 58, 569–572

Cox, D.R., Hinkley, D.V. (1974): *Theoretical statistics.* London: Chapman and Hall

Damiani, S., Alverez, A., Ricardo, J., Pascual, R., Valdés, P. (1987): Analysis multivariado del electroencefalograma en escolares cubanos de 7 a 11 años. In: *Estudios avanzados en neurociencias.* Havana: Ciudad de la Habana, CENIC. 142–157

Davies, R.B. (1983): Optimal inference in the frequency domain. In: *Handbook of statistics: vol 3.* Brillinger, D.R., Krishaiah, P.R. (eds.). Amsterdam: Elsevier, pp. 73–92

Defayolle, M., Dinand, J.P. (1974): Application de l'analyse factorielle a l'etude de la structure de l'EEG. *Electroencep. Clin. Neurophysiol.* 36, 319–322

Elashoff, R.M., Afifi, A.A. (1966): Missing values in multivariate statistics-I. Review of the literature. *J. Am. Statist. Assoc.* 61, 595–604

Elmgren, J., Lowenhard, P. (1973): Un análisis factorial del EEG humano. *Rev. de Psicologia General y Aplicada (Madrid)* 28, 255–271

Galan, L., Bosch, J., Biscay, L., Szava, S., Valdés, P. (1989): Statistical technology of neurometric norm construction. *Neuroscience 89.* La Havana, Cuba. February

Gasser, T. (1977): General characteristics of the EEG as a signal. In: *EEG informatics. A didactic review of methods and applications of EEG data processing.* Rémond, A. (ed.). Amsterdam: Elsevier Biomedical Press, pp. 37–55

Gasser, T., Mocks, J., Bacher, P. (1983a): Topographic factor analysis of the EEG with applications to development and to mental retardation. *Electroenceph. Clin. Neurophysiol.* 55, 445–463

Gasser, T., Mocks, J., Lenard, H.G., Bacher, P., Verleger, R. (1983b): The EEG of mildly retarded children: Developmental, classificatory, and topographic aspects. *Electroencep. Clin. Neurophysiol.* 55, 131–144

Gersch, W., Yonemoto, J., Naitoh, P. (1977): Automatic classification of multivariate

EEGs using an amount of information measure and the eigenvalues of parametric time series models features. *Comput. Biomed. Res.* 10, 297–318

Gibbs, F.A., Knott, J.R. (1949): Growth of the electrical activity of the cortex. *Electroenceph. Clin. Neurophysiol.* 1, 223–229

Herrmann, W.M. (1982): Development and critical evaluation of an objective procedure for the electroencephalographic classification of psychotropic drugs. In: *Electroencephalography in drug research.* Herrmann, W.M. (ed.). Stuttgart: Fischer, pp. 249–351

Herrmann, W.M., Roehmel, J. Streitberg, B., Willmann, J. (1983): Example for applying the COMSTAT algorithm to electroencephalogram data to describe variance sources. *Neuropsychobiology* 10, 164–172

Hjorth, B. (1970): EEG analysis based on time domain properties. *Electroenceph. Clin. Neurophysiol.* 29, 306–310

Ilmoniemi, R.J., Williamson, S.J., Hostetler, W.E. (1988): New method for the study of spontaneous brain activity. In: *Biomagnetism '87.* Proceedings of the Sixth International Conference on Biomagnetism. Tokyo, Japan, August 27–30. Atsumi, K., Kutani, M., Veno, S., Katila, T., Williamson, S.J. (eds.). Tokyo: Denki University Press, pp. 182–185

Jansen, B., Jasman, A., Lenten, R. (1981): Piecewise analysis of EEGs using AR-modelling and clustering. *Int. J. Neurosci.* 14, 168–178

John, E.R., Ahn, H., Prichep, L., Trepetin, M., Brown, D., Kaye, H. (1980): Developmental equations for the EEG. *Science* 210, 1255–1258

John, E.R., Prichep, L., Easton, P. (1987): Normative data banks and neurometrics. Basic concepts, methods and results of norm constructions. In: *EEG Handbook, revised series: vol 1. Methods of analysis of brain electrical and magnetic signal.* Gevins, A.S., Rémond, S. (eds.). Amsterdam and New York: Oxford, Elsevier, pp. 449–495

Khatri, C.G. (1966): A note on a MANOVA model applied to problems in growth curve. *Ann. Inst. Statist. Math.* 18, 75–86

Kingma, Y.J., Pronk, C.N.A., Sparreboom, D. (1976): Parameter estimation of power spectra using Gaussian functions. *Comput. Biomed. Res.* 9, 591–599

Krishnaiah, P.R., Lee, J.C., Chang, T.C. (1983): Likelihood ratio tests on covariance matrices and mean vectors of complex multivariate normal populations and their applications in time series. In: *Handbook of statistics: vol. 3.* Brillinger, D.R., Krishnaiah, P.R. (eds.). Amsterdam: Elsevier, pp. 439–476

Lagerlund, T.D., Sharbrough, F.W. (1988): Computer simulation of neuronal circuit models of rhythmic behavior in the electroencephalogram. *Comput. Biol. Med.* 18, 267–304

Lagerlund, T.D., Sharbrough, F.W. (1989): Computer simulation of the generation of the electroencephalogram. *Electroenceph. Clin. Neurophysiol.* 72, 31–40

Lehmann, D. (1986): Spatial analysis of EEG and evoked potential data. In: *Topographic mapping of brain electrical activity.* Duffy, F.H. (ed.). Boston: Buttersworth

Lindsley, D.B. (1939): A longitudinal study of the alpha rhythm in normal children: Frequency and amplitude standards. *J. Genet. Psychol.* 55, 197–213

Lopes da Silva, F., Storm van Leeuwen, W. (1977): The cortical sources of the alpha rhythm. *Neurosci. Lett.* 6, 237–241

Lopes da Silva, F., Storm van Leeuwen, W. (1978): The cortical alpha rhythm in dog: Deep and surface profile of phase. In: *Architectonics of the cerebral cortex,* IBRO Monograph Series, vol 3, Brazier, M.A.B., Petsche, H. (eds.). New York: Raven Press, pp. 319–333

Lopes da Silva, F., van Lierop, T.H., Schrijerr, C.F., Storm van Leeuwen, W. (1973): Organisation of thalamic and cortical alpha rhythms: Spectra and coherences. *Electroenceph. Clin. Neurophysiol.* 35, 627–639

Lopes da Silva, F., Hoeks, A., Zetterberg, L. (1974): Model of the brain rhythmic activity: The alpha rhythm of the thalamus. *Kybernetic,* 15, 27–37

Lopes da Silva, F., van Rotterdam, A., Barts, P., van Heusden, E., Burr, W. (1976): Models of neuronal populations: The basic mechanisms of rhythmicity. *Progr. Brain Res.* 45, 281–308

Lopes da Silva, F., Vos, J.E., Mooibroek, J., van Rotterdam, A. (1980): Relative contributions of intracortical and thalamo-cortical processes in the generation of alpha rhythms, revealed by partial coherence analysis. *Electroenceph. Clin. Neurophysiol.* 50, 449–456

Matoušek, M., Petersén, I. (1973a): Frequency analysis of the EEG in normal children and in normal adolescents. In: *Automation of clinical electroencephalography.* Kellaway, P., Petersen, I. (eds.). New York: Raven Press, pp. 75–102.

Matoušek, M., Petersén, I. (1973b): Automation evaluation of the EEG background activity by means of age dependent EEG quotients. *Electroenceph. Clin. Neurophysiol.* 35, 603–612

Niedermeyer, E. (1987): Maturation of the EEG: Development of the waking and sleep patterns. In: *Electroencephalography: Basic principles, clinical applications and related fields,* 2nd ed. Niedermeyer, E., Lopes da Silva, F. (eds.). Baltimore: Urban and Schwarzenberg

Nuñez, P. (1981): *Electrical fields of the brain: The neurophysics of EEG.* New York: Oxford Univ. Press

Pascual, R., Valdés, P., Alvarez, A. (1985): Multivariate spectral modelling of the EEG: The $\xi\alpha$ model. Neuroscience Branch, Technical Report No. NC-001, Havana, National Center for Scientific Research

Pascual, R.D., Valdés, P.A., Alvarez, A. (1988): A parametric model for multichannel EEG spectra. *Int. J. Neurosci.* 40, 89–99

Petersén, I., Selldén, U., Eeg-Olofsson, O. (1975) The evolution of the EEG in normal children and adolescents from 1 to 21 years. In: *Handbook of electroencephalography and clinical neurophysiology: vol 6. The normal EEG throughout life. Part B. The evolution of the EEG from birth to adulthood.* Lairy, G.C. (ed.). Amsterdam: Elsevier, pp. 31–59

Praetorius, H.M., Bodenstein, G., Creutzfeldt, O.D. (1977): Adaptive segmentation of EEG records: A new approach to automatic EEG analysis. *Electroenceph. Clin. Neurophysiol.* 42, 84–94

Serfling, R.J. (1980): *Approximation theorems of mathematical statistics.* New York: Wiley

Thatcher, R.W., Krause, P.J., Hrybyk, M. (1986): Corticocortical associations and EEG coherence: A two compartmental model. *Electroenceph. Clin. Neurophysiol.* 64, 123–143

Valdés, P.A. (1984): Statistical bases. In: *Functional neuroscience: vol. III. Neurometric assessment of brain dysfunction in neurological patients.* Harmony, T. (ed.). Hillsdale, NJ: Lawrence Erlbaum Associates, pp. 141–254

Valdés, P., Biscay, R., Pascual, R., Jimenez, J.C., Alvarez, A. (1985): A quantitative description of the development of neurometric parameters. Neuroscience Branch, Technical Report No. NC-003, Havana, National Center for Scientific Research

Valdés, P.A., Pascual, R.D., Jimenez, J.C., Biscay, R., Carballo, J.A., Gonzalez, S.L. (1987): Functionally based statistical methods for the analysis of the EEG and event related potentials. In: *Progress in computer-assisted function analysis.* Willems, J.L., Bemmel, J.H., Michel, J. (eds.). Amsterdam: North Holland, pp. 91–96

Vvedensky, V.L., Gelman, E.B., Gurtovoy, K.G., Naurzakov, S.P., Ozhogin, V.I., Yu

Shabanov, S. (1988a): Study of spontaneous activity of the human brain with a tangential neuromagnetometer. 1. Occipital alpha-rhythm. In: *Biomagnetism 1987.* Proceedings of the Sixth International Conference on Biomagnetism. Tokyo, Japan, August 27–30. Atsumi, K., Kutani, M., Veno, S., Katila, T., Williamson, T.J. (eds.). Tokyo: Denki University Press, pp. 190–193

Vvedensky, V.L, Gurtovoy, K.G., Ilmoniemi, R.J., Kajola, M.J. (1988b): Determination of sources of the human magnetic alpha rhythm. *Hum. Physiol.* 13, 400–404

Vvedensky, V.L., Gurtovoy, K.G., Gribenkin, A. P. (1989): Phase relationships in synchronous magnetic and electrical alpha activity in the parieto-occipital area of the human brain. Seventh International Conference on Biomagnetism. New York, August 14–18, Conference Digest New York University, pp. 365–366

Williamson, S.J., Wang, J.Z., Illmoniemi, J.R. (1989): Method for locating sources of human alpha activity. Seventh International Conference on Biomagnetism. New York City, August 14–18 Conference Digest New York University, pp. 365–366

Wennberg, A., Zetterberg, L.H. (1971): Application of a computer-based model for EEG analysis. *Electroenceph. Clin. Neurophysiol.* 31, 457–468

Wright, J.J., Kydd, R.R. (1984a): A linear theory for global electrocortical activity and its control by the lateral hypothalamus. *Biol. Cybernetics* 50, 75–82

Wright, J.J., Kydd, R.R. (1984b): A test for constant natural frequencies in electrocortical activity under lateral hypothalamic control. *Biol. Cybernetics* 50, 83–88

Wright, J.J., Kydd, R.R. (1984c): Inference of a stable dispersion relation for electrocortical activity controlled by the lateral hypothalamus. *Biol. Cybernetics* 50, 89–94

Wright, J.J., Kydd, R.R., Lees, G.J. (1984): Amplitude and phase relations of electrocortical waves regulated by transhypothalamic dopaminergic neurones: A test for a linear theory. *Biol. Cybernetics* 50, 273–283

Zetterberg, L.H. (1969): Estimation of parameters for a linear difference equation with application to EEG analysis. *Math. Biosci.* 5, 227–275

5

Strange Attractor EEG as Sign of Cognitive Function

EROL BAŞAR, C. BAŞAR-EROGLU, J. RÖSCHKE, AND J. SCHULT

5.1 Aim of the Chapter: Two Approaches to Show that EEG Is a Quasi-Deterministic Sensory-Cognitive Activity

In our earlier approaches, we used the algorithm of Grassberger and Procaccia (1983) to determine the dimensionality of brain waves. The SWS-sleep electroencephalogram (EEG) of intracranial structures of the cat brain was embedded into phase space, and the dimensions of the attractors of the cat's *cortex, hippocampus,* and *reticular formation* were computed. The results confirmed the pioneering findings of Babloyantz et al. (1985) concerning special structures of the cat brain. They also confirmed our long-standing assumptions that the EEG represents an integrative signal stemming from deterministic processes and is a strange attractor (Başar et al., 1988; Röschke and Başar, 1988, 1989). Several authors also used this approach for the human EEG during the waking stage and especially during alpha activity (Babloyantz, 1989; Saermark et al., 1989). For a collection of papers covering this topic, see Başar (1988) and Basąr and Bullock (1989). However, we did not encounter in these references any description of the correlation dimension over long periods of time and from simultaneous recordings over several locations of the human scalp EEG. Since the aim of the present study is to demonstrate that the EEG reflects properties of a quasi-deterministic brain process, we performed analysis of the correlation dimension of the human awake EEG during long periods from various locations. If the EEG is a signal stemming from quasi-deterministic neuronal processes, it should be possible to find replicable EEG states if repeatable initial conditions for a sensory-cognitive input to CNS can be realized experimentally.

In the first section of this chapter, α activity is analyzed using the algorithm of Grassberger and Procaccia (1983). It is then analyzed to show that, with a paradigm in which repeatable cognitive targets are presented, the human brain can emit replicable EEG patterns.

5.2 EEG and Evoked Rhythmicities and Internal Evoked Potentials

Analyses of the EEG, of evoked potentials (EPs), and of endogenous potentials (P300 family) are among the most fundamental research tools for understanding sensory and cognitive information processing in the brain. Since Berger's discovery of the EEG and Adrian's measuring of cortical field potentials, important applications of these powerful neurological techniques have been described in several books (Başar, 1980; Berger, 1938; Freeman, 1975).

Some authors take the view that the spontaneous electrical activity of the brain is an expression of incessant, irregular background neural firing. A contrary opinion to this view was extensively discussed by Başar (1980), who posed the following question: Do we have the right to consider the spontaneous EEG activity of the brain as a background noise in the sense of ideal communication theory, or rather is the EEG a most important fluctuation, which controls the sensory evoked and event-related potentials? We have assumed that regular patterns of the EEG reflect *coherent states* of the brain during which cognitive and sensory inputs are processed and should not be considered as simple noise (Başar, 1980; Başar, 1983a and b).

A preliminary phase portrait analysis of the spontaneous and evoked α activity of the brain indicated that the EEG might reflect properties of a strange attractor. Therefore, an analysis of the EEG has been undertaken in order to compare it with the so-called Roessler-Attractor and attractors similar to those of the Navier-Stokes perturbation system (Başar and Röschke, 1983). In comparing the EEG and evoked potentials, the following tentative formulation has been used to show a stimulus-induced transition of the EEG, which is called the "evoked potential." If a brain structure under study is in a desynchronized state, then excitation (sensory stimulation) would put its activity into a temporary attractor. This attractor was called an "instantaneous attracting cycle," rather than "a limit cycle."

5.3 What is Meant by a "Strange Attractor" and an "α Attractor"?

5.3.1 General Remarks

The theory of nonlinear dynamic systems states that nonlinear systems are able to generate deterministic chaos under selected conditions. Chaos in the sense of nonlinear dynamics means that the behavior of a system is not predictable over longer periods of time, but nevertheless there exists a prescription (i.e., in terms of differential equations) for calculating the future behavior from given initial conditions.

An *attractor* is the quantity of selected states of a dynamic system that are determined under various but delimited conditions. In some cases the sequence of theses states, called a trajectory, seems to be chaotic or unpredictable. Nevertheless, the attractor is always reproducible if the initial conditions are exactly reconstructable (Abraham and Shaw, 1983). Main classes of attractors are defined according to their trajectories: the *point attractor,* the *limit cycle* showing almost periodic sustained behavior, the *torus,* and the *chaotic* or *strange attractor.* The manifestation of a strange attractor is given by a sensitive dependence on initial conditions. (For a mathematical definition of a strange attractor, see Röschke, 1986; and Röschke and Başar, 1988.) *Transition* from one attractor to another is called a *state change* or *bifurcation.* (For a number of definitions adapted to physiological systems, see Freeman and Skarda, 1985.)

5.3.2 *Procedure for Analysis of Chaotic Dynamics*

In order to describe periodic, aperiodic, or even chaotic behavior of nonlinear systems, several approaches have been used. Lorenz (1963) applied concepts of nonlinear dynamics to the convection phenomenon of hydrodynamics in order to describe atmospheric turbulence (Navier-Stokes equation). He demonstrated that the unpredictable or chaotic behavior observed in such an infinite–dimensional system might be caused by a three-dimensional (deterministic) dynamic system.

In order to understand these arguments, we have to consider some recent tools from the theory of nonlinear dynamic systems. Systems behavior (in our case, the EEG from different brain structures) must be analyzed not only in the time domain or frequency domain but also in the *phase space.* In general, a phase space is identified with a topological manifold. An n-dimensional phase space is spanned by a set of n independent linear vectors. This requirement is generally sufficient. There are several possibilities for defining a phase space. We use a proposal of Takens (1981) and span a 10-dimensional phase space by $x(t)$, $x(T = \tau)$, ..., $x(t = 9\tau)$, where τ means a fixed time increment. Every instantaneous state of a system is therefore represented by a set $(x_1, ..., x_n)$, which defines a point in the phase space. The sequence of such states (or points) over the time scale defines a curve in the phase space, called a *trajectory.* As time increases, the trajectories either penetrate the entire phase space or converge to a lower-dimensional subset. In this latter case, the set to which the trajectories converge is called an *attractor.*

In relation to the topological dimension of the remaining attractor, one can deduce various properties of the investigated system. If the dimension of an attractor is a noninteger, called a fractal, the attractor is a *strange attractor* and can be identified with the properties of deterministic chaos. A characteristic phenomenon of deterministic chaos is a sensitive dependence on initial conditions. Similar causes do not produce similar effects. This is a very extensive statement, which apparently refutes the causality principle of natural philosophy. However, by examining the properties of a strange attractor more precisely, one finds that it may have a strong conformity, called self-similarity, which is an invariance with respect to scaling.

One of the oldest concepts of dimension is that of a topological dimension D_T.

For a point, $D_T = 0$; for a line, $D_T = 1$; and for a surface, $D_T = 2$. A first generalization is the *Hausdorff dimension* or *Fractal dimension* D_F. For simple sets—for example, a limit cycle or a torus—the fractal dimension F_f is an integer and is equal to the topological dimension D_T. For an n-dimensional phase space, let $N(\varepsilon)$ be the number of n-dimensional balls (or cubes) of radius ε required to cover an attractor. Then the fractal dimension D_F is defined as

$$D_F = \lim_{\varepsilon \to 0} \frac{\log N(\varepsilon)}{|\log \varepsilon|}$$

A generalization of the fractal dimension is introduced in information theory. The Renyi information of order q is defined as

$$I_q = \frac{1}{1-q} \log \sum_{i=1}^{N(\varepsilon)} p_i^q \quad (q \neq 1)$$

$$I_q = - \sum_{i=1}^{N(\varepsilon)} p_i \log p_i \quad (q = 1)$$

Let p_i be the probability that an arbitrary point (of an attractor) falls into ball i with radius ε, and let $N(\varepsilon)$ be the number of nonempty balls. The generalized dimension D_q of order q is given by

$$D_q = \lim_{\varepsilon \to 0} \frac{I_q(\varepsilon)}{\log 1/\varepsilon}$$

For $q = 0$ we find $D_0 = D_F$. D_1 is called the information dimension, and D_2 is called the correlation dimension. In practice, the correlation dimension D_2 is the generalized dimension that is easiest to estimate from attractors generated by experimental data because

$$I_2 = - \log \sum_{i=1}^{N(\varepsilon)} p_i^2 = - \log C(R)$$

where $C(R)$ is a measure of the probability that two arbitrary points—x_i, x_j—will be separated by distance R. $C(R)$ is called the correlation integral and can be easily computed:

$$C(R) = \lim_{N \to \infty} \frac{1}{N^2} \sum_{\substack{j, i=1 \\ i \neq j}}^{N} \theta(R - |x_i - x_j|)$$

where θ is the heavyside function.

It then follows that

$$D_2 = \lim_{R \to 0} \frac{\log C(R)}{\log R}$$

or
$$C(R) \sim R^{D_2}.$$

The main point is that $C(R)$ behaves as a power of R for small R. By plotting $\log C(R)$ versus $\log R$, one can calculate D_2 from the slope of the curve. For an attractor, the dimension of which is unknown, it is necessary to calculate $C(R)$ for several embedding dimensions. In fact one should choose the maximal dimension n of the embedding phase space two times the dimension of the attractor. The evaluation of the calculated correlation dimension D_2 should converge toward a saturation value. That means $\Delta D_2 / \Delta n = 0$ for some $n > n_0$.

5.3.3 Correlation Dimension of α Activity

Technical considerations on the evaluation of the correlation dimension and its use were explained in detail by Röschke and Başar (1989). Correlation dimensions during various stages have been measured and published by several investigators. Babloyantz (1988) did not find any saturation of EEG activity during the waking stage. Layne et al. (1985) estimated occipital and central α dimensionality during waking stage as ranging from 5.5 to 6.6 (central) and 6.5 to 7.7 (occipital). Dvorak and Siska (1986) estimated the α activity as being between 3.8 and 5.4. Saermark et al. (1989) indicated that it could reach dimensions up to 11. Rapp (1986) described the correlation dimension during two different measurement conditions and found much lower correlation dimensions, ranging from 2.4 to 3. In Table 5.1 several evaluations of correlation dimensions are presented.

Our central aim was to describe the correlation dimension continuously over long periods and simultaneously over several different locations as already described in a preliminary communication (Röschke and Başar, 1989). We performed our investigations with six subjects during "eyes closed" periods and also in several scalp locations and found similar results in all subjects.

Figure 5.1 shows power spectra that were evaluated from four different locations simultaneously (vertex, parietal, occipital, and frontal) during the waking stage of a subject with closed eyes. During the analysis of such compressed arrays of power spectra, the following observations were made. In the central electrode (vertex) the power was usually centered between 7 and 10 Hz, with large peakings in frequencies lower than 10 Hz. In occipital locations the subjects usually showed high-amplitude α activity centered at 10–12 Hz. In frontal electrodes, the 10-Hz component had usually lower amplitudes; high-amplitude activities were mostly centered on lower frequencies, including theta band. Before computing the correlation dimension of the EEG, the data were digitally filtered in the frequency range between 5 and 15 Hz.

In Figure 5.1 the correlation dimension D_2 is also given during corresponding

TABLE 5.1. Human EEG/MEG data

Reference	Parameters	Results
Babloyantz et al., 1985	Δt = 10 msec N = 4000 τ = 20 msec EEG	Sleep stage 2: D_2 = 5.03 Sleep stage 4: D_2 = 4.0 – 4.4 Awake, α activity: D_2 = 6.1 Beta: D_2 – no saturation
Rapp et al., 1986	Δt = 2 msec N = 1000-4000 τ = 10-20 msec EEG	Eyes closed, relaxed: D_2 = 2.4 (N = 1000) D_2 = 2.6 (N = 4000) Eyes closed, counting: D_2 = 3.0 (N = 4000)
Layne et al., 1985	Δt = 2 msec N = 1000–15000 τ = 20 msec (occip.) τ = 40 msec (vertex) EEG	Awake, occipital: D_2 = 5.5 – 6.6 Awake, vertex: D_2 = 6.5 – 7.7
Başar et al., 1989b, 1990, and Chapter 5	Δt = 30 msec N = 16384 points (segments of 3 min) sampling frequency Δf = 100 Hz EEG	Eyes closed, occipital/ vertex/parietal/frontal: D_2 = 5.5–8 finite dimension only when data prefiltered between 5 and 15 Hz
Dvorak et al., 1986	Δt = 5 msec N = 1000–12000 τ = 40 msec EEG	Eyes closed: D_2 = 3.8–5.4 (N = 1000) D_2 = 8–10 (N = 12000)
Van Erp et al., 1987	Δt = 5–10 msec N = 1000–10000 τ = 15–75 msec EEG	α rhythm: D_2 = 5–6 (N = 1000) D_2 = 7–8 (N = 10000) Beta rhythm: D_2 = no saturation
Babloyantz et al., 1986, 1989	Δt = 0.83 msec N = 6000 τ = 16–60 msec EEG	Creutzfeld-Jacob disease: D_2 = 3.7–5.4 Epileptic attack: D_2 = 2.05
Saermark et al., 1989 and personal communication	Δt = 10 msec N = 4000–8000 τ = 100 msec MEG	Healthy subject: D_2 = 11 Epilepsy (2 patients): D_2 = 7 Epilepsy (2 patients): D_2 = no saturation

D_2, correlation dimension; N, number of data points; t, time shift; Δt, sampling time; MEG, magnetoencephalography

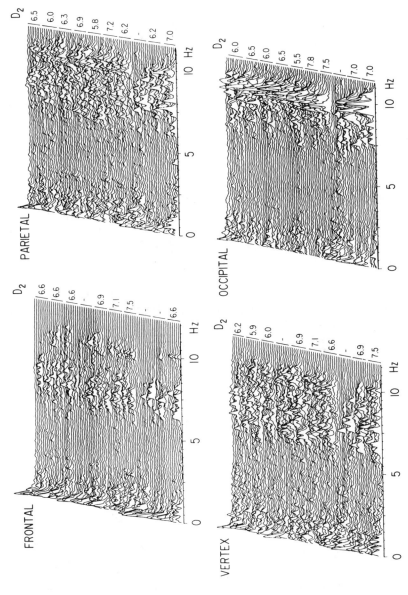

FIGURE 5.1. Power spectra (compressed spectral arrays) and the correlation dimension D_2 belonging to the time segment near D_2-number. Simultaneous recordings from the same subject in frontal, central, parietal, and occipital derivations.

time segments. For evaluation of D_2, EEG segments of a duration of 3 minutes (number of points N = 16384) were used. The sampling frequency was f_s = 100 Hz and the frequency resolution D_f = 0.006 Hz. The correlation dimensions D_2 of the occipital region showed fluctuations between 5.5 to 7.8. Only during a short period of measurement did the correlation dimension not reach any saturation (−) so that this interval cannot be distinguished from "noise."

In vertex the correlation dimensions were observed also with fluctuations between 5.9 and 7.3. In this case there are only two segments showing no saturation and with undefined correlation dimensions. In the parietal location, similar to the occipital region, there was a time window with no saturation (−), whereas the values usually varied between 5.8 and 7.2. Table 5.2 shows fluctuations in correlation dimension D_2 during an experiment lasting about 30 minutes with two different subjects. It is important to note that (1) the correlation dimensions do not vary in a similar way in all locations and that (2) left and right hemispheres may show big differences.

The correlation dimension in the frontal region differed from that in the other locations. Although the correlation dimension had a similar range of fluctuations (between 6.6 to 7.5), time segments with noise-like activity and undefined correlation coefficients were more abundant. Although it is not possible here to discuss all the implications that follow from an analysis of Table 5.2, it is important to emphasize the following: in the existing publications about the description of α activity and its correlation dimension, there are usually no concomitant spectral descriptions demonstrating qualitative changes in α activity and the emergence of "noisy states"(Röschke, 1986; Röschke and Başar, 1989). In addition to the power spectra shown in Figure 5.1, another subject did not show any prominent α activity during eyes closed periods (specially in the frontal region). In this case, shown in Figure 5.2, there are very few time periods where analysis of the EEG shows a finite correlation dimension.

The important implication of the analysis of Figure 5.2 is that two different locations in the brain may show completely different behavior (see also Table 5.2). In this case during a long time period a nonfinite correlation dimension was observed in the frontal region, whereas the simultaneously recorded occipital region showed finite correlation dimensions in almost all segments. (Note that we used a digital filter of 5–15 Hz Röschke and Başar, in preparation).

5.4 Replicable α Patterns in the Human Brain

5.4.1 Why This Paradigm?

In the first sections of this chapter, we presented an analysis of the correlation dimension D_2 of brain α activity. We found a converging D_2 by application of a filter between 5–15 Hz, in contrast to the results presented in the literature, which have shown greater fluctuations (Table 5.1). Therefore, we now consider the most

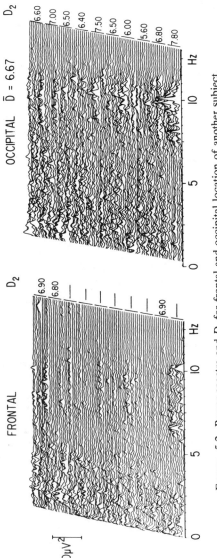

FIGURE 5.2. Power spectra and D_2 for frontal and occipital location of another subject.

important quality of strange attractors. If the input and initial conditions can be replicated, the activity of the strange attractor, which appears to be random, is deterministic and *reproducible.*

In our earlier publications we presented evidence of reproducible and phase-ordered α patterns prior to a cognitive target (Başar, 1988; Başar et al., 1988). In our most recent analysis we presented statistical results from 20 subjects during experiments using sound and light targets. In this chapter, we present the reproducible and phase-ordered α patterns, (phase locking of alpha patterns to the target, that is, omitted stimulation), for one subject only, since the results are representative of the ensemble of experiments already reported in detail elsewhere (Başar et al., 1989).

5.4.2 The Paradigm: Analysis of Prestimulus Activity Prior to a Cognitive Target

The experiments were carried out with volunteer healthy subjects, mostly students between the ages of 19 to 21. The EEG was recorded in vertex, parietal, and occipital locations against an earlobe reference (C_z, P_3, P_4, and O_1, O_2 in the 10/20 system). The EEG signals were amplified by using a Schwarzer EEG machine. The subjects sat in a soundproof and echo-free room that was dimly illuminated. As visual stimulation we used a light step of 800 msec generated with a 20-watt fluorescent bulb that was electrically triggered. The light stimulation was applied at regular intervals of 2000 msec. Every fourth light stimulation was omitted. The subjects were asked to predict and to *mark mentally* the time of occurrence of the omitted signals. One second of the EEG before the omitted stimulation was also recorded with the event-related potential following the omitted stimuli.

We also used as control experiments a control paradigm that is more difficult. In this case the occurrence probability of the omitted signal (target) was drastically decreased: every fourth to seventh stimuli was omitted (the occurrence of target signal is 25% when the subject heard the third tone).

Comparison of the experimental results shows that the subjects emitted coherent and phase-ordered pretarget EEG signals in almost all cases of the easiest paradigm, whereas as the rule, the same subjects did not show the same highly coherent and phase-ordered pretarget EEG during the most difficult paradigm.

5.4.3 Examples of Experiments with Varied Probabilities of Stimulus Occurrence

As mentioned above we present here the results from the experiments on only one of our subjects. J.K. is a medical student who quickly learned the goal of the experiments and was very cooperative during the experiments. Figure 5.3 illustrates samples of the filtered resting EEG as a control before the experiment with the cognitive task. There are three plots of the filtered EEG segments, with ten sweeps in each plot. The three plots present samples during the same recording session. The mean correlation coefficient ($\bar{C}$) of each ensemble of sweeps in a time scale from − 500 to 0 msec is also shown in Figure 5.3.

The subject was told to be attentive to repetitive light stimuli. Every fourth light stimulation was omitted (the easiest paradigm). J.K. reported that at the beginning of the experiment he could easily mark the target signal; however, after approximately 10 omitted signals or after the first 40 sensory stimulations he could not concentrate as well; and toward the end of the measurement he had enormous difficulties in concentrating.

A clear rhythmicity and a high congruency in almost all sweeps were observed. Figure 5.4A shows the first ten filtered sweeps together with the filtered mean value and wide-band mean curves (1-30 Hz). In the following sessions of the experiments (Figures 5.4B and 5.4C), the rhythms were less regular, and the congruency among sweeps almost disappeared. Also during this stage a 10-Hz EEG with larger amplitudes was observed in comparison to the resting EEG that is shown in Figure 5.3. The correlation coefficient decreased drastically toward the end of the experiment. At the beginning, when J.K. reported a good performance, the mean correlation coefficient was high—$\bar{C}$ = 0.38. Later it was diminished ($\bar{C}$= 0.13 and $\bar{C}$ = − 0.01).

Figure 5.5 illustrates a similar experiment with the subject J.K. performed a few months later. Again he reported that at the beginning of the experiment he was able, with ease, to mark the target mentally; during the experiment he lost his ability to follow the target, but near the end he gained better control in marking the target. Figure 5.5B shows the decrease in congruency and the diminishing of the correlation coefficient. In Figure 5.5C the congruency was better ($\bar{C}$ = 0.28).

On the following day we began with the most difficult paradigm (every fourth to seventh stimulation omitted) and proceeded to the easiest paradigm. During the most difficult paradigm J.K. reported that at the beginning of the experiment he felt unsure whether he could follow the rhythm of the light signals. However in the last two-thirds of the experimental period he was able to mark a larger number of the target signals. Figure 5.6A illustrates the beginning, Figure 5.6B the middle stage, and Figure 5.6C the end of the experiment.

The amplitudes of the EEG increased during the experiments, but not the correlation coefficient. During the easiest paradigm, J.K. reported this time that he had not performed well at the beginning. However, toward the end of the experiment he exercised definitely better control in marking the target.

In five subjects measurements of the EEG with the easiest paradigm using light signals were carried out after those with the most difficult paradigm. Yet, during experimental sessions with the most difficult paradigm, congruency of single rhythms, such as the epochs in Figures 5.4 and 5.5, were never observed. Further-

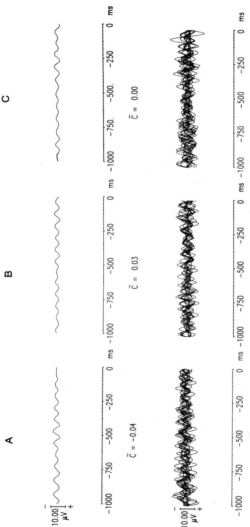

FIGURE 5.3. Resting EEG of the subject J.K. At the bottom ten sweeps of EEG segments that were digitally filtered in the frequency range of 8–13 Hz are shown. Time "0" is arbitrarily chosen. EEG samples were recorded at the beginning (A), middle (B), and end (C) of the recording session. *Top,* Mean value results of ten sweeps. Correlation coefficients C evaluated from three ensembles of ten sweeps.

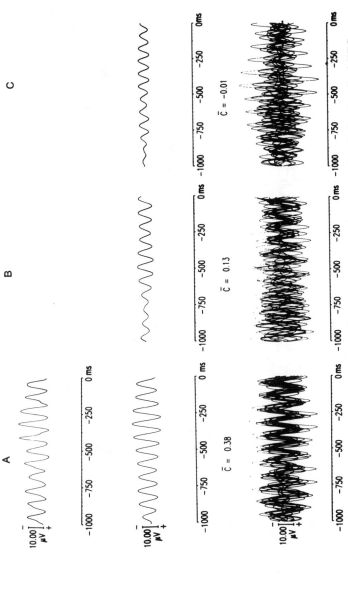

FIGURE 5.4. Pretarget EEG of subject J.K. (experiment No. 3) during the easiest paradigm—every fourth signal omitted. EEG segments were filtered in the frequency range of 8–13 Hz. The time scale from −1000 msec to "0" indicates 1-sec recording time prior to the target (omitted tone). A, Ten single EEG samples at the beginning of the experimental session (*bottom*). Mean value curves of ten sweeps (*middle*). Broad band mean value curve from ten sweeps (*top*); filter range, 1–30 Hz. B, Ten EEG samples in the middle of the experimental session (*bottom*). Mean value curve from ten sweeps (*top*). C, Ten EEG samples at the end of the experimental session (*bottom*). Mean value curve from the sweeps (*top*). The correlation coefficients $\bar{C}$ evaluated from three ensembles of ten sweeps are shown at the top of each ensemble. $\bar{C}$ here describes only the period between −500 to 0 msec; that is 500 msec prior to target. Subject's report: A, good performance; B, and C, bad performance.

more, the mean correlation coefficient, which was calculated during four stages of the experiment, remained around 0.05 in every measurement. It never attained values around 0.4.

5.4.3.1 Comparison of the Easiest and Most Difficult Paradigms

Comparison of the results obtained from using the easiest and the most difficult paradigms for the same subject is important in determining the existence of event-related pretarget rhythms. For the same subject we have the possibility of increasing the probability of the occurrence of the target up to 100%. The increase in the EEG amplitude and the tendency to regularity and phase-ordering are reflected in correlation coefficients. If the probability of the occurrence of a target is then decreased, one would expect a worse or even bad performance. In the case of a bad performance, the phase-ordering of the EEG and the tendency to a repeatable pattern would be expected to diminish. For this reason we applied both paradigms to five subjects on the same days, and we always obtained comparable results, which were similar to the results from the subject J.K.

The increase in correlation coefficient means there is an increase in similarity of single epochs. The fact that subjects who had reported good performance reached a mean correlation coefficient up to $C = 0.4$ shows that the EEG can attain good phase-ordered patterns, which cannot be observed during recordings with less probability of occurrence. We must emphasize that the recording of almost repeatable EEG patterns during defined experiments with cognitive targets requires a large number of experiments and good cooperation on the part of the subjects.

We have described here experiments on one subject in which the evaluation of the correlation coefficients was carried out in a time window between −500 and 0 msec. The use of several time windows and rigorous statistical evaluation, including experiments with 20 subjects and acoustical targets, has been previously reported (Başar et al., 1989).

5.5 Discussion

5.5.1 Quasi-Deterministic EEG, Cognitive States, and Dynamic Memory

The experiments described in this chapter, as well as our earlier work (Başar, 1988), have shown that during cognitive tasks it is possible to measure almost reproducible EEG patterns in subjects expecting defined repetitive sensory stimuli. The use of modern computer techniques makes it possible to search speedily for coherent states during mental tasks. During such coherent states single sweeps of EEG are time-locked to a target signal, at least, for periods of 10 minutes during which the subjects are able to mark a cognitive target mentally in a recurrent manner. [A recording session with 10 target signals (10 omitted stimuli and 30

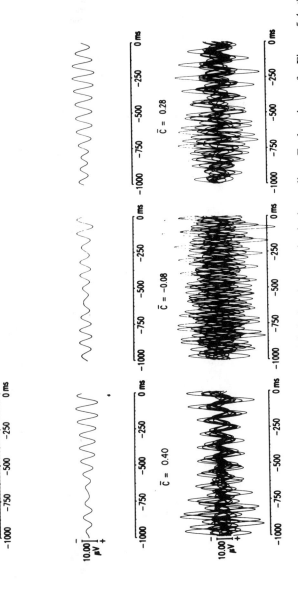

FIGURE 5.5. Pretarget EEG of subject J.K. (experiment No. 15) during the same (easiest) paradigm. Explanation as for Figure 5.4; this shows a repetition after a few months. Subject's report: (A) and (C), good performance; (B), bad performance.

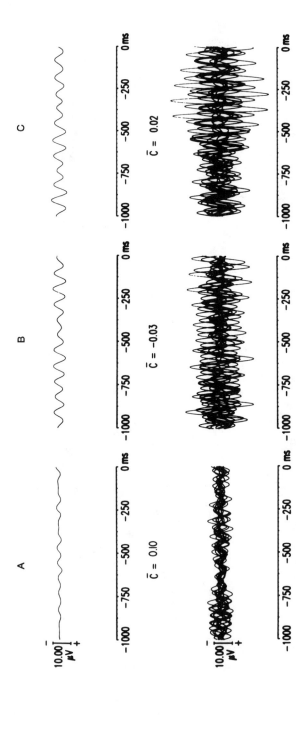

FIGURE 5.6. Pretarget EEG of subject J.K. (experiment No. 16) during the most difficult paradigm. Every fourth to seventh signal was omitted. EEG segments were filtered in the frequency range of 8–13 Hz. The time scale from −1000 msec to "0" indicates a 1-sec recording time prior to the target (omitted tone). A, Ten single EEG samples at the beginning of the experimental session (bottom); mean value curve (top). B, Ten EEG samples in the middle of the experimental session (bottom); mean value curve from ten sweeps (top). C, Ten EEG samples at the end of the experimental session. The correlation coefficients $\bar{C}$ here describe only the period from −500 to 0 msec. Subject's report: tried to do well.

physical stimulations) requires at least 200 seconds or about 3 minutes. Some of the subjects can reach the same performances during 30–40 omitted stimuli; in other words, around 10 minutes.] We use the term "quasi-deterministic EEG" for the recurrently emitted, almost reproducible EEG patterns. The specific brain function is related in this case to a type of short learning process and short-term memory that we have termed *dynamic memory* (Başar, 1988).

In this chapter we focused on the 10-Hz frequency range, although other frequency ranges were briefly commented on in section 5.4. Tentatively, the term "dynamic memory" could be extended to cover all relevant EEG frequency ranges and most probably combinations of several patterns. The use of this term finds legitimacy in our work, as well as that of Babloyantz (1988), Başar et al. (1988), Röschke and Başar (1988), and Freeman and Skarda (1985), which showed that the EEG reflects the properties of a strange attractor mathematically. However, we consider that the description of the EEG with the algorithm of Grassberger and Procaccia (1983) alone is not sufficient to show that the EEG is a strange attractor. What are needed are sensory cognitive paradigms similar to the one in this chapter that can corroborate the position that the waking EEG is a sign of deterministic cognitive processes.

Hillyard and Picton (1979) used the term "selective attention" or simply "attention" as a construct that has a rather broad but circumscribed set of meanings, being clearly distinguished from such nonselective CNS processes as arousal or alertness. Attentional processes are those CNS functions that enable perceptual or motor responses to be made selectively to one stimulus category or dimension in preference to others. Irrelevant stimuli that are not being attended are either partially or completely rejected from *perceptual* experience, entry into long-term memory, and control over behavior. Furthermore, these authors state that *attention* refers to selective aspects of sensory processing. Accordingly, all experimental demonstrations of attention must measure the responsiveness of the organism to more than one category of stimulus. The initial stages of sensory processing are generally thought to proceed rather inflexibly and to consist of an initial afferent registration and feature analysis of incoming sensory data. This information persists in accessible form for about 1 second in a high-capacity sensory memory that has been termed the buffer stores, "iconic" (visual) memory, and "echoic" (auditory) memory.

During our experiments the subjects of the experiments had to pay attention to an omitted stimulus. If they managed to be attentive enough to make the target mentally, they anticipated with 10-Hz waves that were time-locked to the target, showing almost reproducible patterns. Depending on the performance of a subject the *coherency time* of such reproducible wave packets ranged usually from 300 to 1000 msecs, most of it before the time the omitted stimulus was due.

In earlier publications Başar (1983a and b) has hypothesized that the evoked potentials manifest the bifurcation of the strange attractor EEG to a limit cycle attractor of short duration. A strange attractor is manifested by its activity, which appears to be random. However, the activity of a strange attractor is *deterministic* and *reproducible* if the input and initial conditions can be replicated. The experiments with human subjects that were described in Section 5.4 showed that, by increased certainty due to the expectation of repetitive sensory signals and accordingly by increased attention stages, the subjects seem to generate internal cognitive inputs to the CNS. These cognitive inputs are probably due to repetitive similar mental efforts. According to our experimental results, if a subject cannot mentally predict the occurrence of the expected target signal (omitted stimulation), there is no average synchronization of the EEG in the 10-Hz frequency range (see Section 5.4.3).

Furthermore, our earlier results showed that the 10-Hz, 40-Hz, and 4-Hz EEGs go from *disordered states* to *ordered coherent states* during defined cognitive input to the CNS. Herein lies a similarity to evoked potentials elicited by exogenous sensory stimuli, which also show a transition from disordered to ordered coherent states. Moreover, the analysis of the present study shows the following phenomenon: when the certainty of the occurrence of a target signal increases, the 10-Hz EEG goes to coherent phase-ordered states without application of physical stimulation. There are replicable phase-ordered (or time-locked) patterns to a target signal (see Figure 5.4).

5.5.2. More Comments about Deterministic EEG and Chaotic α Attractor

The aim of this chapter is not to discuss the correlation dimension of human EEG and its changes, but rather to link the concept of the strange attractor with physiological and behavioral changes of the CNS. Babloyantz (1989) recently wrote a critical review, emphasizing sources of errors in the estimation of the correlation dimension. According to Babloyantz [and also criticism by Röschke and Başar (1989)] it is not yet possible to carry out an exact comparison between the data published by several investigators since 1985. In Table 5.1 are shown several sets of results that are not always in agreement. Since the D_2 is very sensitively dependent on the use of amplifiers, filter limits, sampling intervals, and recording lengths, the comparison becomes more and more difficult. Accordingly, we suggest that investigators should try to interpret their own data by emphasizing the *transitions* and relative changes in pathological conditions. In other words, it remains necessary to compare pathological data only with data on healthy persons measured and evaluated in the same laboratory, using the same parameters. Important questions remain: What is the meaning of the function of the D_2 of α activity between 5.5 and 8.8, and then why does it suddenly show no saturation, i.e., noise behavior? In this context, Lopes da Silva et al. (1990) and Röschke and Başar (1989) have commented on the activities of intracranial structure (Table 5.3). These authors state that stable attractors were not found during all behavioral stages in animal experiments.

For the time being, it seems most appropriate to work globally by using D_2 as a nonlinear descriptor together with the linear descriptor power spectrum in order to measure important transitions and to try to globally distinguish *noisy* and *deterministic* states. For the interpretation of our data the following observation is also a relevant one. There are stages of EEG activity where the D_2 does not possess saturation. However, the same EEG filtered does suddenly converge and reaches finite values of D_2 after only a few minutes. Another important observation is that two different brain structures may show various degrees of freedom. It is also possible to show that fluctuations of D_2 might be completely independent at various scalp locations. We therefore assume in this chapter that comparison by using the same experimental setup and simultaneous recordings from the same brain can give rise to new interpretations: there are a number of independent strange attractors simultaneously recorded (Babloyantz, 1989).

Lopes da Silva et al. (Chapter 6) studied the correlation dimension of EEG signals recorded during epileptic seizures. Their results indicated that the dimension of the signals varies as a function of brain site in the hippocampus and of time during the course of a kindled epileptic seizure. They further showed that it is not possible to state that EEG signals can be represented, in general terms, as generated necessarily by low-dimension deterministic chaotic systems; however, during given states, such as during epileptiform discharges, this might be the case.

Table 5.2. Changes of correlation dimension D_2 in the course of 30 minutes*

Location	Correlation dimension D_2										Mean value
					Subject A						
O_1	7.0	7.0	—	7.5	7.8	5.5	6.5	6.0	6.5	6.0	6.6 ± 0.7
O_2	6.2	6.6	—	6.5	7.1	6.1	6.0	6.2	7.5	7.1	6.5 ±0.5
P_3	6.2	6.6	—	6.5	6.6	5.8	7.5	5.9	6.0	—	6.3 ± 0.5
P_4	7.0	6.2	—	6.2	7.2	5.8	6.9	6.3	6.0	6.5	6.4 ± 0.4
C_1	7.5	6.9	—	6.6	7.1	6.9	—	6.0	5.9	6.2	6.6 ± 0.5
F_3	6.6	6.6	—	6.1	7.1	5.0	—	5.5	6.0	—	6.1 ± 0.7
F_4	6.6	—	—	7.5	7.1	6.9	—	6.6	6.6	6.6	6.8 ± 0.3
					Subject B						
O_1	6.9	6.8	6.6	6.7	6.5	5.9	6.4	6.0	6.5	6.5	6.4 ± 0.3
O_2	7.1	6.9	6.8	6.5	5.9	6.1	6.2	6.1	6.3	6.6	6.4 ± 0.4
P_3	7.0	6.9	—	6.5	6.3	6.2	6.2	6.2	6.5	6.7	6.5 ± 0.3
P_4	7.5	—	—	6.5	6.0	6.5	7.1	6.2	6.7	—	6.6 ± 0.5
C_1	7.3	6.9	—	6.6	6.5	6.6	6.4	7.0	7.3	6.4	6.7 ± 0.3
F_3	6.1	6.1	6.2	5.9	6.0	5.9	5.9	6.0	6.2	6.1	6.0 ± 0.1
F_4	—	6.9	—	—	—	6.9	—	—	—	—	6.9

*Successive 3-minute samples during good α after filtering the EEG through a 5–15 Hz digital band pass filter (no phase shift).

TABLE 5.3. Intracranial EEG: animal experiments

Başar et al., 1988; Röschke and Başar, 1985, 1988	Cat, SWS, cortex (epidural) $D_2 = 5.0 \pm 0.1$ Hippocampus: $D_2 = 4.0 \pm 0.07$ Reticular formation (mesencephalon) $D_2 = 4.4 \pm 0.07$ (the most stable data)
Röschke and Başar, 1989	Cat, inferior colliculus: $D_2 = 6.7$ Reticular formation (mesencephalon) $D_2 = 7.05$ (unstable attractor, waking state, attractor properties only in 25% of recording time, "high frequency"attractor," data filtered between 100 and 1000 Hz
Röschke and Başar, 1989	Cat, waking state, hippocampus $D_2 = 4.00$ (during synchronized hippocampal theta activity
Lopes da Silva et al.,1989	Rat, hippocampus: $D_2 = 2-3$ or higher (unstable depending on location and on existence of epileptic discharge)
Skinner et al., 1989	Rabbit, olfactory bulb: $D_2 = 5-6$ (event-related shifting from 5 to 6 in evoked activities with odor targets)

SWS, slow wave sleep

Bullock (1989) expressed his hope that we would soon see the dimensionality of many parts of the brain at the same time, second by second, in cats, catfish, and octopus, as rest changes into arousal, directed attention, and recognition. The results given in Figures 5.1 and 5.2 and in Table 5.2 point out the feasibility of gathering some of the data desired by Bullock. It seems that there are a number of brain state changes that might be recorded by using such an ensemble of data.

Is it correct to analyze filtered data? Because the usual embedding dimension is limited, we assume that the description of the absolute value of D_2 during waking stage is limited. Accordingly, the values of D_2 ranging between 5 and 8 should be evaluated only relatively. It is worth emphasizing that for many of the time segments there is no saturation and when it is reached saturation is only temporary. If saturation could be obtained only by means of filtering, all the data would show finite dimensions. Therefore, we reject the possibility of filter-induced strange attractors in the α frequency range. The *reproducible* α is the most convincing demonstration of this finding. Furthermore, the greatest power in studied cases is in the α frequency range so that from the physiological viewpoint the analysis in the α range can be considered as a first approach.

5.5.3 Coherent States During Cognitive Processes

It is largely recognized that the endogenous ERP components are related to cognitive processing of stimulus information or to the organization of behavior, rather than being evoked by the presentation of the stimulus. The event-related rhythms that have been presented in this chapter reflect engagement of the effort performed by the brain in the expectation and prediction of an event. They are purely endogenous because they are emitted in relation to a mental task and not after a physical event or before a physical motion since the subjects cooperated in avoiding every finger or glossokinetic artifact. The cognitive task consisted of mental marking of an omitted signal and not a physical stimulation. Our results greatly differ from reports describing cognitive EEG changes in frequency and amplitudes, because the experiments indicate that the EEG of a subject can reach such patterns as a *constant template* during a constant mental task. This pattern has a defined phase-reordering and alignment during the execution of the mental task, which start approximately 500 msec before the event. Accordingly, we want to emphasize that the EEG might play a highly active and defined role in processes of cognition, being involved especially in generation of percepts and short-term memory.

In this chapter, we focused on EEG rhythmicity in a frequency range of around 8 to 13 Hz during the described cognitive task. Our earlier studies found that activities of slower (1–7 Hz) and higher (40 Hz) frequency with phase-reordering could also be observed during cognitive tasks. We also assumed that synchronization of the electrical activity in delta (1–3 Hz), theta (3–8 Hz), and alpha (8–13 Hz) frequency ranges seems to occur during operative stages of the brain in which the brain processes information coming from sensory and cognitive signals (Başar et al., 1984). It is also important to emphasize the results of Freeman (1983) who showed that the spatial pattern of the 40-Hz EEG appears to be related to the stimulation (odor) that an animal (rabbit) expects to receive. In recent studies, Freeman and Skarda (1985) showed that, by training rabbits to discriminate odors, a new spatial pattern appears with each odor, manifesting a learned regular pattern in the olfactory cortex. The "motor potentials" preceding movements described first by Kornhuber and Deecke (1965) (also called readiness potentials) attracted tremendous interest since these studies showed that changes in slow EEG activity could be interpreted as an indicator of future motions to be directed by the cerebral cortex. The analysis of the EEG before cognitive tasks may open new important aspects in the understanding of cognitive tasks and dynamic memory. Due to the increased speed and memory of computers, several new applications will be made possible by the use of various analytical methods of single EEG-EP trials, which are now being described by several authors (see Başar, 1988).

5.5.4 Two Different Approaches Lead to One Conclusion: The EEG is a Quasi-Deterministic Signal

In this chapter we first presented a nonlinear approach in order to demonstrate that the EEG in various structures of the cat brain is due to deterministic processes. Babloyantz (1985) already formulated that in human subjects the Slow-wave-sleep EEG is also a deterministic signal. We then used a linear approach by digital filtering and were able to show that by using a sensory-cognitive paradigm the EEG can go over to *coherent* and *ordered states* that are of a *repeatable* nature. The purpose of bringing together a *nonlinear* and a *linear analysis* is to try to find coherent and repeatable states of the brain's electrical activity. In a series of experiments that are now in press (Başar et al., 1988, 1989) we have been able to show that the transition from a disordered state to an ordered state also occurred in the frequency range of 40 Hz and 4 Hz. When the probability of occurrence of a target signal is increased, the 10-Hz, 40-Hz, and 4-Hz EEG goes over to coherent phase-ordered states without the application of physical stimulation. These coherent states depict almost phase-ordered patterns. The manifestation of a strange attractor has an activity that appears to be random. However, this activity is deterministic and reproducible if the input and initial conditions can be replicated.

By expecting repetitive sensory signals the subject seems to generate cognitive input due to repetitive mental effort. If the subject cannot predict mentally the occurrence of the expected target signal, there is no *average synchronization* of the EEG in the 8–13 Hz or 40-Hz frequency channels.

The algorithm of Grassberger and Procaccia (1983) permits one to find reproducible patterns of strange attractors in the EEG. By applying the methodology of Section 5.4, we are in fact able to demonstrate that several coherent reproducible patterns can be found in the EEG before a defined target signal. Both approaches, which are from the theoretical viewpoint very different, yield the conclusion that the brain might produce similar EEG patterns during defined experimental conditions days and weeks later. There are many possibilities of using these coherent states of the brain manifested in the EEG to elucidate several sensory-cognitive mechanisms.

This chapter and an earlier study by Röschke and Başar (1989) show that the brain does not operate always with strange attractor states (or deterministic states). In the waking stage there are also "noisy states" during long periods of time. Accordingly, we emphasize the importance of using the expression "quasi-deterministic EEG" since all EEG microstates are not due to deterministic processes. One of the tasks in behavioral brain research is to correlate deterministic EEG stages with behavior and higher mental activity.

References

Abraham, R.H., Shaw, C.D. (1983): *Dynamics. The geometry of behavior.* Santa Cruz, CA: Ariel Press

Babloyantz, A. (1988): Chaotic dynamics in brain activity. In: *Dynamics of sensory and cognitive processing by the brain.* Başar, E. (ed.). Berlin: Springer, pp. 196–202

Babloyantz, A. (1989): Estimation of correlation dimensions from single and multichannel recordings—a critical view. In: *Brain dynamics. Progress and perspectives.* Başar, E., Bullock, T.H. (eds.). Berlin: Springer, pp. 122–130

Babloyantz, A., Nicolis, C., Salazar, M. (1985): Evidence of chaotic dynamics of brain activity during the sleep cycle. *Phys. Lett. [A]* 111, 152–156

Başar, E. (1980): *EEG-brain dynamics. Relation between EEG and brain evoked potentials.* Amsterdam: Elsevier/North-Holland

Başar, E. (1983a): Synergetics of neuronal populations. A survey on experiments. In: *Synergetics of the brain.* Başar, E., Flohr, H., Haken, H., Mandell, A. (eds.). Berlin: Springer, pp. 183–200

Başar, E. (1983b): Toward a physical approach to integrative physiology. I: Brain dynamics and physical causality. *Am. J. Physiol.* 245, R510–R533

Başar, E. (1988): EEG-dynamics and evoked potentials in sensory and cognitive processing by the brain. In: *Dynamics of sensory and cognitive processing by the brain.* Başar, E. (ed.). Berlin: Springer, pp. 30–55

Başar, E., Bullock, T.H. (1989): *Brain dynamics. Progress and perspectives.* Berlin: Springer

Başar, E., Röschke, J. (1983): Synergetics of neuronal populations. A survey on experiments. In: *Synergetics of the brain.* Başar, E., Flohr, H., Haken, H., Mandell, A.J. (eds.). Berlin: Springer, pp. 199–200

Başar, E., Başar-Eroglu, C., Rosen, B., Schütt, A. (1984): A new approach to endogenous event-related potentials in man: Relation between EEG and P300-wave. *Int. J. Neurosci.* 24, 1–21

Başar, E., Başar-Eroglu, C., Röschke, J. (1988): Do coherent patterns of strange attractor EEG reflect deterministic sensory-cognitive states of the brain? In: *From chemical to biological organization.* Markus, M., Müller, S., Nicholis, G. (eds.).Berlin: Springer, pp. 297-306

Berger, H. (1938): Das Elektroencephalogramm des Menschen. *Nova Acta Leopold.* 6, 38

Bullock, T.H. (1989): Signs of dynamic processes in organized neural tissue: Extracting order from chaotic data. In: *Brain dynamics. Progress and perspectives.* Başar, E., Bullock, T.H. (eds.). Berlin: Springer, pp. 537–547

Dvorak, I., Siska, J. (1986): On some problems encountered in the estimation of the correlation dimension of the EEG. *Phys. Lett.* A 118, 63–66

Freeman, W.J. (1975): *Mass action in the nervous system.* New York: Academic Press

Freeman, W.J. (1983): Dynamics of image formation by nerve cell assemblies. In: *Synergetics of the brain.* Başar, E., Flohr, H., Haken, H., Mandell, A.J. (eds.). Berlin: Springer, pp. 102–121

Freeman, W.J., Skarda, C.A. (1985): Spatial EEG patterns, non-linear dynamics and perception: The neo-Sherringtonian view. *Brain Res. Rev.* 10, 147–175

Grassberger, P., Procaccia, I. (1983): Measuring the strangeness of strange attractors. *Physica* [D] 9, 183–208

Hillyard, S.A., Picton, T.W. (1979): Event-related brain potentials and selective information processing in man. In: *Cognitive components in cerebral event-related potentials and selective attention.* Desmedt, J.E. (ed.). *Prog. Clin. Neurophysiol.* 6, 1–52

Kornhuber, H.H., Deecke, L. (1965): Hirnpotentialänderungen bei Willkürbewegungen und passiven Bewegungen des Menschen: Bereitschaftspotential und reafferente Potentiale. *Pflügers. Arch.* 284, 1–17

Layne, S.P., Mayer-Kress, G., Hozfuss, J. (1985): Problems associated with dimensional analysis of electroencephalogram data. In: *Dimensions and entropies in chaotic systems.* Mayer-Kress, G. (ed.). Berlin: Springer, pp. 246

Lorenz, E.N. (1963): Deterministic nonperiodic flow. *Atmos. Sci.* 20, 130

Rapp, P.E. (1986): Oscillations and chaos in cellular metabolism and physiological systems. In: *Chaos.* Holden, A.V. (ed.). Manchester, UK: Manchester Univ. Press, pp. 197–202

Röschke, J., Başar E. (1985): Is EEG a simple noise of a "strange attractor"? *Pflügers Arch.* 405: R45

Röschke, J. (1986): Eine Analyse der nicht-linearan EEG-Dynamik. Dissertation, Universität Göttingen

Röschke, J., Başar, E. (1988): The EEG is not a simple noise: Strange attractors in intracranial structures. In: *Dynamics of sensory and cognitive processing by the brain.* Başar, E., (ed.). Berlin: Springer, pp. 203–216

Röschke, J., Başar, E. (1989): Correlation dimensions in various parts of cat and human brain in different states. In: *Brain dynamics. Progress and perspectives.* Başar, E., Bullock, T.H. (eds.). Berlin: Springer, pp. 131–148

Saermark, K., Lebech, J., Bak, C.K., Sabers, A. (1989): Magnetoencephalography and attractor dimension. Normal subjects and epileptic patients. In: *Brain dynamics. Progress and perspectives.* Başar, E., Bullock, T.H. (eds.). Berlin: Springer, pp. 149–157

Schütt, A. (1989): The EEG is a quasi-deterministic signal anticipating sensory-cognitive tasks. In: *Brain dynamics. Progress and perspectives.* Başar, E., Bullock, T.H. (eds.). Berlin: Springer, pp. 43–71

Skinner, J.E., Martin, J.L., Landisman, C.F., Mommer, M.M., Fulton, K., Mitra, M., Burton, W.D., Saltzberg, B. (1989): Chaotic attractors in a model of neocortex: Dimensionality of olfactory bulb surface potentials are spatially uniform and event-related. In: *Brain dynamics. Progress and perspectives.* Başar, E., Bullock, T.H. (eds.). Berlin: Springer, pp. 158–173

Takens, F. (1981): Detecting strange attractors in turbulence. In: *Dynamical systems and turbulence. Warwick 1980.* Rand, A., Young, L.S., (eds.). Berlin: Springer, pp. 366–381

van Erp, M.G. (1988): On epilepsy: Investigations on the level of the nerve membrane and of the brain. Dissertation, University of Leiden, The Netherlands

6

Cellular and Network Mechanisms in the Kindling Model of Epilepsy: The Role of GABAergic Inhibition and the Emergence of Strange Attractors

Fernando H. Lopes Da Silva, Willem Kamphuis, Jan M.A.M. Van Neerven, and Jan Pieter M. Pijn

6.1 Introduction

The kindling model of epilepsy offers the possibility of studying a number of general properties of the central nervous system: namely, (1) how plasticity or malleability in neuronal networks can be expressed and (2) how the system may become unstable such that the ongoing activity is disrupted. The first point pertains to the question of synaptic plasticity, and the second point is related to the question of the maintenance or interruption of the flow of consciousness.

In the first part of this chapter we consider some recent developments regarding the establishment of the kindled state in a brain structure. This is characterized by the progressive decrease of threshold for the occurrence of afterdischarges or seizure activity. In the second part we analyze the question of how seizure activity can propagate from an initial focus to other brain areas using a nonlinear correlation method. In the third part we examine whether the EEG signals recorded during a seizure may give information about the possibility that the underlying systems may behave as strange attractors. In the concluding remarks we try to derive some general principles from the experimental data that may further an understanding of how the networks in the brain are functionally organized.

6.2 The Kindling Model of Epilepsy

6.2.1 Definitions and Relevance of the Model

The term "kindling" was introduced by Goddard et al. (1969) to indicate the progressive development of EEG and behavioral epileptiform activity in response to repeated administration of weak electrical tetanic stimuli to a given brain area.

The stimuli are initially subthreshold in relation to the capacity of eliciting after-discharges and convulsions. The behavioral expression of kindling has a number of characteristic stages (Racine, 1972), and ultimately tonic-clonic seizures that may last several minutes are elicited.

Kindled animals that are kept unstimulated for several months retain the ability to respond with a full-blown seizure if stimulated again (Goddard et al., 1969; Wada et al., 1974). The cumulative process that is characterized by the progressive reduction of the threshold both for afterdischarges and for behavioral convulsions may be called the "kindling process." The end stage of this process may be called the "kindled state," which is characterized by a long-lasting low threshold for eliciting epileptic seizures by electrical stimulation.

Kindling can be induced in a rather similar way in a variety of species, such as frog, rat, rabbit, cat, dog, rhesus monkey, and baboon (for review, see McNamara, 1986). The kindling process appears to take more time as the brain volume of the animal increases. Evidence for kindling in humans is indirect and has primarily been derived from the observation of the formation of a secondary focus an appreciable time after the appearance of a primary focus, as shown by the group of Morrell (1983) and by Rosén et al. (1984). Furthermore, a case has been described where seizures of progressive severity appeared in a male patient after daily tetanic stimulation (Schramka et al., 1977). The establishment of post-traumatic epilepsy may be due to a phenomenon of the same type as kindling, and the use of antikindling drugs immediately after a severe cranial trauma may have a prophylactic value in preventing the subsequent development of epileptic sei-zures (Servit and Musil, 1981). Therefore, it appears that kindling may offer a good model for at least certain aspects of human focal epilepsy.

6.2.2 Kindling of the Rat Hippocampus: In Search of the Underlying Mechanism at the Cellular Level

In this section are reviewed our more recent experimental findings obtained regarding kindling of the rat hippocampus. We choose the hippocampus because it is a laminated structure and thus appropriate to interpret electrophysiological signals that may indicate changes in local neuronal networks related to the estab-lishment of kindling.

6.2.2.1 Experimental Procedure

The experimental procedures are summarized as follows (for details, see Kam-phuis et al., 1988). Wistar male rats were chronically implanted under pentobarbi-tone anesthesia with two bundles of stainless-steel electrodes (diameter 100 μm sharp) in the left dorsal hippocampus, one bundle for recording and the other for stimulation. The stimulation electrodes were aimed at the Schaffer collaterals that stem from the axons of CA3 pyramidal cells and make synapses with the apical dendrites of CA1 pyramidal cells in the stratum radiatum (Lopes da Silva and Arnolds, 1978). The recording electrodes were placed in such a way that they straddled the stratum pyramidale as shown in Figure 6.1. Reference electrodes were screws placed on the skull. Single-pulse stimulation of the Schaffer collaterals evoked a field potential that reached maximal negativity at the level of the stratum radiatum and was positive at the stratum pyramidale/oriens. The electrodes were cemented in place and connected to a socket that was fixed to the skull. On the plug FET amplifiers were mounted, functioning as impedance transformers and connected via a cable to a commutator and amplifiers. The signals were sampled by a microcomputer that was programmed to run the stimulating program and to average the evoked potentials (EPs). Single- or double-pulse stimulation was carried out using biphasic pulses of 0.1 msec duration that were delivered at intervals of 4 to 10 seconds. Usually double-pulse stimulation was carried out in order to obtain estimates of the short-term state of excitability of the local networks. Routinely we used an interpulse interval of 20 msec at which a clearcut double-pulse depression was apparent at high stimulus intensities: this means that the amplitude of the response to the second pulse of the pair (E2) was smaller than that of the first (E1). In the following discussion the ratio E2/E1 is called the *excitability ratio*. At the interpulse interval of 20 msec, a series of intensities was used to obtain input-output relations as indicated in Figure 6.2.

The kindling procedure consisted of stimulating the Schaffer collaterals with 1-sec trains of pulses (50 Hz, 0.1 msec duration) two or three times a day at intervals of at least 3 hours. The procedure was carried out every day until

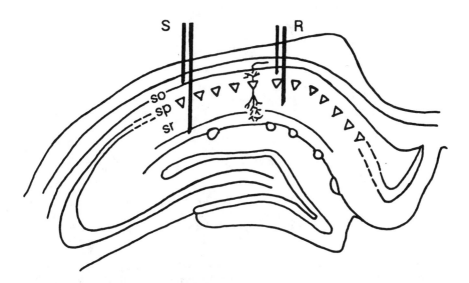

FIGURE 6.1. Section of the hippocampus with an indication of the position of the stimulating (S) and recording electrodes (R). So = stratum oriens; sp = stratum pyramidale; sr = stratum radiatum. A typical CA1 cell is drawn. The cell bodies of CA1 pyramids are indicated by the triangles. The localization of the Schaffer collaterals is indicated by the line in sr.

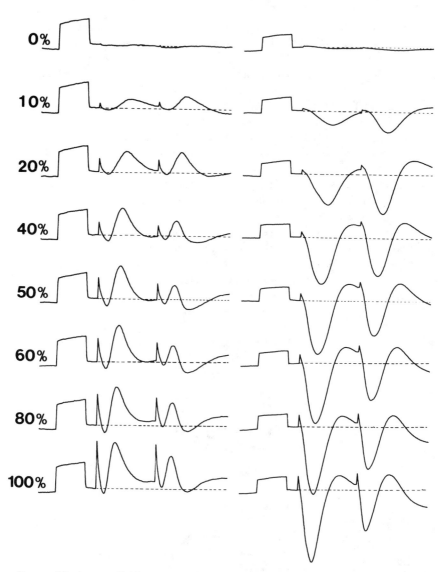

FIGURE 6.2. Average field potentials (F) recorded from the stratum oriens/pyramidale (*left*) and stratum radiatum (*right*) to paired-pulse stimulation of the Shaffer collaterals for different intensities. The maximal intensity, set at 100%, yielded a response to the first or conditioned stimulus (CS), which showed amplitude saturation (control measurements at the first session); the minimal intensity (indicated as 0%) yielded an EP at threshold level for the stratum radiatum. Note that at low intensities there is a paired-pulse facilitation and at high intensities a depression. The calibration pulse represents an amplitude of 1 mV and a duration of 10 msec, with positivity upward. Reprinted from Kamphuis et al. (1988): Changes in local evoked potentials in the rat hippocampus (CA$_1$) during kindling epileptogenesis. *Brain Res* 440: 205–215 with permission of Elsevier Science Publishers BV.

generalized convulsions occurred approximately after 25 stimuli. The stimulus intensity for kindling was usually that intensity necessary to obtain a field potential with an amplitude about equal to half the saturation level of the input-output relations.

A group of control rats was implanted as the experimental animals, but no tetani were applied. In this group, series of evoked potentials (EPs) were investigated under comparable experimental conditions as for the kindled rats. Here we consider first the electrophysiological and then immunocytochemical and some preliminary biochemical findings.

6.2.2.2 Electrophysiology

The electrophysiological results obtained in the course of kindling can be subdivided into those observed immediately after the tetani as discussed in Section 6.3 (afterdischarges), the occurrence of interictal epileptiform transients (Wadman et al., 1983), and the lasting changes in the neuronal networks. Here we concentrate mainly on the lasting changes since they are characteristic of the kindling process. In order to study these changes, we probed the state of the neuronal networks using double-pulse stimulation at the longest possible interval after each kindling tetanus. The main changes of the field potentials were the following:

Only five of eight rats showed long-term potentiation (LTP) of the amplitude of E1, but in two of these rats this effect was limited to the first four kindling stimuli. In these cases the responses did not increase further, as if a saturation level was reached. In the other three rats the amplitude of the field potentials increased up to session 16.

The waveform of the EPs changed in the course of kindling in a consistent way, as illustrated in Figure 6.3. The main change was that the decaying phase of the EP became gradually less steep. Statistical analysis (Mann-Whitney U test) showed that this decrease in decaying slope was significantly larger than that of the control group for session 8 ($p < = 0.032$) and for session 16 ($p < 0.001$).

The excitability ratio or ER (ratio E2/E1) changed also in the course of kindling. Before kindling tetani were delivered, the intensity of the stimuli was chosen in such a way that the baseline value of ER was $79 \pm 7\%$ (paired-pulse depression). In the course of kindling the ratio E2/E1 increased gradually and reached the value $98 \pm 5\%$. A significantly increased ER in comparison with controls (Mann-Whitney U test) was found from session 11 onward ($p < 0.05$). The relation between the mean ER values and kindling session (from session 1 to 11) had a correlation coefficient of $r = 0.97$ ($p < 0.01$). The control animals did not show any significant trend in this respect.

The most conspicuous change in the EPs at the level of the stratum oriens/pyramidale during kindling was the progressive appearance of a population spike at about sessions 7–10. For low-stimulus intensities, the population spike was clearly seen in the second response (E2), indicating an enhanced paired-pulse facilitory effect. At high-stimulus intensity the population spike was seen in the first response.

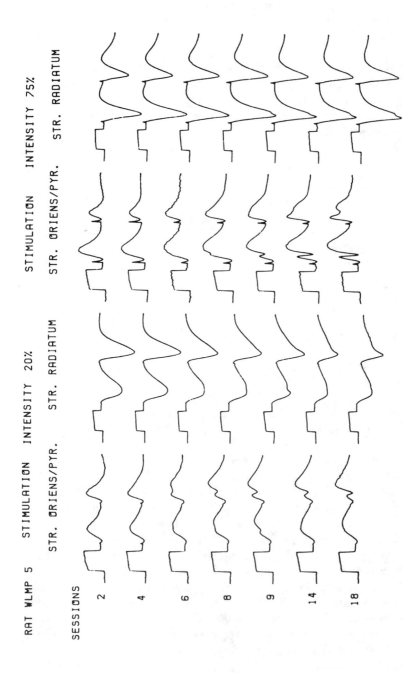

FIGURE 6.3. Average field potentials recorded simultaneously from the stratum oriens/pyramidale and the stratum radiatum to stimulation of Schaffer collaterals for different sessions as indicated at the side of the curves and for two stimulus intensities (20 and 75%). In addition to the changes in the responses recorded from the stratum radiatum, note the changes in the responses from the stratum oriens/pyramidale. Here the gradual emergence of a population spike can be seen. At low-stimulus intensity, this is particularly clear in the second or test response; at high intensity it is clear in the first response. Reprinted from Kamphuis et al. (1988): Changes in local evoked potentials in the rat hippocampus (CA_1) during kindling epileptogenesis. *Brain Res* 440: 205–215 with permission of Elsevier Science Publishers B.V.

These findings need further discussion. First, we have to consider the current interpretation of the EPs studied here. The early rise and peak of the EP have been described (Leung, 1979, 1982) as generated by the extracellular EPSPs of the pyramidal cells, although this component may also include a contribution of a feed-forward inhibitory potential (IPSP) (Buzsaki and Eidelberg, 1982). After the initial excitation, a late component (> 15 msec) appears that has been interpreted as a population "early" inhibitory postsynaptic potential (IPSP) (Schwartzkroin, 1986).

A decrease in the decaying slope of the field EP, as encountered during kindling, may be due to a number of processes. Two main possibilities are the following: an attenuation of the IPSP, which would be reflected in widening of the field EP since the EPSP would not be counteracted by the subsequent recurrent IPSP, and an enhancement and/or prolongation of the depolarization. However, in the present experiments we do not dispose of detailed current-source density analysis nor of intracellular recordings in order to clarify this issue further. Our results indicate that the balance between excitatory and inhibitory processes changed in favor of the former either because of a decrease in the recurrent IPSP and/or because of a prolongation of the depolarization. This conclusion is supported by the following additional findings: the progressive decrease in paired-pulse depression (PPD) and the emergence and relative increase of a population spike at relatively late kindling stages. The former finding could be due to a failure in recurrent inhibition. However, we may speculate that a similar change in the form of the field EP and a decrease in PPD could also be due to a strong and prolonged depolarization mediated by increased activity of NMDA receptors as known to occur in hippocampal slices of kindled rats (Collingridge et al., 1983; Mody and Heinemann, 1987; Wadman et al., 1985). This possibility has to be further tested experimentally. We should note that both these effects—decrease of inhibitory processes and prolonged depolarization—may occur simultaneously and could even reinforce each other. The late emergence of a population spike corresponds probably to a decrease in threshold for the generation of action potentials in the pyramidal cells, because the ratio between population spike amplitude and slope of the field EP at the level of the soma increases steeply around sessions 8–12. This factor would enhance the probability of synchronous firing of these neurons (Traub and Wong, 1983). In regard to a possible role of the long-lasting hyperpolarization or "late" inhibitory process (Knowles et al., 1984), we can only say that a decrease in such a process cannot explain the conspicuous changes observed at a latency of about 15–17 msec. Whether such processes also change in the course of kindling is a matter that merits further investigation.

In contrast to these conclusions, King et al. (1985) reported an increased synaptic inhibition in the dentate gyrus of hippocampal slices in rats that had been kindled in the entorhinal cortex up to 1 day before the in vitro measurements. In a preliminary series of experiments, we studied simultaneously the EPs of CA1 and of the dentate gyrus during kindling. The EPs of the CA1 region were recorded as indicated above. Those of the dentate gyrus were obtained by stimulating the perforant path and recording from around the granular layer. In the CA1 region, we found also that the paired-pulse depression tended to disappear as described above, but in the dentate we found the opposite effect, an increase in paired-pulse depression (Kamphuis et al., unpublished data). This means that different brain areas may react differently to the same kind of stimulation. These results also allow us to emphasize that the balance between excitability and inhibitory processes in a neuronal network is a very sensitive parameter that may change depending on the pattern of stimulation. The direction of change, however, depends probably on local factors that are as yet not well known.

6.2.2.3 Immunocytochemistry

A series of GABA-immunocytochemical studies was undertaken to answer the question of whether the changes in GABA cells that occur during kindling might explain the decrease in the balance between excitatory and inhibitory synaptic processes in the CA1 region. The rats were prepared with electrodes as indicated above. The fixation of the brain for immunocytochemical localization of GABA and parvalbumin (PV) has been described in detail elsewhere (Kamphuis et al. 1987, 1989a and b). Parvalbumin is a specific Ca^{2+}-binding protein that is present in certain neuronal populations. The spatial distribution of PV- and GABA-immunopositive cells shows a close correspondence (Celio, 1986; Kosaka et al., 1987; Stichel et al., 1988), and therefore PV may be considered a marker for a subpopulation of GABAergic interneurons in CA1 stratum oriens and pyramidale (Kosaka et al., 1987).

The main results can be summarized as follows. At the long term in the kindled state (31 days after the last generalized seizure) there was no reduction of PV immunoreactivity nor in the number of GABA-immunopositive cell bodies that colocalized with PV. In contrast, the number of GABA-immunoreactive somata in stratum oriens and pyramidale that did not colocalize with PV was reduced by 50%.

The alterations in GABA-IR at long term were further analyzed by manipulating the system responsible for the breakdown of GABA (Kamphuis et al., 1989b). This was done by injecting the GABA-transaminase inhibitor, amino-oxy-acetic acid (AOAA), which in control rats leads to an increase in the number of cell somata that are immunoreactive for GABA. This increase most likely results from the accumulation of GABA, reflecting a GABA synthesis by L-glutamate decarboxylase (GAD) activity in the somata of interneurons. In kindled rats 31 days after the last seizure, the number of GABA-immunoreactive cells was significantly lower (Mann-Whitney U test $p < 0.001$) than in AOAA-treated controls. This indicates that in the kindled state a GAD-dependent increase in GABA content of the interneurons does not take place in at least a subpopulation of GABAergic interneurons. These results lead to the conclusion that in the *kindled state* the long-term enhancement of seizure susceptibility may be related to the decreased GABAergic inhibitory control of the neuronal population in the CA1 region of the hippocampus.

The changes in GABA and PV immunoreactivity found in the early phases of kindling (i.e., during the development of the kindling process) are more complex.

Instead of finding a monotonous decrease of GABA immunoreactivity during the kindling process, as expected in view of the long-term changes, Kamphuis et al. (1987) found an initial transient increase. Namely, after 14 afterdischarges there was a significant increase (Mann-Whitney U test, $p < 0.037$) in GABA immunoreactive cell density ipsilateral to the side where the kindling stimulus was applied, as compared to controls. The complex time course of GABA immunoreactive cell density (CD) for the ipsilateral side is indicated in Figure 6.4 (in this case, the long-term group was sacrificed at 28 days after the last tetanus). From these results, it may be concluded that during kindling epileptogenesis there is a transient increase in GABA immunoreactive cells in the early phase of kindling acquisition, which is thereafter followed by a process of gradual reduction that finally leads to the long-term decrease of GABA immunoreactivity in the kindled state.

The finding of an increase in GABA immunoreactivity in the early phases of the kindling process presents a paradox since in this same phase we found a decrease in paired-pulse depression and the threshold for epileptiform seizures. One possibility that could explain this apparent paradox would be that the increase in GABA immunoreactive cell density reflects GABA accumulation in the cells and that less GABA is being synaptically released. This hypothesis was tested in a series of biochemical studies carried out in slices of kindled rats. Preliminary results in-

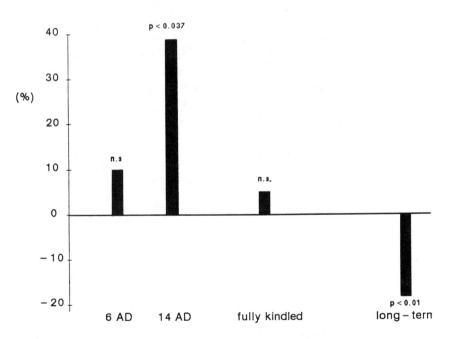

FIGURE 6.4. Relative changes in mean cell density of GABA immunoreactive cell somata along the kindling process and at long term (28 days) after the final seizure on the ipsilateral-stimulated side in CA1 region of the hippocampus. Changes are expressed as percentages compared to a matched control group.

dicate that this possibility does not hold, since in the early phases of kindling there exists an increase of Ca^{2+}-dependent GABA release (Kamphuis et al. 1990).

6.2.2.4 Concluding Remarks: The Cellular Mechanisms Underlying Hippocampus Kindling

In the conclusion of Section 6.2, we may state that the kindled state of the hippocampal CA1 area is characterized by a reduced number of GABA immunoreactive cells. Therefore, it appears that this decrease in GABA cells may explain the change in the balance between excitatory and inhibitory synaptic processes in favor of the former which is characteristic of the kindled state. The underlying mechanism for this change in long-term GABA immunoreactivity is still unknown. It has been suggested that a toxic overload of the neurons with Ca^{2+} may play a role in this respect (Kamphuis et al. 1989a, 1989c). This suggestion is supported by the finding that no loss of GABA immunoreactivity could be found in cells that also contain the calcium-binding protein PV.

It is likely that the repeated application of the tetanus and subsequent afterdischarges depolarize the membrane of the neurons in such a way that NMDA receptors may be activated (Harris et al., 1984; Melchers et al., 1988). This activation may be related to the increase in a dendritic calcium conductance that was found in hippocampal slices of kindled rats (Kamphuis et al., 1989c; Wadman et al., 1985).

However, during the early kindling process, we have the paradox of the existence, at the same time, of a decreased paired-pulse inhibition and an increased number of GABA immunoreactive cells and increased GABA release. A possible way to explain this paradox is that in this phase of kindling there might be a reduction in the sensitivity of the GABA receptor. Such a reduction was noted to occur in vitro after repeated activation of NMDA receptors (Stelzer et al., 1987), resulting in an increase in intracellular Ca^{2+} concentration. Hamon and Heinemann (1986) demonstrated that a reduced inhibition in turn increases the influx of Ca^{2+}. In this way, a positive feedback may become established between Ca^{2+} influx and a reduction of GABAergic inhibition. This point is further discussed in Section 6.5.

6.3. How Does Seizure Activity Propagate from a Kindled Focus to Other Brain Areas?

6.3.1 Seizure Propagation: Methods of Analysis

The kindling model of epilepsy gives one the opportunity to study how epileptiform seizure activity spreads from an initial focal area and progressively involves other brain areas. By applying signal analytical methods it is possible to quantify the degree of relationship between EEG signals recorded from different areas. In this way one can determine whether activity in one area depends on that of other areas. This was done initially by Brazier (1972, 1973) who determined coherency between pairs of EEG signals. To determine the direction of spread, she calculated a time delay (Δt) according to $\phi_0/360°$. f_0 where ϕ_0 is the phase angle at the frequency for which coherency is maximal (f_0 in Hz). Gotman (1983, 1987) also used coherency as a measure of coupling between brain areas, but he estimated the delay not merely on the basis of one frequency but within the range of frequencies (Δf) for which coherency was significant. He tested whether the phase-shift as a function of frequency was linear (as it should be in the case of a pure delay), and if so he calculated the time delay by $\Delta t = \Delta\phi/f. 360°$.

The same approach has been used earlier by Mars et al. (1977) with regard to EEG signals recorded during kindled seizures in dogs, but these authors noted that time delays could be estimated in this way in some cases but not in all. A possible reason for the failure could be that the relationship between the EEG signals might not be linear, whereas measurements based on coherency and phase spectra assume a linear relation between the signals. In our experience in kindled rats, it is clear that nonlinear relationships are present.

This failure led to the development of a new method of analysis based on the average amount of mutual information (AAMI), as described in detail elsewhere by Mars and Lopes da Silva (1987). The average amount of mutual information is a measure of how well one can, on average, predict the values of signal X given signal Y, regardless of whether the signals are linearly or nonlinearly related. The computer program written by Mars and Van Arragon (1982) to calculate AAMI was rather cumbersome. Recently, Moddemeijer (1987) has developed a simpler and quicker algorithm, which was used in the present work. The AAMI (X,Y) can be normalized by dividing it by its theoretical maximal value, which is defined by the properties of the two signals. The normalized AAMI is called the transmission coefficient T.

In this study, we used this T coefficient and also two other measures. The first is the squared cross-correlation (r^2), which measures the similarity in waveform of two signals using linear regression (this is a measure in the time domain that is comparable with the coherency in the frequency domain). The second is a related measure using nonlinear regression: the cross-association (h^2) (Guilford and Fruchter, 1985). All three are calculated as a function of time shift between a pair of signals. The value of the maximum gives the maximal association, and the corresponding time shift gives an estimate of the time delay between the two signals. We use all three methods because differences between r^2 and h^2 give information about the kind of relation between X and Y, whereas T helps in interpreting when several maxima are found for the other two (for details, see Pijn et al., 1989). We should add that the concept of a pure delay in relation to the propagation of epileptiform EEG activity does not necessarily apply in all cases. In the present investigation we considered only those estimates of delay times that could be considered unambiguous (Moddemeijer, 1989).

6.3.2 Experimental Procedures

We applied these methods to the analysis of kindled seizures obtained in rats. The experimental procedures were similar to those described in Section 6.2.2.1, except that in these experiments we implanted additional electrodes for stimulation and recording besides those aimed at the hippocampus. It is of importance to note that the stimulation electrodes used for administering the kindling stimulus were placed in the entorhinal cortex (EC), a structure from where the perforant path fibers arise that innervate the dentate gyrus and CA fields of the hippocampus. The reason for administering the kindling stimulation to the EC was that in this way we could extend the length of the period of spread of the epileptiform activity from this structure to the ipsi- and contralateral hippocampal formation.

The electrode bundle in the EC consisted of four wires with the tips separated by 300 μm. It was positioned with two tips above and two tips below the pyramidal cell layer. This made local stimulation and bipolar recording from the same bundle possible. In both hippocampi, we placed electrodes consisting of three wires (350 μm separated tips) in the CA2-CA3 region and electrodes consisting of four wires more medially. The reference electrode was placed on the frontal bone. By measuring local evoked field potentials, it was possible to identify the approximate positions of the electrodes. Histological verification of the exact position of the tips took place after the experiment. In order to measure only local activity, bipolar derivations between adjacent electrodes were used. Fifteen EEG derivations were recorded simultaneously from the freely moving rat. The signals were digitized and stored (by PDP 11/34 computer) on a magnetic tape. The rat was stimulated daily in the EC with a tetanus consisting of 100 pulses of 200 msec and 500 μA during 2 seconds.

As an example, four EEG signals recorded during an afterdischarge evoked in an animal are shown in Figure 6.5. As described above, we calculated the squared cross-correlation (r^2), the average amount of mutual information (T), and the cross-association (h^2) as a function of time shift between pairs of signals for successive epochs. In Figure 6.5, the maximum of the cross-association function for each epoch is shown. Also shown is the estimated time delay for those epochs. We can see in the figure that the association between the EC and the ipsilateral hippocampus (FD) increases in the first 7 seconds after the end of the tetanus, reaching a value of 85%. During this time, the afterdischarge evoked in the EC spreads to the FD with a delay of about 20 msec. Then the association between these two areas decreases gradually; the hippocampus becomes more and more independent of the EC. During the afterdischarge, the association between the right and the left hippocampus ($CA1_l$ and $CA1_r$) also changes but in a different way. At the beginning, we see also a rather high association between the CA1 areas. A positive delay of about 30 msec was found for this period, meaning that the CA1-left leads the CA1-right. Later, during the second period of the afterdischarge lower values were found for the association. Finally, a third period could

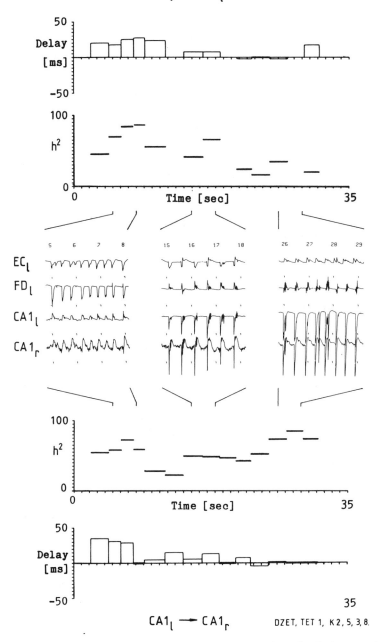

FIGURE 6.5. Examples of the evolution during 35 seconds of an afterdischarge and of the association between EEG signals recorded in the entorhinal cortex (EC) and the hippocampi (CA_1 and FD areas) in a rat. For the combination EC (*left*) toward ipsilateral hippocampus (FD) and the combination CA_1-left toward CA_1-right the value of the maximum of h^2 (as function of time shift between the signals) is shown for successive epochs. For each epoch, the estimated time delay is also shown. Three stages can be distinguished. For each stage examples of the EEG records (4 seconds long) are shown. Reprinted from Pjin et al. (1989): Evolution of interactions between brain structures during an epileptic seizure in a kindled rat. *Adv. Epilep.* 17: 67–71 with permission of Raven Press, New York.

be distinguished during which the association between the CA1 areas increased again. This was found only for the combination of the two CA1 derivations, and thus it is specific for those areas. A clear zero delay was found in this period, meaning that at this stage, the CA1 areas are working as a tightly coupled system (or they are both driven by a third area).

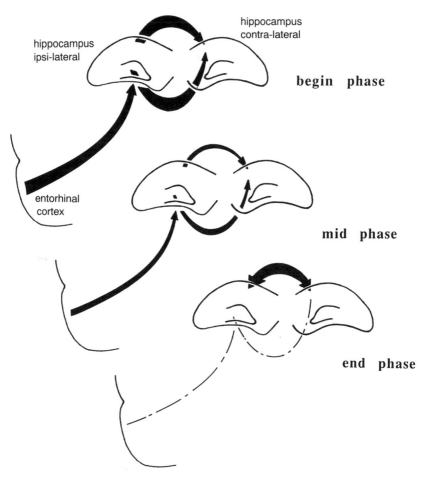

FIGURE 6.6. Main features of the evolution of the afterdischarge of Figure 6.5. The thicker the arrow connecting two brain areas, the larger the association between them. The arrowhead points in the direction of the area lagging. Two arrowheads indicate that zero delay was estimated. Three phases can be distinguished. In the first phase (about 7 seconds long) the activity is dominated by the EC; the midphase (about 12 seconds) the relationships become weaker; in the end phase (about 11 seconds), the association between both hippocampi increases whereas that with the EC becomes negligible.

6.3.3 Concluding Remarks: Coupling and Uncoupling of Different Brain Areas

The analysis of kindled seizures, an example of which was examined in detail here, reveals two important properties: (1) the EEG activity during a seizure is nonstationary and (2) the dynamics of the coupling between different brain areas is rather complex. The fact that the EEG signals recorded during a seizure are nonstationary implies that the analysis must be carried out in a piece-wise fashion, i.e., that the signals must be subdivided into quasi-stationary segments.

The dynamics of the interaction between different brain areas reveals a complex pattern (Figures 6.5 and 6.6). In the initial phase it appears that the activity of the focal area—in this case, the EC—leads the activity of the hippocampal areas that are directly connected to the EC by way of a dense system of fiber projections. After this phase the degree of coupling between the focal and the secondary areas decreases substantially. Meanwhile, there is a pronounced increase of the coupling between homologous sites of the two hippocampi. Under these conditions the delay between the two CA1 signals is not significantly different from zero. This means that the two CA1 neuronal networks produce oscillatory EEG signals that are synchronous. The synchrony in activity may be due to a third source driving both hippocampi, or it may result from the interaction between both hippocampi working as one tight coupled system just as do two coupled oscillators. In order to find out whether a third source may be responsible for this synchrony, we carried out a series of experiments (Fernandes de Lima et al., 1990) in which we applied a tetanus to one CA1 region and recorded the subsequent afterdischarge from both sides in intact rats and after sectioning of the commissural fibers or of the septal area. Sectioning the septal area, the corpus callosum, or the dorsal commissure had no influence on this phenomenon of synchronous coupling. However, sectioning the ventral hippocampal commissure resulted in the uncoupling of the two CA1 signals. According to these experimental results we may conclude that the existence of a strong commissural system of fibers connecting the two CA1 regions in both directions is a necessary and sufficient condition for the establishment of synchronous coupling between both hippocampi during a seizure.

6.4 Does the Epileptiform EEG Activity Indicate that the Underlying Neuronal Networks Behave as Strange Attractors?

6.4.1 Definitions and Methodology

In the past few years an effort has been made by several investigators (Basar, 1988) to investigate whether EEG signals may provide information about the dynamics of the underlying nonlinear neuronal networks. In this type of investigation a new mathematical form of analysis has been applied that was developed to analyze nonlinear dynamic systems that may generate chaotic behavior while the time or space dependence of the system may be deterministic. According to Schuster's description (1984), "Deterministic chaos denotes the irregular or chaotic motion which is generated by nonlinear systems whose dynamical laws uniquely determine the time evolution of a state of the system from a knowledge of its previous history." One of the interesting features of this form of analysis is that it enables one to examine whether a seemingly random signal, such as an EEG, may be generated by a complex nonlinear system that may be described by means of a strange attractor. An attractor can be considered as representing the behavior of the system after a transient period. In mathematical terms an attractor is a region in the phase-space to which all the trajectories having certain initial conditions will be attracted (Serra et al., 1986). Examples of attractors are the point attractor (equilibrium point), the limit cycle (periodic oscillation), or the strange or chaotic attractor (irregular oscillation).

In this approach the first step is to determine whether an apparently random signal, such as an EEG, may contain information about a strange attractor that describes the underlying generator system. This is done by computing the *correlation dimension* of the attractor (Grassberger and Procaccia, 1983). In practice one disposes of a set of time samples describing a sampled signal x_n, with $n = 1 \ldots N + K_{m-1}$. From this series one can construct Nm-dimensional vectors V_m defined by:

$$V_m (i) = (x (i), x(i+ k_1), x(i + k_2), \ldots x(i + k_{m-1})$$

where $k \ll N$ and no $k = k_j$ if $i \neq j$

We may define the correlation integral $C(r,m)$, where r is the radius of a sphere with center $V_m(i)$ in R^m, as follows:

$$C(r,m) = \sum_{n=1}^{N} \sum_{i=1}^{N-n} h(r - d(v_m(i + n), v_m(i)))$$

where h is the heavyside or step function

$$(h(x) = 0 \text{ if } x \leq 0, \text{ and } = 1 \text{ if } x \geq = 0).$$

$C(r,m)$ is the number of pairs $V_m(i)$, $V_m(j)$ with distance $d \leq r$. If an attractor is present in the time series, according to the theorems of Takens (1981), the values of $C(r,m)$ should satisfy the following relation:

$$C(r,m) \sim r^{D_2(x)}$$

For small $r \ll$, $D_2(x)$ is the correlation dimension of x. The computer procedure used by us to estimate $D_2(x)$ was described by Van Neerven (1988). In a log-log plot the above expression defines a straight line with slope D_2, the correlation dimension. In the figures presented here, we plot the slope D_2 against $^2\log(r)$ for increasing values of the embedding dimension m. In this way we may state that there is a strange attractor if the curves for increasing embedding dimensions have an asymptote that tends to a plateau for relatively small values of r. The height of the plateau gives an estimate of the value of the correlation dimension (Figure 6.7). Below we discuss the control procedures used to validate these estimates.

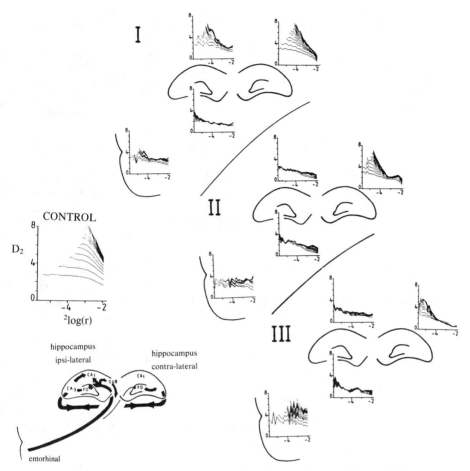

FIGURE 6.7. Results of the correlation dimension analysis of a series of EEG epochs recorded from four sites; EC, FD, and CA₁ on the left and CA₁ on the right hippocampus. Each series of four corresponds to one phase of the kindled seizure, the association analysis of which is shown in Figure 6.6. Each frame represents the plot of the correlation dimension D_2 against $^2\log r$. The value of the asymptote corresponding to a plateau in the plot gives an estimate of the dimension. On the left side, a control plot is shown for the EC signal of phase I after the phase has been randomized. Note that the dimension becomes very large. Similar results were obtained with the other control signals. The values of the dimension D2 are low for the signals recorded from EC and FD in Phase I; from EC, FD, and CA₁ in Phase II; and from FD and both CA₁ in Phase III.

The algorithms used to estimate the correlation dimension are viable to a number of artifacts, originating mainly from the fact that every time series is finite (N samples). Thus, the series is embedded in only a finite number of dimension m, whereas the correlation dimension is defined in the limit for N and $m \to \infty$ and $r \to 0$. Therefore, $C(r,m)$ is calculated only for the finite set of values of r from r_0 to r_e. In practice a reliable estimate of the correlation dimension can only be obtained by minimizing the artifacts. In this respect a few considerations about choices that were made are summarized as follows:

The initial choices of value of the radius r_o and of the ratio between successive radii $f = (r_i/r_{i+1})$ have been done in such a way that these values cover the straight part of the log-log plot and give as many points as possible. A ratio $f = 0.9$ and a value r_o between 0.1 and 0.3 worked well in most cases for the signals analyzed.

An important property of the signal to be taken into account is the autocorrelation function. If the time series is strongly autocorrelated one finds an underestimation of the correlation dimension. To avoid this one should take for embedding only those samples that are uncorrelated. Therefore, in the expression of the correlation integral the variable n should start from a value $n = a \neq 1$. This means that the first a-1 components of each $V_m(i)$ should not be used for the calculation of $C(r,m)$. If we take a of the order of the autocorrelation time, T, this artifact can be avoided (Theiler, 1986).

The choice of the delays k_i should be such that after embedding all components of V_m are as weakly as possible correlated to each other. We used a recursive algorithm such that the first i delays filled the interval (0, WT) as homogeneously as possible. The window W should be a small integer. We used here $W = 3$.

For these analyses the EEG signals were sampled at 512 Hz after filtering with a 48 dB/octave anti-aliasing filter with cut-off frequency at 150 Hz. Low-frequency cut-off was 0.5 Hz.

6.4.2 Analysis of the Correlation Dimension of EEG Signals Recorded During Epileptic Seizures

The correlation dimension was determined for the EEG segments recorded during the kindled seizure illustrated in Figures 6.5 and 6.6. The corresponding plots of the correlation dimension D_2 against radius r are shown in Figure 6.7. These plots demonstrate how the dimensions of the signals may vary as a function of brain site and of time in the course of a kindled epileptic seizure. In the initial phase the EEG signals from EC and the ipsilateral dentate gyrus or fascia dentata (FD) show a rather low dimension, in the order of 2 to 3. To test the reliability of this result we carried out a control analysis as follows: the amplitude and phase Fourier spectrum of the signal were computed, and the phase angle of each component was then changed at random. Thereafter, an inverse Fourier transformation was carried out resulting in a new signal. This new signal has the same power spectrum as the original signal, but the components have random phases. The correlation dimension of this phase-randomized signal equals that of bandlimited noise. We used a control signal with the same autocorrelation function as that of the original signal because, in this way, we could check the dependency of the estimator of the correlation dimension on the autocorrelation function. If the correlation dimension of the original signal is much lower than the dimension of the randomized signal using the same calculation procedure, one can conclude that an attractor of low dimension is present. For the EC and FD signals in the initial phase of the seizure a low dimension was estimated that was clearly lower than that of the corresponding control signal. The EEG signal from the ipsilateral CA1 had a much larger dimension than that of the EC. The dimension of the contralateral CA1 signal was still higher, approximating that of random noise. This shows that in this phase of the kindling seizure an attractor is only clearly present in those brain areas where the activity has its focus. In the midphase of the seizure the dimension of the CA1 areas decreases and becomes similar to that of the other brain areas. This tendency is even clearer in the last phase of the seizure. At this time the estimation of the dimension of the EC signal becomes rather unstable, indicating that in this structure the attractor becomes less pronounced.

6.4.3 Concluding Remarks: EEG Signals as Manifestations of Strange Attractors or as Random Noise?

The analysis of the EEG signals during a kindled seizure demonstrated clearly that the neuronal network responsible for the generation of seizure activity may be considered to act as a strange attractor. A low dimension of the attractor corresponds to a pronounced form of epileptiform activity. However, the analysis of the spontaneous or ongoing hippocampal EEG, either presenting theta rhythm or irregular activity, showed consistently a much higher dimension (data not shown). These findings indicate that it is not possible to state that EEG signals can be represented, in general terms, as generated by low-dimension deterministic chaotic systems. However, under given circumstances, such as during epileptiform discharges, this can be the case.

Babloyantz and Destexhe (1986) have reported that epileptiform EEGs may be described as having a low dimension, whereas the presence of a chaotic attractor could not be distinguished during the awake stage (alpha waves) and REM sleep. We found similar results in the rat. Also Röschke and Başar (1988) found that the spontaneous activity of the cat cortex has a relatively high dimension, of order 8-9, whereas that of the acoustical evoked potential was much lower (between 3.5 and 5).

In conclusion, we may note that in the awake state the ongoing rat hippocampal EEG signals present a high correlation dimension. This has also been found for the human EEG (Babloyantz and Destexhe, 1986) in the awake state and during REM sleep. This means that under these conditions the underlying system has a large number of degrees of freedom. Similarly, Dvorak and Siska (1986) found for the human EEG correlation dimensions between 5 and 10 in the alert state. They stress the need to be cautious in making any assertions about the correlation dimension of the EEG in general terms. However in slow-wave sleep in humans (Babloyantz et al., 1985) and during epileptic seizures (Babloyantz and Destexhe, 1986, in humans and present results) the correlation dimension decreases in a spectacular and consistent way, indicating that the dynamics of the underlying neuronal networks manifests the properties of a strange attractor with a limited number of degrees of freedom.

6.5 General Remarks About Kindling and the Functional Organization of Neuronal Networks

At the cellular level, we may draw a general conclusion and formulate a new hypothesis about the mechanisms of epilepsy. The conclusion is that the kindled state of the hippocampus is characterized by a decrease in the order of 20–30% in the density of GABA-immunoreactive interneurons. This decrease appears to be sufficient to impair the stability of the hippocampal neuronal networks, resulting in a decrease in the threshold for eliciting epileptiform afterdischarges.

A new hypothesis has been formulated to account for the paradox that, during the development of hippocampal kindling (i.e., during the early kindling process), a decrease in inhibitory activity is found at the same time as there occurs an increase in GABA immunoreactivity and in GABA-synaptic release. This hypothesis, illustrated in Figure 6.8, is that during this phase of the kindling process there is a gradual decrease in the sensitivity of GABA receptors induced by repeated depolarizations and a subsequent increase in intracellular Ca^{2+} due to an enhancement of the permeability of the neuronal membranes for Ca^{2+} ions (Kamphuis et al., 1989a and c; Wadman et al., 1985). This hypothesis is being tested experimentally.

At the level of neuronal networks a general remark may be made about the mechanisms of epilepsy. During the development of a kindled seizure, a complex series of changes in the state of the underlying networks takes place. At those brain sites (EC) where the focus starts, a transition from a mode of activity characterized by a large number of degrees of freedom into a strange attractor of low dimension takes place. This is also reflected in the fact that a high degree of coupling between related brain areas (EC, FD, ipsilateral CA1), which behave as strange attractors, is found in the same phase of the kindling seizure. As the seizure progresses, brain areas (contralateral CA1) that are more distant from the focus undergo a similar transition as previously described, with the appearance of strong degrees of coupling with the homologous areas. At a late stage where the strange attractor of the focal area becomes less stable, an uncoupling of the initially related areas takes place, whereas those where a strange attractor of low correlation dimension is present stay strongly coupled.

These observations suggest that it may be useful to follow the development of epileptiform seizures both by measuring the correlation dimension of the local EEG signals and the degree of association between signals recorded from different areas. In practice this implies that if one disposes of a set of EEG signals recorded

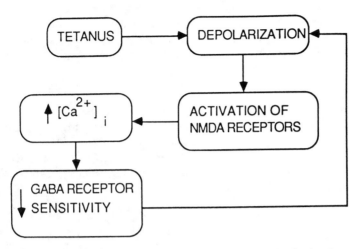

FIGURE 6.8. Positive feedback loop that is probably responsible for the development of the decrease of GABA-receptor sensitivity underlying the kindling process according to the hypothesis proposed here.

from different brain areas during an epileptiform seizure, the signal that, at first, can be considered to contain a strange attractor of low dimension is probably generated in the area closer to the focus. This may have an useful application in focus localization.

In more general terms, it may be said that a neuronal network that contains a strange attractor of low dimension behaves as a complex dynamic system with a small number of degrees of freedom. This is the case during an epileptic seizure in contrast to the high dimensions found in an alert behaving animal or human. Of course, during a seizure, the consciousness state is impaired, and the animal is not able to react to external stimuli. Therefore, we may speculate that there may be a correspondence between the state of a neuronal network characterized by a strange attractor of low dimension on the one hand and a limited capacity for processing of information. In this respect, the observation of Xu Nan and Xu Jinghua (1988) that the correlation dimension of the human EEG tends to increase mainly on the contralateral side to the dominant hand and thus becomes asymmetric during active mental tasks, may be relevant. The state where the EEG signals can be accounted for by attractors of high dimensions or may not be distinguishable from random noise appear to correspond to alert states during which the animals are capable of information processing and of reacting adequately to the environment. Therefore, we may conclude that a neuronal network may switch from one mode of activity to another or from an attractor to another, depending on changes in the parameters or in input state as shown for the olfactory bulb by Freeman (1988).

It makes little sense to try to define a single attractor or a single correlation dimension for a general EEG signal. A correlation dimension varies with the type of underlying neuronal network and therefore with the brain structure and with the conditions under which it operates. The latter are directly related to the behavioral state.

Acknowledgments. The authors thank André Noest for the computer analyses, Evelien Huisman for the experimental work on immunocytochemistry, Wytse Wadman for his advice and criticisms, Simon van Mechelen for the photographic work, and Cristine Cabi and Liesbeth Laan for the secretarial help. This work has been partially supported by grant numbers A49 and A67 from CLEO (Commissie Landelijk Epilepsie Onderzoek)-TNO, Den Haag.

References

Babloyantz, A., Salazar, J.M., Nicolis., C (1985): Evidence of chaotic dynamics of brain activity during the sleep cycle. *Physics Lett.* 111A, 152–156

Babloyantz, A., Destexhe A. (1986): Low-dimensional chaos in an instance of epilepsy. *Proc. Natl. Acad. Sci. USA* 83, 3513–3517

Basar, E., ed. (1988) *Dynamics of sensory and cognitive processing by the brain.* Berlin: Springer

Brazier, M.A.B. (1972): Spread of seizure discharges in epilepsy: Anatomical and electrophysiological considerations. *Exp. Neurol.* 36, 263–272

Brazier, M.A.B. (1973): Electrical seizure discharges within the human brain: The problem of spread. In: *Epilepsy: Its phenomena in man.* Brazier, M.A.B. (ed.). New York: Academic Press, pp. 153–170

Buzsaki, G., Eidelberg, E. (1982): Direct afferent excitation and long-term potentiation of hippocampal interneurons. *J. Neurophysiol.* 48, 597–607

Celio, M.R. (1986): Parvalbumin in most gamma-aminobutyric acid containing neurons of the rat cerebral cortex. *Science N.Y.* 231, 995–997

Collingridge, G.L., Kehl, S.J., McLennan, H. (1983): The antagonism of aminoacid-induced excitation of the hippocampal CA1 neurons in vitro. *J. Physiol.* 334, 19–32

Dvorak, I., Siska, J. (1986): On some problems encountered in the estimation of the correlation dimension of the EEG. *Physics Lett. A.* 118, 63–66

Fernandes de Lima, Pijn, J.P., Nunes, F.C., Lopes da Silva, F.H. (1990): The role of hippocampal commissures in the interhemispheric transfer of epileptiform afterdischarges in the rat: a study using linear and nonlinear regression analysis. *Electroenceph. Clin. Neurophysiol.* (In Press)

Freeman, W. (1988): Nonlinear neural dynamics in olfaction as a model for cognition. In: *Dynamics of sensory and cognitive processing by the brain.* Basar, E. (ed.). Berlin: Springer, pp. 19–29

Goddard, G.V., McIntyre, P.C., Leech, D.K. (1969): A permanent change in brain function resulting from daily electrical stimulation. *Exp. Neurol.* 25, 295–330

Gotman, J. (1983): Measurement of small time differences between EEG channels: Method and application to epileptic seizure propagation. *Electroenceph. Clin. Neurophysiol.* 56, 501–514

Gotman, J. (1987): Interhemispheric interactions in seizures of focal onset: Data from human intracranial recordings. *Electroenceph. Clin. Neurophysiol.* 67, 120–133

Grassberger, P., Procaccia, I. (1983): Measuring the strangeness of strange attractors. *Physics* 9D, 189–208

Guilford, J.P., Fruchter, B. (1985): *Fundamental statistics in psychology and education.* New York: McGraw Hill

Hamon, B., Heinemann, U. (1986): Effects of GABA and bicuculline on NMDA- and quisqualate-induced reductions in extracellular free calcium in area CA1 of the hippocampal slice. *Exp. Brain Res.* 64, 27–36

Harris, E.W., Ganong, A.H., Cotman, C.W. (1984): Long-term potentiation in the hippocampus involves activation of N-methyl-D-aspartate receptors. *Brain Res.* 323, 132–137

Kamphuis, W., Wadman, W.J., Buijs, R.M., Lopes da Silva, F.H. (1987): The development of changes in hippocampal GABA immunreactivity in the rat kindling model of epilepsy: A light microscopic study with GABA antibodies. *Neuroscience* 23, 433–446

Kamphuis, W., Lopes da Silva, F.H., Wadman, W.J. (1988): Changes in local evoked potentials in the rat hippocampus (CA_1) during kindling epileptogenesis. *Brain Res.* 440, 205–215

Kamphuis, W., Huisman, E., Wadman, W.J., Heizmann, C.W., Lopes da Silva, F.H. (1989a): Kindling-induced changes in parvalbumin immunoreactivity in rat hippocampus and its relation to long-term decrease in GABA-immunoreactivity. *Brain Res.* 479, 23–34

Kamphuis, W., Wadman, W.J., Huisman, E., Lopes da Silva, F.H. (1989b): Decrease in GABA immunoreactivity and alteration of GABA metabolism after kindling in the rat hippocampus. *Exp. Brain Res.* 74, 375–386

Kamphuis, W., Wadman, W.J., Huisman, E., Lopes da Silva, F.H. (1989 c): Transient increase of cytoplasmic calcium concentration in the rat hippocampus after kindling-induced seizures. An ultrastructural study with the oxalate-pyro-antimonate technique. *Neuroscience* 29, 667–674

Kamphuis, W., Huisman, E., Dreijer, A.M.C., Ghijsen, W.E.J.M., Verhage, M., Lopes da Silva, F.H. (1990): Kindling increase the K^+-evoked Ca^{2+}-dependent release of endogenous GABA in the area CA_1 of rat hippocampus. *Brain Res.* 511, 63–70

King, G.I., Dingledine, R., Giachino, J.L., McNamara, J.O. (1985): Abnormal neuronal excitability in hippocampal slices from kindled rats. *J. Neurophsyiol.* 54, 1295–1304

Knowles, D.W., Schneiderman, J.H., Wheal, H.V., Stafsrom, C.E., Schwartzkroin, P.A. (1984): Hyperpolarizing potentials in guinea pig hippocampal CA3 neurons. *Cell Mol. Neurobiol.* 4, 207–230

Kosaka, T., Katsumar, H., Hama, K., Wu, J-Y, Heizmann, C.W. (1987): GABAergic neurons containing the Ca^{2+} binding protein parvalbumin in the rat hippocampus and the dentate gyrus. *Brain Res.* 419, 119–130

Leung, L.S. (1979): Orthodromic activation of the hippocampal CA1 region in the rat. *Brain Res.* 176, 49–63

Leung, L.S. (1982): Nonlinear feedback model of neuronal populations in hippocampal CA1 region. *J. Neurophysiol.* 47, 845–868

Lopes da Silva, F.H., Arnolds, D.E.A.T. (1978): The physiology of the hippocampus and related structures. *Annu. Rev. Physiol.* 40, 163–191

Lopes da Silva, F.H., Mars, N.J.I. (1987): Spread of epileptic seizure activity in experimental and clinical epilepsy: The use of mutual information analysis. In: *Presurgical evaluation of epileptics.* Wieser, H.G., Elger, C.E. (eds.). Berlin: Springer, pp. 209–214

Mars, N.J.I., Van Arragon, G.W. (1982): Time delay estimation in non-linear systems using average amount of mutual information analysis. *Signal Processing* 4, 139–153

Mars, N.J.I., Lopes da Silva, F.H. (1987): EEG analysis methods based on information theory. In: *Methods of analysis of brain electrical and magnetic signals.* Gevins, A.S., Rémond, A. (eds.) Amsterdam: Elsevier, pp. 297–307

Mars, N.J.I., Lopes da Silva, F.H, Van Hutten, K., Lommen, J.G. (1977): EEGs during seizures; Localisation of an epileptogenic area. *Electroenceph. Clin. Neurophysiol.* 43, 575

Melchers, B.P.C., Pennartz, C.M.A., Wadman, W.J., Lopes da Silva, F.H. (1988): Quantitative correlation between induced decreases in extracellular calcium and LTP. *Brain Res.* 454, 1–10

Moddemeijer, R. (1987): Estimation of entropy and mutual information of continuous distribution. Internal Report, University of Twente, (The Netherlands), Nr. 080–87–33

Moddemeijer, R. (1989): Delay-estimation with application to electroencephalograms in epilepsy. Ph.D. thesis. University of Twente, Enschede (The Netherlands)

Mody, I., Heinemann, U. (1987): N-methyl-D-aspartate (NMDA) receptors of dentate gyrus granule cells participate in synaptic transmission following kindling. *Nature* 326, 701–703

Morrell, F., Rasmussen, T., Gloor, P., De Toledo-Morrell, L. (1983): Secondary epileptogenic foci in patients with verified temporal lobe tumors. *Electroenceph. Clin. Neurophysiol.* 54, 26P

McNamara, J.O. (1986): Kindling model of epilepsy. In: *Advances in neurology:* vol 44. Delgado-Escueta, A.V., Ward Jr. A.A., Woodbury, D.M., Porter, R.J. (eds.). New York: Raven Press, pp. 1033–1044

Pjin, J.P.M., Vign, P.C.M., Lopes da Silva, F.H. (1989): Localization of epileptogenic foci using a new signal analytical approach. *Neurophysiol. Clin.* 20, 1–11

Racine, R.J. (1972): Modification of seizure activity by electrical stimulation. II. Motor seizure. *Electroenceph. Clin. Neurophysiol.* 32, 295–299

Röschke, J., Basar, E. (1988): The EEG is not a simple noise: Strange attractors intracranial structures. In: *Dynamics of sensory and cognitive processing by the brain.* Basar, E. (ed.). Berlin: Springer

Rosén, I., Salford, L., Starck, L. (1984): Sturge-Weber disease—neurophysical evaluation of a case with secondary epileptogenesis, successfully treated with lobectomy. *Neuropediatrics* 15, 95–98

Schramka, M., Sedlak, P., Nadvornik, P. (1977): Observation of kindling phenomenon in treatment of pain by stimulation in thalamus. In: *Neurosurgical treatment in psychiatry, pain and epilepsy.* Sweet, W.H., Obrador, S., Matin-Rodriguez, J.G. (eds.). Baltimore: University Park Press, pp. 651–654

Schuster, H.G. (1984): *Deterministic chaos.* Weinheim West Germany: Physik-Verlag

Serra, R., Andretta, M., Zanarimi, G., Compiani, M. (1986): *Introduction to the physics of complex septens.* Oxford: Pergamon Press

Servit, Z., Musil, F. (1981): Prophylactic treatment of posttraumatic epilepsy: Results of long-term follow-up in Czechoslovakia. *Epilepsia* 22, 315–320

Stelzer, A., Slater, N.T., Ten Bruggencate, G. (1987): Activation of NMDA receptors blocks GABAergic inhibition in an in vitro model of epilepsy. *Nature* 326, 698–701

Stichell, C.C., Singer, W., Heizmann, C.W. (1988): Light and electron microscopic immunocytochemical localization of parvalbumin in the dorsal lateral geniculate nucleus of the cat: Evidence for coexistence with GABA. *J. Comp. Neurol.* 268, 29–37

Schwartzkroin, A. (1986): A regulation of excitability in hippocampal neurons. In: *The hippocampus:* vol. 3. Isaacson, R.L., Pibram, K.H. (eds.). New York: Raven Press, pp. 113–136

Takens, F. (1981): Dynamical systems and turbulence. In: *Lecture notes in mathematics* vol. 898, Rand, D.A., Young, L.S. (eds.). Berlin: Springer, pp. 365–381

Theiler, J. (1986): Spurious dimension from correlation algorithms applied to limited time-series data. *Physiol Rev* A, 34, 3427–3432

Traub, R.D., Wong, R.K.S. (1983): Synchronized burst discharge in disinherited hippocampal slice. II. Model of cellular mechanism. *J. Neurophysiol.* 49, 442–558

Van Neerven, J.M.A.M. (1987): Determination of the correlation dimension from a time series: Applications to rat EEGs in sleep, theta rhythm and epilepsy. Master's thesis, University of Amsterdam, Amsterdam

Wada, J.A., Sato, M, Corcoran, M.E. (1974): Persistent seizure susceptibility and recurrent spontaneous seizures in kindled cats. *Epilepsia* 15, 464–478

Wadman, W.J., Lopes da Silva, F.H., Leung, L.S. (1983): Two types of interictal transients of reversed polarity in rat hippocampus during kindling. *Electroenceph. Clin. Neurophysiol.* 55, 314–319

Wadman, W.J., Heinemann, U., Konnerth, A., Neuhaus, S. (1985): Hippocampal slices of kindled rats reveal calcium involvement in epileptogenesis. *Exp. Brain Res.* 57, 404–407

Xu, N., Xu, J., (1988): The fractal dimension of EEG as a physical measure of conscious human brain activities. *Bull. Math Biol.* 50, 559–565

Part III

Visual Information Processing

7

A Synergetic Approach to Visual Perception

H. HAKEN

7.1 Introduction

The title of this book, *Machinery of the Mind,* may be interpreted either that we search for the material substrates of mental processes or that we study mechanisms underlying these processes. When we look for the building blocks of the material substrate in brains, we are readily led to neurones. Although properties of individual neurones, including those involved in vision, have been studied in detail, not as much is known about their joint action because studies based on multi-electrode derivations are just in their beginning stages, yet there is no doubt that perception is a process in which a very great number of neurones are involved. In order to obtain a deeper understanding of processes going on in the material substrate when perception takes place, one may follow two different approaches. In the bottom-up approach we start with some idealized properties of neurones, which may be called model neurones. With their help, we try to build up networks that perform specific functions. This way is followed in the rapidly developing field of neurocomputers. There is yet another way; namely, the top-down approach where we start with the known or required functions of a system, in particular of the brain, and then we ask what kind of properties are required so that the system fulfills specific tasks. In this way one may hope to obtain insights into the operation and perhaps also in the properties of the elements of networks that are required to fulfill the prescribed tasks.

This chapter follows the top-down approach that is suggested by synergetics. Although there have been some reports on the application of synergetics to brain research (Basar et al., 1983), not everybody may be familiar with this approach so that a few introductory words may be in order.

7.2 What Is Synergetics About?

Synergetics is an interdisciplinary field of research that studies the cooperation of individual parts of a system that may produce spatial, temporal, or functional structures on macroscopic scales. The spontaneous formation of structures via self-organization is a widespread phenomenon found not only in biology but also in physics, chemistry, and other fields of research. Physics and chemistry provide us with comparatively simple examples, such as a fluid layer in a vessel heated from below. When the temperature difference between the lower and upper surface is small, the fluid is still at rest. Beyond a critical temperature difference, however, a macroscopic motion in form of rolls occurs. In other words, a macroscopic pattern arises in the system. When the vessel has a circular geometry and we look from above at it, the rolls may be oriented in different directions. In other words, the system shows multistability. Which orientation of the rolls will develop depends on the initial "preparation" of the motion of the fluid.

Another famous example is provided by the light source called laser. In a conventional lamp, such as in an excited gas, the individual atoms emit individual wavetracks so that the total light appears as random noise. When the energy pumped into the laser is increased beyond a critical pump rate, the emission may acquire the form of a well-ordered coherent wave, which means that the electrons of the atoms have acquired a coherent action so as to produce that wave. This is again a process of self-organization. Actually when the pump rate is still further increased, the coherent wave is replaced by regular short pulses, and under different pump and loss conditions, so-called deterministic chaos occurs. Individual neurones can show precisely the same firing patterns: namely, noisy, regular, spiking, and chaotic. This shows that the same system can exhibit quite different kinds of behavior if seemingly unspecific control parameters are changed. At the same time this may serve as a paradigm for the behavior of nonlinear networks.

Further examples of self-organization are provided by a number of chemical reactions in which, under specific conditions, the reactants and their products may form patterns in the form of outgoing concentric rings or spirals. The formation of such spatial patterns may be interpreted as the outcome of parallel computation. To this end let us divide the total volume into small volume elements, in each of which the reactions may take place. One may show mathematically that these reactions can be described as addition or multiplication. Because diffusion of chemicals takes place between the individual volume elements, information is transferred between them. Thus, the chemical processes giving rise to patterns can be considered as a parallel computer network. As is shown in synergetics, these and many other processes of spontaneous pattern formation are governed by the same principles. Namely, when a control parameter, such as the temperature difference in a fluid, is changed beyond a critical value, the old state (e.g., the homogeneous state of the fluid) becomes unstable. Close to the transition point one or several collective modes (the fluid rolls) may become unstable while other collective modes are still stable. The amplitudes of the unstable modes serve as so-called order parameters and determine the evolving structures. The behavior of the stable modes or elements is uniquely determined by that of the order parameters in a unique fashion via the so-called slaving principle of synergetics. When an initial state is given, the nonlinear dynamics of the system pulls the total system into a state that is described by one or several order parameters so that a specific pattern occurs. This chapter focuses on those situations where just one order parameter wins a competition between several order parameters.

7.3 Our Goal in Pattern Recognition

This chapter examines the properties of an abstract system that can do pattern recognition. This model has been tested on a serial computer and has proved successful in the recognition of faces and other objects. Our approach has the following characteristics: recognition can be done regardless of the size, position, and orientation of the object, or in other words, the recognition is invariant against translation, scaling, and rotation. The recognition mechanism is strongly resistant against noise, and it may recognize scenes, e.g., group photographs. Furthermore, it can deal with ambiguous patterns, such as the famous vase/face ambiguity, and it takes into account bias. The results are in excellent agreement with psychological (or psychophysical) findings on the perception of ambiguous and biased patterns. A number of conclusions that may be of interest to the neurosciences can be drawn.

The simultaneous invariance against translation, rotation, and scaling is achieved when we subject the patterns first to a Fourier transformation from ordinary space to the wave-vector space and then apply the complex logarithmic map to the wave-vector space. By contrast, it is usually assumed that the map from the retina to the visual cortex is done by means of the complex logarithmic map directly applied to the spatial coordinates. Thus, our own approach comes closer to the point of view of Pribram who conceives of the brain as a hologram.

Our abstract system can be realized by networks of model neurones in various ways, which have properties that are rather distinct from those of model neurones considered so far, e.g., two-state neurones.

Our approach sheds new light on the concept of neural nets and their relationship to grandmother cells. Actually, our algorithm can be realized both by a neural net with distributed information processing and by means of a network of grandmother cells where each cell is responsible for the recognition of a specific object. Thus, our approach does not rule out the existence of grandmother cells, some of which have been probably identified in sheep and monkeys. On the other hand, even when grandmother cells exist, they do not recognize pattern by themselves, but only in interaction with all the other grandmother cells.

The problem of ambiguous patterns that results in oscillations in perception can be resolved by a temporal change of a parameter that may be called the attention parameter and that undergoes saturation. This fatigue can be very simply described with respect to grandmother cells. In the case of a distributed neural network, this kind of fatigue is discerned within the synaptic strengths, and it may be very difficult to discover it experimentally.

7.4 Pattern Recognition by Synergetic Systems

We base our approach to pattern recognition of visual structures on the following concepts. First, we adopt the concept of associative memory, i.e., that in the process of pattern recognition an even uncomplete pattern may be automatically completed and later a specific identification is made, for instance by attaching a name to a face. We also assume that pattern recognition is done by a synergetic system whose parts cooperate so that the incoming pattern is treated as a state of the system. By the dynamics of the system this state is then automatically pulled into a stable state coming closest to the offered pattern, whereby the stable state corresponds to the complete pattern, including, for instance, face plus name.

Third, we use the duality between pattern recognition and pattern formation that had been previously studied by synergetics. In pattern formation the evolution of the order parameters determines the evolving structure whereby the order parameters govern the behavior of the subsystems by means of the slaving principle. If initially a part of the subsystems forms some pattern, these subsystems can generate their order parameter, which in turn enslaves the total system so that the pattern is completed. In pattern recognition we are first provided with features that may be incomplete but determine the order parameter. By its action it may supplement all the missing features. Therefore, the restoration of patterns in pattern formation can be interpreted as an associative memory.

Leaving aside a number of important physiological processes in vision, we shall proceed in the following manner. When we are given a number of prototype patterns, such as faces, we put a grid on these pictures and take the gray values v_j in each cell j as a component of a vector $\mathbf{v} = (v_1, v_2, \ldots, v_j, \ldots)$ describing that pattern. Although we can deal equally well with specific colors, we shall adopt the simpler notion of gray values. Any offered pattern is encoded in a vector $\mathbf{q} = (q_1, q_2, \ldots)$, which changes in time according to the following equation:

$$\dot{\mathbf{q}} = \sum_u \lambda_u \mathbf{v}_u \mathbf{v}^u \cdot \mathbf{q} - \sum_{u' \neq u} (\mathbf{v}^u \mathbf{q})^2 \mathbf{v}_u (\mathbf{v}^u \mathbf{q}) - |q|^2 \mathbf{q} + \mathbf{F}$$

attention	discrimination	saturation	fluctuation
parameter			forces
	learning		
	matrix : association		

[1]

The individual terms are denoted with respect to the task they have to fulfill. The equations that look rather complicated have a very simple meaning. They describe in a high-dimensional space the motion of a particle that slides downhill in some landscape represented by a potential function $V(\mathbf{q}, \mathbf{q}^+)$:

$$\dot{\mathbf{q}} = -\frac{\partial V}{\partial \mathbf{q}^+} + \mathbf{F}$$

[2]

The equations of motion are constructed in such a manner that an initially offered pattern described by $\mathbf{q}(0)$ undergoes a specific temporal change, which eventually leads into one of these originally stored prototype patterns encoded by $\mathbf{v}_u$.

$$\mathbf{q}\,(0) \rightarrow \mathbf{q}\,(t) \rightarrow \mathbf{v}_u \qquad [3]$$

$$\text{offered} \qquad \text{recognized}$$

The motion occurs in such a way that the system approaches that prototype pattern $\mathbf{v}_u$ that comes closest to the offered pattern $\mathbf{q}$. As is explained below, Eq. 1 can be interpreted as a network. Using the projection of $\mathbf{q}$ on $\mathbf{v}$, we may introduce new variables $\mathbf{d}_u = (\mathbf{v_u}\,\mathbf{q})$ and transform Eq. 1 into the following equations:

$$\dot{d}_u = (\lambda_u - D)\,d_u + d_u^{\,3} + F_u(t) \qquad [4]$$

$$D = \sum_u d_u^{\,2}$$

In the parlance of synergetics, d_u is the order parameter, whereas the λ serves as a control parameter. The sliding downhill of a particle can be visualized by means of Figure 7.1. To construct the vectors $\mathbf{v}_u$ of the prototype patterns, faces

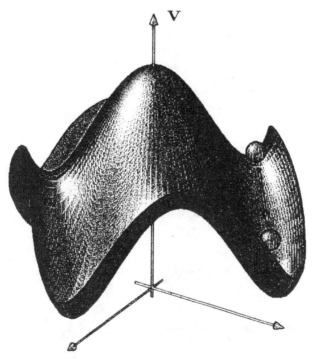

FIGURE 7.1. Dynamics described by Eq. 1 with two prototype patterns and a two-dimensional feature space. The figure shows the landscape of the potential in which the particle, which represents the state of the system, moves. The valleys represent the prototype patterns that are stored in the computer.

FIGURE 7.2. Example of faces and names stored as prototype patterns in the computer.

FIGURE 7.3. When the attention parameter is negative, the corresponding face, even if originally fully offered to the computer, disappears in the course of time.

jointly with their names were digitized on a grid of 60×60 pixels (Figure 7.2). Then in a first test the attention parameters λ were put negative, and part of a face was offered (Figure 7.3). The system "dissolved" the face; that is, with no attention present a face cannot be recognized. Then two further tests with positive-attention parameters (that were equal to each other) were made where either the name (D) or part of a face was offered. Each time the system restored the whole pattern (Figure 7.4).

The process could be made invariant against simultaneous translation, rotation, and scaling by the following procedure. Replacing the components q_j of $\mathbf{q}$ by continuous functions in the x,y plane (i.e., by $q(x,y)$), we subject $q(x,y)$ to a Fourier transformation with coefficients $c(k_x, k_y)$. Since a displacement $x, y \rightarrow x - x_o, y - y_o$ causes a multiplication of $c(k_x, k_y)$ by exp $(ik_x x_o + ik_y y_o)$, by taking the absolute value of c, we may eliminate the exponential function that contains the displacement. Thus, $|c|$ is invariant against translation. In the next step we introduce polar coordinates in the complex plane of

$$\mathbf{k} = k_x + ik_y = r\, e^{i\phi}$$

and apply a logarithmic map
$$\mathbf{k}' = \ln \mathbf{k} = \ln r + i\phi.$$

Evidently, a scaling $r \rightarrow r/r_o$ and a rotation $\phi \rightarrow \phi + \phi_o$ correspond to a translation by $- \ln r_o$ and $i\,\phi_o$. Fourier transformation of $|c|$ in the $\mathbf{k}'$ – space and taking the absolute value of the Fourier coefficients provide us with a representation that is invariant against scaling and rotation. In this way we obtain invariant quantities that now take the role of the $\mathbf{q}$s in our original Eq. 1. The individual steps are shown in Figure 7.5.

The robustness of our procedure against noise is shown in Figure 7.6. We offered complex scenes to the computer (Figure 7.7). The computer first identified the face of the woman. Then we put the corresponding attention parameter equal to 0 or to a negative value. In the next step the computer was able to identify the face of the male person. Identifications were made also of groups and are shown in Figure 7.8.

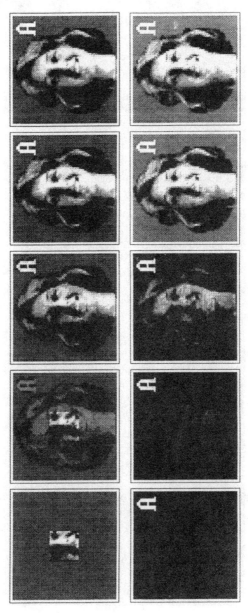

FIGURE 7.4. *Upper row*, If a name is offered to the computer, it can restore the whole pattern, i.e., the face belonging to the corresponding name. *Lower row*, If part of a face is offered, the computer can restore the whole face and the name belonging to it.

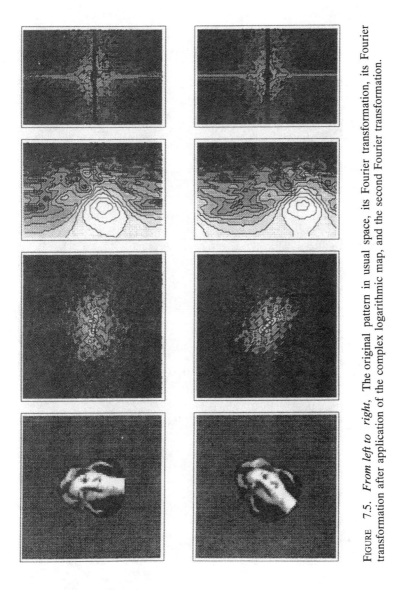

FIGURE 7.5. *From left to right*, The original pattern in usual space, its Fourier transformation, its Fourier transformation after application of the complex logarithmic map, and the second Fourier transformation.

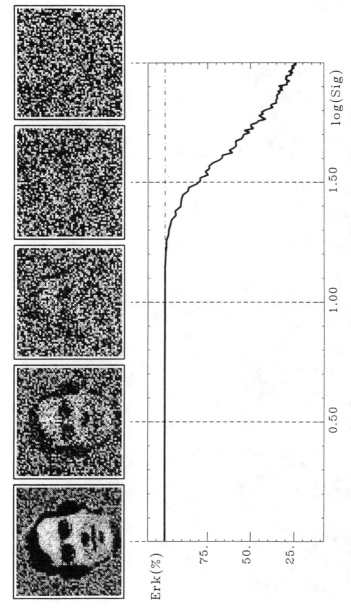

FIGURE 7.6. *Upper row, from left to right,* The original face is subjected to an increasing noise level. *Lower row,* Percentage of faces recognized as a function of the increasing noise level.

FIGURE 7.7. A complex scene that was recognized by the computer.

FIGURE 7.8. Same as in Figure 7.7 with several faces.

7.5 Ambiguous Patterns

To deal with ambiguous figures, such as in Figure 7.9, we used the assumption that the attention parameter becomes saturated.

To illustrate our general procedure, consider the case of two patterns described by the order parameters, d_1 und d_2. The order parameter Eq. 4 were generalized in the following manner. In order to take into account saturation, in addition to Eq. 4, we assumed that the attention parameters λ_1, λ_2 obey the following equations that describe saturation.

FIGURE 7.9. An ambiguous figure (vase or faces).

$$\text{Order parameters} = d_1, d_2$$

$$\text{Attention parameters} = \lambda_1, \lambda_2$$

$$\dot{d}_1 = (\lambda_1 - D)\, d_1 + d_1{}^3 - \frac{\partial V_1}{\partial d_1} \quad D = d_1{}^2 + d_2{}^2$$

$$\dot{d}_2 = (\lambda_2 - D)\, d_1 + d_2{}^3 - \frac{\partial V_1}{\partial d_2}$$

[6]

$$\dot{\lambda}_1 = \gamma\, (1 - \lambda_1 - d_1^2)$$

$$\dot{\lambda}_2 = \gamma\, (1 - \lambda_2 - d_2^2)$$

[7]

In order to describe a bias, we assume that the watershed that separates the two valleys of the potential of Figure 7.1 can be shifted. This has been taken care of mathematically by the additional terms containing V_1 in Eq. 6 where V_1 is given by

$$V_1 = C\, d_1{}^2\, d_2{}^2 \left(1 - 4\alpha\, \frac{(d_1{}^2 - d_2{}^2)}{d_1{}^2 + d_2{}^2} \right)$$

[8]

The parameter α, which is the rotation angle of the watershed against its middle position, can be directly correlated to experimental data because psychological tests were done on test persons who were offered ambiguous patterns. Let us assume that there are n test persons. Then in these tests n_1 persons recognize pattern 1 first, whereas n_2 persons recognize pattern 2 first. Then, the angle α is given by

$$\alpha = \alpha_c \, (p_1 - p_2) \qquad [9]$$

where the quantities p_1 and p_2 are defined as ratios

$$p_1 = \frac{n_1}{n} \, , \, p_2 = \frac{n_2}{n} \qquad [10]$$

The computer calculations show that, for an α smaller than a critical value α_c, oscillations in perception occur. In this way pattern 1 is perceived over time $T_1(\alpha)$ and pattern 2 over time $T_2(\alpha)$. In accordance with the experimental data our model shows that the duration of the perception of each pattern increases or decreases while the total time $T = T_1 + T_2$ increases with increasing α (Figure 7.10). Our results on the dependence on T_1 and T_2 on α are in excellent agreement with the experimental data. For α bigger than α_c, no oscillations occur any more, and the system relaxes to one of two fixed points; that is, the system becomes bistable. We recognize either one or the other pattern for an infinitely long time.

7.6 Interpretation as Network

Eqs. 1 or 4 can be visualized in several ways as a network. When we decompose the vector q into its components q_j, we may attach to each component j a model neurone j. When written down in components, Eq. 1 means that each model neurone receives inputs from other model neurones at an earlier time-step (Figure 7.11). When the connections are fixed by the coefficients of the products of q_j on the right-hand side of Eq. 1, it is evident that the attention parameters λ enter all the connections between neurones. If there are many neurones, evidently these temporal changes of λ as described in our model, which takes care of oscillations, may be rather minute. The behavior of the network can be described equally well by the order parameter Eq. 4. This network is depicted in Figure 7.12. The inputs q_0 are projected by means of $\mathbf{v}$s on the cells denoted by d_j , which refer to a specific pattern and which therefore correspond to grandmother cells. Therefore, the network has only two steps: one is a projection of the incoming inputs onto the grandmother cells, and then the grandmother cells interact within their network. The simplicity of Eq. 4 allows us to replace the network of Figure 7.12 by a still simpler network with considerably fewer connections, as exhibited in Figure 7.13. Here the grandmother cells communicate via a general reservoir; the attention parameters λ refer to each individual grandmother cell so that fatigue takes place

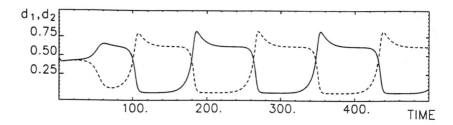

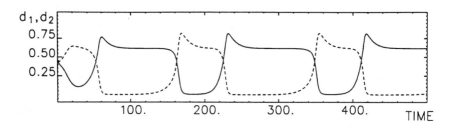

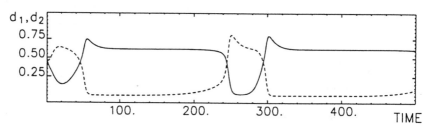

FIGURE 7.10. Examples of the perception of patterns depending on the original bias. The bias was increased from top to bottom.

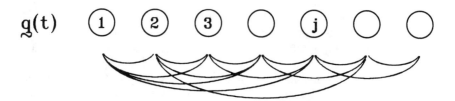

FIGURE 7.11. Interpretation of Eq. 1 as parallel network of specific "neurones."

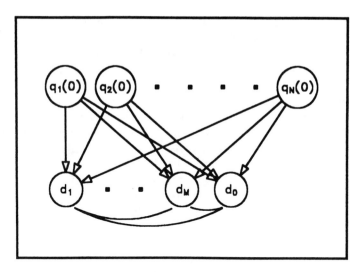

FIGURE 7.12. Network with an input layer and a middle layer containing "grandmother" cells.

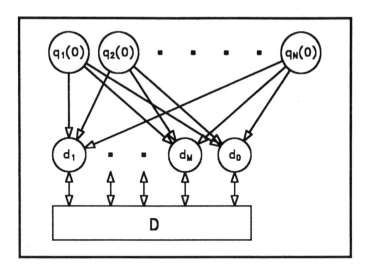

FIGURE 7.13. Reduction of the network of "grandmother" cells to individual "grandmother" cells that are coupled via a common reservoir.

in each cell individually and is not influencing the interconnections between them or with their common reservoir D.

It is surprising that the same behavior of a system can be realized by quite different configurations. This quite naturally leads to the question of how nature proceeds. Indeed, here we have conflicting experimental results. On the one hand, there is strong evidence of the dislocalization of the stored information; on the other hand, other experiments seem to indicate that grandmother cells indeed exist. These experiments shed light on our ideas on learning by means of the Hebb mechanism. Although in the network of type 1 learning proceeds by changing the individual strengths of synaptic connections, in the concept of the grandmother cells two learning mechanisms may occur. Namely, the projections v may be changed during the learning process, and in addition it may be necessary that new grandmother cells are generated or are dying out (in the case of an initial abundance of them).

It should be mentioned that in our discussion on various realizations by networks we considered the simple case without preprocessing, i.e., where the spatial data encoded by $q(x,y)$ were processed directly. In the more complicated case in which we take into account the invariance against translation, rotation, and scaling, first the incoming signals must be preprocessed so that there are two or three layers above the input layer shown here

7.7 Discussion and Outlook

The most important question is how relevant our above considerations are for brain research. In this author's opinion, because of the enormous complexity of the brain, an understanding of its function can be obtained only by a close cooperation between experiment and theory. Although we possess a great amount of information on the action of individual neurones, very little is known about their operation within a network. Yet, when multielectrode experiments on several or many neurones are made, we are confronted with a new difficulty; we are overflooded by an enormous amount of data, i.e., information that is to be processed. Such a processing can be made only if we have some guidelines at hand for which effects and correlations to search. In other words, we need models for possible interpretations of these data. Whether the theoretical model presented in this chapter already provides us with such a means is open for discussion. Yet, I think it strongly shows the direction in which one kind of research may proceed further; namely to study models that can be realized on various kinds of networks that are conceptually still simple and that can perform tasks in a similar way to how they are done by the brain.

Of course, one must not overlook the difficulties still ahead of us, e.g., when two systems fulfill the same task it is by no means guaranteed that the processes underlying the corresponding functions are identical or that even there is a one-to-one correspondence. Indeed, the field of synergetics has shown that the same patterns in the inanimate world can be produced by quite different underlying mechanisms. On the other hand, one may derive from such models specifications for the general frame of the properties of the basic elements. Our network as realized by grandmother cells indicates quite clearly that the function of the network is not achieved by individual logical steps, but rather by majority decisions represented by D in Eq. 4. It is an interesting question to discuss which building blocks of the nervous system may make such decisions. Is it an individual neurone, a group of neurones, or a bigger array of them? Another comment may be in order: to achieve simultaneous invariance against translation, rotation, and scaling we performed several transformations. When we drop the requirement of simultaneity, other realizations become possible. Indeed, for instance it is known that human perception is not entirely rotation invariant, i.e., for instance, faces can no more be clearly recognized when they are upside down. Yet the question of how our perception works regardless of position, orientation, and scale remains unanswered.

At any rate I believe that there is an urgent need for bold new ideas on networks that go considerably beyond the line so far followed; namely, sticking too closely to some idealized properties of individual neurones. I hope that this chapter may serve as a first step toward that goal.

References

Basar, E., Flohr, H., Haken, H., Mandell, A.J. (1983): *Synergetics of the brain.* Berlin: Springer Verlag

Haken H. (1990): *Synergetic Computers and Cognition.* Berlin: Springer Verlag

8

The Eye of Man, the Brain of the Physicist: A System's Approach to Vision

L. Henk van der Tweel

8.1 Introduction

Among the human senses, vision ranks high in its importance for the individual. It can be safely stated that no natural or artificial system in which information is handled surpasses human vision in complexity and efficiency.

When the image of the visual scene is projected on the fundus of the eye, it encounters a structure that in reality is part of the brain, i.e., the retina. A schematic layout of the retina is given in Figure 8.1 (Kuffler and Nichols, 1977). In the few square centimeters of the retina, one finds some 200 million photoreceptors, tiny organs that are able to sense the smallest units in which light energy can be packed (i.e., light quanta) and transfer them into "electricity." The ensuing signal is strong enough to be transported to and through a number of neural stations.

The receptors can be divided into two main classes: cones and rods. The cones, which take their name from their shape in the peripheral retina, serve daylight vision and form the basis of color vision. There are about 2 million cones in the human retina. They are actually present in several distinct shapes, but the specific functions belonging to the various types are not exactly known. The other class of receptors is that of the rods, also named after their shape. The system in which they function (the scotopic system) is extremely sensitive; only a few light quanta suffice to give a visual impression. There is but one kind of rod, and all rods are filled with only one photopigment, rhodopsin; therefore, under conditions of night vision no colors are seen. At the end of this chapter in Section 8.7.3, the condition of total colorblindness is described, which is caused by the absence of functioning cones.

Signals from both cones and rods are transported along very complicated networks through the retina to the retinal ganglion cells. Here, for the first time in the sequence of processes, nerve spikes are generated and sent along the optic tract to the higher centers of the brain.

The human and primate retinas have many elements in common with those of lower animals, but as a special property only the most central part of the retina—

160

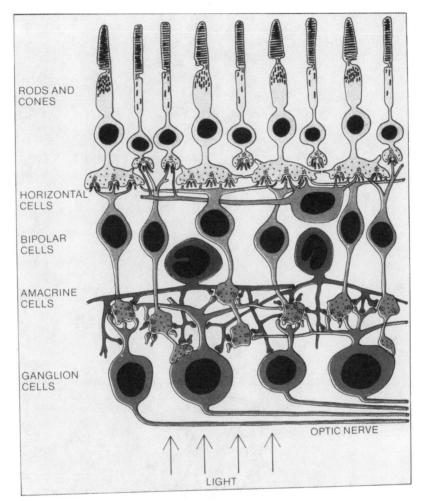

RODS AND
CONES

HORIZONTAL
CELLS

BIPOLAR
CELLS

AMACRINE
CELLS

GANGLION
CELLS

OPTIC NERVE

LIGHT

FIGURE 8.1. Schematic representation of the human retina. Reprinted with permission from Kuffler, S.W., Nichols, J.G. 1977: *From neuron to brain*. Sunderland, M.A.: Sinauer. After Dowling and Boycott, 1966.

the fovea, representing 1–2 degrees of the visual scene—is gifted with sharp vision. Nevertheless, our visual world gives the impression of being homogeneous and sharp. This is caused by continuous "scanning" of the scene. Yarbus (1967), a Russian scientist, recorded the eye movements when a picture is viewed. In Figure 8.2, the fixation points are projected back on the picture; in this case a lion drawn by Delacroix. It is evident that there is a concentration of fixation points on the eyes, mouth, and ear of the lion: the most important and/or complex features of a visual scene attract the main attention.

Even though we scan continuously, we perceive the outside world as stable. MacKay (1974) has paid much attention to this stability of the visual scene (see

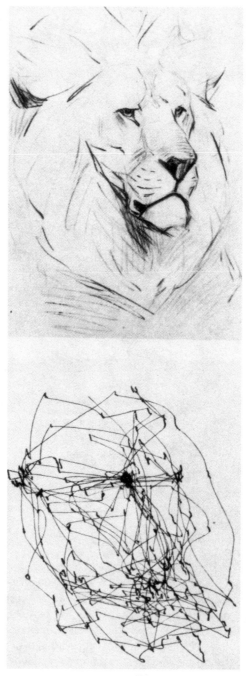

FIGURE 8.2. Recording of the gaze of a spectator viewing a drawing of a lion by Delacroix. Fixation points concentrate on eyes, mouth, and ear. When squinting, the graphic recording starts to resemble the drawing. Reprinted with permission from Yarbus, A.L., 1967: *Eye movements and vision*. New York: Plenum Press.

MacKay and Mittelstaedt, 1974). In addition to the compensatory controls relating to eye movements and the equilibrium sense, MacKay pointed out that an efficient "technique" for stabilizing our representation of the outer world would be the perceptual supposition that this world does not change and that therefore only true changes in the visual scene are consciously noticed. Such a procedure would also result in a kind of data reduction.

8.2 Visual Pathways

Figure 8.3 shows schematically how the visual signal is transported along various stations to the cortex. In the retina, a light stimulus influences the electrical potentials of the cells at many places, and already in this early state various interactions take place. Only when the electrical activity reaches the ganglion cells, however, do spikes become the bearers of information. One can say that transformation from continuous to discrete takes place at this site. Note that the term "discrete" and **not** "digital" is used, because "digital" would mean that true counting is performed as in a computer. It is nearly certain, however, that further handling of the optic information does not occur by calculating (counting), but by sampling the neural activity over periods of approximately 20–50 msec. The result serves then as a measure of stimulus strength.

After having passed the optic chiasm (see the Section 8.7.2 on albinism), the spikes in the optic tract impinge on the next station—the lateral geniculate body— where the spike-trains evoke electrotonic activity that results in new action potentials. In the six layers of this relay station many new interactions and transformations take place. These continue in the cortex where a surprising variety of representations of the visual world can be found.

8.3 Flicker Fusion and System Analysis

Whenever we view TV we become a "living proof" of the slowness of the visual system. The inherent alternation of 50/60 pictures per second is not or only slightly perceived, an effect due to "flicker fusion." From the mid-1800s on there was great interest in this phenomenon, especially from psychologists. Before 1950 about 1500 publications on flicker vision appeared (Landis, 1953), producing a chaotic picture. No unified viewpoint was developed until a Dutch engineer, H. de Lange (1957, 1958), returned to the theory that had been developed by Fourier in 1807 (see Regan, 1989). De Lange's use of modulated light was preceded by Ives (1922), but his analysis was not as deep and there was no climate for this type of approach; it was totally forgotten.

To appreciate the importance of this step, it is necessary to explain at least some aspects of system analysis. A system of the kind we are interested in can be defined as performing a transformation onto its input signals. In our case this may concern the way in which light distributions varying in time are processed electrophysiologically, but also static pictures may be considered.

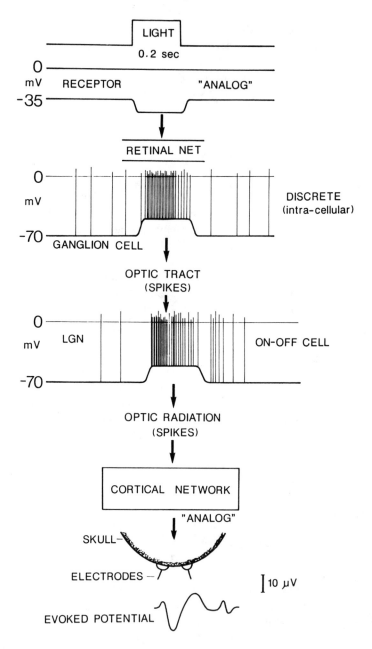

FIGURE 8.3. Schematic representation of signal transport along visual pathways. At each neural station spikes (discrete elements) are transported in electrotonic potentials ("analog"), which in their turn generate new spikes. The term "analog" is colloquial; in reality the information in the spike trains is also analog (without quotation marks).

The simplest system, in a mathematical-physical sense, is a linear one. Mathematically, it is defined by the superposition principle. Suppose that an input A—for instance, a voltage changing in time—is transmitted through such a linear system, which results in an output C that generally will have another magnitude and time course than the input. The same applies for an input B resulting in an output D. The superposition principle holds that the combination from A and B will cause an output C + D. One of the consequences is that in such a system there is proportionality between input and output. For our purpose the following is essential: in any linear system the harmonic function A sin (2πft + Φ) of Figure 8.4 remains harmonic, although practically always the amplitude and phase of the output will deviate from those of the input, depending on the frequency.

It should be noted that the sinusoidal wave shape is the only one that preserves its character. The change of shape to which all other functions are subjected is denoted by the term "linear distortion," as distinct from nonlinear distortion that is so profusely present in physiology. Furthermore, a linear system is exhaustively defined by the input-output relations of a continuum of harmonic functions as a function of frequency. In practice often a series of closely spaced discrete frequencies is employed.

Until de Lange's approach, practically all flicker research in vision was performed by interruption of light with strong and sudden transitions, and it was impossible to comprehend. Moreover, as is well known, the visual system has a logarithmic character, which of course is far from linear. De Lange, however, hypothesized that if one would use sinusoidally modulated light near threshold, even the complicated visual system might show linear properties (the "small signal" approach). The assumption proved very fertile. Figure 8.5, adapted from de Lange's thesis (1958), shows that the temporal properties of the visual system

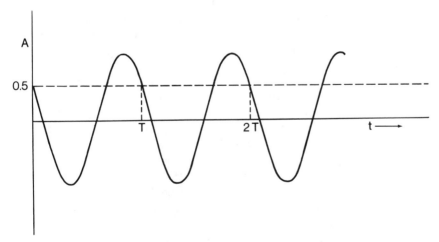

FIGURE 8.4. Sinuoidal wave shape as employed in system analysis. It is illustrated by the function sin(2πft+Φ) with frequency $f=1/T$. It is characterized by its amplitude and its phase at $t=0$; in this case Φ $=150°$.

can indeed be described by a frequency characteristic in a similar way as is done for an audio amplifier. Therefore, it became possible to define flicker sensitivity accurately, and previous ambiguities were avoided. At the same time, de Lange proved that even the mathematical theory of Fourier has its application in vision. The different waveshapes depicted in the inset in Figure 8.5 appear to be evaluated by the system according to their fundamental component only, in any case at the high-frequency side.

Figure 8.5 illustrates also how threshold modulation depth depends on retinal illumination. Modulation depth (MD) is defined as M% = 100×(Lmax − Lmin)/(Lmax + Lmin). 100% MD is the deepest modulation that can be realized technically, because Lmin will then be zero. The curves are presented with decreasing modulation depth plotted logarithmically upward. Since the reverse of threshold is a measure of sensitivity, the curves give a direct impression of the sensitivity of flicker as a function of illumination and frequency. For example, at 430 trolands the threshold at 40 Hz lies at 40% modulation depth, whereas it is (virtually) 150% for 43 trolands. One is justified to say that the difference in sensitivity is approximately a factor of 4, a figure that never could be deduced from conventional flicker fusion measurements.

It was immediately evident that it would be most interesting to try to apply this technique electrophysiologically; Visser and I (1959) started by measuring ERGs in cat with sinusoidally modulated light. Figure 8.6, taken from the original publication, shows that this effort was quite successful. The sinusoidal shape of the stimulus was well preserved, and interestingly, much bigger phase shifts were encountered than were expected. An explanation can be found in the duality of the retina, i.e., the existence of two systems: that of the rods and of the cones. At low frequencies, the ERG of the scotopic system dominates because, under our conditions, it is mainly produced by (abundant) stray light. At higher frequencies, however, the photopic system (mediated by the cones) has the largest impact. The two systems exhibit at low frequencies opposite polarity and have different frequency characteristics. Therefore, the resulting polarity (and/or phase) depends on their relative weights. Large phase-shifts between nearby frequencies can then be expected.

The linearity of the responses proved good enough to apply Fourier techniques; the ERG to a step of light corresponded indeed with that calculated from the amplitude and phase characteristics under the supposition of linearity. Subsequently we could show that the signal transport from eye to brain, objectively measured as VEPs, also obeys laws of linearity at several frequencies.

One of the advantages of the modulation technique is demonstrated in Figure 8.7 (Spekreijse and van der Tweel, 1972) in which the cortical responses to 60 Hz modulation of a homogeneous field at various modulation depths are presented. They were to a high degree proportional to modulation depth (strength of stimulation)!

It should be realized that all these responses were obtained when flicker was in no way perceived any longer by the subject. Apparently there was (physical) information in the brain of which the subject could make no use. That it was yet

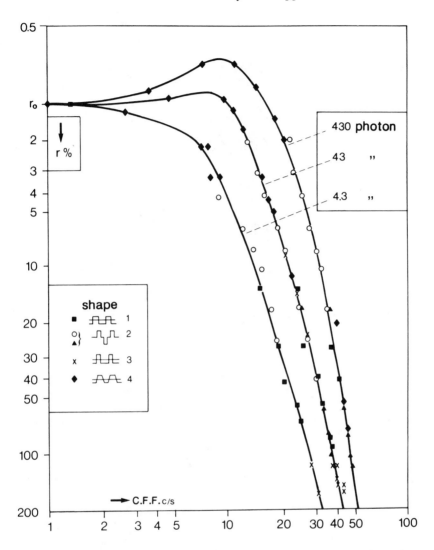

FIGURE 8.5. Original curves from de Lange of human flicker sensitivity. Several waveforms were employed; the curves show that the threshold is exclusively determined by the amplitude of the fundamental, calculated by Fourier analysis (as indicated by the symbols). Modulation percentages > 100% are virtual ones, but fit into the general curve. The attenuation at high frequencies is very steep. The unit of retinal illumination "photon" is identical to the "troland." Adapted from de Lange, H., 1957: *Attenuation characteristics and phase-shift characteristics of the human fovea-cortex systems in relation to flicker-fusion phenomena*. Thesis.

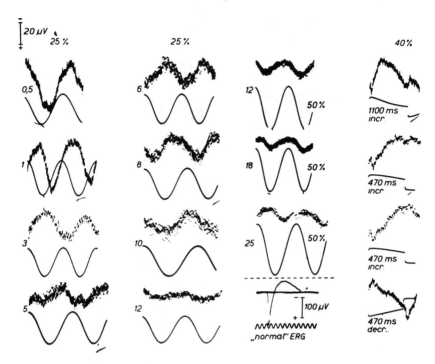

FIGURE 8.6. Electroretinogram of anesthesized cat to sinusoidally modulated light. *Upper curves*, Electroretinogram; *lower curves*, photocell signal of light modulation. High illumination (50.000 trolands). Note large phase-shifts and approximate sinusoidal shape of responses. ERGs to conventional flash and to incremental and decremental steps are included. Reprinted with permission from van der Tweel, L.H., Visser, P., 1959: Electrical responses of the retina to sinusoidally modulated light. *Electroretinographia.* Acta Facultatis Medicae Universitatis Bruneusis Lekarska Fakulta Brne.

possible to record responses to a nonperceived stimulus is due to the infinitely long memory of the computer, as well as to the large retinal and cortical area that is covered by the recording electrode, whereas visual discrimination has a time window of only a few seconds and a spatial extension of 5° to 10°. Another important phenomenon is also shown in Figure 8.7: at stimulation with 30 Hz a strong 60 Hz response is recorded. This represents a fundamental property of the visual system, i.e., the presence of rectifying elements. Spekreijse (1969) has shown that rectification is already found in the retinal ganglion cells of the goldfish; many properties that had been found in humans proved to have their counterpart there.

A very interesting finding at this stage of research was that the signal transport for homogenous fields follows three pathways, one for frequencies up to 15 Hz with latencies of 150-200 msec, a high-frequency one with latencies on the order of 40 msec (Spekreijse and van der Tweel, 1972), and one around 20 Hz (Regan, 1966) with latencies of approximately 100 msec. This division is not reflected in

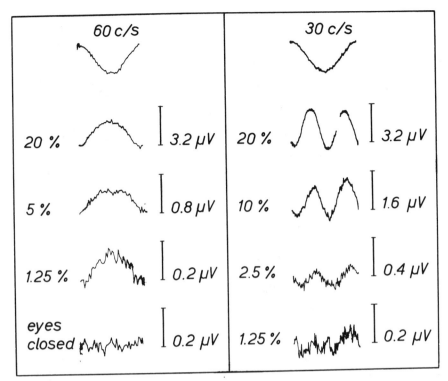

FIGURE 8.7. Occipital VEPs to sinusoidally modulated light at 30 and 60 Hz. Sensitivity of recording is inversely proportional to modulation depth. Reprinted with permission from Spekreijse, H., van der Tweel, L.H., 1972: System analysis of linear and nonlinear processes in electrophysiology of the visual system. *Proc. Kon. Ned. Akad. van Wetensch. 75, 77–105.*

psychophysics. The electrophysiological representations of these temporal frequency "channels" on the skull have definitely different locations, whereas characteristic features, such as saturation, integrative fields and nonlinearities, are also frequency-specific (Spekreijse, 1966).

Because the system analysis technique enables quantitative comparison of frequency characteristics at the various stages of the visual system, their temporal properties (transfer functions) can be determined.

There appeared to be an intriguing parallelism of the low frequency system and the alpha-rhythm: both show a comparable frequency selectivity in a given subject. The selectivity measured in this way bears no known relation to visual or psychological properties and exhibits a large interindividual variation. Figure 8.8 (van der Tweel and Verduyn Lunel, 1965) shows two different subjects with, respectively, a low and a high selectivity at approximately 10 Hz. In analogy, subject B has a very sharply tuned and persistent α rhythm, whereas it is much less outspoken and more reactive in subject A.

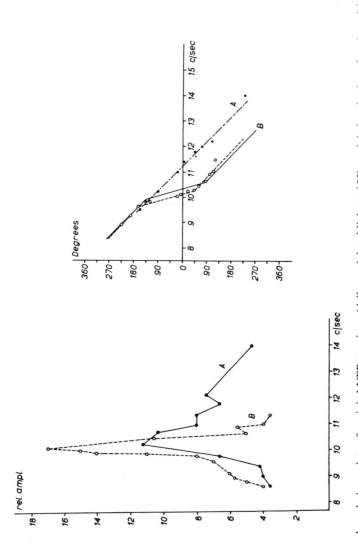

FIGURE 8.8. Amplitude and phase plots of occipital VEPs to sinusoidally modulated light at 10% modulation depth as function of frequency. Subject A shows a broad maximum at 10.5 Hz. Subject B (two identical runs) exhibits high selectivity, which is also reflected in the steep course of the phase plot around 10 Hz. Note the linear frequency scale. Reprinted with permission of Elsevier Science Publishers B.V., from van der Tweel, L.H., Verduyn Lunel, H.F.E., 1965: Human visual responses to sinusoidally modulated light. *Electroenceph. Clin. Neurophysiol.* 18, 587–598.

8.4 Checkerboards

Until now only the processing of unstructured fields was treated, but in the real visual world blank fields are noninformative. During our research we came naturally to experiment with checkerboards. The reason was that we wanted to measure the size of the integrational (perhaps retinal) fields that should precede rectification. If one considers that primitive retinal receptive fields are more or less circular, squares are the best fit for a reversal stimulus with constant average luminance. (One might think that a hexagon would be more suitable, but it is impossible to make an alternating pattern with hexagonal elements.)

To our surprise, the size and quality of the cortical responses to such a stimulus surpassed in most subjects all our expectations; they could in no way be explained by the responses to luminance changes. Figure 8.9 (Spekreijse and van der Tweel,

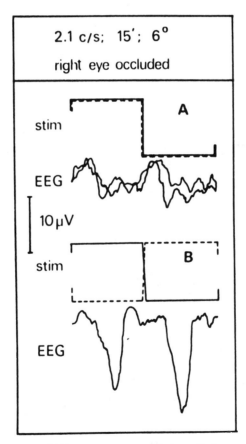

FIGURE 8.9. Occipital VEPs to 20% square wave modulation of a homogenous field of 6°. A, Small and similar responses to increase and decrease of illumination. B, Checkerboard reversal with identical modulation; 15' checks. Pattern responses prove stable and much larger. Reprinted with permission from Spekreijse, H., van der Tweel, L.H., 1972: System analysis of linear and nonlinear processes in electrophysiology of the visual system. *Proc. Kon. Ned. Akad. van Wetensch.* 75, 77–105.

1972) is representative of the results obtained with most subjects; square wave modulation of a homogeneous field shows small and equal responses to both the increase and decrease of illumination. If the field is split up in checks, however, reversal excites crisp responses, which are many times larger than those to homogeneous fields.

Figure 8.10 (van der Tweel, 1979) demonstrates that strong responses are not a privilege of checkerboards. In this case also the appearance-disappearance responses (with constant average luminance) were recorded. The famous Escher picture (by intuition of the artist?) reflects important properties of the retina: Escher's fishes become larger and larger to the outside, conforming with the property that visual acuity diminishes rapidly toward the periphery.

Irritatingly enough, there is no good theory to explain why checkerboards and not black and white stripes, for instance, are so effective in evoking VEPs. Nevertheless, VEP research has provided new tools in electrodiagnosis. Halliday (Halliday and MacDonald, 1977) deserves special mention for his contribution to the diagnosis of multiple sclerosis; in Section 8.7 the significance of checkerboard VEPs for the study of congenital and hereditary visual disorders is demonstrated.

8.5 Single-Cell Approach

It may be too much of a generalization, but one could say that the smaller the brain, the greater the importance of the retina. In an epoch-making series of experiments, Lettvin and co-workers (1959) have shown that the retina of a frog is preprogrammed to catch a fly. A revealing illustration is given in Figure 8.11; it is adapted from the publication, "What the frog's eye tells the frog's brain."

Lettvin's frog was positioned in front of a screen on which optical stimuli could be presented. In the optic tectum a microelectrode was placed that picked up the reactions of a cell "looking" at the site on the screen corresponding with the cell's receptive field. When a light spot was flashed on and off at this place, there was only a slight reaction, and even when a black strip of 15° was moved through the field little activity was found. If, however, a fly-sized spot was shifted through the receptive field, a strong burst of spikes was recorded. In the natural situation the discharge will trigger a jumping reaction of the frog, probably before it has "seen" the fly.

Parallel with the earlier sketched development in gross electrophysiology, much information was obtained from the expansion of Lettvin's work. Especially Hubel and Wiesel (1959) have shown that there are well-organized specialized structures in the cortex that, for instance, respond selectively to line elements of a certain length and orientation. However, more complicated patterns have their counterpart in single-cell responses as well. It may suffice here to quote the research by Bruce et al. (1981). Figure 8.12 shows specialized cells in the superior temporal polysensory area of the macaque that react exclusively to faces. The magnitude of the spike response depends upon the type of face; the monkey makes apparently not so much difference between a bearded man and his fellow monkey. Removal of the eyes, however, depresses strongly the spike bursts.

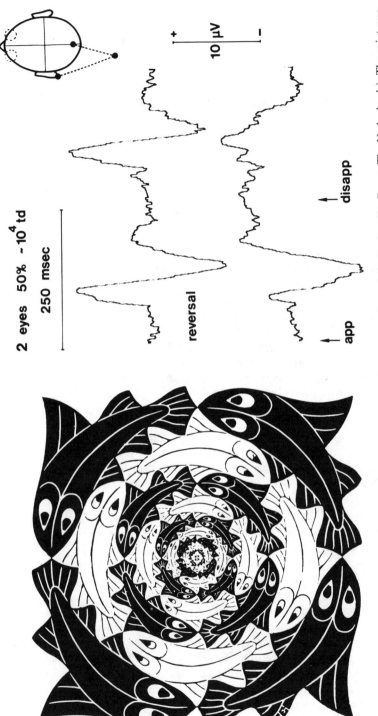

FIGURE 8.10. VEPs to the woodcut, "Ever Smaller", by the famous artist, M.C. Escher (copyright Cordon Art, Baarn, The Netherlands). The print was presented in the appearance-disappearance and reversal modes in the technique of constant average luminance. Reversal occurs twice each period. Reprinted with permission from van der Tweel, L.H., 1978: Pattern evoked potentials: facts and considerations. *Jpn. J. Ophthalmol. Proc. XVIth Symp ISCEV*, Tazawa and Ikeda, (eds).

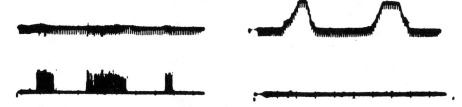

FIGURE 8.11. Spike discharges of a tectal cell of frog to the movement of a black disk of 1° with three different speeds (*left*), and a dark strip of 15° (*right*) with two different speeds through the receptive field of the cell. *Upper trace*, Photocell signal to light changes in the field. Darkness upward. Time calibration 5 Hz. The disk (simulated fly) evokes a large discharge, but only a small response from the photocell. Reprinted with permission from Lettvin, J.Y., Maturana, H.R., McCulloch, W.S., Pitts, W.H., 1959: What the frog's eye tells the frog's brain. *Proc. IRE* 47, 1940–1951. © 1959 IRE (now IEEE).

Analogous to the methods described before, spatial Fourier techniques have been applied to investigate which physical properties of a visual scene are essential for the response of an organism. A classical example in psychophysics is the famous demonstration by Harmon and Julesz (1973). A portrait was coarsely quantisized spatially in squares and in 16 brightness levels (Figure 8.13) and consequently became unrecognizable. Yet, if it is blurred by computer manipulation, the face at once emerges. By squinting or looking from a far distance this effect can be mimicked. One can be assured that this was by no means an incidental finding by the authors, but the result of deep theoretical insight.

8.6 Visual Function and Fourier Analysis

At about the same time that de Lange introduced the use of sinusoidally modulated light for flicker research, Schade (1956) began using sinusoidal spatial grids to define the optical (and physiological) quality of visual perception. This work was followed by that of Campbell and Robson (1968) and many others. One of the practical advantages of sinusoidal grids is that, because of their extension, there is not much need for fixation control.

Just as with the frequency characteristics of flicker vision and electrophysiology, contrast sensitivity functions have been defined and measured employing spatial grids of varying frequencies. The underlying suppositions are that the processing of spatial patterns near threshold is linear, but that at the same time adaptation effects (i.e., nonlinearities) can adequately be used.

In case of linearity, Fourier theory holds that data obtained with more conventional light distributions (space domain) predict those obtained with sinusoidal grids considered in their periodicity (frequency domain) and vice versa. To facilitate the calculations involved, edges, luminous lines or points, (equivalent to step, respectively impulse functions) are commonly chosen as visual stimuli in the spatial domain.

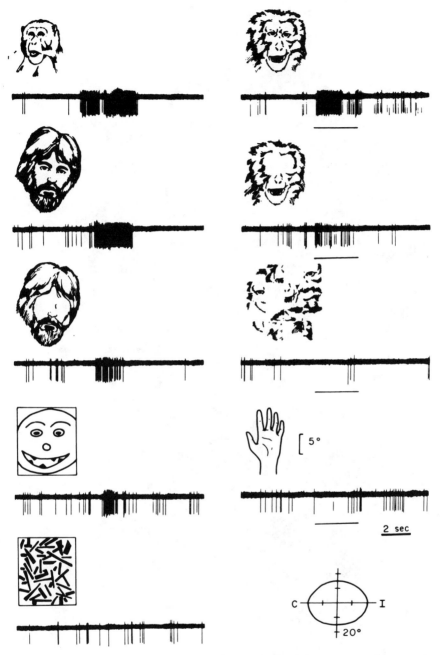

FIGURE 8.12. Response of cell in the superior temporal polysensory area of macacque. Response is better to faces than to other stimuli. Removing eyes or caricaturizing the image reduces response, whereas, cutting the picture into 16 pieces and rearranging it eliminates response. Pictures to the left were swept across the fovea at 10°/sec. Pictures to the right were projected during 3 sec on fovea. Receptive field illustrated on the lower right. Reprinted with permission of the American Physiological Society from Bruce, C.J., Demisone, R., Gross, C.G., 1981: Properties of neurons in a visual polysensory area in the superior temporal sulcus of the macacque. *J. Neurophysiol.* 46, 1057–1075.

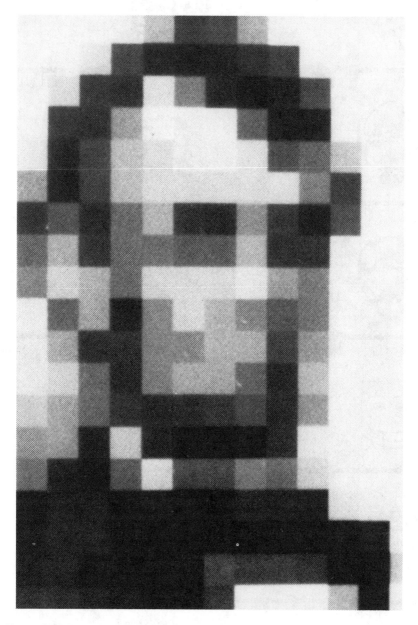

FIGURE 8.13. Portrait of Lincoln quantisized spatially in squares and in 16 brightness levels. When blurred, the face becomes recognizable. Blurring can be mimicked by squinting or looking from afar. Reprinted with permission from Harmon, L.D., Julesz, B., 1973: Masking in visual recognition. *Science* 180, 1194–1196 (Copyright 1973 by the AAAS).

In principle the use of sinusoidal grids is just as legitimate an experimental procedure as that described in this chapter for temporal events. However, one should be well aware of the intricate problems involved, among which that of retinal inhomogeneity is by no means negligible.

The popularity of sinusoidal grid techniques has resulted in a number of publications based on the hypothesis that the visual system itself is organized according to Fourier principles: the visual world should first be transformed to the spatial frequency domain, and the ensuing Fourier spectrum would then be transported along separate spatial frequency channels. Following this, at higher (cortical) centers Fourier synthesis would enable more or less faithful reconstruction of the visual scene (see, for instance, Ginsburg, 1980). Such a model has been proposed also to be analogous to the auditory system where indeed frequency analysis (often somewhat loosely described as Fourier analysis) lies at the basis. It is not the place here to discuss why such a hypothesis is most unlikely, although in a strictly linear system it can never be disproved (see van der Tweel, 1988). It may suffice to note that periodicity in hearing leads to the unique experience of pitch in which the pressure oscillations have lost their individuality, whereas a sine wavegrid is never seen as a global entity but every period (or part of it) is separately perceived. Because there have risen also other objections to visual Fourier functioning, recently more complicated grids (Gabor functions) have been proposed as basic elements in vision, but they are essentially subject to the same criticism.

In conclusion, system analysis scores high in the visual-evoked response system if primitive signals, such as whole-field modulated light, are employed. In particular, manipulation of certain nonlinear properties enables us to determine the sequence of such processes as resonance and rectification. In Figure 8.14 (Spekreijse and van der Tweel, 1972) a scheme is given of the build-up of the human evoked response system for low-frequency luminance variations according to system analytical principles.

As soon as structure is involved, even without cognitive implications, system analysis of VEPs is much less straightforward, although there are reports on the determination of distal transfer functions in cases of checkerboard stimulation (Pijn et al., 1985; Spekreijse and Reits, 1982). On the other hand, as already mentioned, structured stimuli are more adequate visual stimuli than pure light. Whereas with homogeneous light electrical responses could be recorded far below critical modulation depth, VEPs to contrast exhibit a threshold and may extrapolate neatly to that obtained with psychophysical methods (Campbell and Maffei, 1970).

8.7 Visual Evoked Potentials and Aberrant Function

It has proven very profitable to perform research with patterned stimuli in cases of deviant visual function. It is probably surprising that relatively simple stimuli such as checkerboards, can throw light on certain aspects of cortical functioning. This is illustrated with three examples from studies of the Amsterdam group.

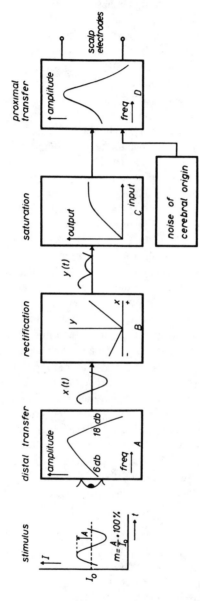

FIGURE 8.14. Schematic model of main processes in the visual evoked response system to luminance modulation. A, Distal "filter"; B, rectification; C, saturation; D, proximal selective processes. Reprinted with permission from Spekreijse, H., van der Tweel, L.H., 1972: System analysis of linear and nonlinear processes in electrophysiology of the visual system. *Proc. Kon. Ned. Akad. van Wetensch. 75*, 77–105.

8.7.1 Amblyopia

The first case is that of a young woman with the affliction of amblyopia aniso-metropia: the "lazy" eye (Spekreijse et al., 1972). This eye had a visual acuity of approximately 1/6. There were no objective signs of any disorder. The retina, including the fovea, appeared normal. Figure 8-15 shows the evoked potentials of the functional and the abnormal eye for patterned and nonpatterned stimuli.

If we look first at the responses to flashes, it is seen that the responses of both eyes are very similar. VEPs to checkerboard appearance stimuli, however, show more interesting features. In the normal eye of the subject the VEP starts with a typical positivity followed by a deep negativity. This is similar to what generally is found in a normal population. In the amblyopic eye the contrast needed to obtain an appreciable response has to be increased. It is only the positive excursion, however, that is present, whereas the negative excursion is practically absent. This negative excursion is known to be directly related to sharp (foveal) vision; it diminishes strongly at blurring. The positive excursion that is less influenced by

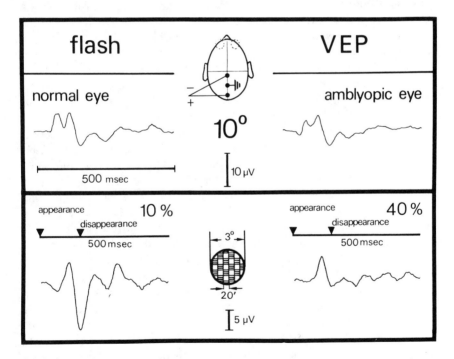

FIGURE 8.15. VEPs to flash and to checkerboard stimulation of normal and amblyopic eye of a young woman. Flash responses are similar; with checkerboard stimulation the amblyopic eye shows only the parafoveal component; it lacks the negative component specific for sharp vision. Reprinted with permission from Spekreijse, H., Khoe, L.H., van der Tweel, L.H., 1972: A case of amblyopia: Electrophysiology and psychophysics of luminance and contrast. In: *The visual system*. Arden, G.B., (ed.), New York: Plenum.

blurring has a more parafoveal origin than the negative peak. Therefore, the result accords with the picture that in the amblyopic eye the most central and detailed functioning is heavily impaired, whereas the coarser parafoveal representation is much less deteriorated.

8.7.2 Albinism

Albinism is a well-known condition occurring in humans and in many animals. The worldwide prevalence of all forms of albinism is estimated at 1:30,000, though prevalence estimates for specific genotypes in nonisolated populations range from 1:17,000, to 1:50,000 (Kinnear, 1985). The typical albino is known as having little pigmentation but this is by no means always the case. The albino phenotype ranges from hypopigmentation of the hair, skin, and eye to only ocular hypopigmentation. It is not uncommon to observe albinos with dark hair, pigmented skin, and even brown iris. Despite the wide variability in pigment expression, all albinos share common ophthalmic features, the most notable of which is foveal hypoplasia with concomitant reduced visual acuity; nystagmus, iris diapheny, fundus hypopigmentation, and photophobia are also common albino features. Because several of these ocular symptoms can be observed in nonalbino disorders (e.g., aniridia, congenital nystagmus, achromatopsia), it is important to find definite properties of the albino visual system that cannot be confused with nonrelated visual syndromes, such as those mentioned above. One of these properties concerns the albino retino-geniculo-cortical organization.

Anatomically, albinism is characterized by misrouting of the optical tracts. As shown in the schematic representation of Figure 8.16, temporal retinal fibers erroneously cross at the chiasm. The checkerboard VEPs of the lower montage of Figure 8.17 (adapted from Apkarian et al., 1983) are illustrative of albinism. In a normal subject (upper montage) both eyes have a mainly symmetrical representation in the two hemispheres. VEPs to 20° central field checkerboard stimulation measured with an electrode configuration as shown are similar for the left and the right eye. This is significantly different in the albino. Identical stimulation yields asymmetric responses for each eye; the left eye projects mainly to the right hemisphere and the right eye to the left one (Figure 8.17, lower montage).

The diagnostic sensitivity of the EP method is practically 100%. Of 198 albino patients whom Apkarian (1989) has tested, there was only one dubious negative judgment of VEP misrouting.

One of the interesting considerations is that misrouting need not incapacitate the subject more than rather superficially. Some albinos have even been shown to have stereopsis (Apkarian and Reits, 1989). On a personal note, this author has played tennis with a typrosinase-positive occulocutaneous albino who proved quite a capable match. Similar interesting observations concerning visually guided behavior can be made of the Siamese cat, an albino with severe visual misrouting

NORMAL   **ALBINO**

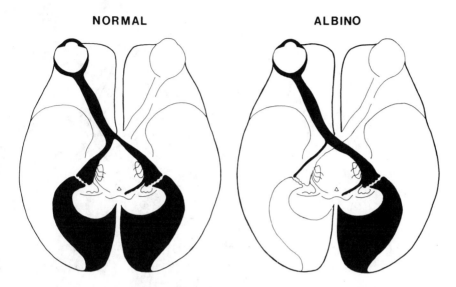

FIGURE 8.16. Routing of the optical pathways of normal and albino. The albino has erroneous crossing at the chiasm. Each eye projects mainly contralaterally. By courtesy of Dr. P. Apkarian, Amsterdam.

but whose jumping belies the condition of disrupted retino-geniculo-cortical organization.

8.7.3 Total Colorblindness (Achromatopsia)

Achromatopsia is a rare but well-documented, mostly hereditary affliction. The subjects are totally colorblind, are photophobic, often have a nystagmus, and are hypermetropic. Their visual acuity is never much better than 1/10, and the maximum flicker fusion frequency is approximately 25 instead of 70 Hz. The condition is totally explained by the absence of (functioning) cones. With respect to perception (leaving aside photophobia and lack of color vision), the main difference with a trichromatic subject is the coarseness of the retinal mosaic and the absence of a true fovea. Most psychophysical tests of structured vision give results very similar to those of the trichromat, with the provision that an enlarging factor of about 7-10 is taken into account.

In Figure 8.18 (van der Tweel and Auerbach, 1975), an achromat's VEPs to checkerboard stimulation with 60' checks are presented together with those to 12' checks of a trichromatic subject. Except perhaps for a somewhat longer latency, the achromat's response falls within the range of those of a normal population, confirming earlier reports (van der Tweel and Spekreijse, 1971).

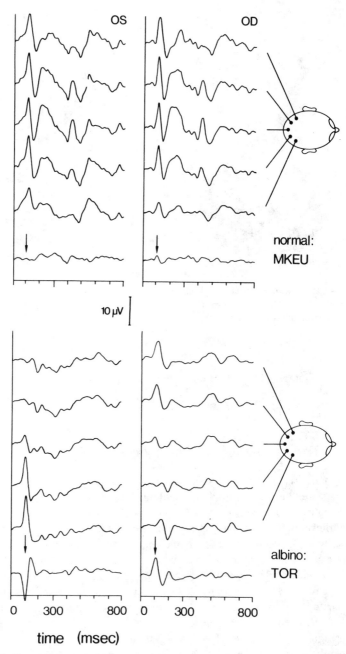

FIGURE 8.17. Comparison between VEPs of the normal and albino eye, in the normal subject VEPs to checkerboard stimulation of the left and or the right eye have a similar and mainly symmetrical distribution on the skull. In the albino the left and the right eye project contralaterally. Adapted from Apkarian, P., Reits, D., Spekreijse, H., van Drop, D., (1983): A decisive electrophysiological test for human albinism. *Electroenceph. Clin. Neurophysiol.* 55, 513–531.

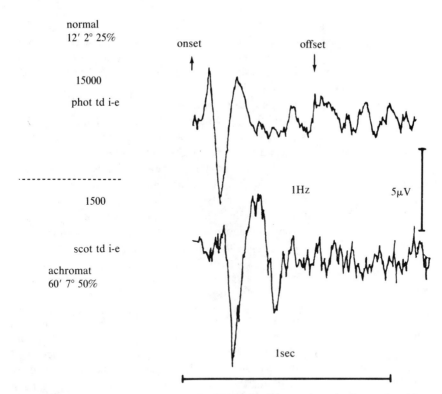

normal
12' 2° 25%

onset

offset

15000

phot td i-e

1Hz

5µV

1500

scot td i-e

achromat
60' 7° 50%

1sec

FIGURE 8.18. VEPs to checkerboard stimulation of trichromatic and achromatic subject. The response of the achromat falls within normal ranges and is unexpectedly large. The checks are used for the achromat are 60' compared to 12' for the normal. Reprinted with permission of Kluwer Academic Publishers from van der Tweel, L.H., Auerbach E., 1977: Achromatopsia, electrophysiological evidence for separate luminance and contrast processing. *Doc. Ophthalmal. Proc.* 2, 163–173.

The results are rather surprising, because in trichromats the checkerboard responses are mainly produced by the inner 2-3 degrees of the retina having the largest representation at the back of the head: 40,000 central cones in contrast vision command a multiple of cortical cells, since such numbers must be involved in order to obtain responses of 10 µV or more. Yet, rods are the only functioning receptors present in the achromat. Possibly to enable the densest packing of cones, the trichromat lacks rods in the inner central fovea. Probably related to this condition, archromats are reported to have a central scotoma; apparently rods do not occupy the empty place of the missing cones. Since the achromat's cortical responses can exhibit considerable amplitudes, the cortical cells that normally are connected to the cones of the fovea via the ganglion cells of the central retina must now get their input from rod receptors. Yet, visual discrimination in the achromat is at its best 10', and therefore the receptive fields of the ganglion cells will largely overlap. From this finding might be concluded, considering the size of the VEPs in the achromat, that the cortical representation is functionally overdimensioned. Could one speculate that the many cortical cells that in the trichromat are connected to the fovea have now been successfully looking for a job, even if this does not contribute to better discrimination?

An interesting point is the comparison of achromatopsia with amblyopia. Amblyopia is the suppression of the function of one eye, originally because of unsurmountable differences between the two eyes. Apparently low vision in itself is not a cause of real malfunctioning because the achromat's vision does not show the properties of the amblyope. The achromat's contrast vision behaves normally within its restricted acuity, and the VEPs do possess the typical negative deflection that is missing in the studied case of amblyopia. A (large) difference between the two eyes seems needed to disrupt the function of the nonfavored eye, as was described in the anisometropic amblyope. This relates well to experimental evidence in cats in which one eye lid was sutured.

8.8 Final Remarks

In this chapter an overview of a main line of nearly 40 years of development of human visual neurophysiology has been presented with an emphasis on system analysis. This emphasis has been at the cost of many intriguing phenomena that have been discovered in the last few years. Topological and neuromagnetic research efforts, for instance, are relatively new fields with already very promising results. Yet, the methodology described in this chapter still has the potential to contribute to neurophysiology, as well as to opthalmology.

References

Apkarian, P., Reits, D., Spekreijse, H., van Dorp, D. (1983): A decisive electrophysiological test for human albinism. *Electroenceph. Clin. Neurophysiol.* 55, 513–531

Apkarian, P., Reits, D. (1989): Global stereopsis in human albinos. *Vision Res.* 29, 1359–1370

Bruce, C.J., Desimone, R., Gross, C.G. (1981): Properties of neurons in a visual polysensory area in the superior temporal sulcus of the macaque. *J. Neurophysiol.* 46, 1057–1075

Campbell, F.W., Maffei, L. (1970): Electrophysiological evidence for the existence of orientation and size detectors in the human visual system. *J. Physiol.* 229, 719–731

Campbell, F.W., Robson, J.G. (1968): Application of Fourier analysis to the visibility of gratings. *J. Physiol.* 197, 551–566

De Lange, H. (1957): *Attenuation characteristics and phase-shift characteristics of the human fovea-cortex systems in relation to flicker-fusion phenomena.* Thesis

De Lange, H. (1958): Research into the dynamic nature of the human fovea-cortex system with intermittent and modulated light. *J. Opt. Soc. Am.* 48, 777–784

Dowling, J.E., Boycott, B.B. (1966): Organization of the vertebrate retina. *Vision Res.* 3, 1–15

Ginsburg, A.P. (1980): Specifying relevant spatial information for image evaluation and display design: an explanation how we see certain objects. *Proc. SID* 21, 219–227

Halliday, A.M., McDonald, W.I. (1977): Pathophysiology of demyelating disease. *Br. Med. Bull.* 33, 21–27

Harmon, L.D., Julesz, B. (1973): Masking in visual recognition: Effect of two-dimensional noise. *Science* 180, 1194–1196

Hubel, D.H., Wiesel, T.N. (1959): Receptive fields of single neurons in the cat's striate cortex. *J. Physiol.* 218, 754–791

Ives, E. (1922): Critical frequency relations in scotopic vision. *J. Opt. Soc. Am.* 6, 254–266

Kinnear, P.E., Jay, B., Witkop, C.J., Jr. (1985): Albinism. *Surv. Opthalmol.* 30, 75–101

Kuffler, S.W., Nichols, J.G. (1977): *From neuron to brain.* Sunderland, MA: Sinauer

Landis, C. (1953): *Annotated bibliography of flickerfusion phenomena, covering the period 1740–1952.* Ann Arbor, MI: Univ. of Michigan Press

Lettvin, J.Y., Maturana, H.R., McCulloch, W.S., Pitts, W.H. (1959): What the frog's eye tells the frog's brain. *Proc. IRE* 47, 1940–1951

MacKay, D.M., Mittelstaedt, H. (1974): Visual stability and motor control (reafference revisited). In: *Cybernetics and bionics.* Keidel, W.D., Handler, W., Spreng, M. (eds.). München: Oldenbourg, pp. 71–80

Pijn, J.P.M., Estévez, O., van der Tweel, L.H. (1985): Evoked potential latencies as a function of contrast: A system analytical approach. *Doc. Ophthalm.* 59, 175–185

Regan, D. (1966): Some characteristics of average steady-state and transient evoked responses evoked by modulated light. *Electroenceph. Clin. Neurophysiol.* 20, 238–248

Regan, D. (1989): *Human brain electrophysiology* New York: Elsevier

Schade, O. (1956): Optical and photoelectric analog of the eye. *J. Opt. Soc. Am.* 46, 721–739

Spekreijse, H. (1966): *Analysis of EEG responses in man.* Thesis. The Hague: Junk

Spekreijse, H. (1969): Rectification in the goldfish retina. *Vision Res.* 9, 1461–1467

Spekreijse, H., Khoe, L.H., van der Tweel, L.H. (1972): A case of amblyopia: Electrophysiology and psychophysics of luminance and contrast. In: *The visual system.* Arden, G.B. (ed.). New York: Plenum

Spekreijse, H., Reits, D. (1982): Sequential analysis of the visual evoked response system in man: Nonlinear analysis of a sandwich system. *Ann. NY Acad. Sci.* 388, 72–97

Spekreijse, H., van der Tweel, L.H. (1972): System analysis of linear and nonlinear processes in electrophysiology of the visual system. *Proc. Kon. Ned. Akad. van Wetensch.* C 75, 77–105

van der Tweel, L.H. (1988): Henkes and the physicist or 40 years of interaction. *Docum. Ophthalmol.* 68, 189–202

van der Tweel, L.H., Auerbach, E. (1977): Achromatopsia, electrophysiological evidence for separate luminance and contrast processing. *Docum. Ophthalmol.* Proc Series, XIIIth ISCERG Symposium, Israel: 105–113

van der Tweel, L.H., Spekreijse, H. (1973): Psychophysics and electrophysiology of a rod-achromat. *Docum. Ophthalmol.* Proc Series 2, 163–173

van der Tweel, L.H., Verduyn Lunel, HFE. (1965): Human visual responses to sinusoidally modulated light. *Electroenceph. Clin. Neurophysiol.* 18, 587–598

van der Tweel, L.H., Visser, P. (1959): Electrical responses of the retina to sinusoidally modulated light. In: *Electroretinographia* Acta Facultatis Medicae Universitatis Brunensis, Lekarska Fakulta Brne 185–196

Yarbus, A.L. (1967): *Eye movements and vision.* New York: Plenum Press.

9

Electrophysiology of Visual Attention

STEVEN A. HILLYARD, GEORGE R. MANGUN, STEVEN J. LUCK, AND
HANS-JOCHEN HEINZE

9.1 Introduction

The ability of human observers to deploy their attention rapidly to selected portions of a visual scene has important consequences for perception. A rapidly expanding research literature has established that stimuli positioned at or near an attended location in the visual field are processed more efficiently than are stimuli at some distance from the focus of attention (for recent reviews, see Eriksen and Yeh, 1985; Prinzmetal et al., 1986). The enhanced processing of attended events may take the form of improved detection of faint stimuli, improved discrimination of stimulus features and patterns, or speeded motor responses to expected targets. This attentional process has been likened to a focal "spotlight" or "zoom lens" that facilitates the processing of stimuli within a circumscribed zone around the attended locus. In some cases, however, the zone of facilitation may take the form of a broader "gradient" of attention that has its peak at the attended location and drops off gradually across the visual field (Shulman et al., 1986). These faciliatory effects are not a consequence of eye movements toward the attended location, which must be strictly controlled in a proper experimental design.

Despite the recent surge of experimental interest in visuospatial attention, many of its fundamental mechanisms remain poorly understood. At the most basic level, there is a controversy regarding the extent to which moving attention to a location results in an actual improvement in the quality of sensory information taken in versus the extent to which it alters the decision threshold or criterion for responding to attended or expected events (Shaw, 1984; Sperling, 1984). This is a version of the traditional "early" versus "late" selection issue—does attention actually modulate the flow of sensory information at a relatively early level of processing, or does it bias higher recognition and decision systems to favor specific inputs?

Electrophysiological studies in human subjects are playing an increasing role in the analysis of attentional processes (for reviews, see Harter and Aine, 1984; Hillyard and Picton, 1987). In particular, recordings of event-related potentials (ERPs) can reveal the precise time course of neural events at successive stages of sensory processing. ERPs are voltage fluctuations recorded from the scalp that are time-locked to sensory, motor, or cognitive processes and are characterized by particular latencies, polarities, waveshapes, and scalp distributions. These voltage fields reflect the synchronous activity of neuronal populations engaged in information transactions in different brain regions. Thus, recordings of ERPs to attended and unattended stimuli can reveal the timing of stimulus selection processes and provide an indication of the anatomical location of the underlying neuronal processes. During visuospatial selective attention, for example, relatively short-latency ERP components (75–150 msec latency) evoked by attended-location stimuli show substantial amplitude enhancements at scalp sites that overlie visual cortical areas. Such information has obvious relevance to the question of whether spatial attention improves visual encoding and perceptual accuracy.

ERPs elicited by visual stimuli consist of a series of positive and negative voltage deflections that have characteristic latencies and morphologies (Figure 9.1). Over the posterior scalp these deflections include the P1 (peaking between 100–140 msec), N1 (160–200 msec), P2 (220–250 msec), and N2 (260–300 msec) waves. These ERP peaks are sensitive to the spatial focusing of attention, with attended-location stimuli evoking larger P1, N1, N2 and sometimes P2 amplitudes, as compared to when attention is directed elsewhere (Eason, 1981; Eason et al., 1969; Harter et al., 1982; Hillyard and Munte, 1984; Neville and Lawson, 1987; Rugg et al., 1987). The early attention effects on the P1 and N1 waves are manifested as amplitude enhancements of the ERP peaks and are usually largest at scalp sites overlying the visual cortex; typically these amplitude modulations do not include any changes in the ERP waveshapes or scalp topographies. Further, the P1 and N1 amplitude changes are essentially equivalent for tasks in which attention is cued to a location on a trial-by-trial basis (Mangun et al., 1987) or is sustained at one location throughout a sequence of stimuli (Hillyard and Munte, 1984; Mangun and Hillyard, 1988).

This pattern of ERP results is consistent with an attentional mechanism of early sensory input modulation or "gain control" that is mediated by descending neural influences upon the afferent sensory pathways (Eason, 1981; Hillyard and Mangun, 1987). This early "gain boost" in the transmission of attended signals presumably underlies the improved perceptual accuracy observed for attended targets and forms at least part of the neural basis of the "attentional spotlight." The precise location in the visual pathway of this selection mechanism is uncertain, but the scalp maximum of the P1 peak (90–140 msec latency) occurs at lateral occipital scalp sites that approximately overlie extrastriate visual cortical areas 18 and 19 (Mangun and Hillyard, 1988, 1990).

The early P1 and N1 amplitude modulations are unique to spatial attention. Different patterns of ERP components are found in association with selective attention to other stimulus features, such as color, spatial frequency, orientation, or shape, as well as to various conjunctions of these features. The predominant ERP response to these nonspatial features when attended is a distinctive broad negative component ("selection negativity") beginning at 150–200 msec and extending until 300–400 msec (Harter et al., 1982; Harter and Aine, 1984). Interestingly, ERP evidence suggests that selection of such features as color or form appears to be hierarchically dependent upon the prior selection for location

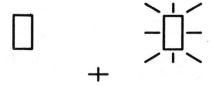

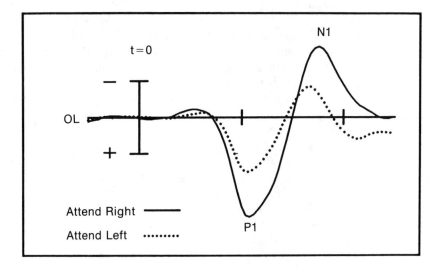

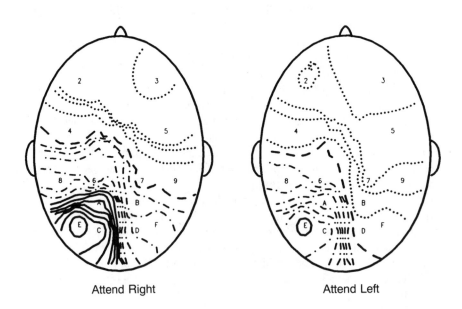

(Hillyard and Munte, 1984). This capability of spatially focused attention to control the analysis of other stimulus features provides strong support for "early selection" models of attention. This result is also in line with recent evidence that spatial attention can gate the registration of simple feature information (Kahneman and Treisman, 1984; Prinzmetal et al., 1986).

During both spatial and nonspatial selective attention, the ultimate identification of task-relevant target stimuli triggers a P300 component with an onset at 200–300 msec, which is absent if the target is "missed" (Coles et al., 1985; Donchin et al., 1986; Hansen and Hillyard, 1983; Wijers et al., 1987). Thus, ERPs provide a window into the entire range of early and late stimulus selections throughout the interval of 75–300 msec, well before the occurrence of the discriminative motor responses that form the main data base for behavioral investigations.

9.2 Current Source Density Analyses

In order to localize the neural generators of attention-sensitive ERP activity more precisely, we have employed two-dimensional current source density (CSD) analysis from multichannel scalp recordings. CSDs can be obtained by placing electrodes in a geometrical grid such that each electrode sits in the center of a square comprised of four neighboring electrodes, i.e., a LaPlacean montage. By subtracting the average voltage of the four neighboring electrodes from the voltage of the central electrode, an estimate can be obtained of the instantaneous brain electrical current flowing perpendicularly to the scalp at that location (MacKay, 1984; Mitzdorf, 1985). By mapping the distribution of the perpendicular currents over the electrode array, the neural generators producing the current flows can be localized within the underlying cortical areas. The CSD analysis is better suited to localizing cortical generators than is simple voltage mapping, since the current flow calculations are reference-free and specifically emphasize superficial rather than deep sources (Pernier et al., 1988; Srebro 1987).

FIGURE 9.1. ERPs (*middle*) and topographic voltage maps (*bottom*) elicited by right visual field stimuli (*top*) under different conditions of attention. Stimuli were flashed to the right and left visual fields in random order, and the task was to attend selectively to one field at a time in order to detect infrequent targets occurring within the stimulus stream on that side. Fixation was maintained on a central point (+). ERPs to the right stimuli are shown when attended (*solid*) and unattended (*dotted*); stimuli onset at t=0, and tick marks every 100 msec. The early P1 and N1 components were of greater amplitude when the eliciting stimuli were attended. Topographic voltage maps show the P1 (108 msec) scalp distribution over the head when the eliciting stimuli were attended (*left side*) and when ignored (*right side*). Note that the location of the scalp voltage maximum remains over lateral occipital scalp sites contralateral to the visual field of the stimulus. Positive voltages represented by solid and dot-dash lines, negative by dotted, and zero potential by dashed line; each contour represents a voltage increment of 0.20 μV. The view is from slightly behind and above; thus, the frontal pole is at the top and the inion at the bottom of each "head."

An example of such a CSD mapping is shown in Figure 9.2 for the P1 component (100–120 msec) elicited by small rectangular flashes of light in the upper left quadrant of the visual field. In this experiment, rectangles were flashed to the four quadrants of the visual fields (about 6 degrees from fixation) in random order at interstimulus intervals of 250–450 msec. The subject's task was to attend selectively to the stimuli at one of the four locations and press a button upon detecting infrequent "target" stimuli (slightly smaller rectangles) at that location. As is usual in this type of experiment, the P1 component showed amplitude enhancement for flashes at attended locations while maintaining its contralateral, posterior scalp distribution (Mangun and Hillyard, 1988, in press).

The map on the left side of Figure 9.2 shows the voltage topography of the P1 wave, and the smaller map on the right shows the topography of the current source density obtained from the 12 Laplacean montages located over the posterior scalp. A sharply focused current source is seen over the right (contralateral) occipital area, with a smaller source appearing in a mirror image location over the left hemisphere. By relating the scalp electrode sites to underlying cortical areas (Homan et al., 1987) we have suggested that the source for the attention-sensitive P1 wave may be in the lateral prestriate cortex (Mangun and Hillyard, 1990).

In monkeys, the lateral prestriate cortex contains areas V2, VP, and V4, which analyze form and color information received from the striate cortex and project outputs to inferior temporal cortex (DeYoe and Van Essen, 1988). There are some intriguing correspondences between the P1 component in humans and neural activity recorded from area V4 in monkeys. The latencies of V4 units in macaques range from 70–90 msec (Robinson and Rugg, 1988), which is in line with onset latencies of 80–100 msec (depending on stimulus brightness and eccentricity) for the human P1. In addition, Desimone and associates have found that unit activity in the monkey area V4 (but not V1 or V2) is sensitive to whether or not attention is being paid to the location of the evoking stimulus (Moran and Desimone, 1985; Spitzer et al., 1988). On the basis of available scalp recorded data in humans, however, it would be premature to conclude that the P1 component reflects neural activity and attentional control in the human homologue of area V4. It is possible that the attentional control manifested in the P1 over prestriate areas is actually imposed at an earlier level of the visual pathway (Eason, 1981). Further work is required to bring together these lines of research in humans and monkeys.

9.3 Selective Attention to Bilateral Stimulus Arrays

The studies of visuospatial attention reviewed above and elsewhere in the literature (Harter and Aine, 1984; Mangun and Hillyard, 1990) all presented random sequences of stimuli, one at a time, to attended and unattended locations. Although paying attention to such stimuli produces robust enhancement of early ERP components (P1 and N1), it is not clear that such conditions are optimal for producing attentional selectivity. Since the appearance of a single stimulus in an otherwise "empty" visual field tends to draw attention to its location whether it is

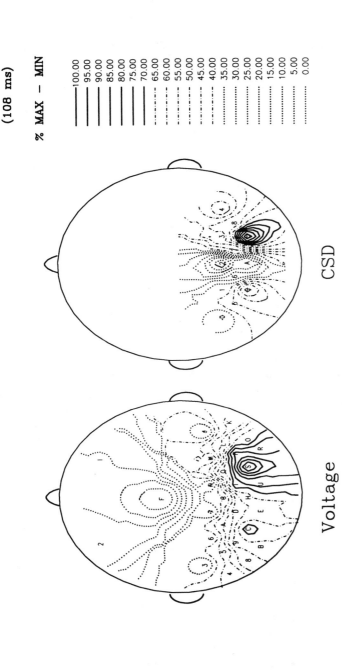

FIGURE 9.2. Topographic voltage map of the P1 component elicited by upper left visual field stimuli (*left*). Data from 29 scalp electrodes were employed using a four-nearest neighbor interpolation algorithm to derive the isopotential contours. The scalp voltage maximum is seen to be over the lateral occipital scalp of the right hemisphere with a weaker second focus visible over the ipsilateral (*left*) occipital scalp. Topographic maps of the CSD are shown at right. CSDs were computed from the voltage data by subtracting the average of the four neighboring electrodes from the value of the central electrode in each of 12 possible Laplecean electrode montages over posterior scalp. Values are scaled as percent of maximum minus minimum values. The CSD maps show a scalp current source at the same contralateral right occipital site as the voltage maximum, a weaker ipsilateral source, and a current sink located between the two at the midline.

attended or not (Jonides, 1981; Posner and Cohen, 1984), it is not surprising that a strong selective attentional set may be difficult to maintain when isolated events are regularly presented to unattended locations (Muller and Rabbit, 1989). Behavioral studies have obtained stronger and more consistent attention effects (i.e., differential processing of attended and unattended elements) in situations where a multielement display, such as a ring of letters, is presented and the subject's attention is precued to one location within the display. Under these conditions, subjects are faster and more accurate at identifying stimuli such as letters in the attended (cued) portion of the display (Eriksen and Yeh, 1985; Jonides, 1981). Further, it should be noted that multielement stimulus arrays are more realistic representations of the visual scenes encountered in everyday experience.

In a recent study (Heinze et al., 1990), we recorded ERPs to bilateral stimulus arrays and unilateral "probes" presented in a rapid, random sequence while subjects attended exclusively to the right or left visual field portions of the display. ERPs to the bilateral arrays exhibited P1 components of enhanced amplitude over the occipital scalp contralateral to the attended visual field. Such a finding was interpreted as evidence that the visual pathways carrying information from the attended portion of the bilateral arrays were relatively facilitated. These findings established that focal attention within bilateral, multielement stimulus arrays can be indexed by electrophysiological measures. Further, these data suggest a means by which ERP and behavioral paradigms can be combined to investigate the relationship between ERP amplitude and perceptual sensitivity in a highly focused attentional task.

Figure 9.3 illustrates the stimulus sequence used in a subsequent study that was designed to investigate the relationship between ERP amplitudes and perceptual sensitivity (d') during focused attention within a bilateral stimulus display. Stimuli consisted of a rapidly flashed sequence (350–700 msec interstimulus intervals) of bilateral arrays of symbols (4.7 degrees eccentricity) that were randomly interspersed with unilateral symbol pairs presented unpredictably to the left or right visual field. A masking stimulus immediately followed the unilateral symbols. While maintaining fixation centrally (verified by the electrooculogram—EOG— and infrared scleral reflective monitoring), subjects were instructed before each run to attend exclusively to the left- or right-side pairs of symbols in the bilateral arrays in order to discriminate and memorize the symbols presented on that side. When the unilateral stimulus was flashed (in either visual field), the subjects' task was to determine whether those unilateral symbols were the same (target) or different (nontarget) from the symbol pair on the attended side of the immediately preceding array. Subjects indicated targets with a right-hand button press. ERPs and detection scores were obtained for the stimuli separately as a function of attend-left or attend-right conditions.

of the display by requiring discrimination and memorization of attended-side symbols and second to compare discrimination accuracy (and its ERP correlates) for the unilateral stimuli falling within versus outside the focus of attention. Significantly, the design allowed us to assess whether or not the subjects' attention was focused at the to-be-attended location immediately before onset of the uni-

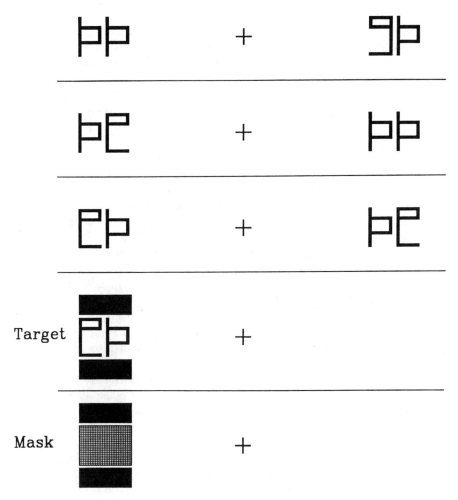

FIGURE 9.3. Diagram of the stimulus sequence viewed by the subjects. Bilateral pairs of symbols were flashed (*top three panels*) and interspersed with unilateral pairs of symbols (*targets, fourth panel*) that were bordered on top and bottom by red rectangles; the unilateral symbols were followed by a mask (*fifth panel*). Subjects were instructed to attend to and discriminate the symbols on one side of the bilateral array for the duration of each run (less than 1 min). Whenever unilateral stimuli were flashed, subjects had to press a button if the unilateral pair was identical to the pair on the attended side of the preceding array. Thus, the unilateral pair in the figure would be a target requiring a response if the attention condition was "attend left" but not if it was "attend right."

lateral stimulus. This was possible because in order for subjects to identify a unilateral stimulus as a target or nontarget they had to know what symbols had preceded it on the attended side. The discrimination of the symbols on one side of the bilateral array was adjusted so as to be too difficult to allow subjects to divide their attention between left and right sides of the display.

The effect of lateralized attention upon the ERPs to bilateral stimuli was to increase the positivity over the occipital scalp contralateral to the attended visual half-field ($p < .01$) (Figure 9.4). The onset of this effect coincided with that of the visual P1 component, and the effect was greatest at the occipital scalp sites. Figure 9.5 shows the ERPs elicited by the unilateral target stimuli that were detected correctly; these data are collapsed over left and right scalp sites to yield waveforms for contralateral and ipsilateral scalp sites relative to stimulus location. Significant attention-related differences included a positive deflection with an onset at 60–80 msec over contralateral occipital scalp sites ($p < 0.01$) and an increased negativity in response to attended-side targets over parietal, central, and frontal scalp sites bilaterally ($p < 0.01$). Attentional difference waves for the unilateral targets were computed by subtracting the unattended ERPs from the attended waveforms (Figure 9.6). The early attention effect over contralateral occipital scalp is seen to be a broad positivity that continues until approximately 300 msec poststimulus. Thus, as in the Heinze et al. (1990) study, the attention effects for both bilateral and unilateral stimuli appear as positive shifts over the contralateral occipital scalp that are not followed at those sites by an N1 attention effect, as is typically seen in experiments using randomized sequences of unilateral stimuli.

The target detection measures (d' and beta) are summarized in Figure 9.7. The d' scores for the unilateral stimuli were significantly higher in the attended versus unattended half-field ($p < 0.001$). Measures of response criteria tended toward larger values for the unattended-side targets, but this difference did not reach statistical significance ($p > 0.10$).

Together, these findings support the hypothesis that spatial selective attention modulates the flow of information in the visual pathways. The ERP data suggest that this effect begins as early as 60–80 msec poststimulus and is localized over lateral occipital scalp sites contralateral to the attended stimulus field. We propose that these attention-related alterations in sensory processing result in improved perceptual representations of the attended location stimuli, as reflected in the higher signal detection scores for those events (Downing, 1988).

9.4 Selective Attention and Feature Detection

Multielement stimulus arrays can also be employed to study the effects of spatial selective attention on feature detection. When an object contains a simple visual feature that distinguishes it from the rest of the visual scene, it appears to "pop

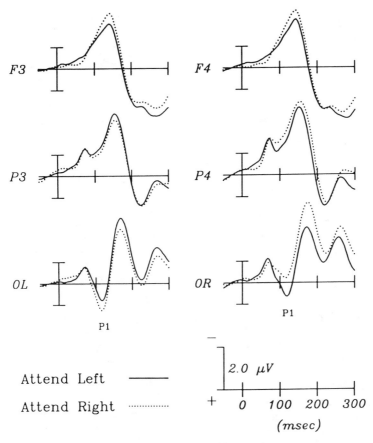

FIGURE 9.4. Grand average ERPs (N = 8 subjects) to bilateral stimuli under attend left (*solid line*) and attend right (*dotted line*) conditions. Recordings are from left (F3, P3, OL) and right (F4, P4, OR) frontal, parietal, and occipital scalp sites. Over the right occipital scalp (OR) the P1 component was largest when the subjects attended the left visual field, whereas over left occipital scalp (OL), the P1 was largest when subjects attended the right visual field.

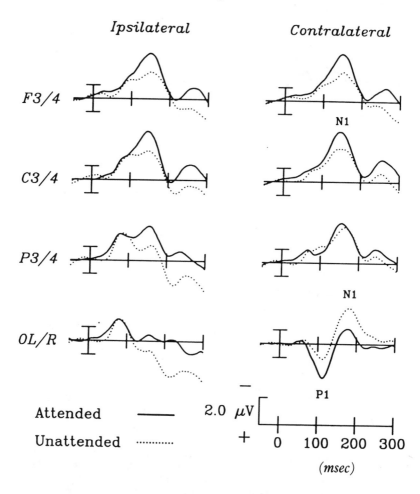

FIGURE 9.5. Grand average ERPs (N = 8) to unilateral stimuli when subjects attended to the side on which the unilateral stimulus was flashed (*solid line*) and when they attended the opposite side (*dotted line*). ERPs are collapsed across the visual field of stimulus and hemisphere of recording to yield contralateral and ipsilateral recordings with respect to the visual field of stimulus. The effect of attention was an amplitude enhancement of the occipital P1 component contralaterally and an increase in the frontal and central N150 component bilaterally. No attentional enhancement of the occipital N1 was observed (compare to Figure 9.1).

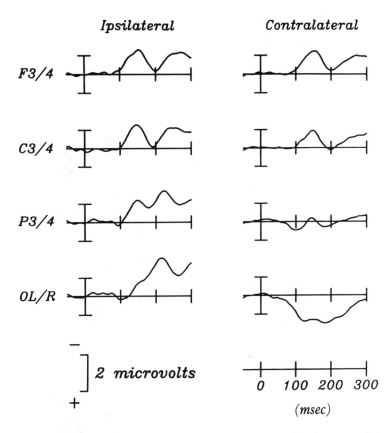

FIGURE 9.6. Attentional difference waves obtained by subtracting the unattended wave-forms from the attended waveforms of Figure 9.5. The effect of lateralized attention is seen as a broad positive deflection over the contralateral occipital scalp. Over the ipsilateral occipital scalp the attention effect can be seen as a slow negative deflection that has a slightly later onset (about 130 msec).

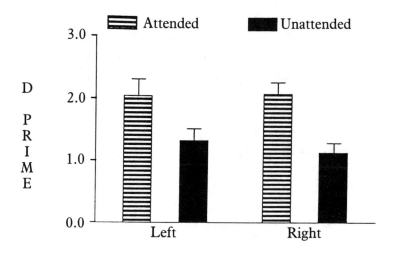

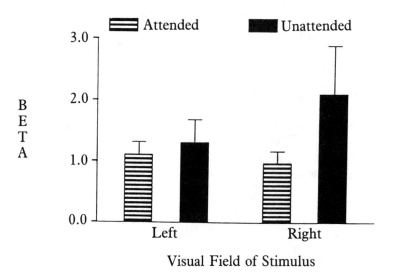

FIGURE 9.7. Bar graphs of d' (*top*) and beta (*bottom*) scores for left and right visual field targets when they occurred on the attended side (*striped*) and on the unattended side (*solid*). The d' scores were significantly higher for the detection of unilateral targets flashed to the attended side. Beta scores were not significantly different for attended versus unattended sides.

out" from the display and can be detected effortlessly. Several authors have suggested that such simple features as color are detected automatically and preatten-
tively and that this automatic feature detection process may provide the basis for
texture segregation (Julesz, 1984; Treisman and Gelade, 1980). It has proven
difficult to assess the automaticity of pop-out detection with behavioral measures,
however, because it is usually necessary to make a stimulus task-relevant in order
to elicit a behavioral response. The ERP technique may be used in this situation
to assess automatic processing in the absence of any overt responses.

Luck and Hillyard (1988) obtained ERP evidence that pop-outs are indeed
detected automatically. In this study, subjects were presented with arrays of eight
randomly positioned items, one of which (i.e., the pop-out) differed from the
others on 50% of trials. The background items were small, blue, vertical rectangles, and the pop-out item differed by virtue of being either green, horizontal,
or larger in size. For each block of stimuli, one of the pop-out types was designated
the target that required a right hand response, and the other two pop-out types were
grouped with the no pop-out stimuli as nontargets (left hand response). Despite the
fact that the nontarget pop-out arrays did not need to be distinguished from the no
pop-out arrays, the former elicited a larger N2 wave over anterior electrode sites,
as shown in Figure 9.8. The presence of this anterior N2 wave provides evidence

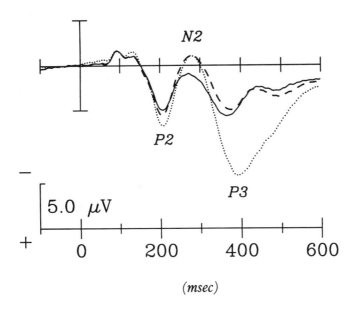

FIGURE 9.8. Grand average ERPs (N = 12 subjects) elicited by no pop-out (*solid*), non target
pop-out (*dashed*), and target pop-out (*dotted*) stimulus arrays in the experiment of Luck and
Hillyard (1988). These ERPs were averaged over left and right frontal electrode sites.

that a pop-out may register automatically even when its discrimination is not explicitly required. The target arrays elicited both an enhanced anterior N2 and a broadly distributed late positive (P3) component.

An additional experiment was conducted to determine whether the detection of pop-out items can be affected by spatial selective attention (Luck and Hillyard, 1989). In this experiment, the stimulus arrays consisted of 12 randomly positioned items, 6 in each visual field. As in the previous experiment, the background items were blue vertical rectangles, and a single pop-out item was present on 50% of trials. The pop-out item could be either a green vertical rectangle or a blue horizontal rectangle and could occur at random on either the left or right side. For each block of stimuli, one of the two types of pop-out stimuli was designated relevant, and the subject was instructed to attend to one side of the display and to respond only when the relevant pop-out occurred on that side. Each block of trials therefore contained some arrays with a relevant pop-out item (25%), some with an irrelevant pop-out item (25%), and some with no pop-out (50%). In order to increase task difficulty and thereby create a need for attentional selectivity, the stimuli were presented at a faster rate than in the previously described experiment (750 msec average interstimulus interval compared to 1500 msec).

The effects of spatial selective attention are displayed for the no pop-out arrays in Figure 9.9. As in the previously described experiment (Figure 9.4), the bilateral arrays elicited a broad positivity that was largest at posterior scalp sites contralateral to the attended side; this asymmetry was present for all stimulus types. The later phase of the attention effect was more anteriorly distributed than the earlier phase, suggesting that it may actually consist of two overlapping positive deflections, the earlier possibly representing a modulation of the exogenous P1 wave. In any case, these data indicate that the attended and unattended sides of the stimulus arrays were differentially processed beginning as early as 65 msec poststimulus.

A key question in this experiment was whether or not pop-out stimuli in the unattended field would be discriminated as indexed by altered ERP components, such as the anterior N2 elicited by task-irrelevant pop-outs in the previous experiment. As shown in Figure 9.10A, the ERP waveform elicited by the irrelevant pop-out stimulus in the unattended hemifield is identical to the ERP to the no pop-out stimulus, whereas the same irrelevant pop-out stimulus elicited an enlarged anterior N2 component when it occurred on the attended side (Figure 9.10B). This finding indicates that spatial selective attention can suppress the processing of task-irrelevant pop-outs, at least to the degree that this processing is reflected by the anterior N2 component.

One possible explanation for this suppression of pop-out information would be that stimulus information from the unattended side is suppressed at an early level to such an extent that the degraded information is not sufficient for later processing systems to detect the pop-out. If this explanation were correct, then the detection of relevant pop-outs occurring in unattended locations should have been suppressed as well. However, relevant pop-outs on the unattended side *did* elicit an

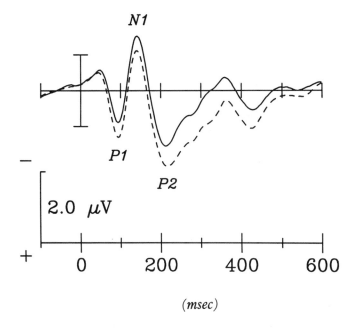

FIGURE 9.9. Grand average ERPs (N = 12 subjects) elicited by the no pop-out stimuli at temporal electrode sites ipsilateral (*solid*) or contralateral (*dashed*) to the attended visual field, averaged over attend-left and attend-right conditions. (From Luck and Hillyard, 1989).

enlarged anterior N2 component (Figure 9.10C), indicating that these pop-outs were detected even though they occurred in an unattended location. Since an early degradation of information from the unattended side should have affected both types of pop-outs, the suppression of the N2 effect only for the irrelevant pop-out suggests that higher, feature-specific selection processes may also be involved in detecting the relevant pop-outs.

Considered together, these results indicate that spatial selective attention may affect several levels of visual processing. The existence of an early selection process based on location alone may be inferred from the finding that all stimuli elicited a positivity over contralateral visual cortex beginning around 65 msec, regardless of whether the stimulus array contained a pop-out item. As described above (Figure 9.7), this spatial filtering process may alter the quality of sensory processing for attended versus unattended stimuli, as reflected in the d′ measure of perceptual sensitivity. In addition, the presence of an anterior N2 enlargement for relevant, but not irrelevant, pop-outs in the unattended half of the array

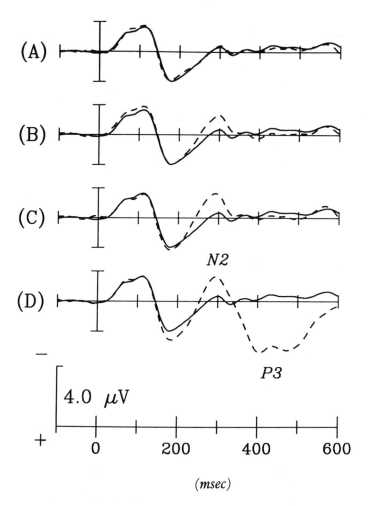

FIGURE 9.10. Comparison of grand average ERPs elicited by no pop-out stimuli (*solid*) and four types of pop-out stimuli (*dashed*) A, Irrelevant pop-out on the unattended side B, Irrelevant pop-out on the attended side C, Relevant pop-out on the unattended side D, Relevant pop-out on the attended side. These ERPs were averaged over left and right frontal electrode sites and collapsed across target pop-out type and direction of attention.

suggests that higher-order feature analyses that lead to pop-out detection may also be affected by attention.

Acknowledgments. This research has been supported by grants from NIMH (MH25594) and NINCDS (NS17778), by ONR Contract N00014-86-K-0291, and by an NSF Graduate Fellowship to Steven J. Luck.

References

Coles, M.G.H., Gratton, G., Bashore, T.R., Eriksen, C.W., Donchin. (1985): A psychophysiological investigation of the continuous flow model of human information processing. *J. Exp. Psychol.* [*Hum. Percept*] 11, 529–553

DeYoe, E.A., VanEssen, D.C. (1988): Concurrent processing streams in monkey visual cortex. *Trends Neurosci.* 11, 219–226

Donchin, E., Karis, D., Bashore, T.R., Coles, M.G.H., Gratton, G. (1986): Cognitive psychophysiology and human information processing. In: *Psychophysiology: Systems, processes and applications.* Coles, M.G.H., Donchin, E., Porges, S.W. (eds.). New York: Guilford Press, pp. 244–267

Downing, C.J. (1988): Expectancy and visual-spatial attention: Effects on perceptual quality. *Psychol. Hum. Percep. Performance* 14, 188–202

Eriksen, C.W., Yeh, Y.Y. (1985): Allocation of attention in the visual field. *J. Exp. Psychol.* [*Hum. Percept.*] 11, 583–597

Hansen, J.C., Hillyard, S.A. (1983): Selective attention to multidimensional auditory stimuli in man. *J. Exp. Psychol.* [*Hum. Percep.*] 9, 1–19

Eriksen, C.W., Yeh, Y.Y. (1985): Allocation of attention in the visual field. *J. Exp. Psychol. Hum. Percept. Performance* 11, 583–597

Hansen, J.C., Hillyard, S.A. (1983): Selective attention to multidimensional auditory stimuli in man. *J. Exp. Psychol. Hum. Percep. Performance* 9, 1–19

Harter, M.R., Aine, C.J. (1984): Brain mechanisms of visual selective attention. In: Parasuraman, R., Davies, D.R. (eds.). *Varieties of attention.* New York: Academic Press, pp. 293–321

Harter, M.R., Aine, C., Schroeder, C. (1982): Hemispheric differences in the neural processing of stimulus location and type: Effects of selective attention on visual evoked potentials. *Neuropsychologia* 20, 421–438

Heinze, H.J., Luck, S.J., Mangun, G.R., Hillyard, S.A. (1990): Visual event-related potentials index focused attention within bilateral stimulus arrays: I. Evidence for early selection. *Electroenceph. Clin. Neurophysiol.*75, 511–527

Hillyard, S.A., Mangun, G.R. (1987): Commentary: Sensory gating as a physiological mechanism for visual selective attention. In: *EEG Supplement 40. Current trends in event-related potential research.* Johnson, R., Jr., Rohrbaugh, J.W., Parasuraman, R. (eds.). New York: Elsevier Publishers, pp. 61–67

Hillyard, S.A., Munte, T.F. (1984): Selective attention to color and locational cues: An analysis with event-related brain potentials. *Percept. Psychophysics* 36, 185–198

Hillyard, S.A., Munte, T.F., Neville, H.J. (1985): Visual-spatial attention, orienting and brain physiology. In: Posner, M.I., Marin, O.S. (eds.). *Attention and performance.* Hillsdale, NJ: Erlbaum Publishers, pp. 63–84

Hillyard, S.A., Picton, T.W. (1987): Electrophysiology of cognition. In: Plum, F. (ed.). *Handbook of physiology: higher functions of the nervous system. Section 1: The nervous system: vol V. Higher functions of the brain, Part 2.* Bethesda, MD: American Physiological Society, pp. 519–584

Homan, R.W., Herman, J., Purdy, P. (1987): Cerebral location of international 10-20 system electrode placement. *Electroenceph. Clin. Neurophysiol.* 66, 376–382

Jonides, J. (1981): Voluntary versus automatic control over the mind's eye's movement. In: *Attention and Performance.* Long, J.B., Baddeley, A.D. (eds.). Hillsdale, NJ: Erlbaum Publishers

Julesz, B. (1984): Toward an axiomatic theory of preattentive vision. In: *Dynamic aspects of neocortical function.* Edelman, G., Cowan, M., Gall, M.D. (eds.). New York: Wiley, pp. 120–148

Kahneman, D., Treisman, A. (1984): Changing views of attention and sutomaticity. In: *Varieties of attention.* Parasuraman, R., Davies, D.R. (eds.). New York: Academic Press, pp. 29–61

Luck, S.J., Hillyard, S.A. (1988): Event-related potentials to visual "pop-out" stimuli. *Soc. Neurosci. Abstr.* 14, 1013

Luck, S.J., Hillyard, S.A. (1989): *On the automatic detection of visual popouts.* Annual Meeting of the Society for Psychophysiological Research, New Orleans

MacKay, D.M. (1984): Source-density mapping of human visual receptive fields using scalp electrode. *Exp. Brain Res.* 54, 579–581

Mangun, G.R., Hillyard, S.A. (1987): The spatial allocation of visual attention as indexed by event-related brain potentials. *Hum. Factors* 29, 195–211

Mangun, G.R., Hillyard, S.A. (1988): Spatial gradients of visual attention: Behavioral and electrophysiological evidence. *Electroenceph. Clin. Neurophysiol.* 70, 417–428

Mangun, G.R., Hillyard, S.A. (1990): Electrophysiological studies of visual selective attention in humans. In: *The neurobiological foundations of higher cognitive function.* Scheibel, A., Wechsler, A. (eds.). New York: Guilford Publishers, pp. 271–295

Mangun, G.R., Hansen, J.C., Hillyard, S.A. (1987): The spatial orienting of attention: Sensory facilitation or response bias? In: *Current trends in event-related potential research.* Johnson, R., Jr., Rohrbaugh, J.W., Parasuraman, R. (eds.). Amsterdam: Elsevier, pp. 118–124

Mitzdorf, U. (1985): Current source-density method and application in cat cerebral cortex: Investigation of evoked potentials and EEG phenomena. *Physiol. Rev.* 65, 37–100

Moran, J., Desimone, R. (1985): Selective attention gates visual processing in the extrastriate cortex. *Science* 229, 782–784

Muller, H.J., Rabbit, P.M.A. (1989): Reflexive and voluntary orienting of visual attention: Time course of activation and resistance to interruption. *J. Exp. Psychol. [Hum. Percept.]* 15, 315–330

Neville, H.J., Lawson, D. (1987): Attention to central and peripheral visual space in a movement detection task. I. Normal hearing adults. *Brain Res.* 405, 253–267

Pernier, J., Perrin, F., Bertrand, O. (1988): Scalp current density fields: Concept and properties. *Electroenceph. Clin. Neurophysiol.* 69, 385–389

Posner, M.I., Cohen, Y. (1984): Components of visual orienting. In: *Attention and Performance X,* vol 5. Bouma, H., Bouwhuis, D. (eds.). Hillsdale, NJ: Erlbaum Publishers, pp. 531–556

Prinzmetal, W., Presti, D.E., Posner, M.I. (1986): Does attention affect visual feature integration? *J. Exp. Psychol. [Hum. Percept.]* 12, 361–369

Robinson, D.L., Rugg, M.D. (1988): Latencies of visually responsive neurons in various regions of the Rhesus monkey brain and their relation to human visual responses. *Biol. Psychol.* 26, 111–116

Rugg, M.D., Milner, A.D., Lines, C.R., Phalp, R. (1987): Modulation of visual event-related potentials by spatial and non-spatial visual selective attention. *Neuropsychologia* 25, 85–96

Shaw, M.L. (1984). Division of attention among spatial locations: A fundamental difference between detection of letters and detection of luminance increments. In: *Attention and performance X: Control of Language processes*. Bouma, H., Bouwhuis, D.G. (eds.). Hillsdale, NJ: Erlbaum Publishers, pp. 109–121

Schulman, G.L., Sheehy J. B., Wilson, J. (1986) Gradients of spatial attention. *Acta Psychologica* 6, 167–181

Sperling, G. (1984): A unified theory of attention and signal detection. In: Parasuraman, R., Davies, D.R. (eds.), *Varieties of Attention*. London: Academic Press, pp. 103-181

Spitzer, H., Desimone, R., Moran, J. (1988): Increased attention enhances both behavioral and neuronal performance. *Science* 240, 338–340

Srebro, R. (1987): The topography and scalp potentials evoked by pattern pulse stimuli. *Vision Res.* 27, 901-914

Treisman, A.M., Gelade, G., (1980): A feature-integration theory of attention. *Cognitive Psychol.* 12, 97–136

Wijers, A.A., Okita, T., Mulder, G., Mulder, L.J.M., Lorist, M.M., Poiesz, R., Scheffers, M.K. (1987): Visual search and spatial attention: ERPs in focussed and divided attention conditions. *Biol. Psychol.* 25, 33–60

Part IV

Cognitive Studies in Humans

10

Brain Electric Microstates and Cognition: The Atoms of Thought

Dietrich Lehmann

10.1 Internal and External Aspects of the Stream of Consciousness: Continuum or Discrete Steps?

A healthy adult human is able to report conscious perceptions and thoughts; they are experienced as an uninterrupted, continuous "stream of consciousness" during wakefulness. Consciousness cannot be thought of as consisting of identifiable subportions that still possess the properties of consciousness, even though it is clear that consciousness occurs only if large numbers of the brain's neural subsystems operate and cooperate adequately.

Yet, discrete neural elements are active in the brain, utilizing all-or-none signals for at least some important information transfer in the network. The macroscopic view always produces a smooth percept from discrete pieces of information as, for example, a rapidly flashing light is perceived as continuous. Accordingly, discrete single unit discharges are obviously compatible with continuous percepts. However, the assumption of an all-knowing homunculus that does the final interpolation between discrete points and that produces continuous perception is not helpful in explaining the internal experience of mentation. It appears that a homogeneous element needs to be the outer aspect of what is internally experienced as continuous stream of conscious experience or as smooth-surface percept. The brain electric field is an obvious candidate for this outer aspect.

Consciousness resembles an all-or-none phenomenon: one is either awake and conscious or asleep with curtailed dream consciousness. Transitional stages in normal life, such as sleep onset or awakening, are brief and fleeting. (Other qualities of consciousness that are experienced in drug states under chemical agents or in meditation after lengthy training are not discussed here). Yet, when considering the stream of awake conscious experiences, different types or modes or strategies of thoughts (disregarding contents) can easily be recognized: a recalled past event or a present mental calculation might be experienced as visual images or as abstract thoughts; associated with such internal experiences there might be strong emotions or neutral feelings. The question thus arises whether these different types of experiences are only highlights in a continuum or whether they indicate the existence of basic mentation packages, of segmentable identifiable building blocks of thought processes, of atoms of thoughts that are concatenated in a stepwise fashion. Our work on brain electric fields suggests that there are quasi-stable microstates in the subsecond range that are associated with different, identifiable behavioral and introspective qualities.

10.2 Brain Information Processing Is State Dependent

All brain work is state-dependent; information treatment strategy, recalled context memory material (Lehmann and Koukkou, 1974), and evaluation of relevance of incoming information all depend on the momentary functional state of the brain (Koukkou and Lehmann, 1983). This dependence is obvious for grossly different states, such as wakefulness and sleep, which are associated with typical differences in brain electric activity (EEG) and in behavior (see Williams et al., 1966 for a classical paper). However, state changes are also evident as fluctuations within wakefulness in which brief EEG changes parallel behavior changes (Bodhanecky et al., 1984; Gath et al., 1983; Keesey and Nichols, 1967; Lehmann et al., 1965). The momentary state is determined by genetic make-up, maturational and external and internal clock-factors, metabolic conditions and disease, and, most important, past experiences that set the goals and provide detailed motivations. The momentary functional state is continuously readjusted via the central mechanisms of the "orienting response" (Rohrbaugh, 1984); this adjustment occurs within the constraints given by the determinants of the present state in response to the demands raised by afferent or memory-retrieved information, as evaluated within the context material of the present state. Different brain global functional states, as manifest in brain electric activity, are associated with different modes of mentation: prime examples are wakeful, reality-oriented thinking and reality-remote dreaming in adults (Koukkou and Lehmann, 1983).

The brain's functional state can be operationalized behaviorally at a relatively simple level as global vigilance. Vigilance varies over time on a gross scale from wakefulness to sleep. On this gross scale of vigilance, such behavioral consequences as differences in reaction time among wakefulness, light, and deep sleep are well known (Williams et al., 1966). Vigilance also varies within wakefulness (or within a conventional "sleep stage") on a much more finely grained scale, in the form of subtle "fluctuations of attention" lasting seconds (Woodworth and Schlosberg, 1954), which account for the well-known variability of wake reaction time performance. However, it should be borne in mind that functional states of the brain are definable in many dimensions, not only in levels of vigilance; a classic example is the wakeful behavior associated with severe memory disturbance during atropine intoxication.

The present challenge is to identify brief epochs of brain activity that belong to different classes as defined by their electric field configurations and to show that different outcomes of information processing depend on the different classes of brain activity, i.e., on the occurrence of different functional states.

10.3 The Brain Electric Field as a Manifestation of Momentary Functional State

At each moment in time, the multitude of the brain's simultaneous electric activities sums into a three-dimensional electromagnetic field, the electric component of which can be recorded noninvasively from the surface of the scalp as electroencephalographic (EEG) or event-related potential (ERP) data. Traditionally, these data are shown as waveshapes of potential differences between electrode locations. If simultaneous recordings are available from a sufficiently large number of electrodes, maps of the momentary distribution of the electric potential on the scalp can be constructed at each moment in time (Lehmann 1971, 1972) as shown in Figure 10.1. For EEG recordings, there typically are 64 or 128 samples/sec (i.e., maps/sec) and up to 512/sec for ERP recordings. The resulting maps are read quite similarly to geographical maps in which higher or lower values exist at different locations. In this way the data can be viewed as an endless stream of momentary maps.

The electric "landscape" of each momentary map reflects the geometry of the momentarily active neural structures. Therefore, different momentary maps must have been generated by different brain activity. This, in turn, implies that different maps manifest different modes or steps of brain information processing. Conversely, similar maps are suggestive of having been generated by a similar mode or step (or content) of information processing. Accordingly, in the pursuit of the atoms of thought the goal is to identify epochs of similar electric landscapes in this continuous stream of momentary maps of brain electric activity and to determine the boundaries in time of these postulated, spatially stable epochs. The advantages offered by this approach is the immense time resolution; in essence, each instant in time could be classed as a particular functional state.

10.3.1 Major Characteristics of Typical Series of EEG Momentary Maps

Spontaneous EEG activity, as is life in general, is basically periodic. In the majority of normal adults during relaxed wakefulness, the activity is dominated by about 18–20 polarity reversals each second, the "9-10 Hz α waves" of the traditional EEG. When examining a series of momentary EEG maps, one finds that they are simple, most often displaying landscapes with one or at most two peaks and troughs (Lehmann, 1971; Lehmann et al., 1987). Equipotential contour lines tend to be concentric around the extrema values; there are no "wavefronts." The locations of the two or three extrema (maximum and minimum) in each map appear to capture the major characteristics of each landscape. Typically, the landscape persists for a certain number of maps and then, within a few maps, changes into a different landscape that again persists for a certain time. There appears to be little continuous propagation of major landmarks (peaks and troughs) over larger distances in space: the classical idea of "traveling EEG

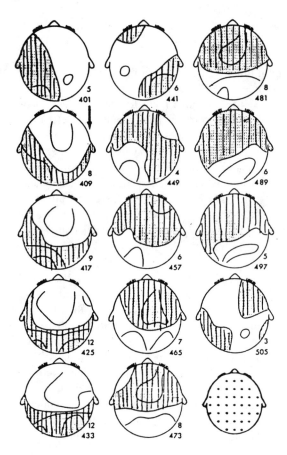

FIGURE 10.1. Series of momentary EEG potential maps recorded from 37 electrodes during relaxation in wakefulness with closed eyes. Sequence from top to bottom, left to right. Time between two maps is 8 msec, total is 104 msec. Negative areas gray, positive white, referred to average reference. Equipotential contours at 10-μV steps. Upper number below each map indicates hilliness, lower number time. Reprinted from Lehmann (1971): Multichannel topography of alpha EEG fields. *Electroenceph. Clin. Neurophysiol.* 31: 439–449 with permission of Elsevier Science Publishers B.V.

waves" is not evident. Periodically, the maps are more pronounced with higher peaks and steeper gradients, corresponding to the polarity reversals.

The multitude of possible EEG waveshapes of potential differences recorded as functions of time between different pairs of electrodes on the scalp cannot be ascribed to single locations (for 20 electrodes, there are 190 possible and different waveshapes). Transformation of waveshape data into a sequence of momentary maps provides no data reduction, but facilitates their understanding (movies are shown as pictures, not as multiple time series of local light intensities!).

As a first survey procedure of map sequences, we assess the momentary "hilliness" (Lehmann, 1971) or "global field power" (Lehman, 1987; Lehmann and

Skrandies, 1980) of each map; this reference-free computation, corresponding to the computation of the standard deviation of all momentary potential values in the map, yields a one-number statement out of any number of simultaneously recorded electrode locations for each moment in time. When applied to an "alpha-type" EEG, a curve with 18-20 peaks/sec results. The momentary maps at these time points of maximal global field power tend to show stable landscapes for several successive maps, as evidenced by the stable locations of occurrence of extreme potential values but with inverted polarities. (Contrary to ERP data, the periodic polarity reversals in the spontaneous EEG have been accepted as a stable-state condition, as in work on time-series-oriented adaptive segmentation (Gath et al., 1983; Lopes da Silva and Mars, 1987). After several successive time moments of maximal global field power during which the polarity-inverting extrema persisted at their original locations, the maps change into a new configuration to then remain stable again for some time.

The maps at times of maximal global field power show optimal signal-to-noise ratios. It can be shown that the mapped standard deviation of the mean of all momentary maps during a given analysis time, corresponding to the band power map versus the average reference, shows a spatial distribution that is very similar to the mean of the maps selected at the time moments of maximal global field power. It can further be shown that the spatial distribution map of the incidence of maximal and minimal extreme potential values during the analysis time is very similar to the power map versus the average reference of the same epoch (Lehmann et al., 1987). These relationships suggested the locations of the momentary extrema potentials as map descriptors for data reduction in space and the selection of maps at the time moments of maximal global field power for data reduction in time.

10.3.2 Recognizing Changes of Momentary EEG Map Landscapes

Space-based adaptive segmentation of spontaneous EEG map series has been developed and applied to EEG data (Lehmann, 1984; Lehmann et al., 1987). In order to investigate the stability or change of landscapes in a series of momentary maps of the spontaneous EEG, some strategy for spatial pattern assessment must be employed. Considering the above-reviewed properties of EEG map series, we have selected the maps at times of maximal global field power for further analysis and used the locations of the maximal and minimal values (the extrema locations) as landscape descriptors (Figure 10.2). This phenomenological description of a map by its most positive and negative locations is related to the concept of accounting for a map by an equivalent electric dipole fitted optimally in three-dimensional space, but does not make assumptions about the electric properties of the media.

The time-selected and space-extracted information consists of a series of "extrema dipole maps" (Figure 10.3), one at each time of maximal global field power.

FIGURE 10.2. Reduction of a momentary map of the brain electric field to the locations of the maximal and minimal potential, the "extrema dipole." *Left*, 19-electrode array where the anterior row was at 40% nasion-inion, the posterior row at the inion and interelectrode spacing was about equidistant on the head. The lateral distortion of the map is produced by the map printing program. *Center*, Interpolated surface as equipotential areas (done with a BioLogic Brain Atlas System); white is more negative, black more positive, and levels in steps of 4 *u*V. The electrodes with the momentary maximal and minimal values are easy to identify; no assumption of an absolute zero value as reference is necessary for this decision. *Right*, Two extrema in the array outline connected by a line that visualizes the orientation of the "extrema dipole." A rough estimate of the location of an equivalent dipole generator accounting for the illustrated field is the center between the two extrema.

For adaptive segmentation of these maps, a spatial window is set around the locations of the two extrema. If in a later map one of the extrema leaves its spatial window, the segment is terminated, and a new segment starts with reset of the windows. We have applied this analysis strategy to 2-minute epochs of 16-channel recordings in six normal subjects during relaxation. The data were digitally filtered (FIR) to the 8-12 Hz (alpha) band and then adaptively segmented into epochs of stable "extrema dipole" landscapes with varying durations. A sample sequence of segments is shown in Figure 10.4; the figure illustrates the varying durations of the individual segments and shows the typical findings that some segment classes as defined by the extrema locations tend to recur more often than others and that the sequencing of the segment classes follows no easily detectable rules. We found a mean segment duration of 210 msec, with 2.6 sec for the longest individual segment. Segments longer than 323 msec covered 50% of the total time, and 25% of the time was covered by those longer than 500 msec. Segments belonged to a mean of 60 classes (defined by the extrema locations) out of the possible 120 classes; the most prominent segment class covered 20% of total time.

10.3.3 Reaction Time Depends on Momentary EEG Microstate at Stimulus Arrival

The functional significance of the formally identified spatially stable segments of brain electric activity was tested in an experiment on selective motor reaction time with eight normal subjects (Lehmann et al., 1987). The subjects had to press a microswitch in response to rare, high-pitch tones that randomly replaced 10% of the lower-pitch tones in a regular series. Each subject responded to a total of 337 rare tones. The classes of the segments that existed at the moment of occurrence

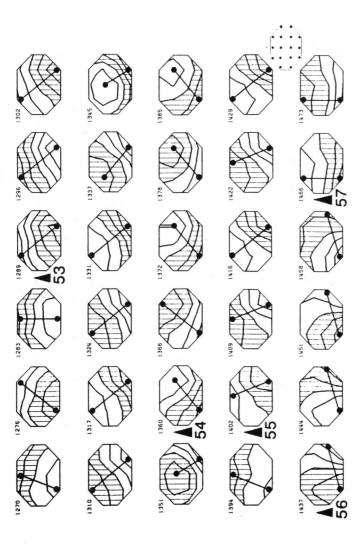

FIGURE 10.3. Space-based adaptive time segmentation of a series of momentary maps of spontaneous EEG selected at successive times of maximal global field power and reduced to the extreme potential locations ("extrema dipoles," black dots connected by a line). A spatial window was set around each extreme, the extrema locations were followed over time, and the segment was terminated when an extreme left its window (*arrowheads* and *large numbers*). White positive, hatched negative to average reference; equipotential lines in steps of 10 uV; the schematic shows electrode array (nose up, left ear left). Note the unequal duration of segments and same segment class for 54th and 56th segment. Reprinted from Lehmann et al., (1987): EEG alpha map series: Brain micro-states by space-oriented adaptive segmentation. *Electroenceph. Clin. Neurophysiol.* 67: 271–288 with permission of Elsevier Science Publishers B.V.

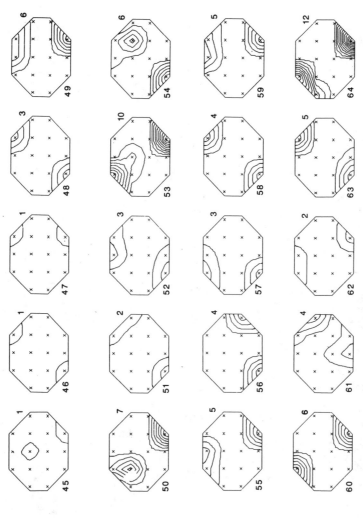

FIGURE 10.4. Sequence of adaptively determined segments of an EEG map series during wakeful relaxation recorded from a normal volunteer. Each map shows the spatial distribution of the extreme potential values in the maps selected at maximal global field power times during one segment. The number of isofrequency-of-occurrence contour lines therefore is related to the times of maximal global field power; the number to the right above each map gives the precise number of global field power maxima in the segment; since these occur at about 50-msec intervals, the number multiplied by 50 indicates segment duration. Total illustrated time is 4.7 sec. Note the repeated occurrence of some segment classes and the long persistence of some segments. Reprinted from Lehmann et al. (1987): Brain microstates by space-oriented adaptive segmentation. *Electroenceph. Clin. Neurophysiol.* 67: 271–288 with permission of Elsevier Science Publishers B.V.

of the rare tones were determined, and the reaction times were measured. Twenty-six different segment classes occurred in all eight subjects. Over subjects, there were similar differences in reaction time associated with these 26 segment classes (Friedman ANOVA chi square 40.6, df = 25, $p < 0.025$). For further data reduction, the centroid (mean) location between the two poles of the extrema dipoles was calculated. The median location over subjects of the dipole centroids of the slowest eight segment classes was over the left parietal area, about 7 cm lateral to the left ($p < 0.05$) of the median location of the fastest eight segment classes (which was over the right parietal area). The mean reaction time difference between the slowest and the fastest eight segment classes was 34 msec, corresponding to a reaction time increase of about 4 msec for each centimeter of lateral displacement of the momentary field centroid to the left.

In summary, the behavioral measure of reaction time in a selective attention paradigm was shown to depend systematically on the class of the momentary EEG field map that accidentally existed at the moment of arrival of the stimulus; the results suggest common characteristics of this relationship over subjects.

These results obtained with spatial analysis of multichannel EEG data in the subsecond time range are in line with results obtained with evoked potential waveshapes. Evoked potentials were shown to depend on characteristics of the prestimulus EEG (Basar, 1980). Further, reaction times were shown to depend on characteristics of the stimulus-preceding EEG time series (Gath et al., 1983). In summary, even on a finely grained scale of state differences within wakefulness, the brain's momentary functional microstate in the subsecond range as reflected in the EEG by a few momentary maps influences the fate of the incoming information.

10.3.4 The Momentary EEG Microstate Reflects Cognitive Mode

In addition to the relationships between the EEG and behavior mentioned above, many workers, including our group, have shown systematic relationships between temporal EEG patterns (reflecting the global brain state) and the kind of mentation; for instance, wakeful thinking, sleep onset reverie, and dreaming (Ehrlichman and Wiener, 1980; Koukkou and Lehmann, 1983; Lehmann et al., 1981; Lehmann et al., 1988) or hallucinogenic drug effects (Koukkou and Lehmann, 1976). In our own studies we have concentrated on spontaneously occurring mentations, avoiding task execution situations (such as "please think about X") since we feel that task execution possibly involves powerful brain subroutines that might overshadow weaker mode-, step- or strategy-specific brain mechanisms that also are reflected in the brain's electric field. In these EEG studies of spontaneous mentation, the subjects are instructed to give a brief report about what just had passed through their mind every time a gentle tone is sounded. The tones were presented at 3- to 6-minute intervals at artifact-free times. The reports were transcribed and scored by raters who used, in different studies, 8 to 23 parameters that refer to mode or strategy of the reported mentation, such as "visual versus abstract mode," "information treatment reality-close versus reality-remote," "emotionally neutral, positive, or negative," "discontinuous ideas versus continuous thought chain," and the like. Fourier-transformation-based characteristics of EEG recordings of 16-20 seconds duration that immediately preceded the alerting tone were then compared with the rating scores. The epoch length for spectral analysis in these studies was sufficient to yield many convincing correlations with the rated mentation characteristics. Typically, these studies were done using two to six simultaneous EEG data recording channels.

We wanted to know whether the method of reporting unconstrained, spontaneous thoughts without specific mentation tasks might still yield reliable correlations with mentation characteristics if the EEG data were radically restricted in time to one sample time point only, but if the spatial configuration of the brain electric field was sampled more extensively from, say, 19 electrodes. In other words, we wondered whether it would be possible to demonstrate that the different functional microstates found in the formal EEG segmentation study not only influence reaction time but also reflect mode or strategy of momentary mentation. After encouraging data had been obtained in pilot studies (Kofmel et al., 1988), we started a systematic experiment that has to date yielded preliminary results.

From each of 20 normal right-handed subjects, who were seated comfortably in a sound-proof and light-shielded room, 30 reports were collected and rated. The EEG was recorded from an array of scalp electrodes located between 40% nasion-inion and inion at about equal interelectrode distances (see Figure 10.2), using a 20-channel Brain Atlas mapping system. The data immediately preceding the report-soliciting signal were recomputed into a time series of global field power (momentary hilliness of the maps as a function of time: Lehmann, 1971; Lehmann and Skrandies,1980). The single map at the last moment of maximal global field power before the prompting signal (on the average this map was found about 30 msec before the prompt) was used for the classification of the momentary functional microstate that preceded the mentation report. (This map involves the highest signal-to-noise ratio in terms of electric landscape; the landscape of a flattish map is more prone to noise distortion than that of a map with high peaks and deep troughs). As map descriptors, the locations of the two extrema potentials (positive and negative) were used. These two extrema locations define a dipole-like figure in the map plane (Figure 10.2). For further data reduction, the median (centroid) location of all dipole extrema was computed for each mentation category of each subject, as illustrated in Figure 10.5. The centroid location of an extrema dipole can be regarded as the surface projection of an estimate of the location of the equivalent model dipole generator that best accounts for the observed momentary map.

Data concerning two pairs of mutually exclusive rating categories of mentations are presented here. Subjects who produced two or more reports on both categories of each pair were included in the analysis. Each subject contributed an average of 10.5 reports to each of the four categories.

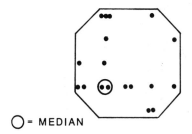

○ = MEDIAN

FIGURE 10.5. Locations of positive and negative extreme potentials (*dots*) extracted from the nine momentary potential maps that preceded the nine reports of the "abstract thought" category in one subject. The circle indicates the spatial median of the nine entries. From a study with B.A. Kofmel and B. Henggeler.

Visual imagery versus nonimagery abstract thought: "I saw that beautiful red-leaved tree which I passed this morning when I drove out of my garage" versus "I wondered whether this recording will continue much longer"; reports from 16 subjects were available

Reality-close versus reality-remote strategy: "I thought about my schedule this afternoon" versus "I was floating over a landscape as if I had wings"; reports from 14 subjects were available

For each category of each subject, the median location of all dipole extrema locations of the momentary maps associated with the reports of that category was computed. Then, grand median locations over subjects were obtained for the four mentation categories (Figure 10.6). The differences between the grand median locations of dipole centroids belonging to the two categories of a pair were tested for statistical significance with the paired Wilcoxon procedure.

Figure 10.6 illustrates the results. The grand median locations of the dipole extrema were more anterior and to the right for visual imagery than for abstract thoughts and also more anterior and to the right for reality-remote than for reality-close mentation. When tested (paired Wilcoxon) along the axis given by the grand median locations, the location differences were significant ($p < 0.005$—visual versus abstract and $p < 0.025$—reality-close versus reality-remote).

The geometries of the two result sets show some similarity. However, when the location differences in the lateral and saggital direction were tested independently, the grand median centroids associated with visual imagery were significantly ($p < 0.025$) more to the right but not anterior than those associated with abstract thoughts; and vice versa, the grand median centroids associated with reality-remote mentation were significantly ($p < 0.05$) more anterior but not to the right than those associated with reality-close mentation, indicative of asymmetries between the results.

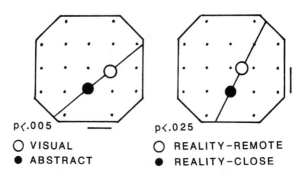

p<.005   ———   p<.025

○ VISUAL   ○ REALITY–REMOTE
● ABSTRACT   ● REALITY–CLOSE

FIGURE 10.6. Grand median locations of the subjects' median extrema locations. *Left*, Grand medians over the 16 subjects who produced visual imagery and abstract reports. *Right*, Grand medians over the 14 subjects who produced reality-remote and reality-close reports. The line across the grand medians for the two mutually exclusive categories in each map is the projection horizon for the paired Wilcoxon statistics that gave the listed p values. Separate tests of lateral and sagittal location differences showed significant results laterally for "visual" versus "abstract," sagitally for "reality-remote" versus "reality-close." From a study with B.A. Kofmel and B. Henggeler.

These preliminary results show clearly that different kinds of spontaneous mentations (as classified by rating of the reports) are associated with significantly different brain functional microstates at the moment of prompting for report of the mentation (microstates as classed by the examination of one single momentary EEG map). Therefore, momentary brain electric microstates do reflect high aspects of cognition and thus indeed are candidates for the role of atoms of thought.

10.4 Interindividual Differences in Brain Mechanisms of Cognition

In a second step of data analysis we examined the possibility that not only the centroid location of the map extrema dipoles differs between mentation classes consistently over subjects but also the orientation of the extrema dipoles differs. For each subject, the extrema dipoles associated with all reports of the same mentation score were averaged, employing an optimizing algorithm that determined that mean dipole which resulted in the least sum of squared deviations of the contributing cases from the mean.

This iterative procedure tests all possible means for the best solution. Given three dipoles (1, 2, 3), each with two extrema (A, B), four combinations are possible: 1A + 2A + 3A; 1A + 2A + 3B; 1A + 2B + 3B; 1A + 2B + 3A. For the general case of N dipoles, there are 2 to the power of (N − 1) possible combinations.

The optimal extrema dipoles were computed for the maps of each mentation category of each subject and then as grand mean extrema dipoles over subjects.

The grand mean dipoles over subjects showed small differences between mentation categories. Examination of the individual subjects' mean dipoles showed that these small differences occurred because the dipole orientations for a given mentation category differed drastically over subjects, whereas the extrema locations of the mean dipoles for a given mentation category within a given subject showed small standard errors (Figure 10.7). This finding is indicative of high intrasubject consistency of the momentary EEG maps associated with a given mentation category; when testing the individual subjects' mean dipoles for differences between mentation categories, about 50% of the mean extrema locations differed significantly.

These results indicate that there are strong interindividual differences in the relative extrema dipole orientations associated with different mentation categories. There are two classical hypotheses concerning such a finding: (1) there might be interindividual differences in the anatomical structures that are available for a given type of higher-level processing or (2) different individuals might use different processing approaches to produce a similar mentation category (they might have found different ways to do a given task). On the other hand, the surprisingly low intrasubject variability of the extrema locations for a given mentation class speaks against the hypothesis that a given subject uses at different times very different approaches to perform a given processing task.

10.5 Conscious Experience

The results presented in this chapter show that the landscape of single momentary brain electric maps reflects different modes or strategies of momentary sponta-

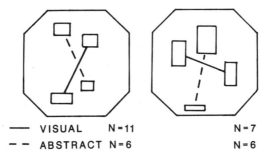

—— VISUAL   N = 11   N = 7
– – ABSTRACT N = 6   N = 6

FIGURE 10.7. Means of extrema dipoles (descriptors of momentary maps) over all cases of two mentation categories in two subjects. The mean locations were computed using an iterative optimization procedure. The mean locations are the centers of the rectangles; the rectangles give the areas of 1 SE. Note the reversed spatial relationships of the dipoles for the two mentation categories in the two subjects (high intersubject variability) and the small SE. of all four means (low intrasubject variability). From a study with B.A. Kofmel and B. Henggeler.

neous thought processes. Our research has demonstrated the possibility of adaptively segmenting series of momentary spontaneous EEG maps into epochs or segments of stable landscapes, which showed durations in the subsecond range. Each segment class reflects a particular functional microstate. We conclude from the results that different segment classes of spontaneous EEG activity might indeed be associated with different kinds of mentation. Accordingly, the different microstates as defined by the segments of the map series could be considered to constitute atoms of thought, units of brain activity states associated with specific steps, modes, or contents of information processing.

That single momentary maps suffice to distinguish successfully the EEG states during different kinds of mentation is accounted for by the fact that a given EEG segment class is defined by stability of the landscape over time. In other words, one typical map classifies an entire segment. Why, on the other hand, would epochs of many seconds in traditional power spectral analysis-based EEG-mentation studies still show the relations between EEG and mentation, if individual building blocks of mentation are so short? We hypothesize that the segment class found right before the prompting might be one that recurred several times over the last several seconds. This would imply that, even though the individual functional microstates are very short, their recurrence over a certain time would permit the brain to re-enter into a given global condition.

Re-entering into a given microstate might be necessary not only to finish the particular processing job but also to provide access to consciousness for a given presently processed content. The work of Libet (1982) on minimal persistence times (500 msec) of stimuli for conscious experience, as well as the studies on (1) visual perception with leading masks (160–250 msec; DiLollo, 1980), (2) interocular intervals for stereo effects (250 msec; Skrandies, personal communication, (3) backward masking (200 msec, Michaels and Turvey, 1979), and (4) time spans for switching of attention (400 msec; Reeves and Sperling, 1986) suggest that brain microstates shorter than some minimum time cannot lead to conscious experience.

Re-entering the same state might provide a second chance to qualify content for consciousness if the first processing occurred in a microstate too short for consciousness access; considering that 25% of total time in our segmentation study (Lehmann et al., 1987) was occupied by states of 500 msec or longer, the chances are good. Re-entering the same state might also conceivably imply a short-term recognition of the very recently treated information: perhaps, multiply recognizing the same information, within the time spans discussed above, is a brain mechanism that is a prerequisite of conscious experience. A structural provision for such multiple recognitions of the same information might be the multiple sensory representation areas in higher mammals.

These considerations also suggest that brain functions that are manifest in early "components" of event-related potentials (ERP) cannot qualify for access to consciousness because of their brevity and nonrepetitiveness. There is no repetitiveness in early ERP components before about 250 msec poststimulus since, contrary to series of spontaneous momentary maps, ERP map series show different localizations of extreme potentials at successive times of maximal global field power (times of maximal global field power are, within map series analysis, the equivalent of the traditional amplitude-based "component latencies" within the waveshape analysis concept); similar results are obtained with adaptive segmentation of ERP map series (Brandeis and Lehmann, 1986; Lehmann and Skrandies, 1984). However, some late ERP segments of relatively long duration need not be associated with conscious percepts, as observed in studies on subliminal stimulation (Brandeis and Lehmann, 1986).

10.6 Microstates and Mental Pathology

We hypothesize that individual global microstates as classed by the brain electric field maps are the electric reflection of the building blocks of conscious mental experiences; each global state that has this potential of consciousness consists of very many parallel local brain substates that, considered separately, would not be sufficient for consciousness. Too, each microstate class would imply a different step or kind or strategy of information processing. The time sequence of occurrence (the concatenation) of the different global functional microstates, the relative occurrence frequency of the different classes, and the mere number of state changes per time unit might vary in normal individuals over gross functional states, such as wakefulness and sleep in adults or immaturity and maturity in wakefulness. Mental dysfunctioning might be associated with abnormal classes of microstates. However, it appears conceivable that mental disease implies more subtle changes that essentially preserve the vocabulary of global microstates but use syntax or rate aberrations; thereby employing erroneous concatenations or aberrant class frequencies or excessive or reduced change rates. Only one of these parameters might be deviant in a given disturbance; even though the "atoms of thought" might still be there, the constructed edifice comes out wrong.

Acknowledgments. The research has been partly supported by grants from NIH, USA, the Swiss National Science Foundation, the Hartmann-Muller Foundation and the EMDO Foundation, Zurich, the Klinik am Zurichberg, Zurich, and the Sandoz Foundation, Basel.

References

Basar, E. (1980): *EEG brain dynamics.* Amsterdam: Elsevier

Bohdanecky, Z., Bozkov, V., Radil, T. (1984): Acoustic stimulus threshold related to EEG alpha and non-alpha epochs. *Int. J. Psychophysiol.* 2, 63–66

Brandeis, D., Lehmann, D. (1986): Event-related potentials of the brain and cognitive processes: Approaches and applications. *Neuropsychologia* 24, 151–168

DiLollo, V. (1980): Temporal integration in visual memory. *J. Exp. Psychol. Gen.* 109, 75–97

Ehrlichman, H., Wiener, M.S. (1980): EEG asymmetry during covert mental activity. *Psychophysiology* 17, 228–235

Gath, I., Lehmann, D., Bar-On, E. (1983): Fuzzy clustering of EEG signal and vigilance performance. *Int. J. Neurosci.* 20, 303–312

Keesey, U.T., Nichols, D.J. (1967): Fluctuations in target visibility as related to the occurrence of the alpha component of the encephalogram. *Vision Res.* 7, 859–879

Kofmel, B.A., Michel, C., Lehmann, D. (1988): Momentary EEG amplitude maps and spontaneous cognitive mode. *Brain Topogr.* 1, 133–134

Koukkou, M., Lehmann, D. (1976): Human EEG spectra before and during Cannabis hallucinations. *Biol. Psychiat.* 11, 663–677

Koukkou, M., Lehmann, D. (1983): Dreaming: The functional state-shift hypothesis, a neuropsychophysiological model. *Br. J. Psychiat.* 142, 221–231

Lehmann, D. (1971): Multichannel topography of human alpha EEG fields. *Electroenceph. Clin. Neurophysiol.* 31, 439–449

Lehmann, D. (1972): Human scalp EEG fields: Evoked, alpha, sleep and spike wave patterns. In: *Synchronization of EEG activity in epilepsies.* Petsche, H. H., Brazier, M.A.B. (eds.). Wien: Springer

Lehmann, D. (1984): EEG assessment of brain activity: spatial aspects, segmentation and imaging. *Int. J. Psychophysiol.* 1, 267–276

Lehmann, D. (1987): Principles of spatial analysis. In: *Handbook of electroencephalography and clinical neurophysiology,* rev. series vol. 1. Gevins, A.S., Remond, A. (eds.). Amsterdam: Elsevier, pp. 309–354

Lehmann, D., Beeler, G.W., Fender, D.H. (1965): Changes in patterns of the human electroencephalogram during fluctuations of perception of stabilized retinal images. *Electroenceph. Clin. Neurophysiol.* 19, 336–343

Lehmann, D., Koukkou, M. (1974): Computer analysis of EEG wakefulness-sleep patterns during learning of novel and familiar sentences. *Electroenceph. Clin. Neurophysiol.* 37, 73–84

Lehmann, D., Koukkou, M., Andreae, A. (1981): Classes of day-dream mentation and EEG power spectra. *Sleep Res.* 10, 152

Lehmann, D., Meier, B., Meier, C.A., Mita, T., Brandeis, D. (1988): Sleep onset mentation characteristics related to lateralized EEG spectral power. *Sleep Res.* 17, 105

Lehmann, D., Ozaki, H., Pal, I. (1987): EEG alpha map series: Brain micro-states by space-oriented adaptive segmentation. *Electroenceph. Clin. Neurophysiol.* 67, 271–288

Lehmann, D., Skrandies, W. (1980): Reference-free identification of components of checker-board-evoked multichannel potential fields. *Electroenceph. Clin. Neurophysiol.* 48, 609–621

Lehmann, D., Skrandies, W. (1984): Spatial analysis of evoked potentials in man—a review. *Progr. Neurobiol.* 23, 227–250

Libet, B. (1982): Brain stimulation in the study of neuronal functions for conscious experience. *Hum. Neurobiol.* 1, 235–242

Lopes da Silva and Mars (1987): Parametric methods in EEG analysis. In: *Handbook of electroencephalography and clinical neurophysiology,* rev. series vol. 1. Gevins, A.S., Remond, A. (eds.), Amsterdam: Elsevier, pp 243–260

Michaels, C.F., Turvey, M.T. (1979): Central sources of visual masking: Indexing structures supporting seeing at a single, brief glance. *Psychol. Res.* 41, 1–61

Reeves, A., Sperling, G. (1986): Attention gating in short-term visual memory. *Psychol. Rev.* 93, 180–206

Rohrbaugh, J.W. (1984): The orienting reflex: Performance and central nervous system manifestations. In: *Varieties of attention.* Parasuraman, R., Davies, D.R., (eds.). New York: Academic Press

Williams, H.L., Morlock, H.C., Morlock, J.J. (1966): Instrumental behavior during sleep. *Psychophysiology* 2, 208–215

Woodworth, R.S., Schlosberg, H. (1954): *Experimental psychology.* New York: Holt, Rhinehart, and Winston.

11

New Prospects in Neurolinguistics

Yu. L. GOGOLITSIN, R.E. KIRYANOVA, AND V.B. NECHAEV

11.1 Introduction

Our drastically extended ability to communicate is unequivocally one of the features most clearly distinguishing us from all other species. It is quite natural, therefore, that the physiological basis of our linguistic abilities attracts attention of researchers working in diverse fields of science, both the social sciences and technical disciplines.

Understanding the mechanisms that enable us to express our thoughts orally or in a written form and to comprehend others by listening or reading is extremely valuable. It will not only open new perspectives in the treatment of various brain disorders, including the most complicated mental ones, in designing new extremely powerful computer systems but will also be of fundamental importance for grasping the nature of consciousness and self-awareness."Indeed, consciousness may be, in large part, a matter of our 'talking to ourselves.' Thus, once we understand the language functions of the brain, we may have gone a long way in understanding how the brain can be conscious of its own existence" (Carlson, 1988, p.10).

This chapter presents a new methodology for studying brain mechanisms of language. It is based on the recording and analysis of neuronal discharges in the human brain during various kinds of verbal processing determined by the tasks of the psychological tests presented to a subject. First, a brief review of methodologies for studying the physiological basis of language is presented. Unique possibilities offered by clinical application of implanted electrodes are considered. Subsequent sections describe various types of neuronal reactions to verbal processing observed in some subcortical nuclei and cortical fields of patients diagnosed and treated with implanted electrodes.

11.2 Methodologies of Research into the Brain Mechanisms of Language

Clinical observations have obviously made the largest contribution to our knowledge of the brain mechanisms underlying human linguistic ability. The first attempts to construct anatomical schemes of the cerebral organization of speech belong to the neurology of the mid-nineteenth century, when for the first time attention was directed to the association between disturbance or complete loss of speech and the localization of brain lesions. Soon after the discovery of the motor and sensory speech areas, the classical neurological scheme of the brain provisions for speech was created, including centers for acoustic patterns of words, for articulation, and for ideas together with connections or associations between these centers. Naturally, this very simplified and pure localizationist scheme was critically revised later. However, its basic principles concerned with the interconnections between speech processes and certain brain zones remained valid.

The methodology based on an analysis of the verbal deficit in association with the localization of brain lesions supplied data of great importance for understanding general features of the anatomical and functional distribution of the brain zones involved in speech processes. Classical works by A.R. Luria (1958, 1966), who founded the neuropsychological approach that combined clinical observations with psychological testing, should be mentioned here first of all.

A closely related discipline with a long history of studies focused on the "brain and language" problem is cognitive psychology. The use of sophisticated psychological tests and simple performance indices, such as reaction time and error score, resulted in the construction of many cognitive models for lexical processing (Mills et al., 1979; Townsend, 1984).

While paying tribute to the contribution of these disciplines to our knowledge of brain provisions for language, it would be improper not to mention here certain natural limitations inherent in both neuropsychological and cognitive-psychological methodologies. Disturbances of verbal function that were sufficiently pronounced to be detected and analyzed by means of neuropsychological methods in the vast majority of cases result from brain lesions, the smallest of which, even though apparently local, have the dimensions of at least several millimeters. It is quite natural, therefore, that the neuropsychological approach primarily supplies valuable information about only the most general, rough anatomical and functional features of the brain system subserving language ability.

Cognitive models of lexical processing, which are of great importance for understanding the basic principles of the cerebral organization of speech, are constructed using the results of analysis of verbal behavior during various tasks. This methodology resembles, to a great extent, the "black box" approach in cybernetics in which output-to-input relations describing the functioning of a system form the basis for a model of its structure. Unfortunately, further verification of many existing and differing models of lexical processing, as well as selection of more adequate ones solely within the framework of the cognitive science methodology, seems hardly possible because of the lack of objective and independently gathered data.

One of the sources of such data is the application of electrophysiological methods. Recording and analysis of the evoked potentials have achieved the status of being one of the effective methods for obtaining information about the activity of brain systems during psychological testing. The very first attempts to use this method showed that characteristic differences between evoked potentials to different types of verbal and nonverbal stimuli could be observed (Buchsbaum and Fedio, 1970; John et al., 1967). In many other studies, a broad spectrum of correlations were obtained between the parameters of evoked responses and different features of the activity performed, though some investigators could find no difference between evoked potentials to meaningful and meaningless stimuli (Shelburne, 1973). The most valuable contribution of the evoked potential studies to our knowledge of the brain mechanisms of speech and mental processes was the discovery of the early (exogenous) and late (endogenous) reaction components with clearly different functional significance, which demonstrated objectively the existence of a certain temporal ordering of the cerebral processes underlying mental activity.

Another important source of physiological data on the brain mechanisms of speech is the clinical application of the brain electric stimulation. Here, by exciting certain areas of the brain with weak electric currents during neurosurgical operations or using long-term therapeutic electrodes, it was possible to observe alterations of verbal behavior and to find regularities in their association with the stimulation site (Ojemann, 1983; Smirnov, 1976).

Recent technological advances—namely, positron emission tomography (PET)—have considerably enriched our ability to study the spatial distribution of the brain areas related to language. The very first applications of PET for measuring regional cerebral blood flow changes during lexical processing (Petersen et al., 1988) already demonstrated the simultaneous activation of widely separated areas of the cerebral cortex, providing evidence in favor of the parallel models of processing the individual words.

Unfortunately, none of the above-mentioned research methodologies has been able to supply strongly desirable data about the physiological events that occur during lexical processing in single neurons and neuronal ensembles located in different areas of the human brain. That is why the introduction into clinical practice of the method of intracerebral electrodes implantation opened unique opportunities for direct investigation and mapping of the most subtle neuronal correlates of human mental activity (Bechtereva, 1978, 1988a), including neuronal reactions to word perception and processing (Gogolitsin et al., 1987a, b, and c).

The results obtained during studies of the brain with the aid of implanted electrodes enabled the formulation and confirmation (Bechtereva, 1978) of the concept of mental activity maintenance by brain systems with cortical and subcortical links of different degrees of rigidity. The pattern of discharge rate of a neuronal population during the performance of a certain task can be considered within the framework of this concept as a multicomponent structure consisting of the elements reflecting the involvement of this population in the functioning of neurophysiological mechanisms formed at different stages of information processing (Gogolitsin et al., 1987d). This idea, in turn, creates a framework for the new psychophysiological approach to the study of neuronal mechanisms of verbal behavior described here, which is based on statistical comparison of the neuronal discharge rate patterns during the patient's performance in a set of specially selected psychological tests. The potentialities of this new approach are illustrated below with the aid of the data obtained in 16 parkinsonian patients with gold electrodes implanted for diagnosis and treatment.

11.3 Neuronal Correlates of Verbal Processing

11.3.1 Subjects and Methods

Strictly on the basis of medical indications, the patients had bundles of intracerebral electrodes implanted unilaterally or bilaterally. Each bundle was formed out of six gold 100-uM wire electrodes. Up to six bundles were usually implanted in one hemisphere for the period of several weeks necessary for diagnosis and treatment. Investigations of neuronal mechanisms of word perception and processing were performed as part of the diagnostic procedures aimed at selection of optimal sites for therapeutic electric stimulation.

A frontal cross-section map of the human brain showing the distribution of neuronal activity recording sites in subcortical structures of the brain hemispheres, for all patients studied, is presented in Figure 11.1. As can be seen, most of the electrodes were located in thalamic and striopallidar nuclei. In 25 cases it was possible to record neuronal activity in the cortex by electrodes placed for monitoring purposes on the cortical surface near the sites of intracerebral electrode penetrations.

Recording of the neuronal population spike activity with the type of electrodes used, the computer system for data acquisition and processing, and specific features of the psychological tests for studying neuronal correlates of mental processes are described and discussed in detail elsewhere (Bechtereva et al., 1985; Gogolitsin et al., 1987a). Here it is sufficient to mention briefly that during the recording sessions patients were seated in a comfortable armchair in front of an opaque screen. Test slides were presented with the aid of a multichannel projection tachistoscope. Data input and storage, as well as stimulus presentation, were controlled by the computer. About a quarter of the electrodes picked up multiunit neuronal spikes with the peak-to-peak amplitude exceeding 50 uV. Only high-amplitude spikes were extracted for further analysis by introducing an amplitude threshold at a level of 0.6-0.7 of the maximal spike amplitude. Data processing routines enabled the calculation and plotting of the peristimulus time histograms (PSHs) of the mean discharge rate across a number of trials. The background mean rate was estimated using 1-sec intervals preceding the presentation of the first stimulus. Significant deviations from the background mean rate were determined by applying the Student t-test for the most general case of unequal sample size and unknown variances.

The psychological tests used in this study were constructed to reveal the neuronal discharge rate responses specifically related to lexical processing. Each test consisted of a series of single trials, with a test-dependent fixed number of slides and order of presentation. In each trial one, two, or three slides containing verbal

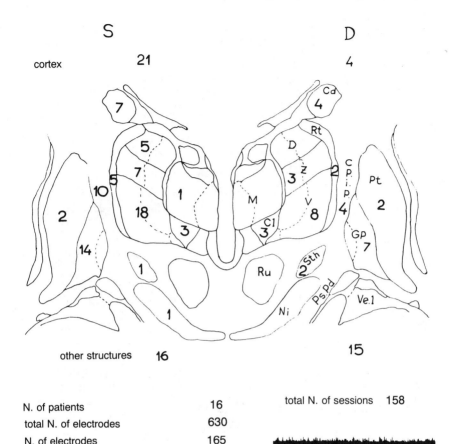

FIGURE 11.1. Frontal cross-section of the human brain, depicting the number of neuronal activity recording sites in the striopallidar and thalamic nuclei of the patients who participated in the study.

or graphic stimuli were presented, depending upon the goal of the test (Figure 11.2). The difference between the trials was determined solely by the type of stimulus presented. For example, in the naming test the stimuli were a single word; a quasi-word (qwrd)—a letter sequence resembling the word but meaningless; a picture of a well-known object (a cup); or an abstract shape bearing no direct resemblance to any well-known object. Single trials were presented in random order. Before testing, the patient was instructed to read aloud each word or to name

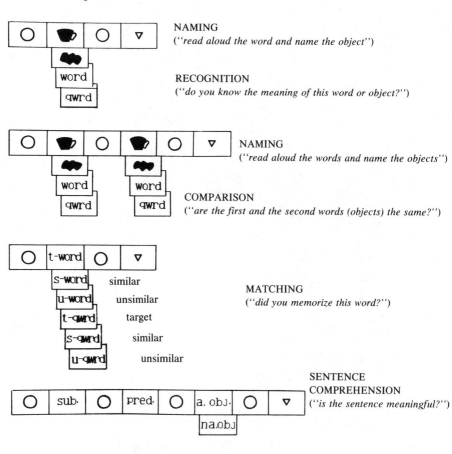

FIGURE 11.2. Structure of individual trials in the psychological tests performed by the patients during recordings of neuronal activity. Each rectangular frame corresponds to a slide with a single verbal or graphic stimulus, a circle used for gaze fixation, or a triangle signaling the patient to pronounce the verbal response. Stimulus slides were presented for 0.1 second and were separated by 1.0 second, in an uninterrupted sequence as shown in the lower part of the drawing. Abbreviations not explained on the figure: qwrd = quasi-word; sub. = subject; pred. = predicate; a.obj. and na.obj = adequate and non-adequate objects.

the object. In the case of unknown objects, the expected response was "no." By comparing the neuronal discharge rate patterns in the above four situations, we hoped to reveal neuronal events related to specific features differentiating verbal stimuli from pictures.

The psychological tests most frequently used in this study were oriented toward differentiation between meaningful and meaningless stimuli (the recognition test); matching the currently presented word or quasi-word to the target stimuli memorized before the beginning of the test; comparison of the two stimuli, and judging the meaningfulness of the sentence (sentence comprehension) that was conferred by an adequate or inadequate object. Comparison tests of two kinds were used. First was a sign comparison test that consisted of trials where two slides with verbal or graphic stimuli were presented with an instruction to respond "yes" only in cases where the stimuli were exactly the same and "no" for all other combinations. In the second, a semantic comparison test, one verbal and one graphic stimulus were shown in each trial, and the subject's task was to respond "yes" only when the presented word appeared to be the proper name for the presented object. In some cases, different versions of these tests were applied. The data presented in the next section are supplied in each case with the diagram of the trial structure for the test performed by the patient.

11.3.2 Results

In many neuronal populations, presentation of the tests evoked complex patterns of discharge rate with single or multiple phasic increases or decreases of firing rate at certain moments during the trial. These patterns were seen in the peristimulus time histograms (PSHs) of average discharge rate across a number of trials. Some of these neuronal populations displayed responses specifically related to lexical processes. Such responses were revealed by comparing patterns from test trials that differed with respect to features of the presented stimuli. Data about populations with particular specific and nonspecific reactions are presented below in a section devoted to the discussion of the association between features of the discharge rate pattern and localization of the neuronal population in a given structure or hemisphere. First, we make an attempt to classify the neuronal reactions revealed in this study and to describe representative samples of each class.

11.3.2.1 Sign Features

The first type of neuronal reactions revealing a specific association with lexical processing consists of an early discharge rate peak in the PSH with a latency of 150–200 msec. In about 10% of the neuronal populations the shape parameters of this peak, in particular the duration or the steepness of its trailing edge, depend upon the graphic contours of the stimulus.

An example of such a reaction, in a class which it seems reasonable to identify as neuronal correlates of the processing of features of word signs, is shown in Figure 11.3. In this particular case, during presentation of a version of the sentence

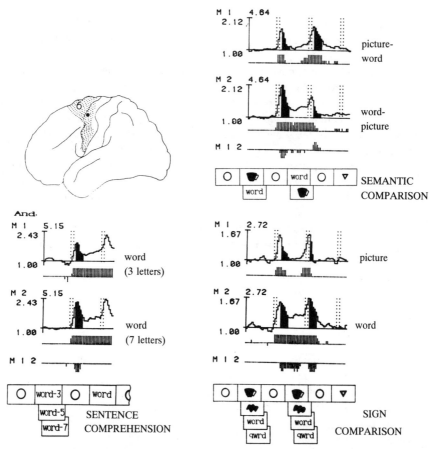

FIGURE 11.3. Discharge rate patterns of a neuronal population in cortical field 6 of the left hemisphere (patient And.) recorded during the presentation of three psychological tests (*sentence comprehension, sign comparison,* and *semantic comparison*). Each graph presented here and in the other figures shows a peristimulus-time histogram (PSH) with 50 msec time bins S, constructed by means of computer averaging of the neuronal discharge rate traces for a group of 40 to 120 trials (depending upon the test) having a common feature (see label to the right of the graph). The deviation in relative discharge rate is plotted along the vertical axis. Unit level corresponds to the mean rate of the population background (absolute values are given in spikes per bin above the graph). The horizontal dotted line under (spikes per 50 msec) each graph shows single time bins and is used to mark those with contents deviating significantly from the background mean value, in the indicated direction based upon student *t*-test for unequal sample sizes and unknown variances. Short, medium, and long bar markers correspond to $p < 0.05$, $p < 0.01$, and $p < 0.001$ levels. The dotted line (M12) located under each pair of graphs is used similarly to represent the results of the pairwise statistical comparison of bin contents based upon student *t*-test for unequal sample sizes and unknown variances. Histogram fragments contributing to the most significant observed differences are also marked by blackening. Double-dotted vertical lines denote the presentation epochs for the slides containing the stimuli to be processed or the instruction to respond.

comprehension test with the first word of variable length, a seven-letter word evoked a more pronounced discharge rate with a lower slope of the trailing edge than was evoked with a three-letter word. Furthermore, during semantic comparison and sign comparison tests, presentation of a word caused a significantly more pronounced response than a picture for all pairwise combinations of verbal and graphic stimuli.

It should be noted that similar correlates of processing of sign features have been found and described earlier (Gogolitsin et al., 1987c) during the matching test. In that case, the correlate consisted of a lower slope of the trailing edge of the early discharge rate peak for nontarget stimuli that were graphically similar to targets than for clearly unsimilar nontargets. In this study, similar neuronal reactions were also observed for other types of lexical processing, which suggests their dependence upon the features of the verbal stimuli but not upon the type of processing.

11.3.2.2 Semantic Features

Another type of neuronal correlate of verbal processing can be frequently observed (Gogolitsin et al., 1987c) during presentation of the recognition test where in each trial a subject has to answer the question, "Do you know the meaning of this word?" A trial consists of presentation of a single slide containing a fixed length letter sequence forming either a meaningful word or a wordlike but meaningless quasi-word. Statistical comparison of the neuronal discharge rate patterns corresponding to these two situations often reveals clear differences, as indicated by an additional discharge rate peak with latency within the range of 300-700 msec (Figure 11.4). Such peaks in the PSH appear more frequently in case of a meaningful word presentation. The opposite situation can, however, also be obingful word presentation. The opposite situation can, however, also be observed.

Demonstrating such pattern differences does not by itself provide sufficient proof of their specific relation to semantic processing. For example, such an effect might also be related to the difference in verbal responses for the two trial types. This interpretation uncertainty can be resolved by comparing the discharge rate patterns in the recognition test with those recorded during the matching tests. During the matching trials, the same words and quasi-words as in the recognition trials are matched to the remembered target stimuli. However, "yes" and "no" verbal responses are now determined by the match or mismatch, but not the meaning of the stimulus.

Taking into account the comparison of patterns observed during different tasks, only those neuronal populations may be considered as specifically associated with semantic processing for which differences such as those shown above are found in a recognition task, but not in a matching task.

11.3.2.3 Operations

The class of neuronal correlates of lexical processing described in this subsection seems to be much less clearly defined compared to the above-described reactions

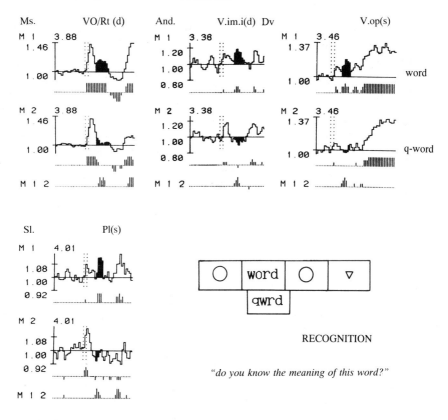

FIGURE 11.4. Discharge rate patterns of neuronal populations in different subcortical brain structures of four patients during the *recognition* test. Patient codes and structure names, abbreviated according to the Shaltenbrandt & Bailey stereotactic atlas, are given near the corresponding graphs.

associated with word sign and semantic features. It includes many different types of reactions with a common feature: they reflect an operation performed on the stimulus according to the test task. It is hoped that future research will provide additional data needed to better understand such reactions and the role played by the corresponding populations of neurons in the brain systems subserving language function.

An example of such a correlate is illustrated in Figure 11.5 by discharge rate patterns recorded during the matching and sign comparison tests. A clear difference is seen between the late components of the peristimulus time histograms (left side of figure) for trials in the matching test, when words and quasi-words similar to the target words and quasi-words were presented, and for trials with presenta-

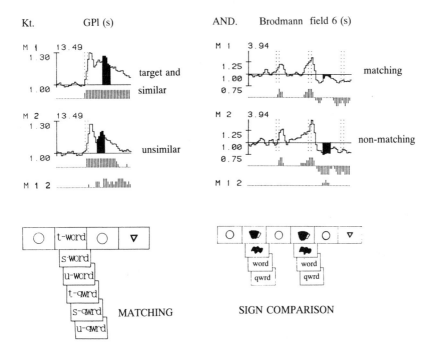

FIGURE 11.5. Discharge rate patterns recorded during *matching* and *sign comparison* tests.

tion of dissimilar stimuli. In a sign comparison test, a population of neurons in the cortical field 6 of the left hemisphere (right side of figure) reveals significant differences between discharge rate patterns for the matching versus nonmatching pairs approximately 400 msec after the presentation of the second stimulus.

Another cortical population (Figure 11.6) displayed the properties of an even more sharply tuned detector. Here additional rate peaks were observed during the matching test, but only for presentation of previously memorized target stimuli, whether quasi-words or quasi-pictures.

During the sentence comprehension test, correlates of deciding whether a sentence was meaningful were revealed both in subcortical and cortical populations of neurons (Figure 11.7). They were manifested in the form of additional rate peaks or higher rate levels approximately 300 msec after the presentation of an object that made the sentence incorrect. In trials with presentation of a correct sentence such peaks or higher rate levels were absent. The observations of such neuronal populations in the cortex confirm the data obtained earlier by N.P. Bechtereva (1988b) in the fields 1-4 of the left hemisphere and in field 4 of the right hemisphere.

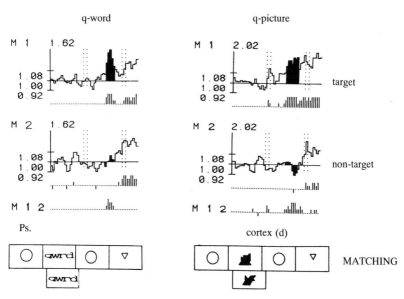

FIGURE 11.6. Discharge rate patterns of the neuronal population in the cortex of the right hemisphere during the test for *matching* quasi-words and quasi-pictures.

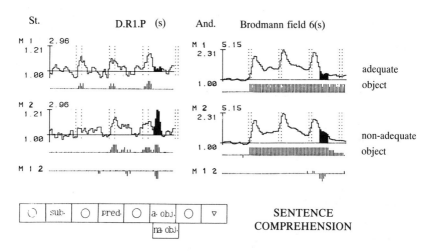

FIGURE 11.7. Discharge rate patterns of the two neuronal populations during the *sentence comprehension* test.

11.3.2.4 Motor Components

Taking into consideration the fact that many of the electrodes used for the diagnosis and treatment of Parkinson's disease are stereotactically aimed at brain structures involved in the regulation of motor activity, it is not surprising that many of the neuronal populations studied demonstrate pronounced reactions during the patient's verbal response. However, the class of neuronal correlates related to motor components of lexical processing includes some other types of neuronal reactions observed long before the pronunciation of the response by the patient.

Some examples are given in Figure 11.8. Populations of neurons revealed a specific reaction in the naming test. Reactions to quasi-pictures were different from reactions to other stimuli that the patient was able to verbalize and pronounce, whether verbal or pictures. In case of quasi-picture presentations, the patients answered "no."

It is clear, of course, that direct evidence for the motor origin of such reactions must be independently obtained, for example, by simultaneous recording of the neuronal activity and the electromyogram of the muscles involved in verbalization. Nevertheless, the effect observed demonstrates a rather high degree of selectivity in reactions of neuronal populations to verbal and graphic visual stimuli.

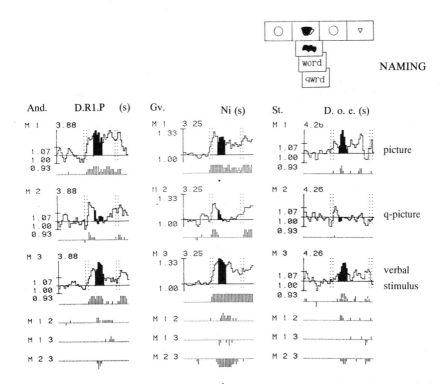

FIGURE 11.8. Discharge rate patterns of the three different populations of neurons during the *naming* test.

11.3.2.5 Mnestic Components

The fifth class of neuronal correlates of verbal processing is comprised of the reactions displayed in multiple stimuli tests in which the subjects required word retention in short-term memory to perform the test correctly. It should be mentioned here that these correlates, presumably related to mnestic processes, are of the tonic type, which differentiates them from reactions described above that were manifested in the form of phasic discharge rate peaks.

In Figure 11.9, discharge rate patterns are presented of cortical neuronal populations recorded during the sentence comprehension test and a short-term memory test. In each trial three words were presented sequentially, followed by a number 1, 2 or 3, which indicated which of the three words was to be repeated aloud. It can be seen that in the short-term memory test sequential presentation of three

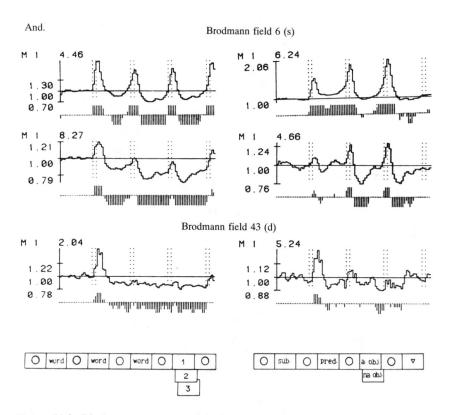

FIGURE 11.9. Discharge rate patterns of the three neuronal populations in the cortex during the *short-term memory* and *sentence comprehension* tests.

words resulted in a pronounced tonic decrease of the discharge rate. This decrease was much less expressed during sentence comprehension when three words were also presented.

11.3.2.6 Hemispheric Asymmetry and Localization-Dependent Features of Neuronal Discharge Rate Patterns

Figure 11.10 presents the distribution of neuronal populations, exhibiting evoked reactions of the above-described types, across all the brain structures where neuronal activity was recorded in this study and separately for thalamic and striopallidar populations.

Although the total number of neuronal populations examined in this study was only 165, which is insufficient for any sound final conclusions, some tendencies can be mentioned that deserve further analysis.

In the left hemisphere, the share of neuronal populations revealing discharge rate evoked responses that were specifically related to lexical processing was considerably higher than in the right hemisphere. This effect can be observed both across all the brain structures and in the thalamic nuclei, but not in the striopallidar system.

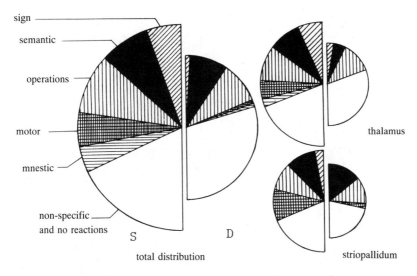

FIGURE 11.10. Relative amounts of different types of neuronal correlates of lexical processing in neuronal populations of the right and left hemispheres, in thalamic nuclei, and in the striopallidum.

General features of the distributions of correlates in the thalamus were similar to those observed when summing the data for all brain structures. In the left hemisphere, motor and mnestic reactions classified as sign processing related were more frequent than in the right hemisphere. The distributions of semantic processing correlates and reactions associated with operations were almost equal for the both hemispheres.

These data confirm the previously reported (Gogolitsin et al., 1988) dependence of the shape of the neuronal evoked reaction pattern upon the localization of neuronal populations in the brain, in addition to their strong dependence upon the psychological task performed by the subject and upon the stimulus features.

11.4 Concluding Remarks

The data obtained in this study demonstrate the potentialities of the illustrated psychophysiological approach in supplying valuable information concerning subtle events that reflect lexical processing in neuronal populations of the human brain. It is evident, of course, that this particular methodology, as is any other single approach, is limited. These limitations are defined first by the strictly followed medicoethical considerations inherent in any invasive technique, which determine solely the sites of electrodes implantation, electrode parameters, and the periods and conditions when the neuronal activity recording and psychological testing can be performed without causing any harm but with the ability to obtain information valuable for diagnostics and treatment of this particular patient.

Another limitation of this methodology is directly associated with its main advantage—the ability to discover local physiological events occurring in the small groups of neural cells, those most probably forming a true physiological basis of our higher functions, including thinking and language (Bechtereva et al., 1985). Accounting for the distributed nature of information processing in the brain, it becomes clear that even simultaneous recording of the activity of dozens of neuronal populations cannot change the situation principally. Yet, when used in parallel with the recording of neuronal activity, other methodologies, among which PET looks extremely promising, are able to supply complementary data necessary to understand the role of both local and spatially distributed processes in the brain systems that maintain higher functions of the human brain (Bechtereva, 1987).

Nevertheless, as it can be seen from the results presented in this chapter, recording of the neuronal activity with the aid of implanted electrodes during psychological tests can be rightfully considered as the source of the unique objective physiological data concerning many principal aspect of organization of the brain systems subserving our linguistic abilities. First, the data obtained during the tests with presentation of both words and pictures clearly demonstrate that visually presented words possess certain yet unknown specific features differentiating them from other visual stimuli, even those ones representing the same objects. Further research into this field may lead to a better understanding of the basic organization of brain mechanisms of language.

Another important conclusion that can be drawn from the analysis of neuronal discharge rate patterns is concerned with the existence of a temporal order of the physiological events in neuronal populations during lexical processing. Though parallel models of this processing are gaining increasing support today, a certain degree of serial organization seems to be also present in brain mechanisms maintaining higher functions. This serial organization is manifested in the existence of the preferred latency ranges of the discharge rate peaks observed during many different psychological tests and resembling to a certain extent the evoked potential components. It should be mentioned, however, that neuronal discharge rate patterns reveal much more complicated and specific reactions compared to the evoked potentials, which opens perspectives of the objective evaluation of information processing stages in the brain and thus linking together psychological and neurophysiological approaches.

References

Bechtereva, N.P. (1978): *The neurophysiological aspects of human mental activity,* 2nd ed. New York: Oxford University Press

Bechtereva, N.P. (1987): Some general physiological principles of the human brain functioning. *Int. J. Psychophysiol* 5, 235–251

Bechtereva, N.P. (1988a): *Healthy and diseased human brain,* 2nd ed. (In Russian). Leningrad: Meditsina

Bechtereva, N.P. (1988b): The neurophysiology of higher cerebral functions. In: *Psychophysiology 88. Proceeding of the Fourth Conference of the International Organization of Psychophysiology.* Prague: 26

Bechtereva, N.P., Gogolitsin, Yu. L., Kropotov, Yu. D., Medvedev, S.V. (1985): *Neurophysiologic mechanisms of thinking* (In Russian). Leningrad: Nauka

Buchsbaum, M., Fedio, P. (1970): Hemispheric differences in evoked potentials to verbal and nonverbal stimuli in the left and right visual fields. *Physiol. Behav.* 5, 207–210

Carlson, N.R. (1988): *Foundations of physiological psychology.* Boston: Allyn and Bacon

Gogolitsin, Yu. L., Melnichuk, K.V., Nechaev, V.B., Pakhomov, S.V. (1987a, b, c): Neuronal correlates of word processing. I. Methodological aspects and methods of investigation. II. General characteristics of evoked changes in discharge rate of neuron populations during word memorizing. III. Discharge rate patterns of individual neuron populations during verbal-mnestic activity. *Hum. Physiol* 13, 531–543; 13, 707–714; 13, 948–956. (English translation of *Fiziolgiya Chelovska* [Russian original data]).

Gogolitsin, Yu. L., Medvedev, S.V., Pakhomov, S.V. (1987d): *Component analysis of neuronal impulse activity.* (In Russian). Leningrad: Nauka

Gogolitsin, Yu. L., Kiryanova, R.E., Nechaev, V.B., Shkurina, N.G. (1988): Reflection of verbal processing in evoked reactions of neuronal populations in striopallidar and thalamic nuclei. In: *Psychophysiology 88, Proceedings of the Fourth Conference of the International Organization of Psychophysiology.* Prague: 93.

John, E.R., Herrington, R.N., Sutton, S. (1967): Effects of visual form on the evoked response. *Science* 155, 1439–1441

Luria, A.R., (1958): Brain disorders and language analysis. *Language Speech* 1, 14–34.

Luria, A.R. (1966): *Higher cortical functions in man.* New York: Basic Books

Mills, R.H., Knox, A.W., Juola, J.F., Salmon, S.J. (1979): Cognitive loci of impairments in picture naming by aphasic subjects. *J. Speech Hearing Res.* 22, 73–87

Ojemann, G.A. (1983): Brain organization for language from the perspective of electric stimulation mapping. *Behav. Brain Sci.* 6, 189–206

Petersen, E.E., Fox, P.T., Posner, M.I., Mintun, M., Raichle, M.E. (1988): Positron emission tomographic studies of the cortical anatomy of single-word processing. *Nature* 331, 585–589

Shelburne, S.A. (1973): Visual evoked responses to language stimuli in normal children. *Electroenceph. Clin. Neurophysiol.* 34, 135–143

Smirnov, V.M. (1976): *Stereotactic neurology.* (In Russian). Leningrad: Nauka

Townsend, J.T. (1984): Uncovering mental processes with factorial experiments. *J. Mathemat. Psychol.* 28, 363–400

12

Event-Related Desynchronization (ERD) Correlated with Cognitive Activity

GERD PFURTSCHELLER, W. KLIMESCH, A. BERGHOLD, W. MOHL, AND H. SCHIMKE

12.1 Introduction

One characteristic feature of the human brain is its ability to generate rhythmic activity within the 7–13 Hz (alpha) and 15–20 Hz (beta) frequency band. These rhythmic activities can be found in intracerebral recordings (Cooper et al., 1965) and in electrocorticograms (Jasper and Penfield, 1949) and were first recorded in the electroencephalogram (EEG) from an intact skull by Berger in 1933. In addition to their sinusoidal behavior and characteristic frequency, alpha and beta band rhythms demonstrate another characteristic feature: the reactivity to exogenous and endogenous events. Well-known examples are the alpha blocking of occipital alpha rhythm after light stimulation (Berger, 1933) and the blocking of central mu rhythm during movement (Chatrian, 1976).

Event-related desynchronization (ERD) is a synonym for amplitude attenuation or blocking of alpha band rhythms. ERD is a phenomenon of the spontaneous EEG and is found parallel to all types of event-related potentials (ERP). For example, visual stimulation results in the generation of a visual evoked potential (VEP) and an occipital ERD (Aranibar and Pfurtscheller, 1978), and voluntary movement is accompanied by a slow negative potential shift (Bereitschafts-potential; Deecke and Kornhuber, 1978) and a central ERD (Pfurtscheller and Aranibar, 1979). This chapter reports on the quantification and topographical display of ERD and its impact on the investigation of cognitive brain functions.

12.2 Prerequisites for ERD Mapping

ERD, a characteristic feature of the EEG, is very often masked by delta, theta, and nonrhythmic EEG components. The application of the averaging technique is therefore a necessary prerequisite for enhancing the signal-to-noise ratio whereby the amplitude change of the alpha band rhythms can be considered as the signal and all other EEG components as the noise. This means that for each ERD study the type of event, the interval between events, and the number of events have to be determined in advance.

An event is defined as a task that a subject has to perform in an experiment. An event can be either externally paced by a stimulus of different modality or can be an internally paced motor behavior or speech. It is assumed that during each event certain brain regions generate specific patterns of activity. For example, before a simple finger movement can be performed, neuronal structures in the basal ganglia, supplementary motor area, thalamus, cerebellum, and motor cortex have to be activated in a predefined temporal sequence. Self-paced voluntary finger movement therefore needs more than 1 sec from the intentional process to movement onset (Deecke and Kornhuber, 1978; Pfurtscheller and Aranibar, 1979; Pfurtscheller and Berghold, 1989).

To study brain functions during such relatively long time intervals (on the order of seconds), the interval between two events must be at least some seconds. From our experiments, we know that a minimum of 30 to 60 events is necessary to obtain satisfying results. This means that with an interevent interval of 6 sec and a total of 60 events, the minimum time for one experiment is about 6 minutes. In these 6 minutes, the general level of vigilance and attention has to be kept constant.

ERD is defined as amplitude attenuation and is therefore dependent on the amplitude of alpha band activity before each event. When there is no alpha band activity at all, no ERD can be measured! ERD can also be negative, meaning that an event is accompanied by an amplitude enhancement or "event-related synchronization" or provocation of alpha band activity. To normalize the ERD, the average amplitude (actually, the squared amplitude) of the alpha band activity some seconds before each event is assumed to be 100%. The interval used for this normalization is called the reference interval. The same procedure that is applied to alpha band activity can also be used to study beta rhythms.

12.3 Quantification of ERD

The procedure of ERD measurement was introduced in 1977 (Pfurtscheller and Aranibar, 1977) and published elsewhere (Pfurtscheller, 1981; Pfurtscheller et al., 1988). The method of data processing together with some factors influencing the calculation of ERD maps are summarized in the flow chart in Figure 12.1. In addition to automatic artifact detection, the calculation of reference-free recordings is an important factor in ERD mapping. The calculation and recording, respectively, of reference-free data (e.g., average reference, bipolar, local average reference, Laplaceian operator) are negligible in experiments with visual information processing—the maps are relatively independent of the type of EEG derivation used—but are of major importance especially in experiments where sensorimotor structures are activated (Pfurtscheller, 1988; Pfurtscheller et al., 1988).

The typical set of data obtained with the ERD mapping technique is displayed in Figure 12.2. A band power map calculated in the reference interval, two ERD

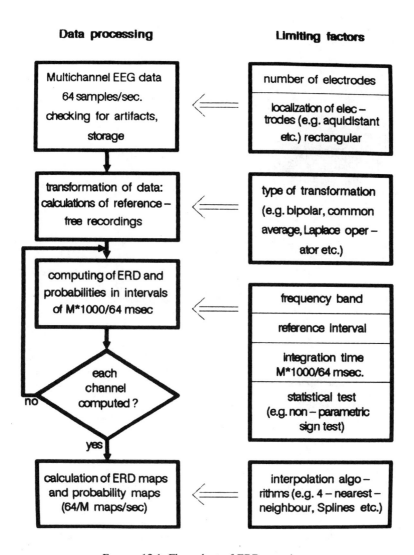

Data processing   **Limiting factors**

FIGURE 12.1. Flow chart of ERD mapping.

maps, and examples of logarithmic power spectra demonstrate most of the important features of ERD mapping. The choice of reference interval, frequency band, and type of derivation can influence the topographical display of the ERD. Referential EEG recordings (with ear reference) or frequencies within the theta range can result in maps without any significant ERD.

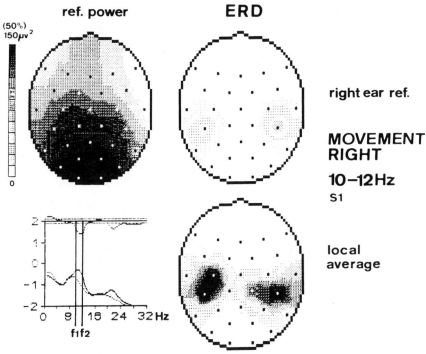

FIGURE 12.2. Examples of an alpha power map calculated in the reference interval (*left upper panel*), logarithmic power spectra with indicated frequency band analyzed (*left lower panel*), and two ERD maps (*right side*) calculated with ear reference and local average derivations. Data from a hand movement experiment. Note the bilaterally localized ERD over both central areas during hand movement.

12.3.1 Classification of Visually Presented Words and Numbers

The subjects tested were 11 right-handed men. The stimuli were 48 words and 48 numbers, presented in two conditions (blockwise and random). Half of the 48 words belonged to the category "tools," and the other half to "animals." The stimuli (words or numbers) were exposed for 250 msec on a video monitor. A warning signal appeared randomly either 1, 2, or 3 sec before the stimulus presentation. The interstimulus interval was 8 sec.

The subjects had to perform a semantic or numerical classification task. They were instructed to indicate to which of two categories (animals/tools or even/odd numbers) an item belonged. Subjects had to respond with "yes" if a word denoted an animal and with "no" if a word denoted a tool. When a number was presented, the subject had to first subtract three and then decide whether the resulting number was odd or even. If the resulting number was even, the subject responded "yes";

otherwise he answered with "no." As an example, consider the number 67. First, the subject performs the subtraction (67 − 3 = 64) and then decides that the result is an even number. Accordingly, the subject gives a yes response. Since the subtraction of three always turns an even number into an odd number and vice versa, subjects could also use an alternative strategy, which consists of responding "yes" to an odd number and "no" to an even number. However, regardless of the strategy used, the purpose of this procedure was to increase cognitive load.

The EEG sampling was started 4 sec before stimulus presentation (word or number). Each sampled trial had a duration of 7 sec. The data presented in this paper are grand averages over all 11 subjects including three different prestimulus intervals (1, 2, and 3 sec) and two stimulus conditions (random and blockwise). Further details about the methodological aspect of stimulus presentation are reported in Klimesch et al. (1988).

12.3.2 ERD Maps During the Visual Task

Grand averages (N = 11) from ERD maps measured before the presentation of a word or number (preparatory state), during semantic encoding and cognitive processing, and during the verbal response are displayed in Figure 12.3. Each map represents a time period of 125 msec. It can be seen that ERD in the 10-12 Hz band is most prominent during semantic encoding and is localized over the occipital region. In contrast, the ERD in the 8-10 Hz band is already present before stimulation, is widespread during cognitive activity, and is more pronounced over the left hemisphere during the verbal response. It is important to note that, depending on the frequency band and time period studied, different ERD maps are found!

12.3.3 Time Course of ERD During the Visual Task

From previous data and similar experiments, we can speculate that specific cortical areas generate their own intrinsic rhythm that is desynchronized when neuronal mass in that area becomes activated (Klimesch et al., 1988; Pfurtscheller et al., 1988). To obtain a better insight into the mechanisms generating localized rhythmic activity, three frequency bands (6–8 Hz, 8–10 Hz, 10–12 Hz) and four cortical regions (occipital, parietal, and left and right centrotemporal) were specified and the time course of ERD analyzed. The ERD values measured on three to four electrodes were averaged for each region specified (see Figure 12.4, lower panel). The mean time courses of the ERD measured in 11 subjects during the visual task are displayed in Figure 12.4 (upper panel). In the occipital region, a brief ERD was found from 4000 to 4875 msec in the 10-12 Hz band. In the parietal region, the 6-8 Hz rhythm was desynchronized from 4250 to 6000 msec.

The intermediate frequency of 8-10 Hz showed the major ERD involving central, parietal, and occipital areas. In contrast to the 6-8 Hz and 10-12 Hz bands, the ERD had already started in the 8-10 Hz band *before* sensory stimulation. Additionally, the 8-10 Hz ERD demonstrated a left hemisphere dominance. In the two specified centrotemporal areas, the ERD increased gradually, starting with the

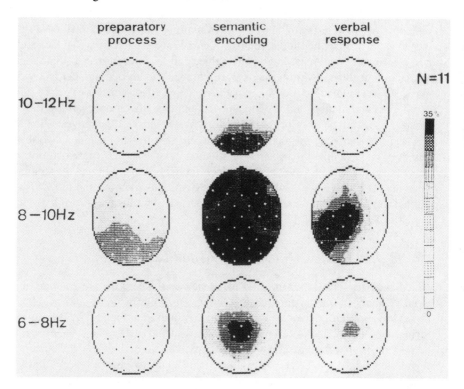

FIGURE 12.3. Grand average ERD maps (11 subjects) from a visual classification task, calculated for three frequency bands. ERD maps calculated before (*left*), during (*middle*), and after (*right*) visual stimulation. Note the occipital ERD before stimulation and the left hemisphere ERD dominance during the verbal response when the 8–10 Hz band is analyzed.

first warning stimulus at 1 sec. The difference of these two time courses displayed a maximum at 4875 msec.

Three characteristic latencies in the ERD time curves were found after stimulus onset:
1. 500 msec: occipital region, 10–12 Hz
2. 625 msec: parietal region, 6–8 Hz
3. 875 msec: centrotemporal left, 8–10 Hz

How can these latencies be interpreted? The semantic encoding process has been predicted for the first few hundred milliseconds (Klimesch, 1988). Perhaps the latency at 500 msec indicates the end of the semantic encoding performed mainly in the occipital region. The parietal cortex with the visual association area shows maximum activation at 625 msec after word presentation. At this latency, the cognitive operation, including the pure long-term memory activation for the

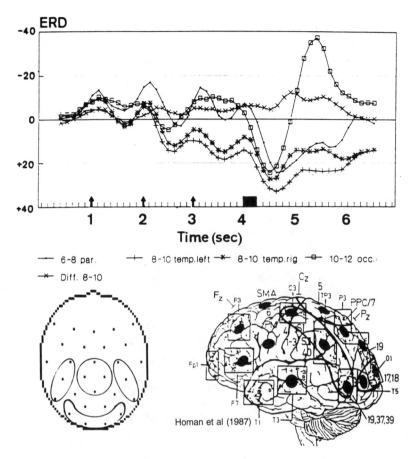

FIGURE 12.4. Time courses of ERD (11 subjects) according to the regions specified. In lower part of figure are maps with electrode location (*left*) and cerebral location of the electrodes (*right*) (Homan et al., 1987). Note the different forms of ERD time course, depending on the frequency band and regions analyzed.

classification process, takes place. The 6–8 Hz ERD in the parietal area can therefore probably be interpreted as the electrophysiological correlate of the cognitive operation.

Both types of ERD in the 6–8 Hz and 10–12 Hz bands were observed mainly after semantic stimulation, whereby the 10–12 Hz band demonstrated the temporally shorter ERD (< 1 sec) and the 6–8 Hz band the longer lasting ERD (> 2 sec). The longest lasting ERD was found in the 8–10 Hz band with a duration of over 2 sec. The preparatory phase for the encoding and cognitive processes began after the warning stimulus and was best reflected in the 8–10 Hz band. Here, the hemispheric asymmetry is of interest: an ERD of larger magnitude was measured over the dominant left hemisphere. The difference in the ERD between both centrotemporal areas was largest at about 875 msec after semantic stimulus presentation; this was about the time when the verbal response was given by the subject. This result seems to indicate that at least one part of the 8–10 Hz band ERD reflects cortical activity associated with the execution of the verbal response.

12.4 Conclusion

The evoked potentials can be interpreted as an on- and off-response made up of components lasting no longer than about 500 msec. Therefore, EPs are only of limited use in the study of cognitive processes and cannot be used to obtain any information about the preparatory stage preceding semantic stimulation. In addition to electric potentials, metabolic measurements can also be used to investigate cortical and subcortical structures involved in cognitive processing. Measurements of regional cortical metabolism using the deoxyglucose method with positron emission tomography (PET) require a steady state of about 40 min, whereas the dynamic method for the determination of oxygen uptake requires 40 sec (Roland and Widen, 1988). Thus, quantitative measurements of brain metabolism, in contrast with measurements of electric potentials and magnetic fields, have a poor time resolution.

The ERD technique is the only method available at this time for studying physiological mechanisms in healthy subjects during cognitive processing with a time resolution of about 100 msec or even less. For the first time, it is possible to discriminate between cortical processes taking place at different time intervals in different cortical areas.

Acknowledgments. The work in this chapter was supported by the Ministry of Science and Research, Austria, and the "Fonds zur Förderung der wissenschaftlichen Forschung" projects S4902-MED and S4905-MED.

References

Aranibar, A., Pfurtscheller, G. (1978): On and off effects in the background EEG activity during one-second photic stimulation. *Electroenceph. Clin. Neurophysiol.* 44, 307–316

Berger, H. (1933): Über das Elektrenkephalogramm des Menschen. *Arch. Psychiat. Nervenkr.* 100, 301–320

Chatrian, G.E. (1976): The mu rhythm. In: *Handbook of electroencephalography and clinical neurophysiology 6.* Remond, A., (ed.). Amsterdam: Elsevier, pp. 46–49

Cooper, R., Winter, A.L., Crow, H.I., Walter, W.G. (1965): Comparison of subcortical, cortical and scalp activity using chronically indwelling electrodes in man. *Electroenceph. Clin. Neurophyisol.* 18, 217–228

Deecke, L., Kornhuber, H.H. (1978): An electrical sign of participation of the mesial "supplementary" motor cortex in human voluntary finger movement. *Brain Res.* 159, 473–476

Homan, R.W., Herman, J., Purdy, P. (1987): Cerebral location of international 10-20 system electrode placement. *Electroenceph. Clin. Neurophysiol.* 66, 376–382

Jasper, H.H., Penfield, W. (1949): Electrocorticograms in man: Effect of voluntary movement upon the electrical activity of the precentral gyrus. *Arch. Psychiat. Nervenkr.* 183, 163–174

Klimesch, W. (1988): *Struktur und Aktivierung des Gedächtnisses: das Vernetzungsmodell: Grundlagen und Elemente einer übergreifenden Theorie.* Bern: H. Huber

Klimesch, W., Pfurtscheller, G., Mohl, W. (1988): ERD mapping and long-term memory: The temporal and topographical pattern of cortical activation. In: *Functional brain imaging.* Pfurtscheller, G., Lopes da Silva, F.H. (eds.). Toronto: H. Huber, pp. 131–142

Pfurtscheller, G. (1981): Central beta rhythm during sensory motor activities in man. *Electroenceph. Clin. Neurophysiol.* 51, 253–264

Pfurtscheller, G. (1988): Mapping of event-related desynchronization and type of derivation. *Electroenceph. Clin. Neurophysiol.* 70, 190–193

Pfurtscheller, G., Aranibar, A. (1977): Event-related cortical desynchronization detected by power measurements of scalp EEG. *Electroenceph. Clin. Neurophysiol.* 42, 817–826

Pfurtscheller, G., Aranibar, A. (1979): Evaluation of event-related desynchronization (ERD) preceding and following self-paced movement. *Electroenceph. Clin. Neurophysiol.* 46, 138–146

Pfurtscheller, G., Berghold, A. (1989): Pattern of cortical activation during planning of voluntary movement. *Electroenceph. Clin. Neurophysiol.* 72, 250–258

Pfurtscheller, G., Steffan, J., Maresch, H. (1988): ERD-mapping and functional topography—temporal and spatial aspects. In: *Functional brain imaging.* Pfurtscheller, G., Lopes da Silva, F.H. (eds.). Toronto: H. Huber, pp. 117–130

Roland, P.E., Widen, L. (1988): Quantitative measurements of brain metabolism during physiological stimulation. In: *Functional brain imaging.* Pfurtscheller, G., Lopes da Silva, F.H. (eds.). Toronto: H. Huber, pp. 213–228

13

Making Sense Out of Words and Faces: ERP Evidence for Multiple Memory Systems

MITCHELL VALDÉS-SOSA AND MARIA A. BOBES

13.1 Introduction

To understand fully the machinery of mind, a blueprint of the human cognitive architecture would be handy. Models of this architecture, based on work in artificial intelligence, cognitive psychology, and neuropsychology, are still in their infancy. More facts are necessary to promote their development. In this chapter, we apply event-related potential (ERP) technology to this endeavor, searching for insights into the organization of long-term memory.

It has been recently argued that at least part of the mind is organized into distinct processing modules (Fodor, 1983; Marshall, 1984; Putnam, 1984; Shallice, 1984). According to this view, since different information domains (i.e., auditory, visual) require different types of analysis, then specialized processors (modules) must exist. They could be found even within the same sensory modality, (i.e., written words and faces). Modules would perform domain-specific computations, possibly mediated by specialized neural hardware, and operate largely independent from each other.

Although the concept of modules is apparently acceptable to many theorists (direct evidence on specialized computations and neural hardware is available), their degree of independence is a matter of some debate. Fodor (1983) adopted an extreme position with his concept of "informational encapsulation": a module will only draw on domain-specific information to perform its computations. Thus, severe limits would exist to top-down influences from higher-level information. However, the data on contextual facilitation of recognition (e.g., Massarro, 1974) provides evidence for top-down influences on the internal operation of modules.

A modular organization of the mind has implications for the organization of memory. An interesting question has been raised recently by Sherry and Schacter (1987). In the face of possibly many distinct processing modules in the brain, does each module draw on its own memory systems, or do different modules use the same general memory system? We may add another question (related to informational encapsulation): if multiple memory systems exist, what rules guide trans-actions between different systems?

Most of the arguments for multiple memory systems in humans lean strongly on dissociations of memory functions found in brain-damaged patients (as indeed do most of the arguments for autonomous processing modules in the cognitive architecture). The peculiar memory losses sustained by patients with medial temporal lobe or medial diencephalic lesions (Squire, 1987), with sparing of other types of memory, support the existence of at least two different memory systems (declarative and procedural memory, according to Cohen, 1984).

Careful single-case neuropsychological studies (Caramazza, 1988) tend to suggest multiple memory systems. On this basis, it has been argued that the mental lexicon for words is subdivided into several distinct compartments, including orthographic, phonological, and semantic components. A recent study (McCarthy and Warrington, 1988) shows impaired recognition of written words from one category ("living things") but not from another ("objects"). Because picture recognition was impaired for neither of the categories, the authors suggest that the semantic system consists of dissociable modality-specific components.

Although the evidence for multiple memory systems appears to be strong, care must be exercised as to how far the subdivision of memory is carried out. Inferences drawn from studies of brain-damaged populations are limited by inherent methodological problems (Beaumont, 1983). Some of the results may be due to artifacts of the test procedures or flaws in the interpretations. The case for a particular memory partition can be considerably buttressed by behavioral and electrophysiological findings in normal subjects. Converging evidence from different sources may reduce the possibility of false conclusions.

Event-related potentials (ERPs), scalp-recorded voltage changes that are time-locked to some observable event, are one such source of converging evidence. ERPs are low-amplitude signals, usually immersed in much larger background noise, and must be extracted by a combination of filtering and averaging procedures (Picton et al., 1984). Potentially, ERP recording offers a unique window into brain function. It is a noninvasive, nonobtrusive technique that permits on-line monitoring of the brain during mental activity in normal human subjects (Kutas and Van Petten, 1987). Its time resolution is much greater than other brain monitoring techniques, such as positron emission tomography, single-photon emission tomography, or cortical blood-flow measurement (Churchland and Sejnowski, 1988).

In this chapter, we present data from ERP experiments that provide converging evidence, together with the brain lesion studies, for the dissociation of two memory systems, one for face-structural information and the other for linguistic-related semantics. In addition, some characteristics of the transactions between the two memory systems, as seen by ERP techniques, are explored. This last problem is of a kind that cannot be examined in brain-lesioned patients with selective memory losses. Brain lesions, by dissociating retrieval, point to independent memory systems, but at the price of drastically isolating them. The interaction between two different memory systems can be examined only when both are functioning.

The tools we use are certain ERP components that can be observed when the eliciting stimulus does not "fit" into the preceding context (Kutas and Van Petten, 1987). These components, negative in polarity, have been related to the contextual facilitation of recognition, also known as recognition priming. The components could well be dubbed "incongruence negativities" if one considers the conditions associated with their appearance; we use this term as a convenient shorthand. The chapter first reviews behavioral and ERP studies of linguistic recognition priming and nonlinguistic priming. Then original data on the priming of memory for faces are presented and theoretical conclusions drawn.

13.2 Linguistic Recognition Priming and N400

Studies of recognition priming have offered valuable clues on the organization of memory. Priming is said to occur when the previous presentation of a stimulus (the prime or context) facilitates the subsequent processing of a related test item, as opposed to unrelated stimuli. The two types of priming most used in studies of memory organization are repetition and associative priming. Most studies have used words (or pseudo-words) as stimulus material. When the same word is repeated in an experiment (repetition priming), its processing is enhanced over nonrepeated words on various performance measures (e.g., tachistoscopic recognition time, Murrell and Morton, 1974; speed in a lexical decision task, Scarborough et al., 1977).

Morton (1969, 1979) proposed that word recognition is mediated by "logogens," memory devices into which sensory information is fed until a threshold is reached and recognition occurs. In a modified version of the logogen model, the threshold could be lowered after the device "fired," returning slowly to the original level. This model could explain facilitated recognition in repetition priming. Higher-order information (context) also feeds into the logogen, reducing the amount of sensory data necessary to reach threshold. This feature makes the model useful in explaining association priming (see below).

Associative priming is the facilitation between related but not identical stimuli. Many studies have found that for different tasks both speed and accuracy are improved when the target item (e.g., "doctor") is preceded by a semantically related word (e.g., "nurse") as opposed to an unrelated word. This facilitation has been demonstrated with various techniques (e.g., speed in lexical decision making, Meyer and Schvaneveldt, 1971; duration of eye fixations during reading, Carroll and Slowiaczek, 1986), and in this case is called semantic priming. Associative priming has also been demonstrated for words (and pseudo-words) related by phonological and/or orthographic similarities (Donnenwerth-Nolan et al., 1981; Hillinger, 1980; Rugg and Barrett, 1987; Shulman et al., 1978).

Possible explanations for associative priming are very interesting. This effect of a different but related word on the recognition of another word necessarily must be mediated through ties existing in long-term memory. This would imply that entries in the mental dictionary are clustered in some meaningful fashion. In many models (Collins and Loftus, 1975; McClelland and Rumelhart, 1981), concepts related to words represent nodes in a network, where all related nodes are linked. In these models, activation may somehow propagate over the links from a stimulated node to related nodes. In this "spreading activation" account, semantic priming reflects basic organizational properties of long-term memory.

Semantic and repetition priming are apparently distinct phenomena (Dannenbring and Briand, 1983; Henderson et al., 1984). Repetition effects last longer (hours) than semantic effects (seconds). Repetition is restricted to within-modality effects, which has been considered evidence for separate auditory and visual logogens (Morton, 1979). Semantic effects can be cross-modal, which agrees with the idea that semantic memory can be reached from different input modules.

The first ERP component (incongruence negativity) related to recognition priming was described by Kutas and Hillyard (1980). When in a sequentially presented sentence, the last word is incongruent with the preceding context (e.g., "I drink coffee and dogs"), a negative component peaking near 400 msec (N400) appears (Kutas and Hillyard, 1980). The N400 component is also evoked in the case of congruent but less probable sentence endings (Kutas and Hillyard, 1984). In this case, the amplitude of N400 is inversely related to the CLOZE probability of the final word. The maximum for N400 is found at centroparietal sites and is only slightly larger over the right side of the scalp (Kutas and Hillyard, 1982).

The amplitude of N400 is not related to the overall truth value of a sentence, but instead seems to reflect the semantic relationship between the content nouns it contains (Fischler et al., 1983). Components similar to N400 are found in tasks in which word series are presented without a sentence structure, if the critical word is semantically unrelated to the preceding ones (Harbin, et al., 1984; Harvey and Marsh, 1983; Polich, 1985a, 1985b).

Components similar to N400 have been evoked by semantic anomalies for many different types of eliciting signals. Words in the visual modality (English, French or Spanish) or even American Sign Language gestures have been used (Kutas et al., 1987; Neville, 1985). Spoken words can also elicit N400 (McCallum et al., 1984).

In a series of papers, Rugg and collaborators (Rugg, 1984a, 1984b, 1985; Rugg and Barrett, 1987) have described a component, peaking around 450 msec after the stimulus onset (N450), that is associated with orthographic/phonologic mismatches in a word pair rhyme detection task. This component, in contrast with N400, is highly asymmetric, with a larger amplitude on the right side of the scalp.

Although establishing the independence or equivalence of two ERP components is a difficult task, few researchers disagree that clearly different scalp distributions point to different components or mixture of components. Different components (or mixtures) also suggest distinct neural generators and associated cognitive processes (for a discussion, see Rugg, 1987b). It is possible that N450 represents a different incongruence negativity, associated with the orthographic/phonological memory domain, in a similar fashion to the association of the more bilaterally symmetrical N400 with the semantic memory domain. Neuropsychological evidence for dissociations between orthographic, phonological and semantic lexicons (Caramazza, 1988) lends support to this interpretation in the case of N400 and N450. Thus, a family of incongruence negativities may exist, reflecting analogous processes in different domains (Rugg and Barrett, 1987).

Rugg (1987b) required subjects to detect the occasional presence of an infrequent nonword in a background of frequent words. Some of the words were preceded by unrelated items, some by a semantic associate, and other were repetitions of the previous word. The effect of stimulus repetition (a positive potential shift for repeats) was widespread and clearly distinguishable from the smaller effect of semantic priming. This last effect has a frontal-central distribution. This finding indicates that the ERP correlates of repetition semantic priming are different, which reinforces the idea that these two types of priming engage different cognitive processes.

According to many ERP researchers, the amplitude of N400 (Kutas and Hillyard, 1984; Kutas and Van Petten, 1987) and related negativities (Rugg and Barrett, 1987) is inversely related to the amount of contextual preactivation of the memory representation of the item under processing. In other words, it serves as an indication of the extent of activation of a memory representation produced by contextual cues. Stuss et al. (1986) have suggested that N400 amplitude is an index of the amount of search in long-term memory. Recently, Stuss et al. (1988) have argued that N400 is associated to postlexical evaluation, rather than lexical access. Although the exact nature of the process associated with N400 awaits definitive identification, it seems clear that N400 and other incongruence negativities can be used to probe associative links between units existing in long-term memory.

13.3 Face Recognition Priming

The study of incongruence negativities for nonlinguistic material could offer information both about associations within the corresponding memory systems and about interactions with other systems. However, in contrast to research on word recognition, relatively few studies of recognition priming have used nonverbal material. Faces seem to impose special computational requirements for their recognition (Bruce and Young, 1986; Young, 1988). This has suggested to some theorists that a face-processing module could exist (Fodor, 1983). Word recognition breakdown in aphasia does not imply failures in face processing. Well-documented impairments in the recognition of familiar faces (prosopagnosia) have been found without disturbances in the linguistic domain (Grusser, 1984; Jeeves, 1984; Young, 1988). This finding is consistent with the existence of independent modules for face processing and words.

In addition to the neuropsychological data, converging evidence for the existence of a specialized face-processing module is also available from unit recordings in monkeys. Monkeys share with humans the ability to discriminate individual faces (Rosenfeld and Van Hoesen, 1979). A great deal of interest has been generated by the description of a population of neurons in the superior temporal sulcus of the monkey that apparently responds selectively to face stimuli (Desimone et al., 1984; Gross et al., 1972; Perrett et al., 1982, 1984; Rolls and Baylis, 1986). The properties of these neurons suggest they could participate in specialized perceptual and memory mechanisms for faces.

An examination of recognition priming for faces must begin by considering the different types of information that can be derived from faces. Bruce and Young (1986; see also Young, 1988) list up to seven different types of face-related information in their model of face processing. At a very basic level there is pictorial information, a source of artifacts in many experiments with photographs. Structural (topographic) processing of face "form" yields several classes of output. The expression, facial speech codes, and visually derived semantic codes (i.e., age, sex, race) can be recognized even in unfamiliar faces and possibly in parallel.

In the Bruce and Young (1986) model, the processing of familiar faces involves several sequential stages. Recognition of a face occurs when the output of the structural encoding process matches a previously stored description. Each structural description corresponds to a known face. In analogy with the logogens, the face descriptions could be contained in face recognition units (FRUs), variable threshold devices feeding on sensory data. The firing of a FRU would permit access to a person identity node containing semantic knowledge about the person (profession, likes, dislikes, family relations, etc.). Eventually, the name code could be activated.

It is important to differentiate effects due to different types of information, because only some of these could be considered domain specific (i.e., belonging to a face-processing module). The person-semantic nodes have been considered part of the central cognitive processes (Bruce and Young, 1986), connected to general semantic memory. It is probable that name codes are similar to other word codes. On the other hand, face structure is evidently domain specific, and the evidence for a face-processing module, mentioned above, essentially concerns this type of information.

Most studies reporting recognition priming for faces have employed variants of repetition (or identity) priming. When the face of a celebrity has to be named or distinguished from unfamiliar faces, previous exposure to the same photograph or a different photograph of the same face produces a priming effect (Bruce and Valentine, 1985; Ellis et al., 1987a). This facilitation has been construed as supporting the existence of FRUs. Previous activation of a unit could lower the threshold for a subsequent firing. However, cross-domain priming, by the celebrity's name, was obtained only if the task required retrieval of the name (Bruce and Valentine, 1985). This last case can be explained by response priming at the level of the name code (in the logogen system), not true priming of FRUs. In the familiarity decision task (which does not require name retrieval), the name presentation did not facilitate face recognition. Also, views of other parts of a known person's body or of the person's clothes do not facilitate recognition of the face (Ellis et al., 1987a) in this task. These results suggest that FRUs are not preactivated by names.

However, Bruce and Valentine (1986) did find associative priming for face recognition when one face of a semantically related pair precedes the other (e.g.,

Paul McCartney and John Lennon) in a familiarity decision task. These authors consider that semantic preactivation could reduce the distance of the corresponding FRU from threshold. This result is interesting because it implies bidirectional interactions between person-semantic nodes and FRUs (see above).

The possible relationships between word and face recognition have been pursued in several ERP studies in which the subjects had to discriminate between repeated and nonrepeated faces. Smith and Halgren (1986) had subjects view a set of unfamiliar faces before recording ERPs. Later, these faces were presented several times each, interspersed with new unfamiliar faces presented only once. They found a negative component (N445) associated with the nonrepeated faces that was much smaller when the repeated (primed) faces were presented. The component was larger at posterior sites and slightly larger at the right posterior temporal site (T6) than the left (T5). Since N445 was not observed in a similar design for meaningless visual patterns, this response possibly is specific for repetition priming of face structure.

In another ERP study of face priming (Barrett et al., 1988), an identity matching task was used. Photographs of familiar faces (celebrities) and unfamiliar faces were viewed in pairs (one face after the other). The subjects had to decide if each pair consisted of the same face (different photographs were used) or different faces. Nonrepeated faces, as opposed to repeated faces, were associated with what seemed to be two different incongruence negativities. The earlier effect (120–160 msec) was confined to the posterior electrodes and familiar faces. The negativity at parietal sites was larger over the left side of the scalp. The later effect was found for both familiar and unfamiliar faces, was more widespread, and was symmetrical over the scalp. However, the effect was larger, especially in the 350–450 msec time window, for the familiar faces. Barrett et al. (1988) conclude that nonrepeated faces can elicit several of what we call incongruence negativities. One of them (the widespread negative modulation around 400 msec) could be related to N400. Both of the ERP studies just described reinforce the idea of a "family" of incongruence negativities, each related to different cognitive processes. If familiar faces are considered analogous to words and unfamiliar faces analogous to pronounceable pseudo-words (Ellis et al., 1987b), then these ERP results are parallels of the repetition priming effect found for verbal material (Rugg, 1987b).

The ERP parallels between word and face recognition may have several explanations. It is necessary to distinguish between effects due to processes within a specialized face processor and those due to processes outside of such a processor, even if these are elicited by faces. Of the intramodule effects, the priming of known faces (activation of FRUs) is of special interest. It would correspond to priming an entry in the lexicon (activating a logogen). In an identity matching task with familiar faces, several types of face-related codes may be involved in the priming effect (structural, person-specific semantic, names, etc.). Even if activation of different codes is not reflected in the behavioral results, overlapping contributions could be made to the ERPs. Thus, part of the ERP results could be due to semantic priming within the central cognitive system (i.e., through person-semantic nodes).

For the reasons mentioned above, it is useful to search for priming of familiar faces (at the structural level) in other paradigms, where it is possible to ensure that the priming effects are not mediated by other types of codes. As seen in the previous section, the ERP correlates of repetition and associative priming may differ. Thus, a paradigm producing associative effects within the structure of known faces would be useful.

13.4 Priming Within Face Structure: The Paradigm

Can associative priming effects, analogous to those described for words, be used to study the internal structure of memories for visual images? More specifically, does priming exist for face structural information? As stressed in the previous section, priming studies using face stimuli have concentrated on repetition effects for whole faces or associative effects between different faces. Yet, many studies have demonstrated that different features within a face influence the evaluation of each other (Haig, 1986a; Parks et al., 1985; Young et al., 1987). This influence is clearly evident when features are interchanged between different face images. The novel configurations are perceived as unfamiliar faces, and the recognition of the constituent parts is hampered. These results underline the role of context in perception and suggest an approach to producing association priming within individual faces.

We have developed a paradigm that probes associative effects within the structure of individual faces that is somewhat analogous to the task used to produce the linguistic N400. In the N400 task an incomplete sentence is read to create a context, after which a congruent or incongruent final word is inserted to complete the sentence (Figure 13.1c). In our paradigm an incomplete face (with missing features) is viewed to create a context, and congruent or incongruent features (see below) are grafted onto the images to complete the face (Figure 13.1A). We present in this section behavioral results and in the next sections ERP data obtained with this paradigm.

Digitized black and white photographs of 70 members from the same department of our research institute were used as stimuli. The images (presented on a microcomputer monitor) consisted of a frontal facial view with deleted eye/eyebrow regions. Several studies have demonstrated the importance of the eye/eyebrow region in face recognition (Haig, 1985; 1986b). The subject pressed a key after freely viewing each incomplete face, and 150 msec later an eye/eyebrow fragment was placed in the correct position (Figure 13.1A). The completed face was displayed for an additional 1 sec.

On a randomly selected half of the trials, these features belonged to the original image (congruent). On the other half of the trials, the fragment corresponded to another face (incongruent), selected in such a way that it was difficult to detect the graft (see Figure 13.2 for examples). In particular, the luminance and race of the incongruent fragments were matched with their new context. After display offset, the subjects classified the face completions as congruent (eye/eyebrows that

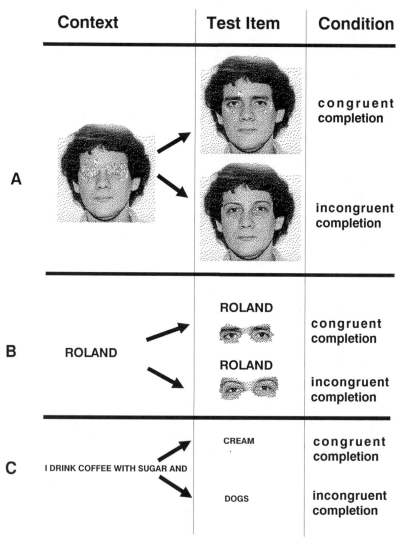

FIGURE 13.1. Three experimental paradigms described in this chapter: A, *Face-processing paradigm.* The stimuli (presented on a CRT monitor) used to produce associative priming between different facial components. On the left, the incomplete face context; on the right, two types of completion: *congruent* face completion; that is, the original eye/eyebrow fragment is put in place, and *incongruent* face completion—an eye/eyebrow fragment originating in another face image. Note that the context precedes and then coexists with test items (eye/eyebrow fragments). B, Paradigm to produce name to face priming. In this case context is a name. C,The classical N400 sentence paradigm.

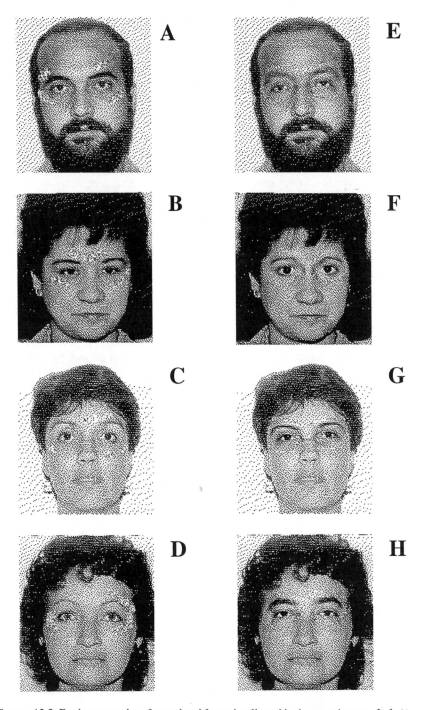

FIGURE 13.2. Further examples of completed face stimuli used in the experiments. *Left,* (A, B, C, and D), Congruently completed face images. *Right,* (E, F, G, and H), The corresponding images with incongruent eye/eyebrow fragments. The resolution of the illustration represents the quality of the images as presented on a EGA CRT monitor. Care was exercised to create incongruent face completions with fragments that were difficult to detect.

belong) or incongruent (eye/eyebrows that do not belong). Two measures of the discrimination behavior were obtained for each subject, using Signal Detection Theory (Swets, 1964): discrimination sensitivity (d') and response bias (Log beta).

Subjects who are very familiar with the persons who posed for the photographs find it easy to recognize the incomplete face images, and to decide if the completing features do or do not belong to each face. A sample of 15 subjects, recruited from the same department as the posers for the face stimuli, were submitted to a session with the task (only half of the faces were used). The discrimination sensitivity scores were high, and no significant response bias was evident (Figure 13.3). Median of the errors over subjects was 9% (ranging from 4 to 13%).

To see if the procedure of previously presenting the face context produced any facilitation on the discrimination task, control trials (using the other half of the faces) were randomly interspersed in the stimulus sequence in the same session described above. On these control trials, no previous context was presented. Instead, the whole face (congruent or incongruent) was presented simultaneously. Although no change in response bias was observed (Figure 13.3), significantly lower ($t = 4.1$, df = 14, $p < 0.001$) discrimination sensitivity scores were found for the control trials. This means that previous viewing of the incomplete face context enhanced the accuracy of subsequent decisions about feature incongruity.

The greater ease of deciding whether features belong to a face when the incomplete face context is viewed alone first, is perhaps related to the difficulty of recognizing constituent parts of a composite face (Haig, 1986a; Young et al., 1987). Spatial displacement of incongruent face parts, preventing fusion into a unitary percept, enhances the recognition of each constituent part (Young et al., 1987). Perhaps the temporal displacement present in our paradigm initially prevents interference due to perceptual fusion and thus enhances subsequent discrimination sensitivity. Also, it is possible that correct recognition of an incomplete, but familiar face allows specific hypotheses to be generated about the missing features. In this sense, the incomplete face could be similar to an incomplete sentence, placing constraints on the expected completion.

To establish if the discrimination behavior just described really depends on the knowledge of specific faces, subjects unfamiliar with the faces used as stimuli were tested (when in doubt, they were asked to guess). Twelve of such individuals produced significantly lower discrimination sensitivity scores (Figure 13.4) than those produced by the subjects familiar with the faces ($t = 6.62$, df = 25, $p < 0.0001$). The mean response bias was not significantly different between the two groups. This suggests that structural descriptions of known faces are useful in performing the task. However, the subjects unfamiliar with the faces performed at a level above chance, with the mean d' significantly different from zero ($t = 3.4$, df = 11, $p < 0.003$).

The modest ability of naive subjects in discriminating between congruent and incongruent face completions could be based on at least two information sources. One source is pictorial mismatch, due to imperfect grafting of incongruent features (differences in texture, luminance, discontinuous borders, etc). The other source

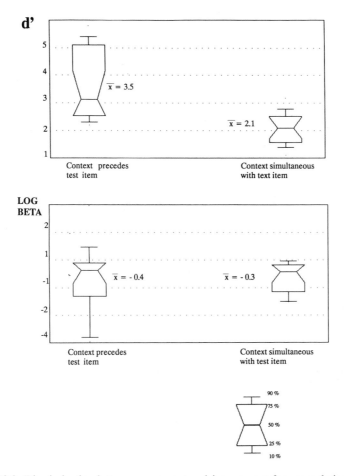

FIGURE 13.3. Discrimination between congruent and incongruent face completion. Box plots of the values of d' and Log beta for 15 subjects familiar with the faces. *Left,* Values when the face context preceded the presentation of the complete face. *Right,* The face context and the eye/eyebrow fragment were presented simultaneously. The mean value for the sample is displayed next to each box plot. *Lower right,* The conventions for the box plots are represented. Note that vertical scales in other figures can be nonlinear.

could correspond to (pictorially consistent) structural incongruities within the grafted images, which are not dependent on descriptions of specific faces. The detection of such incongruities could depend on schematic knowledge structures about faces (Freedman and Haber 1974; Light et al., 1979). This kind of knowledge could permit detection of residual incongruities in age, expression, permissible feature configurations, or even membership in "families" of similar faces. It was possible to select among these alternatives by testing the same group of subjects with inverted, instead of upright face images (turn Figure 13.2 upsidedown). Pictorial mismatch is equally detectable in inverted or upright face images.

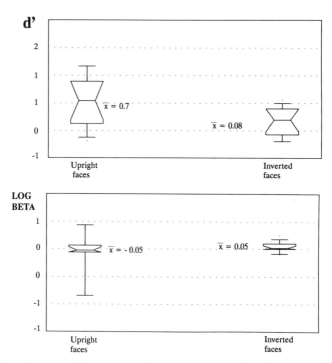

FIGURE 13.4. Box plots of the values of d' and Log beta for 12 subjects unfamiliar with the faces. *Left,* Values when the faces were presented in an upright view. *Right,* Faces were presented in an inverted view (upside down).

In contrast, inverted presentation hampers face recognition more than the recognition of other stimulus classes (Carey and Diamond, 1977; Diamond and Carey, 1986; Yin, 1969). When the subjects unfamiliar with the faces were tested with inverted faces (in the same session and with a different subset of the faces, randomly interspersed with the upright faces), the discrimination sensitivity scores decreased to near zero (the change is significant, $t = 2.74$, df = 11, $p < 0.01$). This implies that the small, albeit significant discrimination ability of naive subjects is not due to pictorial mismatch, but instead to incongruities of face structure.

The previous results suggest that some form of associative priming for face structural information is produced in our paradigm. Face feature processing is facilitated by previous exposure to other parts of a face, perhaps through preactivation of the corresponding memory code. This priming could depend on cumulative contributions of both general schematic knowledge and knowledge of specific faces, although more from the latter. Success in the task does not depend on purely pictorial cues. The priming effect would seem to occur within the FRUs and not between different units, as Bruce and Valentine (1986) suggest for their results with semantic priming.

The internal organization of FRUs (or other devices for face-structural descriptions) is usually not specified in detail in most models. One account is based on the studies in monkeys of neural units responsive to faces that apparently are sensitive to identity (Perrett et al., 1986). Such cells respond selectively to one face, generalizing across expression, viewing distance, and lighting. Some respond to specific parts of the face, i.e., nose, eyes. Perrett et al. (1986) propose that a hierarchical pooling of the output of such cells creates FRUs. A very interesting property of some of these units is, though they only respond to one part of a face, a different but incongruent face component can prevent the cell from responding (Perrett et al., 1984). The mechanisms reflected in the firing of these context-sensitive neurons could underlie the priming effects we have described.

13.5 Priming Within Face Structure: Electrophysiological Data

Further analogies between priming within face structure and semantic priming for words were explored with ERP recordings. Does our face-processing paradigm elicit any component similar to N400? To answer this question, we recorded the ERPs associated with congruent and incongruent face completions.

To increase the number of trials available for averaging, the series of incomplete faces was repeated in two blocks. In each block, half of the faces were completed congruently, the other half incongruently. To prevent the subjects from predicting the type of completion on the second presentation, half of the incomplete faces repeated in the second block were completed in the same way as in the first block. For the other half, the completion type (congruent or incongruent) was inverted.

A sample (Group 1) of ten healthy, right-handed subjects was recruited from the same department to which the individuals who posed for the photographs belonged. Obviously, this group was highly familiar with the faces used as stimuli and had additional associated knowledge, including the corresponding names. The type of completion for each face was counterbalanced across subjects. The performance on the task was accurate and similar to that of the previous group, who were also familiar with the faces (Figure 13.5).

Electroencephalographic activity was recorded after the subject had viewed the incomplete face beginning with the subject's key-press. A prestimulus period of 150 msec was recorded, after which the eye/eyebrow fragment was inserted. Data acquisition ended with the completed face offset 1.024 sec after the key-press. The verbal response was delayed until face offset to reduce movement artifacts. Disk electrodes (Ag/AgCl) were placed on the F7, F8, Cz, Pz, O1, and O2 derivations of the international 10/20 system and referred to linked earlobes. The filter cut-off points were 0.05 and 30 Hz (3 dB down). The vertical and horizontal EOG were monitored. Trials with artifacts, excessive EOG activity, or incorrect responses were eliminated.

For each subject, averaged evoked responses for each electrode were obtained for each stimulus condition. The baseline was corrected by subtracting the average prestimulus amplitude value. Difference waveforms were obtained by subtracting ERPs associated with congruent facial completions from those associated with incongruent completions. In this chapter, when repeated-measure ANOVAs were performed, the Greenhouse-Geisser correction for degrees of freedom was per-

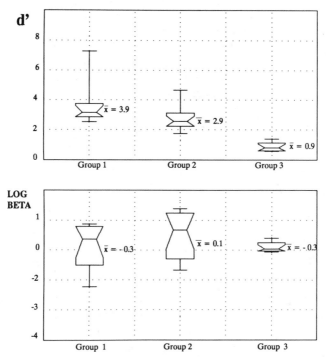

FIGURE 13.5. Box plots of the values of d' and Log beta for three groups of subjects with different degrees of familiarity with the faces and corresponding names (see text for identification of the groups).

formed. The type of face completion (congruent or incongruent) produced no effect on the earliest components (before 250 msec) although it was associated with modulations of the later components at several sites (Figure 13.6). This association is evident in the overlays of congruent and incongruent ERPs where an enhanced negativity is observed for incongruent face completions. The negativity presents a well-defined peak near 374 msec at Cz and is larger over central and posterior regions. We consider this negativity as a component (N374). The results were also analyzed by blocks of trials. Highly similar (although noisier) ERPs and difference waveforms were obtained for the two repetitions of the incomplete face series.

A statistical test of these effects was performed on the average amplitude in the time window from 250 msec to 400 msec as measured for each individual in the difference waveforms. A within-subjects MANOVA indicated a significant main effect of type of face completion ($F(6,4) = 7.69$, $p < 0.034$). The univariate ANOVAs were significant at all derivations except F8. The most significant effects were at Cz ($F(1,9) = 14.2$, $p < 0.004$) and O2 ($F(1,9) = 13.7$, $p < 0.005$). A three-way repeated-measures ANOVA showed that the main conditions of congruity and electrode site were significant, but that the main condition of repetition block (first versus second) was not. Also, the magnitude of the N374 in the difference waveform was not significantly different at Cz ($t = 0.186$, df = 9,

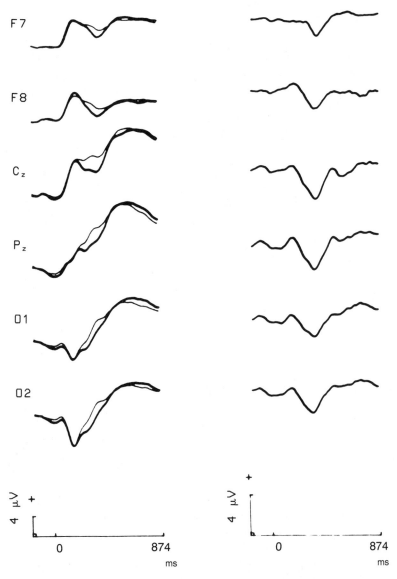

FIGURE 13.6. Group ERPs for subjects who were familiar with both the faces used as stimuli and the corresponding names. Subjects were required to report the names, in addition to indicating type of completion for each trail. In this and subsequent figures, the grand average ERP over all subjects is plotted with positive deflections pointing up. The six derivations used are indicated on the far left. *Left,* ERPs associated with congruent face completions (*thin line*) are overlaid on the ERPs associated with incongruent face completions (*thick line*). *Right,* Difference waveforms (note expanded amplitude scale) obtained by subtracting the ERPs from the two conditions.

$p \leq 0.857$) for first and second presentations. The error rates in both blocks were also similar (6% and 6.2%).

These ERP results indicate that, in our paradigm, structural incongruities within faces are indeed capable of eliciting a negative endogenous component, possibly analogous to N400. The analogy is based on the hypothesis that incongruent completing features (in contrast to congruent features) are not preactivated by the incomplete face. Thus, N374 amplitude, like that of N400 (Kutas and Hillyard, 1984), could be inversely related to memory preactivation. However, the pre-activation is possibly due to interactions within the description of individual faces (FRUs).

However, it is necessary to establish whether other types of incongruities, occurring outside of face structure, contribute to N374. The (easily recognizable) incomplete face context can provide access (via structural encoding) to other memory codes, such as the correct name. If the correct name is retrieved by the context, then the perception of the incongruent face may create a conflict. The incongruent composite could activate other FRUs and the associated person-identity and name codes. Thus, an incongruence at a semantic or verbal level could be produced. If this alternative explanation is valid, then N374 would be linguis-tically mediated, even if not evoked by words. In this case, N374 would resemble the N400 found in studies of sign language (Neville, 1985). It is possible to select between these explanations by examining subjects with different degrees of knowledge about the faces.

13.5.1 Locating the Incongruity Related to N374

Using the same procedures applied to Group 1, two additional samples were studied; they were designed to comprise subjects with different levels of knowl-edge about the face stimuli. Group 2 consisted of 15 subjects who were familiar with most of the faces (>75%) but not with most of the corresponding names (< 40%). These subjects were tested for face familiarity and name knowledge with photographs before the experiment (they work in the same institute as the posers). Those trials in which the stimuli consisted of unknown faces or faces with known names were excluded from further consideration. Group 3 consisted of ten in-dividuals who, before the experiment, had never seen the faces used as stimuli.

The discrimination performance of the two groups is displayed in Figure 13.5. The subjects of Group 3, unfamiliar with the faces, were the least successful. The mean d', though low, was significantly different from zero ($t = 8.6$, df $= 9$, $p < 0.0001$). The subjects in Group 2, who were familiar with the faces, were more accurate. A one-factor ANOVA on d' (including Groups 1, 2, and 3) showed the main effect of the experimental group to be highly significant ($F(2,31) = 14.6$, $p < 0.00001$). Posthoc Tukey tests ($\alpha = 0.05$) indicate that the values of d' were smaller in Group 3 than in Groups 1 and 2, which did not differ significantly. In a one-factor ANOVA on the values of Log beta, the main effect of experimental group was not significant ($F(2,31) = 0.7$, $p = 0.51$). The mean value of Log beta was not significantly different from zero in any group. This means that the differences in accuracy between groups were not due to changes in decision criterion, but to differences in discrimination sensibility.

In Group 2 the incongruent ERPs, compared with the congruent ERPs, presented an enhanced late negative-positive sequence (Figure 13.7). The characteristics of these enhanced components are more evident in the difference waveforms. The negative component had a clearly defined peak, with latency again near 374 msec (N374), and was present in all derivations. The negativity was smallest in the frontal regions and larger for central and posterior regions of the head. The MANOVA for the time window of 250–400 msec shows that the main effect of face completion type was significant ($F(6,8) = 3.95$, $p < 0.039$). The univariate ANOVAs were significant at all derivations and most significant at O2 ($F(1,13) = 22.7$, $p < 0.0004$), and O1 ($F(1,13) = 19.53$, $p < 0.0007$).

The enhanced positivity following N374 was of largest amplitude at central and parietal sites. The peak of this component was near 538 msec. A within-subjects one-factor ANOVA for Cz, on the mean amplitudes for the 450–600 msec time window (which includes the positivity), indicates that the main effect of face completion type was significant ($F(1,14) = 10.02$, $p < 0.007$).

In Group 3 (Figure 13.8), the difference between ERPs elicited by congruent and incongruent completions was small. Slightly enhanced late negativities, associated with the incongruent completions, were observed in several derivations (most clearly seen at F7, O1, and O2). In the corresponding difference waveforms, a slight negative peak (with latency near 328 msec) is present only at O2. This observation was confirmed by a within-subjects MANOVA, indicating that the main effect of completion type was significant ($F(6,4) = 7.43$, $p < 0.036$), with the univariate ANOVAs significant only at O2 ($F(1,9) = 7.04$, $p < 0.026$) and just missing significance at F7 ($F(1,9) = 5.07$, $p \leq < 0.051$).

A four-way repeated-measures ANOVA was performed on the pooled measurements of all groups for the 250–400 msec time window (with groups as a between subjects factor). The only significant main effects were due to congruity (reflecting the N374 effect), group (reflecting the amplitude modulations of ERPs between groups), and site. Neither repetition blocks nor any interaction was significant.

Comparisons between the three groups were performed. The main effect of the experimental group was not significant in a between-subjects one-way MANOVA ($F(10,53.3) = 1.26$, $p < 0.28$) performed on the measurements of the N374 time window (250–400 msec). It was significant only for the univariate ANOVA at Cz ($F(2,30) = 4.14$, $p < 0.026$). Posthoc Tukey tests ($\alpha = 0.05$), show that, for the 250–400 msec time window, the mean amplitude did not differ for groups 1 and 2, but was significantly more negative than for Group 3. Planned comparisons show that the measures over all groups were significantly more negative in O2 than in O1 ($F(1,30) = 6.63$, $p < 0.015$) and that the difference between F7 and F8 was not significant ($F(1,30) = 0.01$, $p < 0.88$). In the P538 time window (450–600 msec), the main effect of experimental group was not significant.

The product-moment correlation between the mean amplitude measures (250–400 msec time window) in each individual at Cz and that individual's efficiency in task performance, as measured by d' was obtained for the pooled data of the

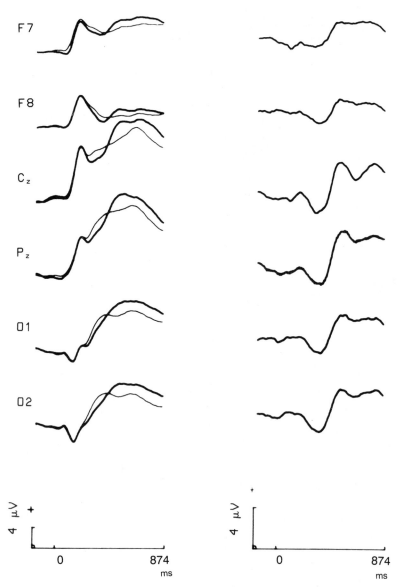

F7

F8

C$_z$

P$_z$

01

02

4 μV +

0   874
ms

4 μV +

0   874
ms

FIGURE 13.7. Group ERPs for subjects who were familiar with the faces used as stimuli, but not with the corresponding names. Subjects were required to indicate type of completion for each trial. *Left,* ERPs associated with congruent face completions (*thin line*) are overlaid on the ERPs associated with incongruent face completions (*thick line*). *Right,* Difference waveforms obtained by subtracting ERPs from the two conditions.

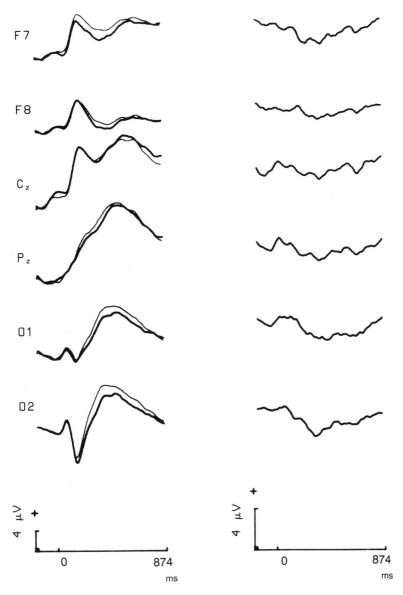

FIGURE 13.8. Group ERPs for subjects who had no previous knowledge of the faces used as stimuli, but were asked to guess if each completion was congruent or incongruent. *Left*, ERPs associated with congruent face completions (*thin line*) are overlaid on the ERPs associated with incongruent face completions (*thick line*). *Right*, Difference waveforms for the two conditions.

three groups. The correlation (r = 0.49) was highly significant (t = −3.14, df = 31, p < 0.004). The intercept of the linear regression of the two variables was not significant. In the P538 time window (450–600 msec), this correlation was not significant r = −0.04).

The use of context (based on previous knowledge about faces) modifies discrimination sensitivity without changing response bias. Studies (Norris, 1985) of context effects on word recognition have found shifts in response bias (with or without discrimination sensitivity changes). This difference is important and suggests that facial context interacts directly with the access to face representations and cannot be explained by postaccess criterion shifts, as posited in some models of word recognition (Norris, 1985).

The results from the group familiar with the faces replicate the findings of the previous section: incongruent face completions are associated with a negative component peaking near 374 msec. The much smaller incongruity effect in the group unfamiliar with the faces indicates that previous knowledge of the specific faces used as stimuli enhances this negativity. This is also suggested by the correlation between discrimination sensitivity and N374 amplitude.

On the other hand, no significant differences in discrimination sensitivity or N374 amplitude were found between the group knowing the face names and the group only familiar with the faces themselves. This finding suggests that naming was not involved in the ERP effects or the discrimination behavior. Although semantic information about the posers for the stimuli was not measured in the two groups familiar with the faces, obvious differences existed in the quality and amount of this knowledge. It follows that differences in person-semantic knowledge probably do not affect the N374 amplitude either. It seems reasonable to assume that N374 is associated primarily with incongruities within the face-structural domain through a mechanism similar to that invoked for N400.

It is pertinent at this point to compare N374 with another endogenous component, N200. One possibility is that N374 is a delayed N200 and part of the complex including P300 (Ritter et al., 1984). The N200/P300 complex is elicited when an unexpected, task-relevant change occurs in an attended stimulus sequence. The more improbable the change, the larger the amplitude of N200. In fact, some debate has existed if the N400 is equivalent to N200 (Ritter et al., 1984). The identification of N200 with N374 is apparently supported by the late positivity following this component in Group 2. In our design, trials with incongruent face completions occurred with the same frequency as trials with congruent completions. If incongruent completions are unexpected, this must be due to subjective probability estimates set up within a trial, as a consequence of viewing the face context. In other words, perhaps congruent face completions are subjectively more probable than incongruent completions.

A particular face context can generate a strong expectancy of the correct (congruent) eye/eyebrows only if the face is familiar. The more familiar the face, the higher the expectancy should be. The possibility of predicting correct face completions should also be related to the discrimination sensitivity. Although N374 amplitude is correlated with discrimination sensitivity, P538 amplitude is not. Also, P538 amplitude is significant in only one group. In fact, Group 1 (the one most familiar with the faces) does not present a significantly larger positivity associated with incongruent face completions.

This dissociation of amplitude modulation of N374 and P538 behavior suggest that they are not part of a N200/P300 complex. It is not clear why P538 is enhanced for incongruent trials in Group 2. This was the only group viewing a mixture of known and unknown faces (Group 1 was familiar and Group 3 was unfamiliar with all faces). Perhaps this could have induced some secondary decision processes leading to a late positive complex.

An argument used to differentiate N200 from N400 is precisely that the latter is always followed by P300 for attended stimuli, whereas in not all experiments is N400 followed by a positivity (Ritter et al., 1984). This property is apparently shared by N374.

The amount of face-structure knowledge could permit different degrees of constraint to be placed on the face completion. If this is so, then the correlation between discrimination sensitivity and N374 amplitude could be analogous to the relation between the CLOZE probability of sentence endings and N400 amplitude found by Kutas and Hillyard (1984).

13.5.2 Scalp Distribution of N400 and N374

Is the N374 component, described in the previous sections, identical with N400 or merely similar? The equivalence of components obtained in different conditions is difficult to establish. Different criteria for comparison have been considered, but no clear-cut decision rules exist (Kutas et al., 1988; Ritter et al., 1984). In addition to waveform morphology, latency, and polarity, aspects to consider are (1) equivalence of task structure and specifications, (2) equivalence of response to specific experimental variables, and (3) equivalence of scalp distributions. The first two criteria are tricky when comparing components elicited by material from different information domains. There is a formal similarity of task structure in the N400 paradigm and in our design. However, the differences in stimulus material probably induce different processing strategies in both tasks. The existence of associations in long-term memory seems to be a prerequisite for both N400 and N374. The strength of the association (or rather the degree of expectancy violation) apparently is related to the amplitude of both N374 and N400.

Qualitative differences in scalp distribution suggest different ERP generator populations. A direct comparison of the topography of N374 and N400 was performed, using 19 sites of the international 10/20 system and constructing interpolated voltage maps. The same recording conditions described above were used, with the exception of the amplifier high-pass cut-off, which was 0.5 Hz.

To obtain the scalp distribution of N400, a replication of the Kutas and Hillyard (1980) experiment was carried out. A group of ten young adults were presented with six word sentences in an analogous manner to that described in the original paper. Half of the sentences ended with a congruent word (high CLOZE probability) and the other half ended with an incongruent word that did not fit (Figure

13.1C). At the end of each trial, the subjects reported if the last word was congruent with the context.

The ERPs associated with incongruent final words presented enhanced negativities compared to the ERPs associated with congruent sentence endings. A well-defined negative component (N400), peaking near 370 msec, was found at almost all sites. A narrow time window, centered around this peak in the difference waveform, was used for the amplitude measurement: from 345 to 395 msec. The effect of congruity was significant in a repeated-measures ANOVA (F(1,9) = 31.3, $p < 0.0003$). Electrode site and the interaction of site and incongruity were also significant.

In a separate session, ERPs were obtained in our face-processing paradigm from another group of subjects, who were familiar with the face stimuli. The recordings corresponding to incongruent face completions again displayed a negative component that corresponded to the N374, although the peak latency was earlier in the conditions of this experiment (325 msec). The time window from 300 to 350 msec was used to measure this incongruity effect, which was significant in a repeated-measures ANOVA (F(1,9) = 5.7, $p \leq 0.04$). Electrode site and the interaction of site and congruity were also significant.

The average amplitudes in these time windows were interpolated and mapped (as percentage of the largest value) over the scalp. The distribution of the two components is very different (Figure 13.9). We found an almost symmetrical distribution for N400, with a centroparietal maximum.

The distribution of N400 is slightly different under different conditions (Kutas

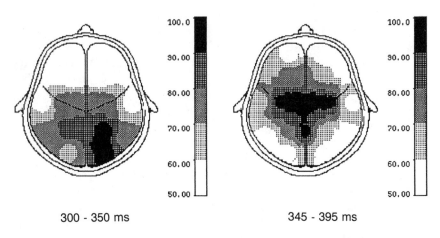

300 - 350 ms   345 - 395 ms

FIGURE 13.9. Voltage maps representing the scalp distribution of N374 (*left*) and N400 (*right*).The measure used to calculate each map was the mean amplitude in the time window indicated below the map. The measure was corrected by subtracting the average prestimulus baseline for each derivation. The values were obtained for the 19 derivations used, and potential values for intermediate points were obtained by triangular interpolations. Each map is scaled in percentages of its maximum value. All maps depicta view of the scalp from above, with the nose up.

et al., 1988). In silent reading it also has a centroparietal maximum, but has substantial amplitude at occipitals (up to 80% of the amplitude at vertex; Kutas and Hillyard, 1983). It is slightly (but significantly) larger over the right side of the scalp (Kutas et al., 1988). However, if an immediate decision to each sentence is required (Fischler et al., 1983), then the distribution is slightly more anterior and the negativity less prolonged (Kutas et al., 1988), perhaps due to the overlap of decision-related positivities (with a parietal maximum). This may cancel part of N400, producing a more anterior distribution. This hypothesis seems to correspond to our situation, where an immediate response was demanded (activity at occipitals was only 60% of the amplitude at vertex).

The distribution of N374 is more posterior and more asymmetrical (Figure 13.9) than that found for N400. The maximum is found at the right occipital site. This is a slightly different distribution for N374 than found in previous groups. The right occipital maximum is also present in the experiment of the next section. The topographical difference with N400 was assessed with an ANOVA on the average amplitude for the time windows mentioned above, measured on each individual difference waveform. The group was considered a between-subject factor and the electrode site a repeated measure. Both main effects were significant, with the negativity in the group with the linguistic paradigm larger than the negativity in the group with the face paradigm ($F(1,18) = 10.04$, $p < 0.005$). To analyze the interaction of site and paradigm type, one of the corrections recommended by McCarthy and Wood (1985) was performed on the data: amplitudes in each individual were scaled to produce a norm equal to 1. After the correction a significant interaction was found between group and site ($F(18,56.7) = 4.18$, $p < 0.05$). Assuming the interaction does not reflect intrinsic difference between subjects in the two groups, this confirms that the scalp distributions of N374 and N400 are different.

It is not possible to distinguish at this point whether N374 and N400 are relatively unmixed components originated by very different generators or if both correspond to a mixture of basically the same components in different proportions. Both explanations indicate activity in different populations of ERP generators. As Rugg (1987b) has pointed out, it is reasonable to assume that these different populations are not related to entirely the same cognitive processes.

13.6 From Names to Faces: Priming and Encapsulation

Fodor (1983), in his discussion of the modularity of mind, proposed that modules were " informationally encapsulated"; that is, the internal operation of a module should not be influenced by top-down processes from the " outside." This notion is related to that of computational autonomy (Marr, 1982). The module does not draw from the whole store of knowledge available to the subject, but only on domain-specific data structures. This view has been contested (Marshall, 1984; Shallice, 1984) and runs counter to many models of perception. Many interpretations of contextual effects conceive interactions at very early stages of perception.

Relevant experimental data would be desirable. If face structure is processed within a module, the degree of encapsulation of that module could be reflected in the requirements for N374 generation.

In this section, we describe an experiment where we substitute the incomplete face context with the name of the face. One possibility is that the name can penetrate the hypothetical face-processing module and activate the appropriate face-structural codes. This will occur if the module is not informationally encapsulated. The other possibility is that the name cannot penetrate the module. If, as we believe, N374 reflects the preactivation of face structural codes, then the presence of this component when a name context is presented is a test of which of the two alternatives is true.

An additional sample of ten subjects who were familiar with both the faces and the names corresponding to the stimulus material participated in the study. Two types of context (Figure 13.1) were used: the incomplete faces used in the experiments described above (Figure 13.1A) and the written names associated with those faces (Figure 13.1B). The subjects again had to decide if eye/eyebrow fragments superimposed on the context were congruent or incongruent. The time between context onset and eye/eyebrow fragment onset was limited to 1 sec. The number of trials was doubled by pairing each incomplete face-eye/eyebrow fragment with a duplicate trial where the incomplete face was replaced by the corresponding written name. Then, the order of the trials was randomly shuffled.

For the incomplete face context, the discrimination performance of this group closely resembled that of similar groups (Figure 13.10). When the name context was used, the mean d' diminished significantly ($t = -4.5$, df = 9, $p < 0.0014$), although it was still significantly different from zero ($t = 15.2$, df = 9, $p < 0.00001$). The response bias shifted significantly ($t = 5$, df = 9, $p < 0.0008$). The subjects in this case were more cautious in responding that a match had occurred. Obviously, the task was more difficult when the prime and test items were in different information formats. In spite of the lower value of d' for verbal context, it is still higher than that obtained for subjects not familiar with the faces.

As before, only correctly classified trials were included in the ERP averages (one subject was excluded because of excessive artifacts). The pattern of ERP results associated with each type of priming is very interesting. In the case of incomplete face context, a clearly defined N374 was again obtained (Figure 13.11). The scalp distribution of this component is similar to that found in the previous groups, with the largest response at the right occipital site.

The ERPs obtained with verbal context displayed a different pattern of results (Figure 13.12). The exogenous components presented a morphology different from that obtained in other groups, a fact obviously related to the gross physical changes in the stimulus background. A large, apparently endogenous negativity was associated with incongruent eye/eyebrow fragments. This negativity had a slower decrease toward the maximum and did not present a sharply defined peak. The maximum amplitude for this negativity was found at vertex. The response was smaller at occipital sites.

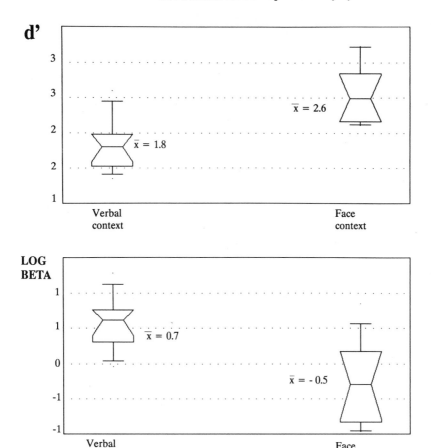

FIGURE 13.10. Box plots of the values of d' and Log beta for the experiment with two types of context (ten subjects familiar with the faces participated). *Left,* Values when a face context were presented. *Right,* Values when a name was presented as a context.

Measurements of the peaks of these negativities in the individual difference waveforms show that the mean latency of the negative peak for the verbal context was 249 msec later than the mean latency for the facial context. This effect was significant ($t = 9.265$, df = 8, $p < 0.0001$). Also, the standard deviation of the latency for the negative peak associated with incongruent face completions was almost twice as large for the verbal context, ($\sigma = 105$ msec) than for the facial context ($\sigma = 58$ msec). The longer latency and increased jitter of the negative peak may reflect the fact that context and test item belonged to different information domains. If the information processing transactions that took place between context and test item required a common representational format, then some sort of translation process was necessary.

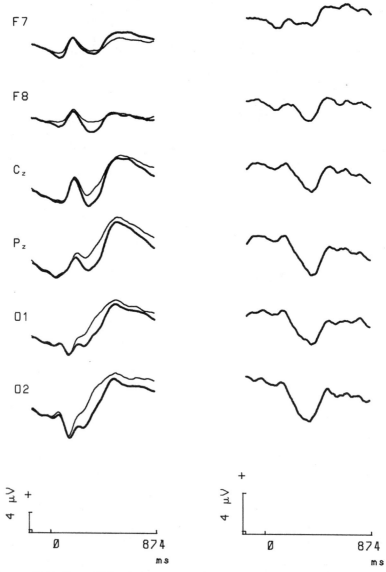

FIGURE 13.11. Group ERPs obtained with facial context. Subjects who were familiar with both the faces and the corresponding names had to indicate the type of completion for each trial. Names were not reported. Context presentation time was limited to 1 sec. *Left,* ERPs associated with congruent face completions (*thin line*) are overlaid on the ERPs associated with incongruent face completions (*thick line*). *Right,* Difference waveforms obtained by subtracting the ERPs for the two conditions.

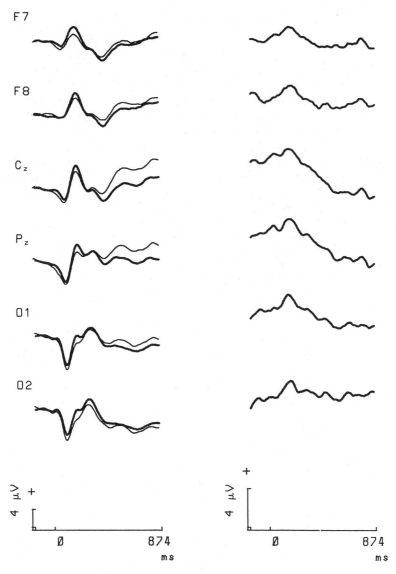

FIGURE 13.12. Group ERPs obtained when verbal context was used in the same group of subjects as in Figure 13.11. *Left,* ERPs associated with congruent face completions (*thin line*) are overlaid on the ERPs associated with incongruent face completions (*thick line*). *Right,* Difference waveforms obtained by subtracting the ERPs for the two conditions.

A three-way repeated-measures ANOVA was performed on the amplitudes for the time window from 250 to 450 msec in the case of face context ERPs and from 500 to 750 msec in the case of face context ERPs. The main effects of congruity ($F(1,8) = 16.9, p < 0.003$) and site ($F(2.2, 17.6) = 3.56, p < 0.05$) were significant. However, the amplitudes of the negativities in the two types of context were not different. The topographical change of the incongruence negativity with context type was tested on measurements of the difference waveforms for the corresponding time windows. Each observation vector (sites were components) was normalized to a length equal to 1.0. The interaction of site and type of context was significant ($F(2.9, 23.3) = 3.1, p < 0.05$), which confirms the conclusion that the scalp distribution of the incongruence negativity changes according to the type of context used.

These results suggest that, at least for the conditions of our experiment, the name cannot produce the same effects as the incomplete face. Although a late negativity is associated with eye/eyebrow fragments that are incongruent with the names, the topography of this negativity does not correspond with that of N374. The amplitude of this component is very small at the right occipital site. In fact, the scalp distribution of the negativity in this case is very similar to that previously described for the linguistic N400 paradigm, with a maximum at vertex. The failure to produce N374 could be due to many factors; however, the really interesting finding is that when an incongruence negativity occurs it is delayed in latency and with topography similar to N400. The occurrence of another type of negativity is more informative than a failure to produce N374.

If the name penetrated the face-processing module and preactivated face-structural codes, then N374 would be expected. Instead, what apparently occurs is that the eye/eyebrow fragments are recoded and person-semantic nodes are activated. In this case, the incongruence created by the experiment would occur within the general semantic system, leading to the generation of N400. This would certainly be consistent with the topography of the negativity found in this experiment.

Thus, the negativity found when names were used as context can perhaps be explained by the "leakage" of the incongruity to the semantic system. This interpretation is consistent with the data on face repetition priming. Faster responses after a congruent name, rather than an incongruent name, were found only when the response involved naming the face (Bruce and Valentine, 1985). When the response was based only on structural information (as in the familiar versus nonfamiliar decision), no priming for faces after name presentation was found (Bruce and Valentine, 1985; Ellis et al., 1987a). This finding points to a locus outside of the face-processing module for name influences. It does not mean that under appropriate conditions, say with mental image generation, a name could not activate facial structure codes. Yet, that activation would probably require time-consuming translation operations. Image generation seems to be a gradual constructive process (Kosslyn, 1988). Encapsulation may be a matter of degree, measured in terms of effort.

13.7 Final Comments

Natural stimuli, such as faces and words (but especially faces), have several undesirable properties as experimental material, of which the most important is the difficulty in establishing a tight control on the information subjects can extract from them. Yet, the advantages outweigh the disadvantages. Current ERP research on long-term memory would not be possible with the simpler and more controlled (but artificial) stimuli of the first decades of cognitive electrophysiology. We believe that the evidence presented in this chapter, together with the Smith and Halgren (1986) and the Barrett et al. (1988) studies, demonstrates that incongruence negativities related with recognition priming for faces exist. Our findings differ from previous studies because they provide evidence that associative priming can be produced for structure within faces. The data indicate that N374 is closely related to knowledge about face structure.

The N374 effect is apparently robust, appearing in the four groups of subjects described in this chapter who were familiar with the faces. The different pieces of evidence presented suggest that N374 is similar to N400, but is a different component. The topographical differences are too great for identity. Part of the second negativity described by Barrett et al. (1988), which is larger for familiar faces and has a widespread distribution and central maximum, is perhaps more similar to N400. The nature of the task used by these authors (identity matching) could certainly lead to activation of person-semantic and name codes for familiar faces. When our subjects were forced to use verbal codes, as in the name context experiment, the negativity associated with incongruent face parts was more similar to N400. The maximum of N374 is located over the right occipital and centroparietal regions. The similar topography of N445 described by Smith and Halgren (1986) suggests that this component is related to N374.

Component scalp topography is not an infallible guide to generator localization. Responses lateralized to one side of the scalp may originate in the opposite hemisphere (Regan, 1989). However, scalp distributions can direct our speculations and generate testable hypotheses. Neuropsychological studies have indicated a special role for the right hemisphere in face processing (reviewed by Rhodes, 1985). Most studies report that unilateral right-sided brain damage impairs face recognition more than left-sided damage, although performance remains above chance. The recognition of familiar faces is most affected by lesions in the right occipitotemporal region. For the complete inability to recognize previously familiar faces (prosopagnosia) to occur, brain damage must affect both hemispheres, always with a right occipitotemporal lesion. In these cases, location of the damage on the left side is more variable (Damasio et al., 1982; Meadows, 1974). Evidence that, in at least some types of prosopagnosia, memory associations, not basic perceptual mechanisms, are affected is presented by Campbell et al. (1986b). This view is also favored by Jeeves (1984).

Studies (Levy et al., 1972) in patients with split brains (disconnected hemispheres) have used chimeric figures consisting of half of one face to the right of the fixation point and half of another face to the left (due to the anatomy of the visual system, each half-face projects to the hemisphere on the opposite side). When the patient must point to the perceived photograph in a sample, the right hemisphere displays a clear advantage. The face reported as seen is that projected to the right. If the name is the required response, the task is much more difficult, and a slight left hemisphere advantage appears, apparently based on salient easily verbalized features.

Studies with lateralized tachistoscopic presentation of faces to different visual hemifields suggest that both hemispheres are capable of processing faces, each applying different strategies. Thus, specific experimental conditions may allow one hemisphere to be more efficient than the other. The right hemisphere has an advantage for processing faces, especially when global properties and face-structural information are important (Marzi et al., 1985; Young et al., 1985b). The advantage may pass to the left hemisphere if emphasis is on local detail and easily verbalized individual features (Sergent, 1987).

In particular, the right hemisphere advantage in this type of study is greater when the face stimuli have to be compared to a memory representation, as in familiar versus nonfamiliar tasks (Marzi et al., 1985; Young et al., 1985b) and when low spatial frequencies dominate the images (Sergent, 1985). Thus, our experimental conditions would seem to favor a right hemisphere advantage, since the congruency judgments were apparently based on memory for specific faces and the images were of relatively low spatial resolution.

In light of these neuropsychological results, we advance the hypothesis that N374 is somehow related to the face-recognition processors, with a postulated location in the posterior regions of the right hemisphere. This hypothesis, suggested by the scalp topography, is consistent with the idea that N374 is associated with priming effects within the facial memory domain.

One problem that has been discussed much in the neuropsychological literature (Diamond and Carey, 1986; Jeeves, 1984; Rhodes, 1985) is whether the perceptual and memory mechanisms for faces are shared with other visual and spatially complex stimuli or whether these mechanisms are special and only dedicated to face processing. The issue has not been resolved, and exactly the same question must be addressed about N374. Discrimination between other objects (kinds of birds, individual cows) that belong to large sets of similar exemplars, differing in subtle features, is sometimes also impaired in prosopagnosia (Jeeves, 1984). If incongruence negativities equivalent to N374 can be produced for other categories of visually complex objects (e.g., interchanging parts of the pictures representing the objects) in individuals with the appropriate knowledge, then perhaps this issue can be experimentally addressed in normal subjects. The scalp topography of such an incongruence negativity can be compared to that of N374, and a discrepancy would suggest independent memory systems.

In fact, a late negative component has been observed when comparing the ERPs of drawings of objects with drawings of nonobjects in an object-reality decision task (Campbell et al., 1986a). The nonobjects were created by tracing parts of drawings of real objects and regularizing the resultant figures. Although these results are not directly comparable to ours since different electrode sites, stimulus generation, and stimulus presentations rules were used, it is interesting to note that the negative component is apparently larger at frontocentral sites.

In conclusion, we return to the organization of long-term memory. One side of the coin is that several parallels between memory for words and memory for faces are worth exploring (Barrett et al., 1988). The evidence from repetition priming experiments (behavioral and ERP) suggests that face recognition units could be the basis of memory for faces. Our results, indicating associative priming between the components of familiar faces, can be construed as evidence that facial memory is organized as an interactive network, with face components as nodes (see Pinker, 1984). The review of data from other sources does lead to the conclusion that neither holistic templates nor vector-like feature lists are appropriate models for facial memory representations (Pinker, 1984; Sergent, 1987).

The other side of the coin is that the different properties of N400 and N374 support the notion of independent memory systems for semantic/linguistic information and face structure. The thrust of this chapter is not to demonstrate that faces and words are processed through the same mechanisms. The analogies between priming in face processing and priming between words should not be drawn too far. In fact, the whole argument for modular organization of mind is linked to considerations of the evolutionary advantages of specialization. Selective evolutionary pressure should provide optimal mechanisms for processing specific classes of stimuli. However, the same evolutionary considerations suggest multiple replication and subsequent divergence of brain systems as a mechanism for the origin of modules. A common ancestor would produce a " family resemblance" among different modules. It is not clear whether the number of modules or separate memory systems is as great as suggested by some neuropsychological studies. Perhaps the use of ERP techniques can test some of the finer partitions proposed for memory.

References

Barrett, S.E., Rugg, M.D., Perrett, D. (1988): Event-related potentials and the matching of familiar and unfamiliar faces. *Neuropsychologia*, 26, 105–117

Beaumont, J.G. (1983): Methods for studying cerebral hemispheric function. In: *Functions of the right cerebral hemisphere.* Young, A.W. (ed.). London: Academic Press

Bruce, V. (1983): Recognizing faces. *Phil. Trans. Roy.* 302B, 423–436

Bruce, V., Valentine, T. (1985): Identity priming in the recognition of familiar faces. *Br. J. Psychol.* 76, 373–383

Bruce, V., Valentine, T. (1986): Semantic priming of familiar faces. *Q. J. Exp. Psychol.* 38A, 125–150

Bruce, V., Young, A. (1986): Understanding face recognition. *Br. J. Psychol.* 77, 305–327

Campbell, K.B., Karam, A.M., Noldy-Cullum, N.E. (1986a): Event-related potentials in a lexical and object decision task. In: *Proceedings of the Eighth International Conference on Event-Related Potentials of the Brain.* Rohrbaugh, J.W., Johnson Jr., R., Parasuraman, R. (eds.)

Campbell, R., Landis, T., Regard, M.(1986b). Face recognition and lipreading. A neurological dissociation. *Brain* 109, 509–521

Caramazza, A. (1988): Some aspects of language processing revealed through the analysis of acquired aphasia: The lexical system. *Ann. Rev. Neurosci.* 11, 395–421

Carey, S., Diamond, R. (1977): From piecemeal to configurational representation of faces. *Science* 195, 312–314

Carroll, P., Slowiackzek, M.L. (1986): Constraints on semantic priming in reading: a fixation time analysis. *Memory and Cognition* 14, 509–522

Churchland, P.S., Sejnowski, T.J. (1988): Perspectives on cognitive neuroscience. *Science* 242, 741–745

Cohen, N.J. (1984): Preserved learning capacity in amnesia: Evidence for multiple memory system. In: *Neuropsychology of memory.* Butters, N., Squire, L.R., (eds.). New York: Guilford Press, pp. 83–103

Collins, A.M., Loftus, E.F. (1975): A spreading activation theory of semantic processing. *Psychol. Rev.* 82, 1–8

Damasio, A.R., Damasio, H., Van Hoesen, G.W. (1982): Prosopagnosia: Anatomical basis and behavioral mechanisms. *Neurology* 32, 331–341

Dannenbring, G.L., Briand, K. (1983): Semantic priming and the word repetition effect in a lexical decision task. *Can. J. Psychol.* 36, 435–444

Desimone, R., Albright, T.D., Gross, C.G., Bruce, C. (1984): Stimulus-selective properties of inferior temporal neurons in the macaque. *J. Neurosci.* 4, 2051–2062

Diamond, R., Carey, S. (1986): Why faces are and are not special: An effect of expertise. *J. Exp. Psychol.* 115, 107–117

Donnenwerth-Nolan, S., Tanenhaus, M.K., Seidenberg, M.S. (1981): Multiple code activation in word recognition: Evidence from rhyme monitoring. *J. Exp. Psychol.* 7, 170–180

Ellis, A.W., Young, A.W., Flude, B.M., Hay, D.C. (1987a): Repetition priming of face recognition. *Q. J. Exp. Psychol.* 39A, 193–210

Ellis, A.W., Hay, D.C., Young, A.W. (1987b): Modelling the recognition of faces and words. In: *Modelling cognition.* Morris, P.E. (ed.). Chichester: Wiley, pp. 269–297

Fischler, I., Bloom, P.A., Childers, D.G., Roucos, S.E., Perry, N.W. (1983): Brain potentials related to stages of sentence verification. *Psychophysiology* 20, 400–409

Freedman, J., Harber, P.N. (1974): One reason why we rarely forgot a face. *Bull. Psychonomic Soc.* 3, 2

Fodor, J.A. (1983): *The modularity of mind.* Cambridge, MA: MIT Press/Bradford

Gross, C.G., Rocha-Miranda, C.E., Bender, D.B. (1972): Visual properties of neurons in inferotemporal cortex of the macaque. *J. Neurophysio.* 35, 96–111

Grusser, O.J. (1984): Face recognition within the reach of neurobiology and beyond it. *Hum. Neurobiol.* 3, 183–190

Haig, N.D. (1984): The effect of feature displacement of face recognition. *Perception* 13, 505–512

Haig, N.D. (1985): How faces differ. A new comparative technique. *Perception* 14, 601–615

Haig, N.D. (1986a): Exploring recognition with interchanged facial features. *Perception* 15, 235–247

Haig, N.D. (1986b): High-resolution facial feature saliency mapping. *Perception* 15, 373–386

Harbin, T.J., Marsh, G.R., Harvey, M.T. (1984): Differences in the late components of the event-related potentials due to age and to semantic and non-semantic tasks. *Electroenceph. Clin. Neurophysiol.* 59, 489–496

Harvey, M.T., Marsh, G.R. (1983): Effect of semantic category typicality on amplitude and latency of N400. *Psychophysiology* 20, 444–445 (abstr.)

Hay, D.C., Young, A.W. (1982): The human face. In: *Normality and pathology in cognitive functions.* Ellis, A.W., (ed.). London: Academic Press, pp. 173–202

Henderson, L., Wallis, J., Knight, D. (1984): Morphemic structure and lexical access. In: *Attention and performance X.* Bouwhuis, D., Bouma, H. (eds.). London: Lawrence Erlbaum Associates

Hillinger, M.L. (1980): Priming effects with phonemically similar words: The encoding bias hypothesis reconsidered. *Memory and Cognition* 8, 115–123

Hillyard, S.A., Picton, T. (1987): Electrophysiology of cognition. In: *Handbook of neurophysiology.* Plum, F. (ed.). Bethesda, MD. American Physiology Society

Hillyard, S.A., Woods, D.L. (1979): Electrophysiological analysis of human brain function. In: *Handbook of Behavioral Neurobiology: vol. 2.* New York: Plenum, pp. 345–378

Jeeves, M.A. (1984): The historical roots and recurring issues of neurobiological studies of face perception. *Hum. Neurobiol.* 3, 191–196

Kosslyn, S.M. (1988): Aspects of a cognitive neuroscience of mental imagery. *Science* 240, 1621-1626

Kutas, M., Hillyard, S.A. (1980): Reading senseless sentences: Brain potentials reflect semantic incongruity. *Science* 207, 203–205

Kutas, M., Hillyard, S.A. (1982): The lateral distribution of event-related potentials during sentences processing. *Neuropsychologia* 20, 579–590

Kutas, M., Hillyard, S.A. (1983): Event-related brain potentials to grammatical errors and semantic anomalies. *Memory Cognition* 11, 539–550

Kutas, M., Hillyard, S.A. (1984): Brain potentials during reading reflect word expectancy and semantic association. *Nature* 307, 161–163

Kutas, M., Van Petten, C. (1987): Event-related brain potential studies of language. In: *Advances in psychophysiology.* Ackles, P.K., Jennings, J.R., Coles, G.H. (eds.). Greenwich, CT: JAI Press

Kutas, M., Van Petten, C., Besson, M. (1988): Event-related potential asymmetries during the reading of sentences. *Electroenceph. Clin. Neurophysiol.* 69, 325–330

Kutas, M., Neville, H., Holcomb, P.J. (1987): A preliminary comparison of the N400 response to semantic anomalies during reading, listening and signing. The London Symposia. *Electroenceph. Clin. Neurophysiol.* Suppl 39, 325–330

Levy, J., Trevarthen, C., Sperry, R.W. (1972): Perception of bilateral chimaeric figures following hemispheric deconnexion. *Brain* 95, 61–78

Light, L.L., Kayra-Stuart, F., Hollander, S. (1979): Recognition memory for typical and unusual faces. *J. Exp. Psychol. Hum. Learning Memory* 5, 212–228

Marr, D. (1982): *Vision.* San Francisco: Freeman

Marshall, J.C. (1984): Multiple perspective on modularity. *Cognition* 17, 209–242

Marzi, C.A., Tassinari, G., Barry, C., Grabowska, A. (1985): Hemispheric asymmetry in face perception task of different cognitive requirement. *Human Neurobiol.* 4, 15–20

Massaro, D. (1975): *Understanding language.* New York: Academic Press

McCallum, W.C., Farmer, S.F., Pocock, P.K. (1984): The effects of physical and semantic incongruities on auditory event related potentials. *Electroenceph. Clin. Neurophysiol.* 59, 477-488

McCarthy, G.A., Wood, C.C. (1985): Scalp distributions of event-related potentials: an ambiguity associated with analysis of variance models. *Electroenceph. Clin. Neurophysiol.* 60, 203–208

McCarthy, R.A., Warrington, E.K. (1988): Evidence for modality-specific meaning system in the brain. *Nature (London)* 334, 428–430

McClelland, J.L., Rumelhart, D.E. (1981): An interactive activation model of context effects in letter perception: Part 1. An account of basic findings. *Psychol. Rev.* 88, 375–407

Meadows, J.C. (1974): The anatomical basis of prosopagnosia. *J. Neurol. Neurosurg. Psychiatry* 37, 489–501

Meyer, D., Schvanevelt, R. (1971): Facilitation in recognizing pairs of words: Evidence of dependence between retrieval operations. *J. Exp. Psychol.* 90, 227–234

Morton, J. (1969): Interaction of information in word recognition. *Psychol. Rev.* 76, 165–178

Morton, J. (1979): Facilitation in word recognition: Experiments causing change in the logogen model. In: *Processing of visible language.* Kolers, P.A., Wrolstad, M.E., Bound, M. (eds.). New York, Plenum Press, pp. 259–268

Murrell, G. A., Morton, J. (1974): Word recognition and morphemic structure. *J. Exp. Psychol.* 102, 963–968

Neville, H. (1985): Biological constraints on semantic processing: A comparison of spoken and signed languages. *Psychophysiology* 22, 576 (Abstr.)

Norris, N. (1986): Word recognition: Context effects without priming. *Cognition* 22, 93–136

Parks, T.E., Coss, R.G., Coss, C.S. (1985): Thatcher and the Cheshire cat: Context and the processing of facial features. *Perception* 14, 747–754

Perrett, D.I., Rolls, E.T., Caan, W. (1982): Visual neurons responsive to face in the monkey temporal cortex. *Exp. Brain Res.* 47, 329–342

Perrett, D.I., Smith, P.A.J., Potter, D.D., Mistlin, A.J., Head, A.S., Milner, A.D., Jeeves, M.A. (1984): Neurones responsive to faces in the temporal cortex: Studies of functional organization, sensitivity to identity and relation to perception. *Hum. Neurobiol.* 3, 197–208

Perrett, D.I., Mistlin, A.J., Potter, D.D., Smith, P.A.J., Head, A.S., Chitty, A.J., Broenniman, R., Milner, A.D., Jeeves, M.A. (1986): Functional organization of visual neurones processing face identity. In: *Aspects of face processing.* Ellis, A.D., Jeeves, M.A., Newcombe, F., Young, A. (eds.). Dordrecht, Nijhoff pp. 187–198

Picton, T.W., Hink, R.F., Perez-Abalo, M., Dean Linden, R., Weins, A.S. (1984): Evoked potentials: How now?. *J. Electrophysiol. Technol.* 10, 177–221

Polich, J. (1985b): N400s from sentences, semantic categories, number and letter strings? *Bull. Psychonomic Soc.* 23, 361–364

Squire, L.R. (1987): Memory: Neural organization and behavior. In: *Handbook of physiology—The nervous system, sec V.* Plum, F. (ed.). Bethesda, MD American Physiology Society

Putnam, H. (1984): Models and modules. *Cognition* 17, 209–242

Regan, D. (1989): *Human brain electrophysiology: Evoked potentials and evoked magnetic fields in science and medicine.* New York: Elsevier

Rhodes, G. (1985): Lateralized processes in face recognition. *Br. J. Psychol.* 76, 249–271

Ritter, W., Ford, J., Gaillard, A.W.K., Russell Harter, M., Kutas, M., Naatanen, R., Polich, J., Renault, B., Rohrbaugh, J. (1984): Cognition and event related potentials: I. The relation of negative potentials and cognitive processes. *NY Acad. Sci.* 425, 24–38

Rolls, E.T., Baylis, G.C. (1986): Size and contrast have only small effects on the responses to faces of neurons in the cortex of the superior temporal sulcus of the monkey. *Exp. Brain Res.* 65, 38–48

Rosenfeld, S.A., Van Hoesen, G.W. (1979): Face recognition in the rhesus monkey. *Neuropsychologia* 17, 503–509

Rugg, M.D. (1984a): Event-related potentials in phonological matching task. *Brain Lang.* 23, 225–240

Rugg, M.D. (1984b): Event-related potentials and the phonological processing of words and non-words. *Neuropsychologia,* 22, 435-443

Rugg, M.D. (1985): The effects of semantic priming and word repetition on event related potentials. *Psychophysiology* 22, 642–647

Rugg, M.D. (1987a): What can ERPs tell us about language laterality? Fourth International Conference on Cognitive Neuroscience, Paris-Dourdan, France

Rugg, M.D. (1987b): Dissociation of semantic priming, word and non-word repetition by event related potentials. *Q. J. Exp. Psychol.* 39A, 123–148

Rugg, M.D., Barret, S.E. (1987): Event-related potentials and the interaction between orthographic and phonological information in a rhyme-judgment task. *Brain Lang.* 32, 336–361

Scarborough, D.L., Cortese, C., Scarborough, H.S. (1977): Frequency and repetition effects in lexical memory. *J. Exp. Psychol. Hum. Perception Performance* 3, 1–17

Sergent, J. (1985): Influence of task and input factors of hemispheric involvement in face processing. *J. Exp. Psychol. Hum. Perception* 11, 846–861

Sergent, J. (1987): Face perception: Underlying process and hemispheric contribution. In: *Perspectives in cognitive neuropsychology.* Denes, G., Semenza, C., Bisiacchi, P., Andreewsky, A. (eds.). Hillsdale, NJ: Lawrence Erlbaum Associates

Seymour, P.H.K. (1987): Word recognition processes. An analysis based on format distortion effects. In: *Cognitive approaches to reading.* Beech, J.R., Colley, A.M. (eds.). New York: John Wiley and Sons, pp. 31–57

Shallice, T. (1984): More functionally isolable subsystems but fewer "modules"? *Cognition* 17, 243–252

Sherry, D.F., Schacter, D.L. (1987): The evolution of multiple memory systems. *Psychol. Rev.* 94, 439–454

Shulman, H.G., Hornak, R., Sanders, E. (1978): The effects of graphemic, phonemic and semantic relationships on access to lexical structures. *Memory and Cognition* 6, 115–123

Smith, M.E., Halgren E. (1986): ERPs elicited by familiar and unfamiliar faces. In: *Proceedings of the Eighth International Conference on Event-Related Potentials of the Brain.* Rohrbaugh, J.W., Johnson Jr., R., Parasuraman, R. (eds.)

Smith, M.C., Theodor, L., Franklin, P.E. (1983): The relationship between contextual facilitation and depth of processing. *J. Exp. Psychol. Learning Memory Cognition* 9, 697–712

Smith, M.E., Stapleton, J.M., Halgren, E. (1986): Human medial temporal lobe potentials evoked in memory and language tasks. *Electroenceph. Clin. Neurophysiol.* 63, 145–159

Squire, L.R. (1987): Memory: Neural organization and behavior. In: *Handbook of physiology—The nervous system, sec V.* Plum, F. (ed). Bethesda, MD American Physiology Society

Stuss, D.T., Picton, T.W., Cerri, A.M. (1986): Searching for the names of pictures: An event-related potentials study. *Psychophysiology* 23, 215–223

Stuss, D.T., Picton, T.W., Cerri, A.M. (1988): Electrophysiological manifestations of typicality judgment. *Brain Lang.* 33, 260–272

Swets, J.A. (1964): *Signal detection and recognition by humans observers.* New York: John Wiley and Sons

Warren, C., Morton, J. (1982): The effects of priming on picture recognition. *Br. J. Psychol.* 73, 117–129

Yin, R.K. (1969): Looking at upside-down faces. *J. Exp. Psychol.* 81, 141–145

Young, A.W. (1988): Functional organization of visual recognition. In: *Thought without language.* Weiskrantz, L. (ed.). Oxford: Oxford University Press

Young, A.W., Hay, D.C., McWeeny, K.H., Flude, B.M., Ellis, A.W. (1985a): Matching familiar and unfamiliar faces on internal and external features. *Perception* 14, 737–746

Young, A.W., Hay, D.C., McWeeny, K.H., Ellis, A.W., Barry, C. (1985b): Familiarity decisions for faces presented to the left and right cerebral hemispheres. *Brain Cognition* 4, 439–450

Young, A.W., Hellawell, D., Hay, D.C. (1987): Configurational information in face perception. *Perception* 16, 747–759

Part V

Cognitive Studies in Animals

14

Spatial Memory in Animals

JAN BURES AND OLGA BURESOVA

14.1 Introduction

We are living in an epoch of fundamental conceptual changes. This statement applies not only to world politics, military doctrines, and economic models but also to scientific concepts. In neurosciences this fundamental change can be best illustrated by the rapidly increasing influence of cognitive theories of behavior. Whereas the founding fathers of modern physiological psychology—Pavlov (1927) and Watson (1919)—purged their theories of such terms as memory, goals, expectations, and the like, present-day neuroscience has returned respectability to the ideas espoused by Beritashvili (1971) and Tolman (1932), who endowed animals with the capability to form a neural representation of the outside world and to use this representation for the control of behavior. The change of attitude was not brought about by persuasive arguments but rather by the different philosophical context in which this research is pursued. The antimentalistic position prevailing at the turn of the century served the noble aim of achieving a strictly objective explanation of behavior, leaving no room for speculative constructs based on introspection. The situation is quite different now when computer sciences modeling the sensorimotor control of robots show that the flexible solution of the task requires the application of such cognitive principles as setting the goal of action and selecting the strategy for achieving it. What appeared to be a misleading speculation a century ago is now a well-justified assumption that has inspired much of the current research into the neural mechanisms of learning and memory.

Although cognitive mechanisms probably participate even in the simplest behaviors, including habituation or classical conditioning, they can be most convincingly demonstrated in situations that cannot be solved by direct response to perceptible stimuli but require additional knowledge, a memory-stored model of the outside world with cognitive maps of the surrounding space, and lists of expected consequences of various actions. Particularly, locomotion through complex environments was always considered a convincing demonstration of goal-directed behavior. This was the reason mazes of various design have formed the standard equipment of psychological laboratories during a century of research.

Their use reached a peak in the 1920s and 1930s, but declined later with the advent of operant conditioning, discrimination learning, and passive avoidance reactions, and with the failure to resolve the controversies connected with the interpretation of maze experiments. Hull (1943) asserted that the animal learns the correct route through the maze, i.e., a sequence of responses to stimuli appearing at different points of the maze, usually represented by a succession of left and right turns. This S-R view was contested by Tolman (1932) who suggested that the animal can solve the problem correctly because it has a cognitive map of the maze and can use it for finding the way to the goal from any point of the maze. In spite of concentrated efforts to prove one of these views by specific experiments, the controversy remained unresolved. The last decade marked a sharp increase in maze studies mainly connected with an increased interest in ethology and naturalistic approaches and with the introduction of new techniques and procedures that considerably strengthened cognitive interpretation of maze behavior.

14.2 Olton's Radial Maze

The first breakthrough was the introduction of the radial maze for the study of spatial working memory by Olton and Samuelson (1976). The radial maze is an elaboration of the earlier Y or + mazes (Figure 14.1), with the number of equidistant arms extending from the central platform increased to 6, 8, 12, 17, or 24. A 12-arm radial maze had been used for the study of egocentric orientation earlier (Potegal, 1969) but more important than the design of the apparatus was the task introduced by Olton and Samuelson (1976). The animal had to collect n food pellets placed in n individual arms of the maze. Optimum performance required the animal to visit each channel only once, i.e., not to revisit the channels from which the food had already been removed. Several distinctive features of the task are immediately obvious.

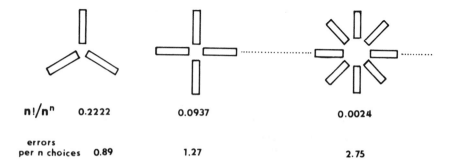

| $n!/n^n$ | 0.2222 | 0.0937 | 0.0024 |

| errors per n choices | 0.89 | 1.27 | 2.75 |

FIGURE 14.1. Evolution of radial mazes: Y maze, + maze, eight-arm maze. The numbers denote the probability of visiting n different arms in n random choices (first row) or the average number of repeated visits of the already visited arms in n random choices (second row).

There is no fixed position of the start and goal in the radial maze. The animal is initially started from the central platform with each arm being a goal, but in the course of the experiment the last entered goal becomes a start, the earlier entered goals turn into blind alleys, and the number of available goals decreases with each correct choice by one.

The task resembles the animal's behavior during foraging or territory patroling and is thus well prepared by the genetically transmitted evolutionary experience of the species.

Mathematically, the task can be modeled as choice with return, and the performance of the animal can be tested against the outcome predicted for random choice. The probability of n successive correct choices is $n!/n^n$, which is 0.0024 for $n = 8$. The expected number of errors (i.e., choices of arms that had already been visited in the particular trial) in the first eight choices in an eight-arm maze is 2.75; the experimentally established values range from 0.2 to 0.4.

Due to the regular geometry of the apparatus, the task can be solved successfully by a rat repeating a definite egocentrically oriented response; that is, turning after leaving the last entered maze arm always into the first arm to the left (right). In fact, a correct solution can be obtained with any integer distance between arms that is not contained in n (Figure 14.2). Already the first radial maze studies showed that rats do not systematically follow such simple or complex routes corresponding to the S–R solution of the task, but that they generate in successive trials different sequences of choices, thereby suggesting that they remember the already visited and/or the yet not entered maze arms. Of course, a maze can be solved by nonmemorial means, exemplified by Ariadne's thread, which helped Theseus find his way through the Kretan labyrinth. Although a rat passing through the maze leaves such a thread in the form of scent trails that may allow the animal to recognize the arms that have been visited, a number of control experiments demonstrated the limited contribution of smell to radial maze behavior (Buresova and Bures, 1981). Also, path integration may contribute to the correct solution of spatial problems (Mittelstaedt and Mittelstaedt, 1982), but only in the case of uninterrupted trials. When the animal is allowed to make the first half of choices, is then removed for several minutes or hours from the radial maze, and is only then permitted to complete the trial, a cognitive explanation remains the only alternative, with extramaze distal landmarks serving as cues for distinguishing individual arms.

14.3 Morris Water Maze

The second breakthrough came 5 years later when Morris (1981) introduced the water maze procedure. The apparently empty circular pool does not look like a maze, but there is a fixed goal represented by the hidden underwater platform and any part of the pool can be used as the start (Figure 14.3). The situation is ideally suited for testing the properties of the cognitive maps postulated by Tolman (1932) and by O'Keefe and Nadel (1978). Such a map is a neural representation of the

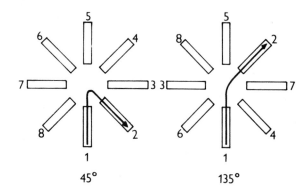

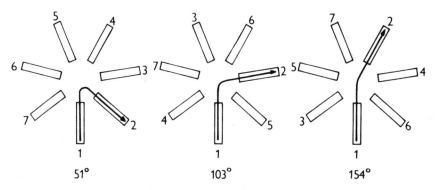

FIGURE 14.2. Noncognitive solutions of the eight-arm (*top*) and seven-arm (*bottom*) radial mazes. The numbers denote the sequences of arm visits with the indicated interchoice angle.

animal's environment, with the current position of the animal and the position of the goal determined by coordinates corresponding to relational properties of distal extramaze landmarks. When the orientation of the rat's cognitive map corresponds to the real situation and when the animal identifies correctly its own position and the position of the target on the map, then it can use the map to find the shortest trajectory between the two points. The efficiency of these cognitive operations is expressed by the length of the actual path measured in multiples of the ideal path. The evaluation of the experiments was greatly improved by the use of computerized videosystems (Buresova et al., 1985; Morris, 1984), which not only plot the animal's trajectory at 40-msec intervals but can also compute the time spent in the target area or in different quadrants of the pool. In interactive systems, the computer also controls the presentation of the escape platform, which is initially collapsed at the bottom of the pool and emerges only after the animal has spent a criterion time (e.g., 1.2 or 2.5 sec) in a predetermined target area (30 cm in

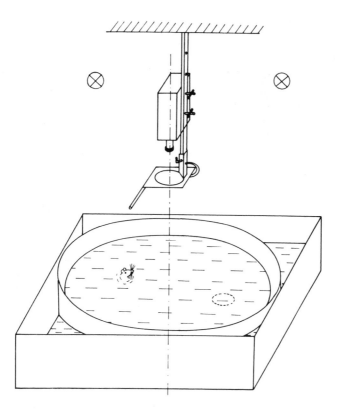

FIGURE 14.3. Morris water maze with the television camera input of the computerized videosystem.

diameter) above the platform. Use of such an "on demand" platform precludes random hitting of the target and requires that the animal start treading water and turning around when it reaches the critical area.

The results show the extreme proficiency of rats in this task, with the swim trajectory parameters (escape latency, path length, initial heading angle) approaching asymptotic values after only 8 to 16 trials. In particular, the ability of the rats to go straight from any start to the goal and to stay in the target area until the platform is raised shows convincingly that the animal knows where the target should be. Yet, it is much less clear how the cognitive map is implemented. O'Keefe and Nadel (1978) distinguished place learning based on the use of cognitive maps from taxon learning, characterized by locomotion toward directly perceptible, usually visible targets. The word "taxon" is unfortunate in this context because it seems to imply that the place navigation cannot be implemented by the mechanism of taxis. Vast literature on the orientation of animals shows that this is not the case. The so-called menotaxis describes locomotion not toward the target but at a definite angle to the orienting field or beacon. As pointed out by Tinbergen

(1952), two simultaneous menotaxes controlled by two orienting vectors may guide the animal toward a hidden target. Such navigation called mnemotaxis or farotaxis is well documented in wasps or bees (Cartwright and Collett, 1982). It is controlled by memory, which is reduced, however, to snapshot memory of the azimuths of at least two remote landmarks as seen from the vantage point of the target. When these landmarks are also seen from the starting point, the animal compares their azimuths with the remembered ones and swims in a direction that would reduce the difference between the perceived and remembered values. The two vectors are added, and the resulting sum points more or less directly toward the target. Similar results are obtained when the animal remembers the distance of at least two remote landmarks from the target (Wilkie and Palfrey, 1987). The distance can be measured by the size of the retinal projections of the landmarks or by their elevation above the horizon. The animal tends to swim toward the landmark that is too far and away from the landmark that is too close. Again, addition of the two vectors leads the animal to the target (Figure 14.4).

Comparison of the memorized target scene with the current sensory input can also be accomplished by cross-correlation, which approaches maximum (= 1)

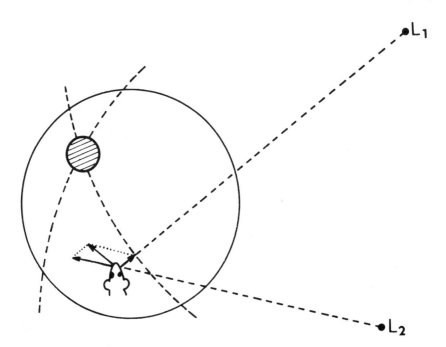

FIGURE 14.4. Distance reduction navigation in the Morris water maze. The interrupted circle segments indicate points equidistant from the visible extramaze landmarks, L_1 and L_2. The rat is too far from L_1 but too close to L_2. The amplitude of the corresponding vectors is proportional to the difference between the remembered target-landmark distance and the currently observed distance. The resulting vector, guiding the animal in the approximate direction of the target, is corrected continuously as the position of the rat changes.

when the animal has reached the target. The cross-correlation product is a scalar value, however, that does not allow direct orientation toward the target. The animal has to emit movements in different directions to assess how they have influenced the cross-correlation product and to select movements increasing the cross-correlation. In many respects, the situation resembles the so-called chemoklinotaxis (Shorey, 1972), i.e., the capability of organisms to find the source of smell, such as a pheromone emitter. The animal samples the olfactory stimulus while exploring the environment, establishes the concentration gradient, and moves against it until the source is found. It must be pointed out, however, that such distance–reduction navigation does not need a cognitive map and represents only an oriented movement in a stimulus–intensity gradient or in a gradient corresponding to the difference between memory image and sensory input (Buresova et al., 1988).

14.4 Kinship of Radial Maze and Water Maze Procedures

The radial maze and water maze are often conceived as devices predestining the spatial memory problem addressed by them. Thus, Olton's radial maze is usually employed in working memory studies and typically features appetitive motivation, multiple perceivable goals, and win-shift strategy. On the other hand, Morris' water maze examines reference memory and is characterized by aversive motivation, a single hidden goal, and win-stay strategy. In spite of the above differences the kinship of the two procedures can be demonstrated by numerous examples of their less conventional applications.

Olton (1983) used the radial maze for simultaneous examination of the working memory and reference memory: half of the arms were baited in the usual way, whereas the other arms never contained food. The rats gradually learned to ignore the empty arms (reference memory) and to visit only the baited ones in a particular trial (working memory). This arrangement made it possible to demonstrate in a single experiment differential effects of lesions or drugs on the two forms of spatial memory.

Buresova et al. (1985a) introduced an adversively motivated version of the radial maze task by placing a radial maze into a water tank (Figure 14.5). The animal was started from a central platform; upon submersion of the platform the animal had to swim to an underwater bench in one of the arms of the maze. After a short rest the bench was collapsed to the bottom of the pool (and remained collapsed until the end of the trial) while the animal returned to the central platform. After the latter had been lowered again, the rat had to find another arm with the bench still available. The trial was terminated after the animal found the last bench.

There are also intermediate stages in the geometry of the radial maze and water maze. By reducing the length of the arms and increasing the diameter of the central platform it is possible to arrive at a situation in which the arms degenerate to a set of circular targets at the circumference of a large open field (Figure 14.6). This approach was used by Barnes (1979) and by Schenk (1988) who trained rats to run

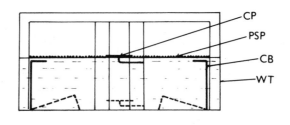

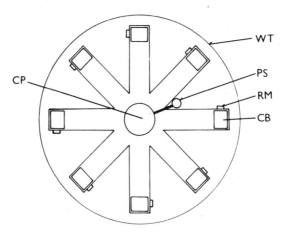

FIGURE 14.5. Radial water maze. *Top,* Cross-section of the maze. *Bottom,* Floor plan. The interrupted lines indicate the low position of the central platform and collapsed benches. CP-central platform; PSP-layer of polysterene pellets; CB-collapsible benches; WT-water tank; PS-pneumatic system for lifting the central platform; RM-releasing magnet controlling the position of the bench. Reprinted from Buresova et al. (1985): Radial maze in the water tank: an adversively motivated spatial working memory task. *Physiol. Behav.* 34: 1003–1005 with permission of Pergamon Press PLC.

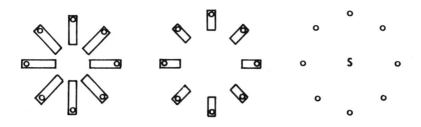

Olton, Samuelson (1976)   Barnes(1979)

FIGURE 14.6. From radial maze to open field maze. A radial maze can be converted into an open field maze by gradually shortening the maze arms and expanding the central platform.

from the center of the field to one of many holes at the circumference of or within the field. The target hole provided access to a tunnel connecting the open field with the home cage. This appetitively motivated dry version of the Morris water maze resembles in some respects the homing test developed by Mittelstaedt and Mittelstaedt (1982). These authors had a lactating gerbil dam with pups in a nest in the center of an open field. In complete darkness, one of the pups was removed from the nest and placed into an open cup at the periphery of the field. The mother was allowed to leave the nest, to search the field, to find the pup, and to carry it back to the nest, which she approached along a straight line obtained by integration of the complex trajectory generated during the search.

The radial maze can be used in a way resembling the single goal situation typical for the water maze: a single channel is baited and serves as the goal while all other channels can be used as starts. This arrangement was employed by Collier et al. (1982), who gave rats two radial maze trials separated by a brief interval. In the first trial, the animal was allowed to explore an eight-arm maze and to find a single baited channel. The animal was then removed from the apparatus, the same channel was rebaited, and the rat was allowed to find it again (win–stay strategy). After several days of training, the animals found the baited channel in the first trial after 4.5 choices on the average (a value corresponding to random choice without return in case of eight possibilities) but needed only 1.5 choices to find this channel in the second trial (Figure 14.7). The radial maze served in this case to implement a spatial delayed-matching-to-sample problem, another version of which was a year later proposed for the water maze by Morris (1983) and further modified by Panakhova et al. (1984) and Whishaw (1985). In the latter case, the rat was allowed to find a randomly hidden underwater escape platform in Trial 1 and to find it in the same place again after a suitable delay in Trial 2. Working memory enters this version of the water maze task at two levels: during the delayed matching-to-sample manifested by the difference between the Trial 1 and Trial 2 escape latencies, and during the search phase of Trial 1. The animal searching the hidden platform must remember the already covered parts of the pool in order to avoid repeated visits of the same location or leaving other areas unexplored. The win–shift principle helps the rat cover the possible platform locations in the most economical way. The same type of win–shift strategy governs the animal's patrolling of a familiar territory (Jucker et al., 1988) and provides the maximum amount of information with the minimum expenditure of time and effort.

The memory trace underlying the spatial delayed matching-to-sample in the Morris water maze can be formed in two phases of Trial 1: (a) during the active search of the hidden platform when the animal is examining the possible target locations and (b) during the 30-sec stay on the platform when the animal actively observes the extramaze cues from the vantage point of the target. Semenov and Bures (1989) assessed the relative contribution of these two components by comparing savings in Trial 2 when Trial 1 was performed under the following conditions (Figure 14.8): SP—conventional combination of active search and subsequent stay on the platform, S—elimination of the observation phase by covering the platform with an opaque cylinder as soon as the animal has climbed

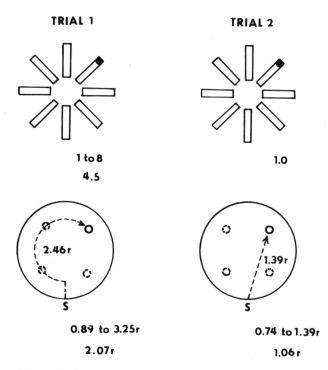

FIGURE 14.7. Spatial delayed matching-to-sample task in the radial maze (*top*) and in the water maze (*bottom*). The lowest, highest, or average number of visited arms (*top*) and the shortest, longest or average path (*bottom*) needed for finding a randomly located target when using the optimum search strategy in Trial 1. Trial 2 illustrates optimum retrieval results. r-radius of the water pool.

upon it, and P—elimination of the search phase by placing the animal directly onto the platform for the 30-sec observation period. The retrieval latency was about 40% of the initial search time under SP conditions, about 75% under S conditions, and about 60% under P conditions. Such result indicates that both active search and latent learning contribute almost equally to the formation of the spatial working memory in this task. The possibility of latent learning in the Morris water tank, first demonstrated by Sutherland and Linggard (1982), shows that entering new information into the cognitive map does not require active exploration of the environment, as postulated by O'Keefe and Nadel (1978).

The above examples show that two extreme forms of spatial tasks can be implemented with either apparatus: to remember a single location out of many (i.e., the spatial matching-to-sample) and to visit all locations from a list. The second task is obviously more complex than the first one because it implies not only recognition of individual locations but also their transient time labeling. An efficient search strategy in the water maze requires this type of spatial memory,

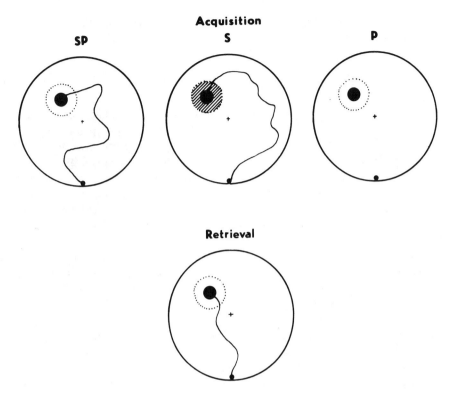

FIGURE 14.8. Two components contributing to the acquisition of the working memory version of the Morris water maze task. SP, search of the platform followed by a 30-sec stay on the platform. S, the platform component is eliminated by covering the rat as soon as it has reached the platform by an opaque cylinder. P, the search component is bypassed by placing the rat directly on the platform. Same retrieval test is used with all three acquisition conditions. Reprinted from Semenov and Bures (1989): Vestibular stimulation disrupts acquisition of place navigation in the Morris water tank test. *Behav. Neural Biol.* 51: 346–363 with permission of Academic Press.

but the task can also be solved in a noncognitive way by adhering to a definite route.

One of the technical problems involved in the tasks of the second type is how to reward completeness of the search. In the conventional radial maze, the reward for completeness is the receipt of all the pellets. The motivation is probably stronger when the animal receives a single pellet upon entering the last arm of the maze. A similar procedure in the water maze requires that the pool be divided into a number of equal segments (e.g., 25), the visits to which would be automatically recorded by the computerized tracking system. The platform is raised only after the last segment has been reached. Another possibility is to have several floating platforms in the pool that will not sustain the weight of the animal (Morris, 1984).

Only when the last of these platforms is visited is it locked and thus turned into a safe escape place.

14.5 Beyond Errorlessness

The solution to the problem of visiting a set of places in a most economic way cannot be restricted to avoiding repeated visits of the same points. Even an errorless trial in which each goal was visited only once can be solved in many alternative ways differing by the length of the path. The next higher step of cognitive efficiency is to generate a correct trajectory that is at the same time the shortest one; that is, it allows the correct solution with the minimum expenditure of energy. This situation, known as the "traveling salesman problem," is well amenable to mathematical formalization and computer modeling (Hopfield and Tank, 1985).

The radial maze is not suitable for the investigation of this problem because the long arms account for more than 80% of the path length, the variability of which is limited to movement over the central platform. Even this variability can be totally suppressed in the two-level radial maze where the animal must return after each choice below the central platform and to climb upon it through a central hole (Magni et al., 1979). Buresova (1980) used a multiple goal maze without the radial maze geometry. Twelve tubular segments equipped with one-way entrance and exit doors were scattered in an enclosed open field (Figure 14.9). The rat was allowed to visit all elements of this random maze and to collect a food pellet from each segment. The acquisition rate was almost the same as in the radial maze, and the small number of errors (one to two errors in the first 12 choices in the 12-goal random maze) indicated that spatial working memory does not depend on the regularity of maze design. The tubular segments are, however, not the ideal goals: the entrance and exit points are far from each other, and the orientation of the long axis of the tube is another confounding variable.

In a new version of the random maze the tubular segments were replaced with closed feeders mounted on heavy plastic cylinders (8 cm high, 6 cm in diameter), the tops of which were formed by a disk-shaped plate (Figure 14.10). Whenever a rat examined the feeder (a 1 cm wide and 0.5 cm deep hollow in the center of the disk covered by a plastic cap that could be lifted by a solenoid-operated lever), it had to lean against the disk and thus close a sensitive contact, indicating a visit to this particular feeder. The sequence of feeders visited during one trial was recorded, repeated visits of the same feeder were classified as errors, and completion of the search was rewarded by automatic opening of the last feeder. The eight cylinders were arranged in an open field to form different geometrical configurations (rectangular, hexagonal, scattered, linear, etc.). The rats were given ten trials per day, first with the rectangular configuration of goals and then with five different arrangements alternating in one daily session. From the very beginning of training the rats displayed good working memory: they made only 1.5 errors on the average in the first eight choices of a trial. Their performance

FIGURE 14.9. Random maze. Tubular segments of a radial maze are scattered over an enclosed open field. Reprinted from Buresova (1980): Spatial memory and instrumental conditioning. *Acta Neurobiol. Exp.* 40: 51–66 with permission of Acta Neurobilogiae Experimentalis.

improved with training and reached the asymptotic level of 1.0 errors per the first eight choices after five sessions. Switching from the regular rectangular configuration of goals to the alternation of several configurations of widely different geometry caused only negligible deterioration of performance, which rapidly returned to the asymptotic level of 1.2 errors per the first eight choices. The yield of errorless trials in the eight- goal random maze was nevertheless lower (below 20%) than in the eight- arm radial maze and made the analysis of corresponding trajectories impractical. Reduction of the number of goals to $n = 6$ increased the incidence of errorless trials to 60% and made it possible to examine them from the point of view of the traveling salesman problem.

Six points can be visited once and only once in 6! = 720 closed tours. However, as each tour can be started from a different point and traversed either clockwise or counterclockwise, the number of tours with different geometry is only n!/2n, i.e., 60 for $n = 6$. Figure 14.11 shows 9 of these 60 routes for a regular rectangular configuration of six points. The tours differ by path lengths, which are measured in arbitrary units, and range from the shortest (6.00) to the longest (10.12). The distribution of the 60 valid tours according to the path length is approximately Gaussian (Figure 14.12), with the average length being 8.30. The rats selected

FIGURE 14.10A. Open-field maze with feeders on cylindrical pedestals. A rat during one of the early choices in a trial. The feeder remains closed.

FIGURE 14.10B. Open-field maze with feeders on cylindrical pedestals. The same rat examining the last unvisited pedestal in the same trial. The food cup cover opens, and the animal is about to collect the pellet.

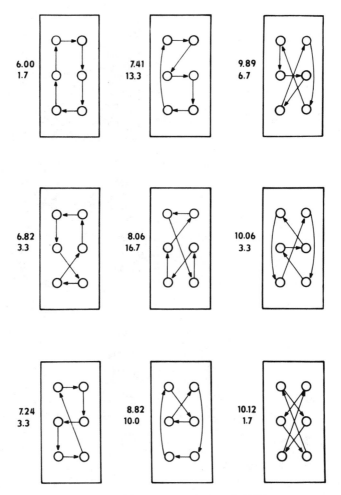

FIGURE 14.11. Examples of the shortest (*left*), medium (*middle*), and longest (*right*) trajectories of errorless trials in an open-field maze with six goals forming a regular rectangle. The upper numbers indicate path length in arbitrary units, the lower numbers the percentage of paths of this length in the set of all possible errorless trials (n = 120).

from this array of valid tours the short ones. The observed tours formed a distribution shifted to the left, with the modal class 7 and the mean path length 7.34. With continued training the tendency to follow the shortest paths further increased, and the mean path length decreased to 6.80.

The above analysis can be applied to any configuration of goals that can always be visited by a multitude of tours with lengths ranging from some minimum to some maximum. The law of effect (Thorndike, 1914) requiring that the task be solved with a minimum expenditure of effort and energy operates in this case at two levels: (1) by avoiding repeated visits of the same points the animal saves the

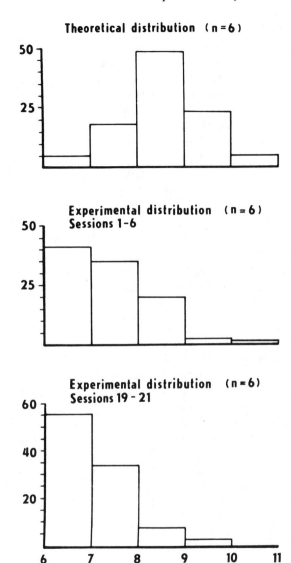

FIGURE 14.12. Comparison of the theoretical and experimental distributions of trajectories according to their length. The results apply to a rectangular configuration of six feeders shown in Figure 14.11. Abscissa-length of the trajectory in arbitrary units; ordinate-percentage of trajectories of the corresponding length. The experimental distributions are based on 452 errorless trials obtained in sessions one to six (13 rats, 10 trials per session) and on 118 errorless trials obtained in sessions 19 to 21 (12 rats, 5 trials per session).

energy needed for the examination of the feeder (rearing, pressing the man-
ipulandum etc.) and (2) by minimizing the length of the correct tour the animal
saves the energy necessary for the locomotion between goals. The relative importance of these two factors varies with the nature of the task. The lengths of the
correct trajectories are not much different in the radial maze where the absence of
errors is the most important criterion of success. In the random maze, path length
may assume a more important role and thus give a new dimension to cognitive
efficiency. Of course, growing experience with the same goal configuration may
lead to a praxis solution of the task, i.e., adherence to a definite route. Yet, only
a cognitive solution is possible whenever a new goal configuration is introduced
or when a different arrangement of goals is used in each trial, as in the way
described by Menzel (1973). The latter author trained chimpanzees to recover
several pieces of food that were hidden before their eyes in different cachets
scattered in a large garden. The hungry animal observed with greatest attention the
movement of the person distributing the food and when released visited the
cachets not in the order in which they were baited but in a way optimizing the
length of the route used by the animal.

14.6 Conclusion

The recent renaissance of spatial memory research was brought about by the
expectation that cognitive maps may provide effective tools for testing the highest
nervous functions in animal models of clinically important degenerative brain
diseases (Alzheimer). Lesion studies and pharmacologic investigations show that
acquisition of spatial memories is critically dependent on the integrity of the
hippocampus and of other prosencephalic structures and on the unimpaired function of cholinergic circuits and of glutamatergic NMDA channels (Barnes, 1988).
The assumption that the neural representation of cognitive maps is topographically
organized prompted attempts to elicit by focal lesions scotoma-like map defects
and to establish principles encoding the map contents in the activity of single
neurons (McNaughton et al., 1983). A better understanding of the behavioral
phenomena studied is a prerequisite for designing experimental paradigms addressing the correspondence between categorized sets of memories and definite
brain loci at the system level and the problem of the nature of the memory record
at the network level.

References

Barnes, C. (1979): Memory deficits associated with senescence: A neurophysiological and
    behavioral study in the rat. *J. Comp. Physiol. Psychol.* 93, 74–104
Barnes, C.A. (1988): Spatial learning and memory processes: The search for their neu-
    robiological mechanisms in the rat. *Trends Neurosci.* 11, 163–169

Beritashvili, I.S. (1971): *Vertebrate memory—characteristics and origin.* New York: Plenum

Buresova, O. (1980): Spatial memory and instrumental conditioning. *Acta Neurobiol. Exp.* 40, 51–66

Buresova, O., Bures, J. (1981): Role of olfactory cues in the radial maze performance of rats. *Behav. Brain Res.* 3, 405–409

Buresova, O., Bures, J., Oitzl, M-S, Zahalka, A. (1985a): Radial maze in the water tank: An aversively motivated spatial working memory task. *Physiol. Behav.* 34, 1003–1005

Buresova, O., Krekule, I., Zahalka, A., Bures, J. (1985b): On-demand platform improves accuracy of the Morris water maze procedure. *J. Neurosci. Methods* 15, 63–72

Buresova, O., Homuta, L., Krekule, I., Bures, J. (1988): Does nondirectional signalization of target distance contribute to navigation in the Morris Water maze? *Behav. Neural Biol.* 9, 240–248

Cartwright, B. A., Collett, T.S. (1982): How honey bees use landmarks to guide their return to a food source. *Nature* 295, 560–564

Collier, T.J., Miller, J.S., Travis, J., Routtenberg, A. (1982): Dentate gyrus granule cells and memory: Electrical stimulation disrupts memory for places rewarded. *Behav. Neural Biol.* 34, 227–239

Hopfield, J.J., Tank, D.W. (1985): Neural computation of decisions in optimization problems. *Biol. Cybern.* 52, 141–152

Hull, C.L. (1943): *Principles of behavior: An introduction to behavior theory.* New York: Appleton-Century-Crofts

Jucker, M., Oettinger, R., Baettig, K. (1988): Age-related changes in working and reference memory performance and locomotor activity in the Wistar rat. *Behav. Neural Biol.* 50, 24–36

Magni, S., Krekule, I., Bures, J. (1979): Radial maze type as a determinant of the choice behavior of rats. *J. Neurosci. Methods* 1, 343–352

McNaughton, B.L., Barnes, C.A., O'Keefe, J. (1983): The contributions of position, direction, and velocity to single unit activity in the hippocampus of freely-moving rats. *Exp. Brain Res.* 52, 41–49

Menzel, E.W. (1973): Chimpanzee spatial memory organization. *Science* 182, 943–945

Mittelstaedt, H., Mittelstaedt, M-L. (1982): Homing by path integration. In: *Avian navigation.* Papi, F., Wallraff, H.G. (eds.). Berlin: Springer-Verlag, pp. 290–297

Morris, R.G.M. (1981): Spatial localization does not require the presence of local cues. *Learning Motiv.* 12, 239–261

Morris, R.G.M. (1983): An attempt to dissociate "spatial mapping" and "working memory" theories of hippocampal function. In: *Neurobiology of the hippocampus.* Seifert, W. (ed.). New York: Academic Press, pp. 405–432

Morris, R.G.M. (1984): Developments of a water-maze procedure for studying spatial learning in the rat. *J. Neurosci. Methods* 11, 47–60

O'Keefe, J., Nadel, L. (1978): *The Hippocampus as a cognitive map.* Oxford: Oxford University Press

Olton, D.S. (1983): Memory functions and the hippocampus. In: *The neurobiology of the hippocampus.* Seifert, W. (ed.). London: Academic Press, pp. 335–373

Olton, D.S., Samuelson, R.J. (1976): Remembrance of places passed: Spatial memory in rats. *J. Exp. Psychol: Animal Behav. Processes* 2, 97–116

Panakhova, E., Buresova, O., Bures, J. (1984): Persistence of spatial memory in the Morris water tank task. *Int. J. Psychophysiol.* 2, 5–10

Pavlov, I.P. (1927): *Conditioned reflexes.* London: Oxford University Press

Potegal, M. (1969): Role of the caudate nucleus in spatial orientation in rats. *J. Comp. Physiol. Psychol.* 69, 756–764

Schenk, F. (1988): A homing procedure for studying spatial memory in immature and adult rodents. *J. Neurosci. Methods* 26, 249–258

Semenov, L.V., Bures, J. (1989): Vestibular stimulation disrupts acquisition of place navigation in the Morris water tank task. *Behav. Neural Biol.* 51, 346–363

Shorey, H.H. (1976): *Animal communication by pheromones.* New York: Academic Press

Sutherland, R.J., Linggard, R. (1982): Being there: A novel demonstration of latent spatial learning in the rat. *Behav. Neural Biol.* 36, 103–107, 1982

Tinbergen, N. (1952): *The study of instinct.* Oxford: Clarendon Press

Thorndike, E.L. (1914): *The psychology of learning.* New York: Teachers College

Tolman, E.C. (1932): *Purposive behavior in animals and men.* New York: Appleton-Century-Crofts

Watson, J.B. (1919): *Psychology from the standpoint of a behaviorist.* Philadelphia: JB Lippincott

Whishaw, I.Q. (1985): Formation of a place-learning set by the rat: A new paradigm for neurobehavioral studies. *Physiol. Behav.* 35, 139–143

Wilkie, D.M., Palfrey, R. (1987): A computer simulation model of rat's place navigation in the Morris water maze. *Behav. Res. Methods Instruments Comp.* 19, 400–403

15

Visual Representation and Short-Term Memory in Inferotemporal Cortex

Joaquin M. Fuster

15.1 Introduction

It is increasingly evident that the cortical processing and representation of sensory data of any modality extend well beyond the primary sensory cortex and occupy substantial parts of the associative cortex. According to electrophysiological and imaging studies, the temporal and spatial characteristics of the cortical representation of a discrete stimulus vary greatly depending on several factors, most notably its modality and its associations with past experience. In the primate, both the processing and the representation of a stimulus tend to follow certain anatomical gradients that conform to the patterns of connectivity originating in primary sensory areas and spreading through association cortex (Pandya and Yeterian, 1985). In the case of vision, those gradients extend into the inferotemporal (IT) cortex.

It has been established that the cortical pathways flowing into and invading IT cortex constitute a hierarchically organized substrate for the analysis of visual properties (DeYoe and Van Essen, 1988). Within that substrate, convergent fibers most probably serve the perception of objects. Divergent and descending fibers, on the other hand, may serve selective attention and the analysis of individual features within objects. That the IT cortex is involved in selective attention has been previously indicated by evoked potential (Newer and Pribram, 1979) and single-unit (Moran and Desimone, 1985; Richmond and Sato, 1987) studies.

This chapter presents some recent evidence indicating that the neurons of the IT cortex take part in the extraction of features from compound stimuli, in selective attention, and in the retention of behaviorally significant stimuli in short-term memory. This evidence has been obtained by recording single-unit activity from the cortex of the IT convexity and the lower bank of the superior temporal sulcus in monkeys performing a visual short-term memory task (Fuster, 1990).

15.2 The Task

The data were obtained from animals fully trained to perform delayed matching-to-sample with complex visual stimuli (Figure 15.1). The task required that the

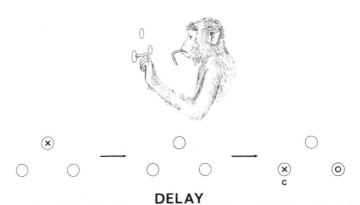

DELAY

SAMPLES   **CHOICES**

FIGURE 15.1. Delayed matching-to-sample task. At the beginning of each trial, the sample stimulus is briefly presented on the top translucid button; that stimulus is a symbol on a colored background (R, red; G, green). If the symbol is =, the subject must retain the background color through a subsequent delay (10–20 sec). At the end of that delay, the subject has to choose the one stimulus—from two in the lower buttons—that contains the distinctive feature of the sample: its color (red or green) or its symbol (X or O). If it makes the correct choice (c), a squirt of fruit juice is automatically delivered to its mouth; if not, the trial is terminated without reward. The sample and the position of its distinctive feature in the lower buttons are changed at random from one trial to the next. Two variants of the task were utilized in the experiments, both designed to ensure that the animal looked at the sample. In one, the sample was turned off by the animal's pressing of the sample (top) button, whereas in the other the sample was of fixed duration (0.5 sec) and appeared 2 sec after an alerting signal—a low-intensity strobe flash that illuminated the entire field briefly and diffusely.

animal notice a particular feature of one such stimulus, retain it for a few seconds in short-term memory, and choose, from two stimuli presented simultaneously, the one that contained that particular distinctive feature. The monkeys performed their task with six circular sample stimuli. Each consisted of a geometric symbol (=, X, or O) on a colored background. If the symbol in the sample was =, the monkey had to retain and, after a delay, choose the color. If the symbol was X or O, the monkey had to retain and select that symbol, ignoring the color. Thus, when the sample appeared at the beginning of each trial, the animal had to attend first to the symbol and, depending on it, attend then also the color. Either a color or a symbol had to be memorized through the subsequent delay for correct performance of the task.

15.3 Feature Extraction

The majority of IT units recorded responded with activation of firing to the sample stimulus. Three categories of sample-activated units were found: (1) nondifferentially responsive to all six samples, regardless of symbol or color; (2) selectively responsive to one symbol (=, X, or O) regardless of background color; and (3) selectively responsive to one color (red or green) regardless of foreground symbol.

Figure 15.2 illustrates responses of a symbol-selective cell to the six samples. Two color-selective cells are shown in Figure 15.3. In some units (Figures 15.2 and 15.3A), the selectivity for a symbol or a color was nearly absolute, at least in terms of the five features (three symbols and two colors) utilized in the task: those units responded exclusively to one stimulus feature. In other units (Figure 15.3B), however, the selectivity was relative in that they responded to more than one feature, although significantly more to one than to the others.

Among the color-selective cells, a subcategory of cells was found that responded more to their preferred color when it was accompanied by = than by the other two symbols. Because color was only relevant (i.e., had to be memorized) if it appeared together with =, the response of those cells to their preferred color was presumed enhanced by its relevance. For this reason they were called *color-attentive* units.

The analysis of the anatomic location of the various types of units revealed substantial differences, especially with regard to cortical depth, although histologic localization in terms of depth was subject to considerable error. Unselective units were found scattered through different layers. The same was true for color units, with the exception of color-attentive units. Symbol units and color-attentive units were most common in subgranular layers (V and VI). Thus, in those deeper layers the cells were found that best responded to the distinctive feature—symbol or color—that the animal had to attend to for correct performance of the task.

The latency of the response to the sample stimulus also varied for the different types of cells (Figure 15.4). Unselective and color cells (the color-attentive cells again excepted) showed the shortest latencies. Symbol cells responded with somewhat longer latency. Color-attentive cells showed the longest latencies of all. This

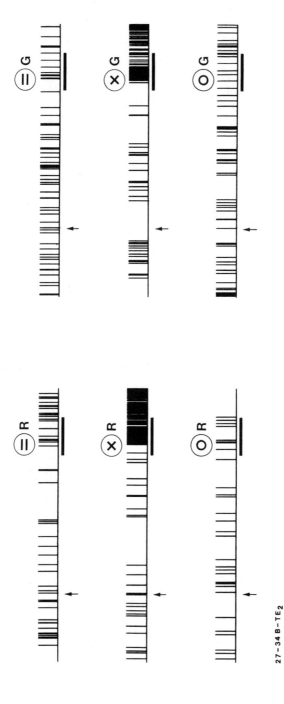

27-34 B-TE$_2$

FIGURE 15.2. Firing of an IT cell at the start of six trials, each with a different sample stimulus (marked by horizontal bar and preceded by alerting signal—a diffuse flash—at the arrow). R (red) and G (green) indicate the color of the sample, which lasts 0.5 sec. The cell responds selectively to X regardless of background color.

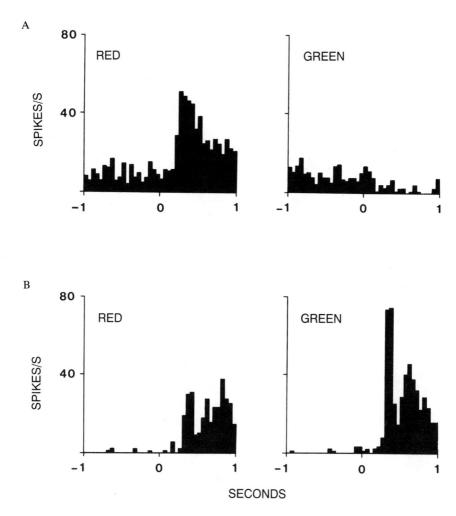

FIGURE 15.3. Average frequency histograms from two units selective for color. Sample stimulus begins at 0 sec on the time scale; bin width, 50 msec. Responses are averaged for all samples of one color regardless of symbol. A, Red-sensitive unit (note inhibition by green). B, Green-selective unit.

order of increasing latency probably reflects the order in which the different types of neurons were sequentially recruited by the sample stimulus: first, the un-selective and some of the color neurons, responding to physical features that were common to several or all samples (shape, brightness, color, etc.); second, the neurons attuned to the symbol; and third, the neurons that were attuned to the color, especially if it was relevant (color-attentive units). This order also reflects the presumed psychophysical order in which the subject attended to the properties of the sample in performance of the task: first the mere presence of the stimulus, then the symbol, and lastly the color.

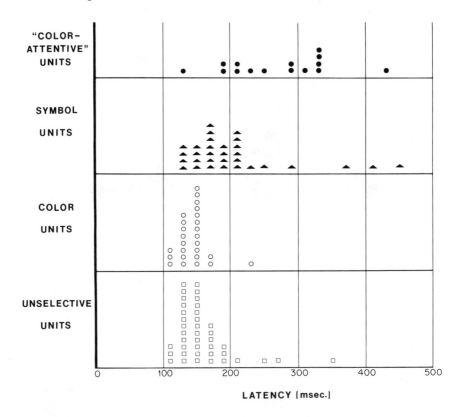

FIGURE 15.4. Latency distributions of four different types of responsive IT units.

15.3.1 Selective Attention to Color

Closer analysis revealed that almost all color units—not only the color-attentive ones—were to some degree susceptible to the relevance of color. In order to quantify the effect of color relevance and therefore the enhancing influence of attention on the response to color, I utilized an attention index (*AI*), which was computed as follows:

$$AI = \frac{Cr - B}{Ci - B}$$

where *Cr* was the mean firing frequency in response to the preferred color when color was relevant (i.e., with =), *Ci* the same when color was irrelevant (i.e., with X or O), and *B* the baseline frequency between trials. The median *AI* in IT cortex was 1.23; by comparison, in a group of color-selective units from V1 cortex, the median *AI* was 0.97. The difference between those two medians was significant ($p \leq 0.01$). These data indicate that the responses of IT neurons to color, unlike those of V1 neurons, are enhanced by attention to color.

The evidence that color-attentive cells were concentrated in deeper cortical layers was substantiated by the finding of a direct relationship between AI and cortical depth (Figure 15.5). The deeper the color cells were in the cortex, the greater was their AI. This suggests a transcortical anatomic gradient of susceptibility to the effect of selective attention on color cells. It also implies that the tuning of such cells to the behavioral relevance of color increases as a function of cortical depth.

A direct relationship was also found between latency and AI, further substantiating the mentioned long latency of color-attentive units (Figure 15.4). Such units, of course, turned out to have the highest AIs.

The apparent correlation between AI and latency suggests that, in the selective

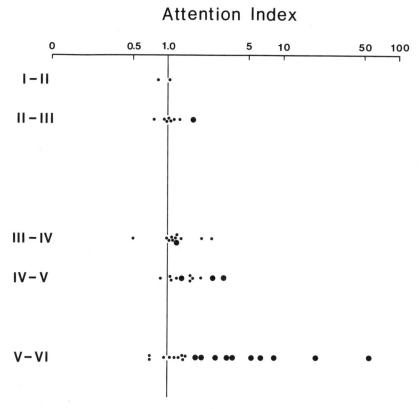

FIGURE 15.5. Distribution of color-selective IT units by their cortical depth and attention index. (Because of limitations in the accuracy of histologic localization, each unit has been plotted between two layers). Larger dots mark the color-attentive units.

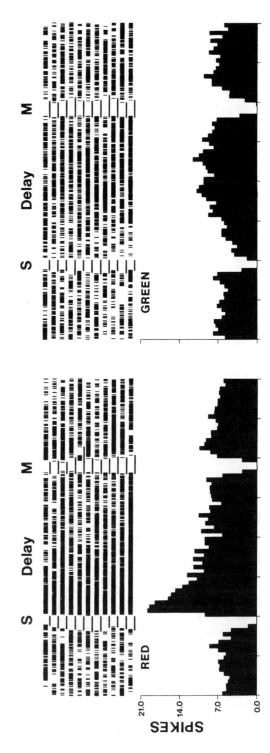

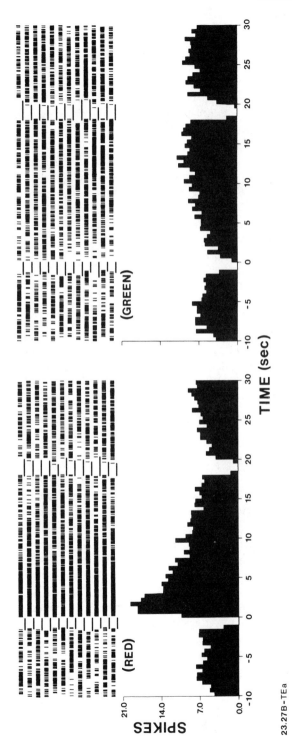

23.27B-TEa

FIGURE 15.6. Rasters and frequency histograms of the activity of an IT unit inhibited by all samples (S) and choice stimuli (M), but activated during the delay of red-sample trials. (The unit was recorded during a variant of the task in which the sample was terminated by the animal's pressing of the sample button—see legend to Figure 15.1; this accounts for variability of sample duration). All records and histograms are aligned with the termination of sample, i.e., beginning of delay. *Upper half,* Records are grouped for trials with color-relevant sample (=). *Lower half,* Records grouped with color-irrelevant sample (X or O).

attention to a color as a function of its context, there is a trend of color information processing toward progressively higher levels of a visual processing hierarchy in IT cortex. That this takes place in the direction of cortical depth remains a tentative inference. The AI-latency correlation per se implies, however, that, in the dynamics of the awake and behaving animal, the selective attention to color information provides that information with greater penetrance into the processing system, whatever its anatomic organization may be.

It is difficult to support that interpretation of the data with what is presently known about intrinsic cortical connectivity. It is also difficult to construe a mechanism by which that filtering or unfiltering of information might take place. Nevertheless, the latency data of color cells further support the general notion (mentioned in the previous section) that selective visual attention, unlike what has been called preattentive or "early" vision, involves serial processing (Koch and Ullman, 1985). Such serial processing seems to take place, at least in part, within the IT cortex.

15.3.2 Short-Term Memory

In this study, as in a previous one (Fuster and Jervey, 1982), I found cells in IT cortex that underwent sustained activation in the delay period of the task trials; that is, during the retention of information contained in the sample stimulus. In some cells that activation was selective, varying in degree depending on the characteristics of the sample stimulus. So, just as there were feature-selective cells in the sample period, there were also feature-selective cells in the delay period. Again, during the latter period, some cells preferred a symbol and others a color. These delay-selective units seemed to participate in the memorization of the feature for which they were selective.

For example, during the delay the unit in Figure 15.6 fired more after red samples than after green ones. The unit may have engaged in short-term retention of the color red, and it seemed to do it equally well whether color was relevant or not. As the reaction of this unit illustrates, the feature selectivity of a given IT cell during the delay was not necessarily the continuation of selectivity in the sample period, but instead could be circumscribed to the delay. In other words, cells that differentially respond to the stimulus when it was present appeared to constitute a separate population from those that did it in the subsequent delay. Thus, neurons involved in the perception of features seem to coexist in IT cortex with neurons involved in the retention of those features in short-term memory.

Unexpectedly, I also found delay-activated units in V1 cortex, although there the incidence of such units was smaller than in IT cortex. Some of them were selective either for symbol or for color. The delay-activated units of V1, however, differed somewhat from those of IT cortex. The most apparent difference was in their discharge after the second presentation of the sample at the time of the choice that ended the trial. Although the firing of delay-activated IT units tended to revert to baseline level after that second presentation (Figure 15.6, upper left quadrant), that of V1 units showed a reactivation similar to the one occurring during the delay. Therefore, V1 units seemed more coupled to the physical properties of the stimulus than IT units; the latter seemed more attuned than the former to the need to retain the stimulus or one of its attributes in short-term memory. At any rate, the finding of selective delay-activated V1 cells indicates that, already at the first stage of the visual cortical processing hierarchy, there is the potential for neuronal retention of information beyond the time of its presence in the environment.

15.4 Conclusions

Cells in the IT cortex discriminate and select the features of compound visual stimuli that the animal discriminates and selects for performance of a delayed matching task.

The latency of the response of IT cells to a compound stimulus depends on their feature selectivity and is related to the presumed psychophysical order of attention to the features of the stimulus according to the rules of the task.

Cells of IT cortex that select behaviorally relevant features tend to concentrate in deep cortical layers.

Selective attention to a color as a function of its context seems to result in the successive recruitment of color-responsive IT cells that are progressively more susceptible to the behavioral relevance of color; this finding suggests serial processing in selective attention.

Cells in IT cortex are apparently capable of retaining individual features of complex visual stimuli in short-term memory.

Different groups of IT cells seem to engage in the perception of a feature and in its retention in temporary memory.

References

DeYoe, E.A., Van Essen, D.C. (1988): Concurrently processing streams in monkey visual cortex. *Trends Neurosci.* 11, 219–226

Fuster, J.M. (1990): Inferotemporal units in selective visual attention and short-term memory. *J. Neurophysiol.* (In press)

Fuster, J.M., Jervey, J.P. (1982): Neuronal firing in the inferotemporal cortex of the monkey in a visual memory task. *J. Neurosci.* 2, 361–375

Koch, C., Ullman, S. (1985): Shifts in selective visual attention: Towards the underlying neural circuitry. *Hum. Neurobiol.* 4, 219–227

Moran, J., Desimone, R. (1985): Selective attention gates visual processing in the extrastriate cortex. *Science* 229, 782–784

Newer, M.R., Pribram, K.H. (1979): Role of the inferotemporal cortex in visual selective attention. *Electroenceph. Clin. Neurophysiol.* 46, 389–400

Pandya, D.N., Yeterian, E.H. (1985): Architecture and connections of cortical association areas. In: *Cerebral Cortex: vol 4.* Jones, E.G., Peters, A. (eds.). New York: Plenum Press, pp. 3–61

Richmond, B.J., Sato, T. (1987): Enhancement of inferior temporal neurons during visual discrimination. *J. Neurophysiol.* 58, 1292–1306

16

Gangliosides, Learning, and Behavior

H. Schenk, U. Haselhorst, N.A. Uranova, A. Krusche, H. Hantke,
and D.D. Orlovskaja

16.1 Introduction

Gangliosides, a complex group of cell-surface sialoglycosphingolipids particularly abundant in neuronal tissues, are assumed to be involved in a variety of cell-surface events, such as synaptogenesis, regulation of cell-growth, neuronal regeneration (for review see Ledeen, 1984), and, last but not least, synaptic transmission (Rahmann, 1983; Wieraszko and Siefert, 1986). Indeed, some investigators have suggested that exogenous gangliosides promote structural repair after brain lesions in vivo (Sabel et al., 1984; Toffano et al., 1983), which may have implications for recovery of function (Dunbar et al., 1986). The facilitated recovery may be due as well to a reduction in neuronal cell loss and axonal/dendritic degeneration and a subsequent neuronal regeneration or both together (Karpiak et al., 1986). Nevertheless, the molecular events that underlie these effects remain unexplained.

In contrast to the experimental knowledge about the effect of applied gangliosides on cell cultures and brain injury, there are no studies describing the effect of exogenous gangliosides on normal animals. In our experiments, we studied the effect of exogenous applied gangliosides on the acquisition of an active conditioned avoidance response and memory formation.

It is well known that modulators of synaptic transmission produce alterations of the ultrastructure of synapses, as was demonstrated both by dopamine blockers and stimulators (Benes et al., 1985; Klinzova et al., 1989; Uranova et al., 1989). Both act also as modulators of learning behavior (Beninger et al., 1981; Sanger, 1985). It has been shown that gangliosides interact with the dopaminergic system of rats (Tilson et al., 1988). Therefore, in addition to learning experiments we looked at the same time for some morphological changes in the ultrastructure of dopamine-rich substantia nigra under chronic ganglioside treatment.

16.2 Method

The study was undertaken in male Wistar rats (VEB Versuchstierproduktion, Schönwalde, GDR) weighing about 200 g at the beginning of the experiments. The animals were housed under standard conditions with a 12-hour dark/light cycle and free access to food and water. All injections were given intraperitoneally in a volume of 1 ml/100 g body weight. The gangliosides used in this study were produced in our laboratory from pig brains with the following pattern: GQ1b = 4.4%, GT1b = 26.8%, FucGD1b = 6.5%, GD1b = 13.4%, GD1a = 38.6%, GD3 = 5.3% and GM1 = 5.0% (Svennerholm nomenclature; Svennerholm, 1964). The content of neuraminic acid was about 18%. The intraperitoneal injection dose was 20 mg/kg. Controls received an injection of isoosmolar sodium chloride.

Three groups of animals were examined.

1. *Single ganglioside injection.* Motor activity (open field) and the acquisition of a conditioned avoidance reaction (CAR) were examined 1 day later. The retention of the CAR was tested twice at 1-day intervals.
2. *Daily injection of gangliosides for 3 weeks.* The acquisition and retention of the CAR were examined after a ganglioside-free interval of 2 days. In addition, morphologic studies of the brains were performed in this group.
3. *Gangliosides injected twice, immediately after the training and the first retention session 24 hours later.* For the behavioral experiments (CAR), we used a computer-controlled two-way shuttle box that we built (for details of the technical design, see Nahrstedt et al., 1987). A 4-sec tone was presented as the conditioned stimulus.

The unconditioned stimulus, 1-mA footshock, was given for a maximum of 15 additional sec. The intertrial interval was randomized with a mean of two trials/min. The training session was stopped after the criterion of three successive avoidance responses was reached (for details see Haselhorst et al., 1988). Retraining followed the original acquisition procedure after 24 and 48 hours. An open-field observation was carried out 5 minutes before the training session in the first group.

For morphologic studies, four animals of the second group and four control animals were used. At the end of the 20 days of ganglioside injection and a drug-free interval of 2 days, the rats were anesthetized with hexobarbital (150 mg/kg body weight) and perfused intracardially with sodium chloride and heparin first and then with a fixative containing 4% paraformaldehyde and 2.5% glutaraldehyde in 0.1 M phosphate buffer. The brains were left overnight in situ at 4°C. On the next day, the brains were removed. Pieces of pars compacta of the substantia nigra were excised. Tissue samples were postfixed in 1% osmium tetroxide in 0.2 M phosphate buffer, dehydrated with ethanol and propylene oxide, and embedded in araldite epoxy resin. Semithin (1-μm) sections were cut by LKB-TV ultramicrotome and were stained by toluidine blue. Ultrathin sections were counterstained on grids with uranyl acetate and lead citrate and viewed with a Philips electronmicroscope.

16.3 Results

In the first group, pretreatment with a single ganglioside injection 1 day before the training session improved the acquisition of the CAR (Figure 16.1). The number of crossed squares in the open-field test (Figure 16.2) and the number of intertrial jumps during CAR did not differ between the control and ganglioside-treated group. In the second group, pretreatment with gangliosides for 3 weeks followed by a 2-day ganglioside-free interval had no effect on the CAR (Figure 16.3). In the third experimental group, the consolidation of the memory was disturbed by gangliosides applied after training. The mean number of trials to reach the criterion in the retraining sessions after 24 and 48 hours was only reduced in controls (Figure 16.4).

Electronmicroscopic study of substantia nigra after chronic ganglioside treatment in Group 2 revealed some differences in the ultrastructure of nerve cells, glial

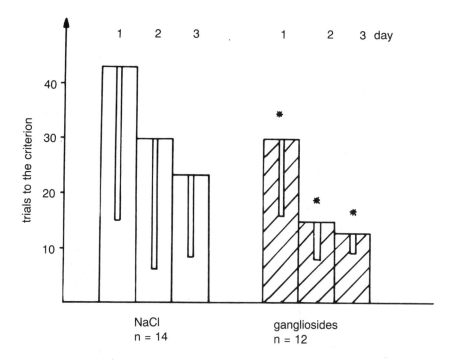

FIGURE 16.1. The effect of gangliosides on the acquisition of the conditioned avoidance reaction (CAR) (day 1) and on the relearning sessions (on day 2 and day 3) in rats. Ganglioside application 24 hours before the CAR (Group 1). Mean number of trials ± SD to reach the criterion. Controls were treated by iso-osmolar sodium chloride. n = number of animals; * = significant differences, assessed by the Mann-Whitney U-test ($p < 0.05$).

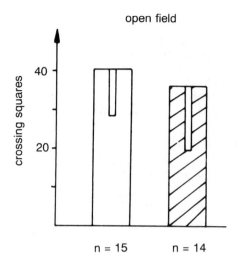

FIGURE 16.2. The effect of gangliosides (*right*) on motor behavior in rats. No significant differences were found.

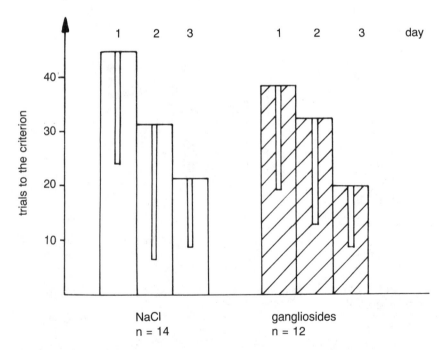

FIGURE 16.3. The effect of gangliosides on acquisition of the CAR (day 1) and on the relearning session on days 2 and 3. Ganglioside application was done daily for 3 weeks followed by a ganglioside-free interval of 2 days before the first CAR (Group 2).

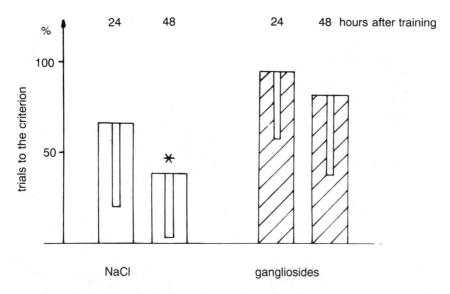

FIGURE 16.4. Effect of posttraining application of gangliosides on memory formation of a CAR. Gangliosides were applied after the first CAR (day 1) and the first relearning session (day 2) (Group 3). Mean number ± SD of trials to reach the criterion in the relearning sessions after 24 and 48 hours in % of the initial training session. * $p < 0.05$ Mann-Whitney U-test; NaCl controls, $n = 15$; ganglioside-treated rats, $n = 15$.

cells, and synaptic contacts between the experimental and control group (Figure 16.5). In the nuclei of some nerve cells, the nucleoli increased in size. In some cells, the size of nucleoli was doubled in comparison with control samples (Figure 16.5A). The cytoplasm of such cells contained a higher density of polyribosomes and mitochondria.

In the neuropil, some dendrites of different diameter contained enormously great number of mitochondria (Figure 16.6). Some of these mitochondria were shrunken in the experimental animals (Figure 16.6B), whereas the ultrastructure of controls (Figure 16.6A) looked quite normal. The presynaptic part of some axosomatic synapses both on large and small cells was increased in size and contained larger mitochondria in experimental animals (Figure 16.7B) than in the control group (Figure 16.7A).

Some axon terminals containing granular vesicles also increased in size. In the experimental group (Figure 16.8B), those terminals contained much more granular vesicles than those of the control group (Figure 16.8A).

Very often, axon terminals in experimental animals that demonstrated such types of changes were surrounded by astroglial cell processes with signs of hypertrophy containing a lot of polyribosomes (Figure 16.9B); these signs were not common for control animals (Figure 16.9A). The cytoplasm of some astroglial and oligoglial cells contained many mitochondria and had a higher density of ribosomes and lysosomes.

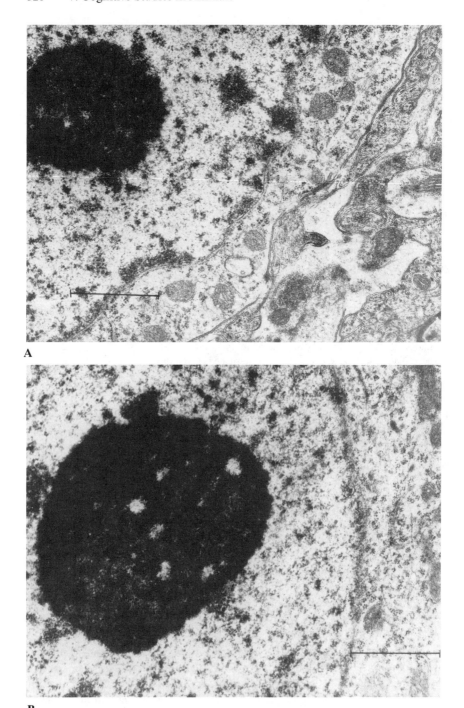

FIGURE 16.5. Cell somata of a small neuron in control and ganglioside-treated animals. Bar = 1 um. *A,* Control and *B,* ganglioside-treated animals.

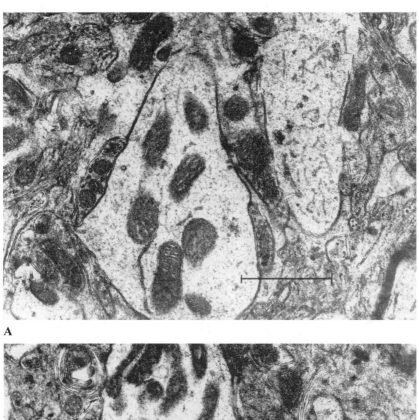

A

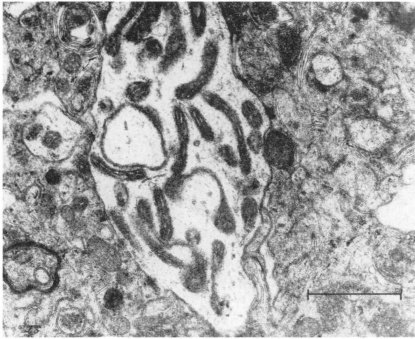

B

FIGURE 16.6. Mitochondria in dendritic profiles. *A,* Control and *B,* ganglioside-treated animals.

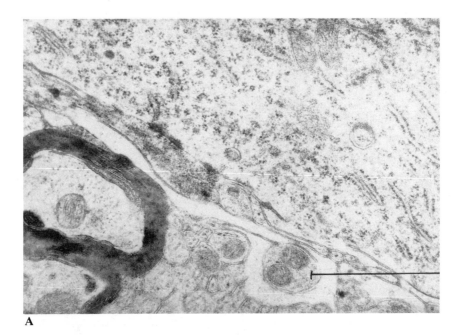

A

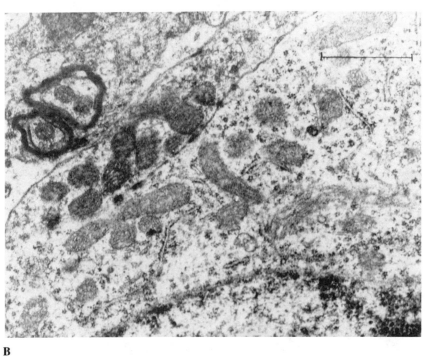

B

Figure 16.7. Axosomatic synapses on large neuron. *A*, Control and *B*, ganglioside-treated animals.

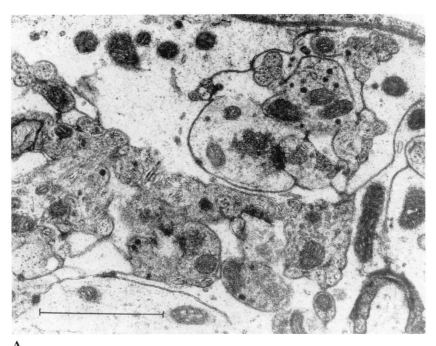

A

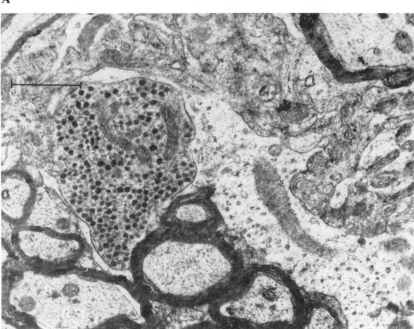

B

FIGURE 16.8. Axon terminals containing a mixture of clear and granular vesicles surrounded by astroglial cell processes. A, Control and B, ganglioside-treated animals.

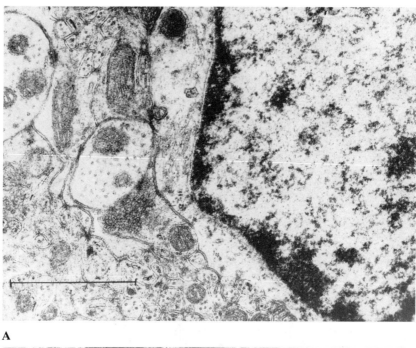

A

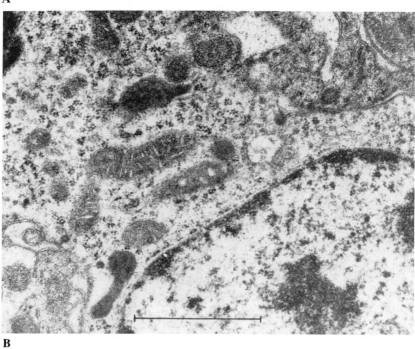

B

FIGURE 16.9. Astroglial cell. *A,* Control and *B,* ganglioside-treated animals.

16.4 Discussion

Knowledge of the fate and metabolic pathways of gangliosides administered to animals is interesting with respect to whether exogenous gangliosides reach the brain. Some investigators have pointed out that a small but definite amount is incorporated into the brain (Ghidoni et al., 1986; Tettamanti et al., 1981). From data reported by Rahmann (1983); Ramirez et al., (1990) and by Wieraszko and Seifert (1986), we know that gangliosides take part in neuronal transmission processes. These findings permit the hypothesis that not only endogenous but also exogenous gangliosides are involved in the modulation of synaptic transmission. Ganglioside pretreatment without a drug-free interval improved the acquisition of the CAR (Figure 16.1). This effect seems to be caused by the ganglioside action directly and not by different motor behavior. The motor activity was equal in both groups, as seen in the open-field experiment (Figure 16.2). The pretreatment with gangliosides for 3 weeks had no effect on the CAR (Figure 16.3). The ganglioside-free interval of 2 days might be the reason for this observation. Perhaps during this ganglioside-free interval, the concentration of gangliosides and its availability for neuronal processes are reduced to a normal level. In contrast, we could demonstrate remarkable changes on the morphologic level after ganglioside pretreatment for 3 weeks (Figures 16.5 to 16.9).

It is well known that dopamine neurons are involved in the modulation of memory processes (McGaugh, 1973). The ultrastructural changes reported in learning resemble those found in our study. Common findings are activation of neuroglial interactions and an increase in the number of polyribosomes in some nerve and glial cells (Figure 16.9); these changes are accompanied by an enhanced energy production that is seen as an increase in the number of mitochondria (Figure 16.6) in the nerve cell processes and plastic changes in the ultrastructure of synapses (Figure 16.7) (Stewart et al., 1984, 1987).

The multiple effects of exogenous gangliosides may be dependent on the production or activation of specific neurotrophic factors, such as nerve growth factor (NGF) (Leon et al., 1984; Stephens et al., 1987). There is some evidence of GM1 interaction with neurotrophic factors. Our data are in agreement with the observation that exogenous gangliosides are involved in trophic interactions of neurons (Fuxe et al., 1986)

The role of gangliosides in the interaction between nerve cell and astroglial cells is known (Flott et al., 1988). Thus, the increased metabolism of some nerve and glial cells revealed in our study might be due to their activation by endogenous trophic factors. The nonspecific effect of exogenous gangliosides has been shown to protect neurons from degeneration, stimulate the process of regeneration, and improve the behavioral deficit after different types of brain lesions, as well as to improve the process of learning. These effects of gangliosides might be due to their activation of metabolism in both neuronal and glial cells, resulting in an increase in the potential of nerve tissue for recovery by the establishment of an optimal environment for processes of neuronal and synaptic plasticity. These changed environments might facilitate the process of learning.

On the other hand, the behavioral data are very similar to the well-known effect of some neuroleptics. For example, clozapine and haloperidol posttraining injections produce a disturbance of memory consolidation (Sanger, 1985). The morphologic changes after chronic administration of gangliosides are also very similar to those observed after chronic treatment with clozapine in the substantia nigra (Uranova, 1985), especially the hyperproduction of mitochondria (Figures 16.6 and 16.7) and the increase in polyribosomes (Figure 16.9).

Recently, it has been shown that a ganglioside mixture can influence the up-regulation of dopamine receptors. Gangliosides can decrease hypersensitivity to apomorphine produced by haloperidol (Tilson et al., 1988). On the behavioral level, we can also find such neuroleptic-like effects as the retardation of memory formation by gangliosides, when applied immediately after the learning procedure (Figure 16.4).

In conclusion, the higher concentration and availability of gangliosides improve the acquisition and increase the consolidation of the CAR in rats. Chronic application of gangliosides induces a neuroleptic-like alteration of the ultrastructure of synapses of the substantia nigra, without influencing learning behavior and memory formation.

References

Benes, F.M., Paskevich, P.A., Davidson, J., Domesick, V.B. (1985): The effect of haloperidol on synaptic patterns in the rat striatum. *Brain Res.* 329, 265–274

Beninger, R.J., Hanson, D.R., Phillips, A.G. (1981): The acquisition of responding with conditioned reinforcement. *Br. J. Pharmacol.* 74, 149–154

Dunbar, G.E., Butler, W.M., Fass, B., Stein, D.G. (1986): Behavioral and neurochemical alterations induced by exogenous gangliosides in brain damaged animals: Problems and perspectives. In: *Gangliosides and neuronal plasticity, FIDIA Research Series: vol 6*, Tettamanti, G., Ledeen, R.W., Sandhoff, K., Nagai, Y., Toffano, G. (eds.). Padova: Livana Press, pp. 365–380

Flott, B., Masco, D., Seifert, W. (1988): Incorporation of ^{3}H-GM1 ganglioside into astrocytes in cell culture. *Neurochem. In.* 13 (Suppl. 1), 111

Fuxe, K., Agnati, L.F., Benfenati, F., Zini, I., Gavioli, G., Toffano, G. (1986): New evidence for the morphofunctional recovery of striatal function by ganglioside GM1 treatment following a partial hemitransection of rats. Studies on dopamine neurons and protein phosphorylation. In: *Gangliosides and neuronal plasticity, FIDIA Research Series: vol 6*, Tettamanti, G., Ledeen, R.W., Sandhoff, K., Nagai, Y., Toffano, G. (eds.). Padova: Livana Press, pp. 347–365

Ghidoni, R., Trinchera, M., Venerando, B., Fiorilli, A., Tettamanti, G. (1986): Metabolism of exogenous GM1 and related glycolipids in the rat. In: *Gangliosides and neuronal plasticity, FIDIA Research Series: vol 6*, Tettamanti, G., Ledeen, R.W., Sandhoff, K., Nagai, Y., Toffano, G. (eds.). Padova: Livana Press, pp. 183–200

Haselhorst, U., Krusche, A., Schenk, H., Hantke, H. (1988): Einfluß von Gangliosiden auf das Erlernen einer bedingten Fluchtreaktion der Ratte an der Shuttle-Box. *Biomed. Biochim. Acta,* 47, 475–480

Karpiak, S.E., Li, Y.S., Aceto, P., Mahadik, S.P. (1986): Acute effects of gangliosides on

CNS injury. In: *Gangliosides and neuronal plasticity, FIDIA Research Series: vol 6,* Tettamanti, G., Ledeen, R.W., Sandhoff, K., Nagai, Y., Toffano, G. (eds.). Padova: Livana Press, pp. 407–415

Klinzova, A.J., Haselhorst, U., Uranova, N.A., Schenk, H., Istomin, V.V. (1989): The effects of haloperidol on synaptic plasticity in rat's medial prefrontal cortex. *J. Hirnforschg.* 30, 51–57

Lankford, K.L., DeMello, F.G., Klein, W.L. (1988): D1-type dopamine receptors inhibit growth cone motility in cultured retina neurons: Evidence that neurotransmitters act as morphogenetic growth regulators in the developing central nervous system. *Proc. Natl. Acad. Sci. USA.* 85, 2839–2843

Ledeen, R.W. (1984): Biology of gangliosides: Neuritogenic and neuronotrophic properties. *U. Neurosci. Res.* 12, 147–159

Leon, A., Benvegnu, D., Daltoso, L., Presti, D., Facci, L., Giorgi, D., Toffano, G. (1984): Dorsal root ganglia and nerve growth factor. A model for understanding the mechanism of GM1 effect on neuronal repair. *J. Neurosci. Res.* 12, 227–288

McGaugh, J.H. (1973): Drug facilitation of memory and learning. *Ann. Rev. Pharmacol.* 13, 229–241

Nahrstedt, K., Weber, P., Kunert, M., Haselhorst, U. (1987): Steuerteil für Verhaltenstestapparaturen. *Medizintechnik* 27, 63–64

Rahmann, H. (1983): Functional implication of gangliosides in synaptic transmission. *Neurochem. In.* 5, 539–547

Ramirez, O.A., Gomez, R.A., Carrer, H.F. (1990): Gangliosides improve synaptic transmission in dentate gyrus of hippocampal rat slices. *Brain Res.* 506, 291–293

Sabel, B.A., Dunbar, G.L., Stein, D.G. (1984): Gangliosides minimize behavioral deficits and enhance structural repair after brain damage. *J. Neurosci. Res.* 12, 429–443

Sanger, D.J. (1985): The effects of clozapine on shuttle-box avoidance responding in rats: Comparison with haloperidol and chloriazepoxide. *Pharmacol. Biochem. Behav.* 23, 231–236

Stephens, P.H., Tagari, P.C., Garofalo, L., Maysinger, D., Piotte, M., Cuello, A.C. (1987): Neural plasticity of basal forebrain cholinergic neurons: Effects of gangliosides. *Neurosci. Lett.* 80, 80–84

Stewart, M.G., Rose, S.P.R., King, T.S., Gabbott, P.L.A., Bourne, R. (1984): Hemispheric asymmetry of synapses in chick medial hyperstriatum ventrale following passive avoidance training: A stereological investigation. *Dev. Brain Res.* 12, 261–269

Stewart, M.G., Csillag, A., Rose, S.P.R. (1987): Alterations in synaptic structure in the paleostriatal complex of the domestic chick, Galeus domesticus, following passive avoidance training. *Brain Res.* 426, 69–81

Svennerholm, L. (1964): The gangliosides. *J. Lipid. Res.* 5, 145–155

Tettamanti, G., Venerando, B., Roberti, S., Chigorno, V., Sonnino, S., Ghidoni, R., Orlando, P., Massari, P. (1981): The fate of exogenously administered brain gangliosides. In: *Gangliosides in neurological and neuromuscular function, developmental and repair.* Rapport, M.M., Gorio, A. (eds.). New York: Raven Press, pp. 225–240

Tilson, H.A., Harry, G.J., Nary, H., Hudson, P.M., Hong, J.S. (1988): Ganglioside interactions with dopaminergic system of rats. *J. Neurosci Res.* 19, 88–94

Toffano, G., Savoni, G.E., Moroni, F., Lombardi, M.G., Calza, L., Agnati, L.F. (1983): GM1 ganglioside stimulates the regeneration of dopaminergic neurons in the central nervous system. *Brain Res.* 261, 163–166

Uranova, N.A. (1985): Action of clozapine on the ultrastructure of the brain. *Z. Nevropatolog. i. Psichi.* 85, 1006–1011

Uranova, N.A., Klinzova, A.J., Istomin, V.V., Haselhorst, U., Schenk, H. (1989): The effects of amphetamine on synaptic plasticity in rat's medial prefrontal cortex, *J. Hirn-forsch.* 30, 45–50

Wieraszko, A., Seifert, W. (1986): Involvement of gangliosides in the synaptic transmission in the hippocampus and striatum of the rat brain. In: *Gangliosides and neuronal plasticity, FIDIA Research Series: vol 6,* Tettamanti, G., Ledeen, R.W., Sandhoff, K., Nagai, Y., Toffano, G. (eds.). Padova: Livana Press, pp. 137–151

17

Neurophysiological Study of Animals' Subjective Experience

V.B. Shvyrkov

17.1 Introduction

Early concepts of brain processes were based on clinical observations and experiments with stimulation and lesions of various brain structures. The data collected by these methods laid the foundation for the morphofunctional understanding of brain activity.

The unit activity recording method was used first to study neural mechanisms of already known functions of various structures. According to the morphofunctional approach, unit activity of sensory structures was compared with stimuli features, and unit activity of motor structures was compared with movement features, etc. These experiments led to the very important discovery of neural unit specialization, which was treated as functional specialization. In particular, various receptive fields of sensory structures units (Mountcastle, 1957; Hubel and Wiesel, 1959) and muscle fields of motor structures units (Fetz and Cheny, 1978) were described.

Contemporary data collected in studies of unit activity in freely behaving animals have demonstrated many other types of unit specialization. In various brain areas there have been discovered units that were active, for example, only during certain movements of the limb (De Long and Strick, 1974) or during the use of "cognitive maps" (O'Keefe, 1976), in hunger (Ono et al., 1981), in aggression (Pond et al., 1977) and attention (Mountcastle, 1978), while reaching certain goals (Shvyrkov, 1980), and during a unique behavioral act (Shvyrkov 1985, 1986).

These data inspired the suggestion of many new brain functions. On the basis of unit activity study, Mountcastle, for example, arrived at the conclusion that:

the neocortex of the parietal lobe is an essential node of a distributed cerebral system which generates and updates a neural image of the body form, the position of the body within the immediately surrounding space, the relation of the body parts to one another and to the gravitational field, the direction of gaze and visual attention, and of dynamic changes in these postural and attitudinal sets. This neural mechanism appears to be correlated with the internal state of the organism in terms of needs and interests such as hunger and thirst and from time to time generates commands for action, for the selective and directed visual attention into the immediate behavioral surround, for the visual grasping of objects, and for skilled coordinated actions of hand and eye (1978, p. 37).

It became clear that the list of brain as well as neural unit functions may be endless and that the morphofunctional approach is unsatisfactory on both cell and brain levels. The reason for the weakness of this approach lies in the subjectivity and arbitrariness of singling out any function.

According to the functional system theory of Anokhin (1974), behavior is not divisible into sensory or motor functions or processes taking place in special structures, but rather consists of behavioral acts, i.e., adaptive changes of the relationships of an organism with its environment. Any behavioral act is performed by activity of a system of elements located in many different structures, but the activity or any element of this system leads only to that same adaptive relationship with the environment. For example, the act of food seizure is performed by activity of some muscle, glandular, or vessel elements in various parts of the body and by some neural elements in different brain structures. All this activity is directed to only one goal—to obtain some food in the mouth. Behavioral acts are established in trial-and-error learning in phylogenesis and ontogenesis, and their totality constitutes an organism's memory. This theory inspires the search for behavioral rather than functional specialization of neural units.

17.2 Experimental Studies

In our studies we correlated unit activity with the behavioral acts established at successive steps of learning in rabbits. In an experimental box (Figure 17.1) with two pedals and two automatic feeders mounted in the corners, rabbits learned first to obtain a food portion from the left feeder, which was given from time to time. At this stage of learning the rabbits simply sat in front of the left feeder and waited for food and took food from the feeder. Then the situation was changed, and the rabbits obtained no food if they simply sat in front of the feeder. They demonstrated searching behavior and left the feeder for other places of the box. This movement actuated the feeder, and the rabbits learned to approach the feeder and take the food from it. At the third stage of learning, being distant from the feeder was no longer sufficient to actuate the feeder. The rabbits again demonstrated searching behavior during which they occasionally found themselves in the left pedal corner. This placement now actuated the feeder. After some trials, the rabbits went to the left pedal corner just after food seizure and stayed there until hearing the sound of feeder actuation. At the last stage of learning, staying in the pedal corner was made ineffective. After many trials, the rabbits learned to press the left pedal that switched on the feeder motor.

The learning process took approximately 2–3 days. Then the rabbits learned to perform the same behavior on the right side of the experimental box. This procedure took 2 more days. Completely learned behavior on each side looked like a

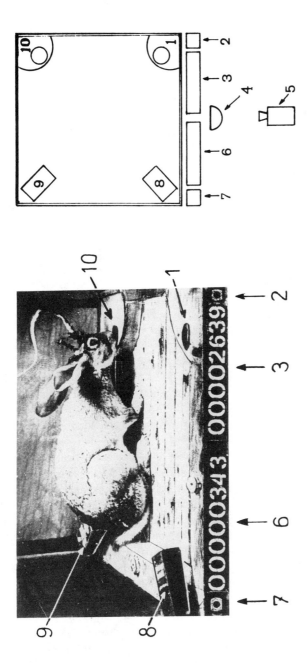

FIGURE 17.1. The experimental box and its scheme. 1, left feeder; 2, marker for the feeder holes encountered by a rabbit's nose; 3, counter of spikes of a unit; 4, flicker-marker; 5, video camera; 6, timer; 7, marker for pedal press; 8, left pedal; 9, right pedal; 10, right feeder.

cycle of a linear sequence of acts: food seizure, turning the head to the pedal, approaching the pedal, pedal pressing, approaching the feeder, food seizure, and so on.

In the main experiments with unit activity recording, the pedals were alternately effective for releasing food. One pedal was effective 10 to 20 times in a row, and then the other one became effective. Accordingly, the rabbits had to "work" on the left and on the right sides of the box in turn. The behavior and unit activity were recorded on a videotape. In addition, unit activity and various marks of behavioral events were simultaneously recorded with a multichannel magnetic tape recorder.

17.3 Discussion

In various cortical areas we found units specialized with respect to each behavioral act and with respect to each stage of learning. In particular, there were units that were only active during food seizure in the left feeder or in the right feeder (the first stage of learning on different sides of the box, (Figure 17.2); units active during the feeder approach and food seizure (the second stage of learning, Figure 17.3); units active while the rabbit remained in the pedal corner (the third stage of learning, Figure 17.4); and units active only during the left pedal press or the right pedal press (the last stage of learning, Figure 17.5).

Although the rabbit might press this pedal with either forepaw and although each time the objective relationship between the rabbit and its environment was different in detail, the unit was active each time that this pedal was pressed and was almost completely silent during the pressing of the other pedal. The behavioral act of left pedal pressing can be one and the same act only in the subjective world of the rabbit. This implies that these kind of units are specialized with respect to elements of subjective experience, i.e., behavioral acts established at successive stages of learning.

Other types of units are specialized with respect to innate behavioral acts, such as food seizure at any place in the experimental box or various movements. Units phenomenologically specialized with respect to some movement—for example, movement of the rabbit to the right—may be interpreted as specialized not with respect to sensory or motor functions, but with respect to an ancient behavioral act, i.e., the change of the relationship between an organism and its environment that gave some useful result at some time during phylogenesis.

Behavioral rather than functional unit specialization can be understood by the fact that during phylogenesis and ontogenesis and in individual learning by trial-and-error, processes—not functions, but specific behavioral acts—are established that are the basis for natural selection. In a problem situation, various trials are performed. All acts, although different in detail, that yield the same result are one and the same for the animal itself. All errors also are equally erroneous. This is the only possible way for animals to differentiate their relationships with the environment. Various sensory and motor functions artificially identified by sophisticated researchers are only variable and unlimited aspects of behavior.

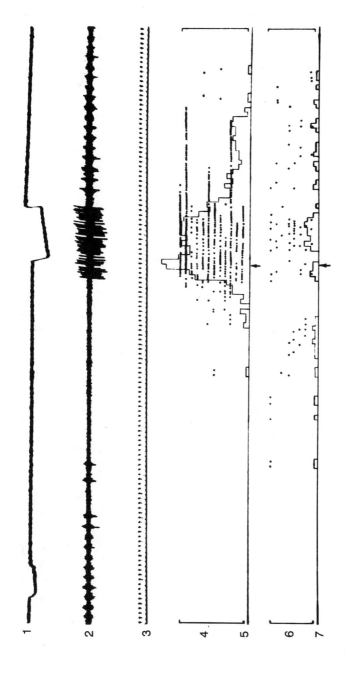

FIGURE 17.2. An example of a motor cortex unit specialized with respect to food seizure in the right feeder (the first stage of learning at the right side of the box). 1, Marks of right pedal press (small deflection) and food seizure in the right feeder (large deflection); 2, unit activity; 3, timer, each 100 msec; 4, raster of the unit activity in 12 food seizures from the right feeder; 5, histogram of this activity; 6, raster of the unit activity in 8 food seizures from the left feeder; 7, histogram of this activity.

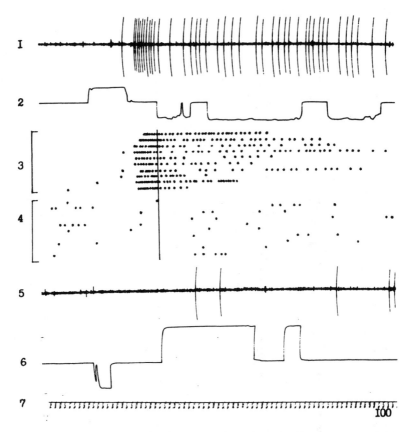

FIGURE 17.3. An example of a limbic cortex unit specialized with respect to the left feeder approach and obtaining food there (the second stage of learning). 1, unit activity; 2, marks of the left pedal press (up) and receiving food in the left feeder (down); 3, raster of the unit activity in ten acts of left feeder approach, accumulated from the beginning of the left feeder marks; 4, raster of the unit activity in ten acts at the right side of the box; 5, unit activity; 6, marks of the right pedal press (down) and receiving food in the right feeder (up); 7, timer, each 100 msec.

Behavioral specialization of neural units (i.e., their specialization in respect to elements of the subjective experience of the animal under study) provides the solution to the brain–mind problem. Description of the totality of specializations of active units at any moment constitutes at the same time the description of the state of the subjective world at that moment. The list of all unit specializations becomes the list of all memory elements established during phylogenesis and ontogenesis, i.e., the list of all subjective experiences.

This identity of brain and mind processes has an evolutional basis. The system-evolutionary approach regards behavior as the fulfillment of the genetic program of an organism's life cycle in dynamic relationships with its environment. This genetic program is expressed mostly in neural cells. In the mouse brain, for

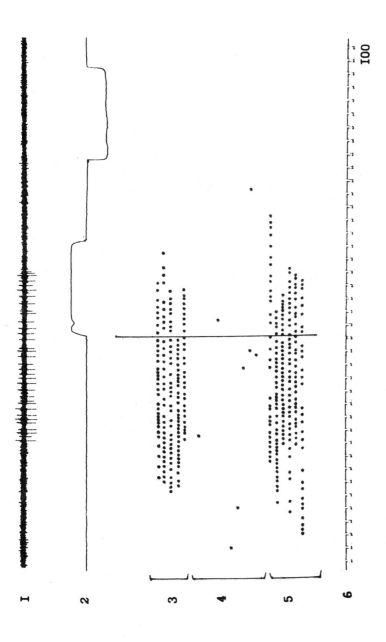

FIGURE 17.4. An example of a limbic cortex unit specialized with respect to the left pedal approach and press or to being in the left pedal corner (the third stage of learning). 1, unit activity; 2, marks of the left pedal pressing (up) and having food in the left feeder (down); 3, raster of the unit activity built up from the beginning of the pedal pressing in 5 behavioral cycles at the left side of the box; 4, the same in 11 cycles at the right side; 5, the same in 6 cycles again at the left side of the box; 6, timer, 100 msec.

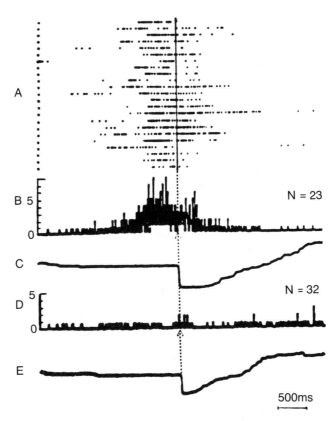

FIGURE 17.5. An example of a limbic cortex unit specialized with respect to left pedal pressing (the last stage of learning). A, raster of the unit activity accumulated from the beginning of the pedal press across 23 behavioral cycles at the left side of the box; B, histogram of this activity; C, averaged mark of the left pedal press; D, histogram of the unit activity across 32 behavioral cycles at the right side of the box; E, averaged mark of the right pedal press.

example, 140,000 genes are expressed and 100,000 of these are expressed exclusively in the brain (Hahn et al., 1982). Like any living cell, a neural cell fulfills its genetic program in metabolic processes and needs some metabolites coming from other cells. Spike activity appears when the metabolic needs of a cell are not satisfied by the metabolites coming from other cells. This activity can influence other cells and bodily processes, which can change the relationships between the body and the environment; that is, can cause some external behavior. This behavior can change the synaptic inflow to the cell and stop its activity.

Because the bulk of genes are expressed in the nervous system, the development of the brain in evolution has changed the formula of relationships between an organism and its environment. Instead of the formula "genome body environment," which is common to all living organisms, it became "genome nervous system body environment." The brain appears to be a special internal screen built up during evolution between the genetic program and its fulfillment in behavior. In the context of the objective external relationships between an organism and its environment, brain activity appears to be the internal subjective reflection of these relationships and at the same time the regulator of bodily processes. The nervous system makes possible the exchange between an organism and its environment not only of matter and energy, which is life, but also of information. Through behavior, the relationships with the environment are incorporated by means of the memory, i.e., the totality of adaptive behavioral acts that occurred in trial-and-error processes and were accumulated in neural cell specializations. With the discovery of behavioral specialization of neural units, a way was found to study mind; that is, the composition and interrelationships of elements of subjective experience, using objective neurophysiologic methods.

First, we tried to ascertain the locus of learning ability or the locus of units specialized with respect to acquired behavioral acts. In order to study whether unit specializations remain constant, we compared the incidence of limbic cortex cells with different specializations in rabbits that had learned to obtain food from two feeders and, with further training, to press the left pedal for food in the left feeder and the right pedal for food in the right feeder. Units active during chewing, food seizures in both feeders, or some particular movement were regarded as specialized with respect to "old" behavioral acts that a rabbit had in memory before its acquaintance with the experimental box. Units activated only during food seizure in one feeder or when a rabbit approached one feeder were regarded as specialized with respect to "new" behavioral acts, which rabbits learned during the first stage of learning. Finally, single pedal press units and single pedal approach units were regarded as belonging to the "newest" elements of memory.

We found that the appearance of units specialized with respect to the "newest" behavioral acts did not change the incidence of other types of units (Table 17.1). This result led us to suggest that newly specialized units were recruited from silent units, but not from previously specialized units.

To test this hypothesis, we compared the amount of active units found in limbic cortex before and after further learning. Before learning we encountered 108 active units in 15 tracks. After learning to press the pedals, we found 123 active

TABLE 17.1. Number of limbic cortex units specialized with respect to different behavioral acts

	Feeders only	Pedals and feeders
Newest	0	13
New	23	17
Old	23	22
Ancient-unspecialized	42	39
Total	88	91

units in 11 tracks. If one accepts that in each track a microelectrode actually met as many as 1600 units (Livanov, 1965) then in the first condition the microelectrode was in contact with 24,000 units and in the second condition with 17,600 units. Comparison of these amounts with the numbers of active units (108 and 123, respectively) shows that new learning increased the pool of active units from 0.45% to 0.69% (X^2=11; p <0.001) and consequently decreased the amount of silent units, which are probably the special reserve for new experience. It seems that we must assume that, during the learning process, newly specialized units come from the pool of silent reserve units.

We also tried to compare unit specializations in brain structures with different functional specializations and found units with identical behavioral specialization in different structures. This whole set would seem to constitute the functional system of the corresponding behavioral act. On the other hand, even these preliminary data show that different brain structures contain units with different behavioral specialization. This probably reflects the projection of subjective experience on the brain.

Table 17.2 shows that units of acquired specializations (behavior at the feeder and pedal places) are located in cortical areas and are almost absent in more phylogenetically ancient structures, such as the hypothalamus, cerebellum, lateral geniculate body, and optic tract. The units of inborn specializations (food seizure at any place, movements) spread over cortical and subcortical areas. Old structures differ in their behavioral specialization: optic tract and lateral geniculate body contain units specialized with respect to old locomotor acts, whereas dorsal hypothalamus and cerebellum also contain some units specialized with respect to ancient acts of feeding behavior. Cortical areas also are not equipotential. Units specialized with respect to inborn locomotor acts are more numerous in visual cortex (35.0%) than in the motor area (12.5%); in contrast, units of ancient feeding behavior are more numerous in motor cortex (31.0%) than in the visual area (4.0%). The number of units specialized with respect to acquired behavioral acts increases from the frontal pole (10.5%) and motor cortex (14.0%) to the visual (20.0%) and limbic cortex (31.5%).

It is well known that nervous system morphology reflects its phylogenesis. In the process of phylogenesis, new structures were added to old ones. The morphologic complication of the nervous system in evolution is the result of genetic changeability; natural selection deals not with the new morphologic formations by themselves, but with the behavioral acts that became possible with the help of units in these new formations. The difference of unit specializations in different cortical areas may reflect the history of these areas' formation in evolution and their different connections with more ancient structures and consequently with different ancient behavioral acts.

This then is the foundation for our suggestion that any external influence on an organism meets first the ancient elements of memory. The increment of unit response latency from periphery to central structures may reflect the movement of excitation through the memory structure from the oldest to the newest elements of the memory.

Table 17.2 Units of acquired specialization

N of units studied	Movement	Food seizure or chewing	Acquired behavioral acts (at one feeder or pedal location)	Unknown specialization
		N and % of units specialized with respect to		
Frontal pole	8	25	9	43
85 units	9.4%	29.4%	10.5%	50.6%
Motor cortex	10	25	11	34
80 units	12.5%	31.0%	14.0%	42.5%
Somatosensory cortex	16	19	14	41
90 units	17.7%	21.0%	15.4%	45.4
Visual cortex	35	4	20	41
100 units	35.0%	4.0%	20.0%	41.0%
Limbic cortex	17	2	37	61
117 units	14.5%	1.7%	31/5%	52.1%
Dorsal hypothalamus	6	2	3	30
41 units	14.6%	4.7%	7.2%	73.0%
Cerebellum (vermis)	10	8	0	19
37 units	27.0%	21.6%		51.3%
Lateral geniculate body	27	2	0	3
32 units	84.8%	6.25%		9.3%
Optic tract	34	0	0	0
34 units	100.0%			
Total	163	87	94	272
616 units	26.4%	14.1%	15.3%	44.1%

Any acquired behavioral act is accomplished not only by activation of cortical units that are newly specialized with respect to this act but also by activation of units specialized with respect to previous stages of learning and by units of many brain structures specialized with respect to ancient evolutionary acts, including spinal cord units directly related with the bodily processes. The learned behavioral act appears to be the recapitulation during some hundreds of milliseconds of all the long history of formation of this subjective experience element, as ontogenesis repeats phylogenesis in a curtailed manner.

In the structure of individual subjective experience, food deprivation also first activates older acquired elements of memory, such as food seizure. These acts cannot be realized in external behavior because there is no food in the feeder. This leads to the activation of more recent elements of experience. One of them—say, the pedal corner approach—is not only desirable but also possible at this moment. The realization of this act makes it possible to press the pedal and receive food in the feeder, which makes possible the realization of the older acts of feeder approach and food seizure. Because older elements are partly activated during the realization of younger acts, there exist various anticipation phenomena in be-havior.

Behavioral specialization of units means that spike activity of a specialized unit is the evidence that this subjective experience occurred. The units of individual acquired specializations may give inconstant and less pronounced activations during performance of other acts of learned behavior. The study of these incon-stant activations could provide us with an understanding of the interrelationships between subjective experience elements. Preliminary data show that there are facilitatory as well as inhibitory interrelationships between them. Figure 17.6 shows five units specialized with respect to the right pedal approach. Variable activations may occur when the rabbit approaches the left pedal (similar goals),

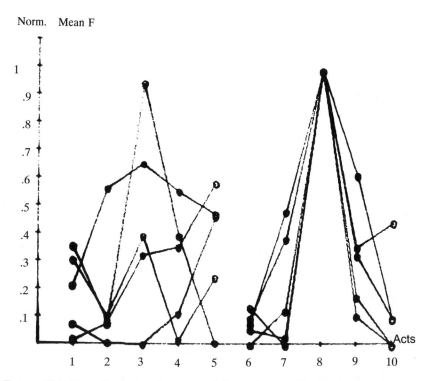

FIGURE 17.6. Normalized mean frequency of five units specialized with respect to right pedal approach in other acts on both sides of the box. Abscissa: 1, food seizure in the left feeder; 2, head turning toward the left pedal; 3, left pedal approach; 4, left pedal press; 5, left feeder approach; 6, food seizure in the right feeder; 7, head turning toward the right pedal; 8, right pedal approach; 9, right pedal press; 10, right feeder approach.

approaches the left feeder (similar movement to the left), and approaches the right feeder (similar position in front of the right wall of the box).

Therefore, ancient elements of innate subjective experience, activated on the basis of needs and goals or on the basis of environment–body relationships, may influence corresponding elements of learned experience. Various reciprocal and synergic relations also exist between innate acts, as we know from neurology. The state of each element of subjective experience (i.e., the number of active units and their frequency of discharging) depends on the state of very many other elements. Variable activations also indicate that during the performance of any behavioral act many unnecessary elements are partly activated. External behavior depends on the state of subjective experience at the moment, which in its turn depends on the metabolic needs of various neural cells, on interelement relationships, and on the environment and bodily processes.

The next steps are to study concretely how animal behavior is determined by these components and how the results of these studies may be applied to human beings.

References

Anokhin, P.K. (1974): Biology and neurophysiology of conditioned reflex and its role in adaptive behavior. Oxford: Pergamon Press, p. 574

De Long, M., Strick P. (1974): Relation of basal ganglia, cerebellum and motor cortex units to ramp and ballistic limb movements. *Brain Res.* 71, 327–335

Fetz, L., Cheney, P. (1978): Muscle fields of primate corticomotoneuronal cells. *J. Physio.* 239–245

Hahn, W.E., Van Ness, I., Chanhard, N. (1982): Overview of the molecular genetics of mouse brain. In: *Molecular genetic neuroscience.* New York: Raven Press, pp. 323–334

Hubel, D., Wiesel, T. (1959): Receptive fields of single neurons in the cat's striate cortex. *J. Physiol.* 148, 579–596

Livanov, M.N. (1965): Neyrokinetica. In: *Problemy sovremennoy neyrophysiologii.* Moscow: Nauka (in Russian), pp. 7–21

Mountcastle, V.B. (1957): Modality and topographic properties of single neurons of cats somatic sensory cortex. *J. Neurophysiol.* 20, 408–434

Mountcastle, V.B. (1978): Some neural mechanisms for directed attention. In *Cerebral correlates of conscious experience.* New York: North Holland Publishing Co, pp. 37–52

O'Keefe, I. (1976): Place units in the hippocampus of the freely moving rats. *Exp. Neurol.* 51, 78

Ono, T., Mishino, H., Sadaki, K., Fucuda, M., Muramoto, K. (1981): Monkey lateral hypothalamic neuron response to sight of food and during bar press ingestion. *Neurosci. Lett.* 21, 99–104

Pond, F., Sinnamon, H., Adams, D. (1977): Single unit recording in the midbrain of rats during shock-elicited fighting behavior. *Brain Res.* 120, 469–484

Shvyrkov, V.B. (1980): Goal as a system-forming factor in behavior and learning. In: *Neural mechanisms of goal directed behavior and learning.* New York: Academic Press

Shvyrkov, V.B. (1985): Toward a psychophysiological theory of behavior. In: *Psycho-physiological approaches to human information processing.* New York: North Holland Publishing Co, pp. 47–71

Shvyrkov, V.B. (1986): Behavioral specialization of neurons and the system-selection hypothesis of learning. In: *Human memory and cognitive capabilities.* New York: North Holland Publishing Co, pp. 599–611

Part VI

Human Development

18

Thoughts on the Role of the Mind in Recovery from Brain Damage

PAUL BACH-y-RITA

This chapter comments on the current status of neurorehabilitation and the role of the mind in rehabilitation. Rehabilitation is a medical specialty that crosses over into the field of education. In fact, in the United States the National Institute of Disability and Rehabilitation Research is part of the Department of Education, and in Mexico, rehabilitation is not part of the Ministry of Health, but of DIF (Desarrollo Integral de la Familia). Rehabilitation cannot be applied to a patient, as can an injection of penicillin for pneumonia, but must be accomplished primarily by the patient with the aid of rehabilitation professionals. Thus, the role of motivation, family support, the environment, and other psychosocial factors is of importance. It must be noted, however, that these factors are also of significance in all aspects of medicine, although their importance may not be appreciated by modern medicine, which is highly procedure-oriented. For example Ulrich (1984), in a retrospective study of all the patients in a suburban Philadelphia hospital who underwent gallbladder removal between 1972 and 1981, noted that the view from the window was correlated with the rate of recovery and the rate of usage of pain medications. Patients assigned to rooms looking out on a natural scene had shorter postoperative hospital stays and needed fewer analgesics than a comparable group of patients assigned to similar rooms with windows facing a brick wall.

Rehabilitation medicine is probably the least scientifically based medical specialty. There is little or no demonstrated scientific basis for most of the procedures, and few of the procedures have been validated by controlled studies. The importance of the intervention of the various therapies (e.g., speech, occupational, physical, recreational, etc.) has not been demonstrated. In spite of all of these facts, patients in rehabilitation settings do improve in function (Lehman et al., 1975).

There are a number of reasons for the delay in progress in neurorehabilitation. Some of these have been discussed elsewhere (Bach-y-Rita, 1981) and include the following:

Problems in determining the location and extent of damage to the brain (although the advent of brain imaging techniques has, in recent years, improved localization): Clinicopathologic correlations in cases with good rehabilitation outcome are rare. When recovery of function occurs, it has been difficult to determine the relative importance of resolution of local factors (tissue debris, edema) and of the development of alternate neural mechanisms (brain plasticity).

Difficulty in obtaining sufficient numbers of patients with similar lesions and characteristics which has hampered the effective use of matched group studies

Difficulty in quantifying progress accurately, which is in part due to inadequate functional evaluation procedures

Long time-scale of recovery: Progress can continue irregularly over many years. This feature makes research in stroke rehabilitation difficult due to cost and data collection factors.

High cost of long-term programs, which has limited the availability and access to rehabilitation, as well as impeding the development of comprehensive and functional rehabilitation programs

There are many factors that relate to the mind that have influenced the present state of neurorehabilitation, which produces little recovery after the first few months. It can be argued that one reason that little recovery occurs is because little recovery is expected to occur, and the absence of recovery may be a "self-fulfilling prophecy" (Merton, 1968). Philosophical and cultural factors are as important as psychosocial and biological factors, and there are changing styles and fads in medicine just as there are in the clothing fashions.

Polanyi (1964) commented that even fundamental science is an art. He stated:

Admittedly, there are rules which give valuable guidance to scientific discovery, but they are merely rules of art. . . . The rules of scientific enquiry leave their own application wide open to be decided by the scientist's judgment.

Since an art cannot be precisely defined, it can be transmitted only by examples of the practice which embodies it. He who would learn from a master by watching him must trust his example. He must recognize as authoritative the art which he wishes to learn and those of whom he would learn it. Unless he presumes that the substance and method of science are fundamentally sound, he will never develop a sense of scientific value and acquire the skill of scientific enquiry. . . . To learn an art by the examples of its practice is to accept an artistic tradition and to become a representative of it. Novices to the scientific profession are trained to share the ground on which their masters stand and to claim this ground for establishing their independence on it. The imitation of their masters teaches them to insist on their own originality, which may oppose part of the current teachings of science. It is inherent in the nature of scientific authority that in transmitting itself to a new generation it should invite opposition to itself and assimilate this opposition in a reinterpretation of the scientific tradition (Polanyi, 1964).

If fundamental science is an art, medicine is even more of an art, and it is thus not surprising that it is subject to changing styles. Thus, although the degree of recovery following damage to the brain depends in part on the extent of the injury, the age of the patient, the rapidity of the damage, and other biological factors, it also depends on the prevailing style, as well as on psychosocial and cultural factors. Home rehabilitation programs that are developed by nonprofessional family members may sometimes be more effective than professional programs, in part because the family members are unaware of the prevailing belief, which is that little or no recovery can be expected beyond some specified time (generally 6 months to 2 years). Frank (1986) has discussed the role of prevailing beliefs or "conceptual substance" of scientific fields. It is surprising how much effect the prevailing belief can have on results.

Elsewhere (Bach-y-Rita, 1988) I have discussed the effect that the very exciting finding of Broca, in 1861, had on the subsequent course of neuroscience: that is that a particular specific region of the brain was related to speech production. Broca's finding, interpreted in the context of all the other exciting scientific findings of the late nineteenth and early twentieth centuries, may have led to the domination of neuroscience by localizationist theories; the concept of rigid localization may have greatly delayed the emergence of plasticity concepts and of an expectation of recovery from brain damage. In turn, this delay may have negatively influenced the way that brain-damaged patients have been rehabilitated.

The human mind is capable of altering physiologic functions. The whole area of psychosomatic medicine is based on this fact. Yet, in general, the role of the mind is not sufficiently appreciated in designing treatment environments or the treatment itself. In traditional medicine, it is primarily in the study of placebos that this interaction is emphasized. For a new drug to be accepted, it must undergo rigorous testing, and the placebo effect must be subtracted from the "real" effect. Yet, a further consideration of this effect emphasizes the enormous therapeutic power of the mind (Bach-y-Rita, 1981).

Cousins (1977) has pointed out that the fact that a placebo will have no physiological effect if the patient knows it is a placebo only confirms the capacity of the human body to transform hope into tangible and essential biochemical change. He states, "The placebo is proof that there is no real separation between mind and body. Illness is always an interaction between both." He further points out that, in the absence of a strong relationship between doctor and patient, the use of placebos may have little point or prospect. In this sense, " . . . the doctor himself is the most powerful placebo of all" (Cousins, 1977).

For many years the brain was considered to be a rigid, unchanging organ. Clemente (1976) has pointed out that, although everyone admitted that an individual could learn and thus show functional plasticity, learning was considered an abstract function related to the mind and not the brain. Once the physical neural patterns of the nervous system had developed (i.e., once the fiber connections had become established), alterations in neuronal geometry were not considered possible. Clemente considers that a change in thinking occurred as the gap between the mind and the body began to close. "The so-called structure of the mind had never been considered rigid, only the structure of the brain. It followed naturally that since the brain was the organ of the mind, and that since the brain was composed of cells, that plasticity could only be the result of cellular events in the central nervous system" (Clemente, 1976). Clemente suggests that the ambient aura for acceptance of the concept of central nervous system plasticity really developed as a post-World War II phenomenon. He considers that this acceptance, together with the recent reevaluation of old anatomic evidence and new experimental evidence (especially anatomic, physiologic, and biochemical) for plasticity, has altered our views (Bach-y-Rita, 1980).

Elsewhere (Bach-y-Rita, 1980) I noted that rehabilitation shares with psychiatry in particular a need to examine and incorporate psychosocial factors, many of which have not been explained and which may be unexplainable by the biomedical model. Engel (1977) has discussed the dilemma of psychiatry in this regard. He pointed out that most of medicine appears neat and tidy, with a firm base in the biological sciences, enormous technical resources, and a record of astonishing achievement in elucidating mechanisms of disease and devising new treatments. He considers that the problem of psychiatry (and much of the rest of medicine) stems from the logical inference that since "disease" is defined in terms of somatic parameters, physicians need not be concerned with psychosocial issues that lie outside medicine's responsibility and authority. He notes that "rational treatment" directed only at the biochemical abnormality does not necessarily restore the patient to health, even in the face of documented correction or major alleviation of the abnormality. Conspicuously responsible for such discrepancies between correction of biological abnormalities and treatment outcome are psychological and social variables. He further notes that, even with the application of rational therapies, the behavior of the physician and the relationship between patient and physician influence powerfully therapeutic outcome for better or worse. These variables constitute psychological effects that may modify the illness experience directly or affect underlying biochemical processes indirectly, the latter by virtue of interactions between psychophysiological reactions and biochemical processes implicated in the disease (Ader, cited by Engel, 1977). Engel (1977) states,

The physician's role is, and always has been, very much that of educator and psychotherapist. To know how to induce peace of mind in the patient and enhance his faith in the healing powers of his physician requires psychological knowledge and skills, not merely charisma. These too are outside the biomedical framework.

Motivation, in the context of this discussion, relates mental activity to action, reflecting the desire to recover and the willingness to work for delayed rewards. Rehabilitation can be a long process, and patients often become discouraged. It is necessary to demonstrate that recovery is occurring by such means as pointing out increased movement control. Using videotapes at various stages in the recovery, graphing certain motor activities or electromyographic evidence of improved control, or other objective means that encourage mental images of improvement are approaches used in the clinic. Family support can be critical. Patients, however, can develop the motivation from their own resources, especially if there is a specific goal.

The major gains in function are usually made early, and small increases in function continue for some time, but generally the efforts cease short of full recovery. The actress, Patricia Neal, for example, was unwilling to proceed to work for the small gains.

I got fed up with working so hard. I felt certain that I was as good as I'd ever be. I was about 80% recovered. Still plenty of problems. But I was ready to take a breather. I wanted to give up and lie back and do nothing. And that is exactly what I would have done if Roald (her husband) hadn't made me go on. I had reached the point where so many people stop work and just cruise along (P. Neal, in Griffith, 1970, p. 89).

The extra effort enabled Patricia Neal, an Academy Award winning actress, to recover that last portion of function that permitted her to resume her acting career. However, it is doubtful that a 100% recovery can be obtained after major lesions, such as those suffered by Ms. Neal (Bach-y-Rita, 1980).

Jokl (1964) analyzed factors leading to superior performance in Olympic champions with major motor deficits acquired either before or after reaching their championship level. He considered that an Olympic champion demonstrated the highest level of human performance, mobilizing all possible mental and physical resources, and thus could offer a model of the capacity to recover. For example, Harold Connolly (1956 hammer-throw champion) had a combined injury of upper and lower left brachial plexus (Erb-Duchenne-Klumpke-Dejerine type) with marked growth defect and pareses of left arm muscles as a result of a birth trauma. Bill Nieder suffered, 1 year before his second Olympic Games, spontaneous tearing in the right pectoral muscle, resulting in reduced force of the main muscle involved in propulsion of the right arm (he is right-handed) during the shotput. Yet, his Gold Medal performance at the second games actually was 5 feet longer than his previous Olympic performance (Jokl, 1964). Among factors in such effective rehabilitation, Jokl considers conceptual originality and creative awareness, appropriate habit formation, intelligence for redesigning motor control strategies, and intensive exercise. The handicapped person is compelled to become aware of sensorimotor features that can integrate for action every part of the motor system that remains available.

A negative example of the capacity of the mind to influence somatic function was provided by Cannon, who published an article in 1942 entitled "Voodoo Death," apparently because of his interest in the ability of mental activity to influence homeostatic mechanisms even to the point of producing death in otherwise healthy persons. More recently, Rosenzweig (1980) has reviewed the literature on animal studies related to the role of the environment on developmental measures and on recovery from brain damage. This review is particularly pertinent to a study of brain rehabilitation, which must be related to the mechanisms of action of environmental and psychosocial factors in influencing recovery. Rosenzweig reviewed some of the few animal studies that demonstrate the capacity of brain-damaged animals to recover. Generally, neuroscientists have been more interested in describing what occurs following a lesion than they are in examining what the animal could do with an excellent training (rehabilitation) program. One excellent cat study that did explore the capacity to recover function was published by Chow and Stewart (1972). Previous studies had revealed virtually permanent amblyopia following lid suture of one eye during a few months of the critical period of development of vision. With an interesting rehabilitation program, Chow and Stewart obtained functional improvement in vision correlated with increased binocular cells. In addition to their brain plasticity findings, their procedures are pertinent to human rehabilitation. For example, they noted that the usual rewards were insufficient; the cats required periods of "gentling" and petting (to establish an effective bond with the experimenters). They developed a demanding, intensive rehabilitation program while avoiding frustration.

When possible, rehabilitation programs tailored to the particular interests of the patient may motivate him or her to spend the time and effort to recover function. In order to develop coordinated eye movements sufficient to permit children with cerebral palsy to learn to read, Gauthier and his colleagues (Gauthier and Hofferer, 1983; Gauthier et al., 1978) developed a functional pendulum. They had noted that the children resisted the usual therapy with a pendulum, becoming bored and disinterested. They used a large projection screen, sat the child in front of it with head immobilized, and projected children's movies (Snow White; Mickey Mouse) from the opposite side of the screen, but in a small dimension (similar to a small TV image). As the child observed the film, the therapist moved the image slowly from side to side in pendular movements. As the children gained more control, the pendular movements became larger and faster, and other types of movements (irregular, interrupted) were introduced. Thus, various types of eye movements could be trained. With this functional training program, sessions could last 2 hours without the onset of boredom or fatigue, and marked improvement in oculomotor control was obtained. In another example of this approach, which I have called functional rehabilitation, a hemiparetic man regained virtually full function with a 5-year home program, designed by his son, that was based on the patient's interest (Bach-y-Rita, 1980). The evidence reviewed here strongly suggests that the mind has an important role in successful rehabilitation following brain damage. At the same time, the role of mental activity in the recovery process may serve as a model of the ability of the mind to influence somatic activity.

References

Bach-y-Rita, P. (1980): Brain plasticity as a basis for therapeutic procedures. In: *Recovery of function: Theoretical considerations for brain injury rehabilitation.* Bach-y-Rita, P. (ed.). Bern: Hans Huber

Bach-y-Rita, P. (1981): Brain plasticity as a basis of the development of rehabilitation procedures for hemiplegia. *Scand. J. Rehabil. Med.* 13, 73–83

Bach-y-Rita, P. (1988): Brain plasticity. In: *Rehabilitation medicine.* Goodgold, J. (ed.). St. Louis: C.V. Mosby

Cannon, W. (1942): Voodoo Death. *Am. Anthropol.* 44, 169–181

Chow, K., Stewart, D. (1972): Reversal of structural and functional effects of long-term visual deprivation in cats. *Exp. Neurol.* 34, 409–433

Clemente, C. (1976): Changes in afferent connections following brain injury. In: *Contemporary aspects of cerebrovascular disease.* Austin, G. (ed.). Dallas: Professional Information Library

Cousins, N. (1977): The mysterious placebo: How mind helps medicine work. *Sat. Rev.* Oct. 1, 9–16

Engel, G. (1977): The need for a new medical model: A challenge for biomedicine. *Science* 196, 129–136

Frank, R. (1986): The Columbian exchange: 1986, American physiologists and neuroscience techniques. *Fed. Proc.* 45, 2665

Gauthier, G., Hofferer, J. (1983): Visual motor rehabilitation in children with cerebral palsy. *Int. Rehabil. Med.* 5, 118–127

Gauthier, G., Hofferer, J., Martin, B. (1978): A film projection system as an eye movement diagnostic and training technique for cerebral palsied children. *Electroenceph. Clin. Neurophysiol.* 45, 122–127

Griffith, V. (1970): *A stroke in the family.* New York: Delacorte Press

Jokl, E. (1964): *The scope of exercise in rehabilitation.* Springfield, IL: Charles C. Thomas

Lehman, J., DeLateur, B., Fowler, R., et al. (1975): Stroke rehabilitation: Outcome and prediction. *Arch. Phys. Med. Rehabil.* 56, 383–389

Merton, R. (1968): The self-fulfilling prophecy. In: *Social theory and social structure.* Merton, R. (ed.). New York: Free Press

Polanyi, M. (1964): *Science, faith and society.* Chicago: University of Chicago Press

Rosenzweig, M. (1980): Animal models for effects of brain lesions and for rehabilitation. In: *Recovery of function: Theoretical considerations for brain injury rehabilitation.* Bach-y-Rita, P. (ed.). Bern: Hans Huber

Ulrich, R. (1984): View through a window may influence recovery from surgery. *Science* 224, 420–421

19

Analysis of Electroencephalographic Maturation

Thalía Harmony, Erzsébet Marosi, Ana E. Díaz DeLeón, Jacqueline Becker, and Thalía Fernández-Harmony

19.1 Introduction

Neurometrics is a methodology born of the necessity to obtain objective parameters for electroencephalogram (EEG) and evoked potentials in order to perform automatic analysis of brain function. Based on the quantitative analysis of electrophysiological signals, its goal is to obtain probabilistic criteria regarding the maturation and functional integrity of the nervous system. This chapter focuses on the maturation of the EEG.

To construct a neurometric procedure for the evaluation of this process, it was necessary to (1) detect the EEG parameters that change significantly with age, (2) acquire data in several samples of normal subjects to establish statistical criteria about the distribution of such parameters, and (3) identify the factors that may alter the behavior of these parameters by function of age.

The first problem was relatively easy to solve by frequency analysis of the EEG. The power in the different frequencies or in a given range of frequencies is obtained by this procedure. The EEG power of the delta, theta, alpha, and beta bands was computed by comparison with what had been reported by visual inspection. The pioneer paper of Matousek and Petersén (1973) demonstrated that absolute power decreases with age and that relative power or the percentage of delta and theta also decreases with age while alpha and beta increase. Later, John et al. (1980) published the regression equations with age of the relative power of each band for the same bipolar derivations used by Matousek and Petersén in a sample of U.S. children. They compared these American norms with those previously published in the Swedish population and found no differences. These results led the authors to state that all children follow the same pattern of EEG maturation, independently of ethnic and sociocultural factors. Once the age-regression equations had been obtained in a sample of normal children, Ahn et al. (1980) and John et al. (1983) applied them to the study of children with slight mental retardation, learning disabilities, and neurological deficits. In every child the parameters derived from frequency analysis of the EEG were Z-transformed, or standardized according to age using previous normative age-regression equations. This procedure ensures the removal of age effect and establishes a common metric of relative probability for each measurement across age. Many of the children from the groups with neuropsychological problems presented deviant values, and for this reason it was proposed that this procedure might be used as an objective criterion for the evaluation of new subjects. Such impressive results moved us to explore the legitimacy of the age-regression norms in other groups of subjects with different economic, psychosocial, and cultural characteristics.

19.2 Maturation of the EEG in Children With Different Psychosocial Characteristics

19.2.1 Mexican Rural Populations

The great majority of children from underdeveloped countries live in the marginal urban areas of large cities or in rural regions. Both places are characterized by poverty. There are inadequate sanitary conditions and poor living environments, which have severe consequences for children: malnutrition, infectious diseases, pregnancy and parturition without medical attention, and deficiencies in the early caregiving environment. Children from these areas rarely achieve the criteria used for the selection of subjects for computation of the EEG age-regression equations used as norms. Our first idea was to study a wide sample representative of Mexican rural areas. The population selected was the one studied by Dr. Cravioto, which included children born between February 1, 1966, and February 28, 1967, as well as their brothers and sisters.

The results described first are based on a study of 228 children (104 girls, 124 boys) between 6.5 and 15.5 years. A questionnaire provided the following information for each child: personal pathological antecedents, socioeconomic status (SES), and academic experience and achievement. Physical development (weight and size), language, and collaborative behavior at the time of the EEG test were also evaluated. The pathological antecedents considered were retarded physical development, head trauma with loss of consciousness, perinatal asphyxia, birth after five parturitions of the mother, and problems in verbal comprehension or expression.

Children were classified into four groups according to their level of education:

1. Those who attended school at a grade level appropriate to their age ($n=106$)

2. Those who attended school at a grade lower than that corresponding to their age and those who left school before completing the elementary grades ($n=82$)

3. Academic underachievers ($n=26$)

4. Illiterate children ($n=14$)

EEG acquisition and analysis followed exactly the specifications of John et al. (1980). Computation of the linear equations in the same bipolar derivations used by John showed that in central (C3Cz and C4Cz) and frontotemporal (F7T3 and F8T4) regions, the slopes were very similar to those described for U.S. children. However, in temporal (T3T5 and T4T6) and especially in parieto-occipital (P301 and P402) regions, the slopes were less marked in Mexican children. As these equations were obtained in the whole sample, we suspected that these results were due to the presence of children with antecedents, as well as of academic under-achievers in the sample; therefore, the equations were recomputed for the group of children with no antecedents from Group 1. The results were very similar to those previously obtained. We concluded that some sociocultural and nutritional factors cannot be discarded as the causes of the differences observed in the EEG maturation of children from rural areas with clear handicaps compared to those of the urban middle class.

We also wanted to know if there were differences in relative power between the different groups of children. In all children, Z-transformed values using American normative equations were computed for all measures. In Figure 19.1 the distribution of abnormal values (greater or lower than 1.96) in the different groups of children is shown. The presence of three or more abnormal values occurred more frequently in the children with antecedents and in the illiterate group. Comparison of Z values between children with and without antecedents demonstrated more delta and theta and less alpha in children with antecedents. The differences were more marked when comparisons were made considering only Groups 2, 3, and 4.

Composite measures computed from multiple variables were also compared among groups. These composite measures were the following: overall delta, theta, alpha, or beta (the values of delta, theta, alpha, or beta across the eight bipolar derivations); left hemisphere (all values for left derivations); right hemisphere (all values for right derivations); anterior region (all values in F7T3, F8T4, C3Cz, C4Cz); posterior region (all values in P301, P402, T3T5, T4T6); and total values for each derivation (delta, theta, alpha, and beta values for the corresponding derivation). Such composite measures also have U.S. norms; therefore we could compute the Z-transformed values according to age. When comparisons of these composites were made between children with and without antecedents, greater significant differences were observed in right hemisphere, posterior region, generalized theta, and in F8T4, T4T6, P301, and P402.

Comparisons among groups with different educational levels showed significant differences between Groups 1 and 4. Table 19.1 shows that the illiterate children had more delta in P402 and more theta and less alpha in almost all derivations. In the same table the differences in the composite measures can be observed. Overall theta and alpha, left hemispere, right hemisphere, anterior region, and posterior region values were significant as were total values for each lead except C3Cz.

This study led to the conclusion that psychosocial handicaps were as important as biological factors in producing abnormalities in EEG maturation.

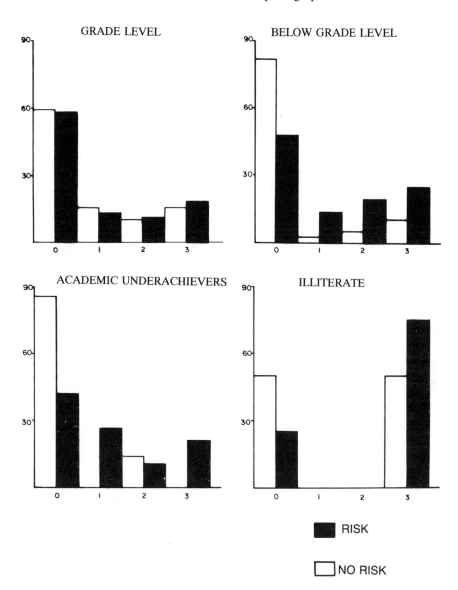

FIGURE 19.1. Percentage of Z abnormal values in children with and without risk of the four groups from a rural area. It is possible to observe that the percentage of children with three or more abnormal values is higher in those with antecedents.

TABLE 19.1. Significant differences between normal academic achievers and illiterate children (Groups 1 and 4) in rural Mexico.

Relative power	t	Combined measures	t
Delta P402	−1.96	Overall theta	−4.11
Theta C3Cz	−3.46	Overall alpha	−4.12
Theta C4Cz	−2.21	Left hemisphere	−3.78
Theta T4T6	−2.46	Right hemisphere	−3.41
Theta P301	−2.33	Anterior region	−3.20
Theta P402	−2.35	Posterior region	−3.79
Alpha C3Cz	3.13	Total C4Cz	−3.38
Alpha C4Cz	2.14	Total T3T5	−2.06
Alpha T3T5	1.98	Total T4T6	−2.87
Alpha T4T6	3.03	Total P301	−3.66
Alpha P301	2.81	Total P402	−2.03
Alpha P402	2.35	Total F7T3	−2.98
Beta T4T6	−1.96	Total F8T4	−2.59

19.2.2 Latin American Urban Samples

As it was very difficult to obtain representative samples of urban areas, our criteria for the selection of children changed. In a collaborative study that included Cuba, Mexico, and Venezuela (Harmony et al., 1988), samples of children selected with different criteria were studied. There were six groups:

1. Cuba: 96 children (40 boys and 56 girls) between 7 and 11 years old from the city of Havana. Children were selected with strict criteria of normality: normal neuropsychological development, absence of neurological and psychiatric symptoms, satisfactory academic achievement according to the school and to an independent pedagogical evaluation, and global IQ score (WISC) not lower than 85. In the case of children with global IQs lower than 90, verbal and performance IQs were required to be higher than 85.

2. Mexico 1: 28 children (13 boys and 15 girls) between 7 and 12 years old from a middle-class school in Mexico City. Children were selected with strict criteria of normality: no personal antecedents of risk factors associated with brain damage, normal psychomotor development, normal pediatric and neurological evaluations, IQ score not lower than 90, and good academic achievement according to the school and an independent pedagogical evaluation.

3. Mexico 2: 28 children (15 boys and 13 girls) between 7 and 12 years old from a marginal (very poor) urban area of Mexico City. Many of these children (20) had known antecedents with risk factors and/or delayed psychomotor development. All belonged to a group with clear socioeconomic and cultural handicaps. They had no neurological symptoms or great behavioral problems, although sleep disturbances and some degree of aggressive behavior were

observed frequently. All had IQ scores no lower than 90, attended school regularly, and had satisfactory academic achievement at school corroborated by an independent pedagogical evaluation.

4. Mexico 3: 30 children (19 boys and 11 girls) between 7 and 12 years old from a marginal urban area of Toluca (State of Mexico). All belonged to the lower socioeconomic class. Eighteen had known antecedents with risk factors associated with brain damage. At the moment of the study they had neither neurological symptoms nor important psychiatric disturbances, although sleep disorders and habit disorders were observed. All had normal IQ scores and attended school regularly with a satisfactory academic achievement.

5. Venezuela 1: 55 children (28 boys and 27 girls) between 4 and 12 years old from Caracas. They belonged to the lower and marginal (poor and very poor) socioeconomic classes. Seventeen children had known antecedents with risk factors associated with brain damage. The pediatric examinations had normal results. In 53% soft neurological signs were observed. Forty children had an IQ (WISC or WPPSI) above 90 and four children above 100. The remaining 11 children had scores between 75 and 90. All attended school regularly.

6. Venezuela 2: 26 middle-class children (14 boys and 12 girls) between 7 and 12 years old whose parents worked at the Central University of Venezuela in Caracas. They had normal pediatric and neurological evaluations, normal IQ scores (WISC), attended school regularly, and had satisfactory academic achievement.

In Groups 1, 5, and 6, direct bipolar EEG recordings were obtained in F7T3, F8T4, C3Cz, P301, P402, T3T5, and T4T6, with eyes closed. In Groups 2, 3, and 4, bipolar recordings were constructed by computer using the monopolar recordings with reference to linked ear lobes. The characteristics of the amplification system in each country were slightly different. The EEGs were visually edited for artifact rejection in all cases.

In Groups 1 to 4 the frequency analysis was performed by fast Fourier transformation (FFT) of 24 segments with a duration of 2.5 sec each, so as to complete 1 minute of EEG. In Groups 5 and 6, FFT analysis was performed for 10 sec of EEG (five segments of 2 sec length).

Power of the following bands was computed: delta (1.5–3.5 Hz), theta (3.5–7.5 Hz), alpha (7.5–12.5 Hz), and beta (12.5–19 Hz). Relative power of each band was calculated as the percentage of total EEG activity. The log $(x/(1-x))$ transformation of relative power was used for statistical analysis because it has been shown that with this transformation Gaussian distributions of relative power are obtained (Gasser et al., 1982; John et al., 1983).

Linear regression equations were calculated for the relative power of all spectral bands as a function of age in every derivation. The values of the intercepts were not considered because the differences in the measurement systems could account for an approximately constant offset. The slopes of the regression equations in each group were compared with those published by John et al. (1980) for the U.S. children using Student's *t* test (Yamane, 1976).

The regression analyses showed that the correlation between age and alpha and theta relative power was significant in almost all derivations in Groups 1, 2, 5, and 6. In the delta band only Groups 2, 5, and 6 presented significant correlations with age. In the beta band both groups from Venezuela had significant correlations between age and relative power. Group 3 showed significant correlations only in the beta band, and Group 4 showed no significant correlations between age and relative power.

In relation to the comparison of slopes of the regression equations as a function of age, Groups 1, 2, and 6 presented very few significant differences from those of the U.S. children. Cuban children only showed significant differences in theta-relative power in T3T5 and P402, Group 2 of Mexican children showed significant differences in theta and alpha-relative power in C4Cz, and Group 6 from Venezuela showed one difference in alpha-relative power in F7T3. The overall results for each of these groups were well below chance level for multiple comparisons using the Bonferroni correction (Morrison, 1976). Therefore, a remarkable degree of congruence was obtained, considering these two methodological differences in procedure: John et al. (1980) used digital filters for the calculation of the power of each band instead of FFT as used in the present work, and the filter characteristics of the amplification systems were different in every country with respect to the system used by John et al. (1980). On the other hand, children from the marginal urban areas of Mexico City and Toluca presented many significant differences in the slopes of the regression equations in comparison with those published by John. Group 5 from Caracas had smaller slopes of theta-relative power in almost all derivations.

From this study it was concluded that children who had grown up with adequate nutritional, sanitary, and cultural environmental conditions followed the same pattern of EEG maturation independently of cultural and ethnic factors. However, children brought up in poor socioeconomic and sanitary environments and who frequently had pathological personal antecedents with risk factors associated with brain damage showed either a slow EEG maturation characterized by smaller slopes of theta-relative power or a great variance of EEG parameters and no relation of these parameters to age.

19.3 Effect of Sex, Socioeconomic Status, and Biological Risk Factors on EEG Maturation

In a sample of 118 children from Mexico City, monopolar eyes-closed EEG recordings were obtained in F3, F4, C3, C4, P3, P4, 01, 02, F7, F8, T3, T4, T5, and T6. This study was done in order to analyze the effects of sex, psychosocial disadvantages, and biological risk factors on EEG maturation. Absolute and relative power for delta, theta, alpha, and beta bands were computed.

19.3.1 Sex Effect

19.3.1.1 Absolute Power

Linear and quadratic polynomial regressions with age were calculated for each parameter in each derivation. Significant correlations with age were found in almost all derivations for delta, theta, and total absolute power. Absolute power decreased with age. Quadratic expressions accounted for slightly more variance than linear models. Regression models were also computed for each sex separately. Boys' values were much better accounted for age than girls' values. Quadratic models accounted for a higher amount of variance in boys. In delta power, between 30 and 40% of the variance was explained by age. Theta power was explained even better: in right central and parietal areas, R-square values reached 0.54. In alpha and beta bands, a significant effect only existed in some leads. However, girls' values had a significant correlation with age only in delta power, with the linear model more adequate than the quadratic regression. In Figure 19.2 the distribution of delta and theta powers according to age is illustrated for boys in the right leads.

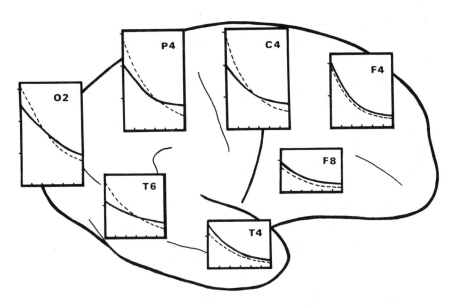

FIGURE 19.2. EEG development in right monopolar leads from 6–12 years (horizontal scale in steps of 1 year) for absolute power in the delta *(solid line)* and theta *(dashed line)* bands in boys.

19.3.1.2 Relative Powers

Linear and quadratic models were also explored in relation to age. Taking both sexes together, the linear model was better for many leads in delta-, theta-, and alpha-relative powers. Beta-relative power was better accounted for in the quadratic model and in more leads. In general, R-square values were lower than those observed with absolute power.

However, when boys and girls were assessed independently, the quadratic model explained more variance in girls. Delta- and alpha-relative powers with significant correlations with age were found in many more leads in girls than in boys. Slopes were always greater in girls than in boys, except in F7. The intercepts of theta-relative power in boys were smaller than in girls, demonstrating that at lower ages boys had a lower percentage of theta than girls. In general, slopes were greater in boys, except in P3, P4, and 01. Beta-relative power was significantly correlated with age in few leads in boys and girls. In Figure 19.3 the regression equations for alpha-relative power in 02 for boys and girls have been plotted. One can see that at age 6 the differences between sexes are greater than at age 12 because slopes are more marked for the female sex.

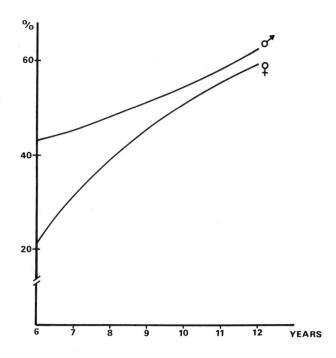

FIGURE 19.3. EEG development of 02 from 6–12 years for alpha-relative power in girls and boys.

19.3.2 Socioeconomic Status Effect

For the analysis of socioeconomic status (SES) affecting EEG, Z-transformed values with age were used. We evaluated SES according to the mother's level of education and income per capita. If either the mother was illiterate or the income per capita was below 25% of the minimum monthly wage, or if the mother had only completed 3 years of elementary school and income was between 25 and 50% of the minimum monthly wage, the SES was considered low. If the income was more than 50% of the minimum monthly wage or if it was between 25 and 50% of the minimum monthly wage and the mother had a vocational school or university education, the SES was considered to be good. An income of between 25 and 50% of the minimum monthly wage and a mother who had completed elementary or secondary school was considered to be intermediate SES.

19.3.2.1 Absolute Power

Differences between SES were assessed by analysis of variance. It was found that total, delta, theta, and beta absolute powers in F4 and F8 and delta and theta absolute powers on F3 and T4 were higher in children with low SES than in children with good SES (Figure 19.4). Delta values in F4 and beta values in F8 were greater in lower than in intermediate SES.

19.3.2.2 Relative Power

When comparisons were made combining data from both sexes, higher delta-relative power in 01, 02, T3, and T6 and lower alpha-relative power in F3, P4, 02, F7, T3, and T6 were observed in low SES than in good SES. Boys with good SES had less relative power in the delta band in T3, T5, and T6 and more alpha-relative power in F8 than boys with bad SES. No differences were observed between girls with different SES values. Computation of age-regression equations by SES demonstrated that, although the slopes were similar, children with good SES had more alpha-relative power, as is illustrated in Figure 19.5.

Canonical correlations analyses between Z-transformed values of relative power and SES were performed. The results suggested that a low SES was related to higher delta-relative power, whereas a good SES had a higher amount of alpha.

19.3.3 Risk Effect

Age-regressed and Z-transformed values were used to compare children with and without antecedents. Comparisons of both sexes together showed significant differences in delta absolute power in F4, P4, F8, T4, T5, and T6; in theta power in F4, F8, and T4; in alpha in F4, C4, P4, F8, T4, T5,and T6; in beta in F4, C4, P4, F8, T4, T5, and T6; and in total absolute power in F4, C4, P4, F8, T4, T5, and T6. Children with antecedents had higher values.

Taking into account that interactions between sex and risk exist (Diaz de León et al., 1988), boys and girls were assessed independently for risk. Using quadratic

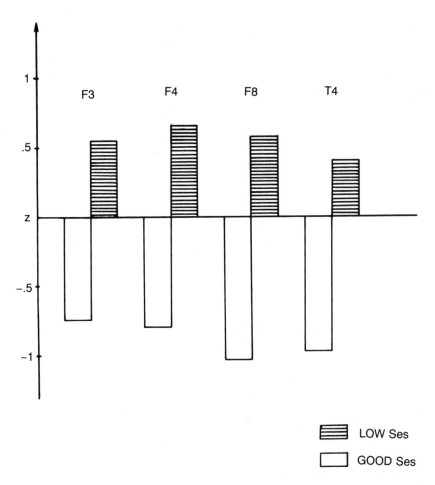

FIGURE 19.4. Z-transformed values of delta-absolute power in F3, F4, F8, and T4 in children with good and low SES.

Z-transformations for age for boys, more significant values were observed in children with risk in all bands in F3, F4, C4, P3, F8, T3, T4, T5, and T6; in delta, alpha, and beta in C3; in theta, alpha, and beta in P4; in theta and alpha in 01; in alpha and beta in 02; and in beta and delta in F7.

When only girls were considered, the differences between children with and without antecedents were not apparent. Only two significant values were observed, which may be due only to chance since multiple comparisons were

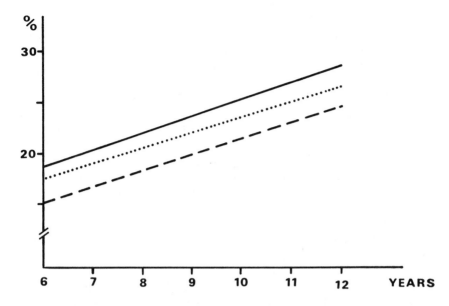

FIGURE 19.5. EEG development at F7 from 6–12 years for alpha-relative power in children with good SES *(solid line)*, intermediate SES *(dotted line)*, and low SES *(dashed line)*.

performed. Absence of significance may also be due to the small sample of girls with antecedents.

MANOVA were performed to analyze the effect of the severity of risk in each sex. Z-transformed values of total power and power in delta, theta, alpha, and beta bands in each lead were used for the multivariate analyses. Wilk's lambda was employed to assess the multivariate significance and Tukey's Studentized Range Tests were used for the comparisons between degrees of severity of risk. The regions that showed significant Wilk's lambda in boys were T4 ($p=0.004$); C4, P4, T5 ($p=0.02$); and F4 and P3 ($p=0.04$). Univariate significant values in all parameters were found in F3 and F8 and in beta and total power in all areas. Comparisons between grades of severity of risk demonstrated great differences between children with no antecedents and children with severe antecedents and between children with slight and severe antecedents.

In girls, no significant Wilk's lambda were observed. In F7, however, all univariate values were significant, as well as in T3 theta and total power. Comparisons among groups of different risk severity showed differences between children without antecedents and with slight antecedents.

In relation to risk antecedents, in samples of children with low SES with and without antecedents, we have observed that relative power had no relation with age in different bands from bipolar derivations (Harmony et al., 1988). We were then interested to know if the values of relative power behaved the same way in monopolar leads. Linear regression equations with age were computed separately in children with antecedents. Significant *positive* slopes were found for delta-relative power in C3, C4, P3, P4, F8, and T4; significant negative slopes were found for theta in F4, F8, and T4; and a *negative* slope for alpha-relative power was found in almost all leads, although it was only significant in C3. These results explained why no correlation with age was observed when samples of children with and without risk were mixed together. However, no differences were found when Z-transformed values were compared in children with and without antecedents. Neither did differences appear when boys and girls were compared separately. Children were then grouped in three different ranges of age: 6.0–8.5, 8.6–10.0, and 10.1–12.9 years. It was found that in the lower age group children without antecedents had more theta and less alpha. In the older group the opposite effect was observed: children with no antecedents had less theta and more alpha. Examination of the composition of the younger group with no antecedents showed that the great majority belonged to the low SES group. Therefore, it was impossible to reach any conclusion.

To determine if there are deviations from normal physical development in children with risk antecedents, the following study was conducted by Fernández-Bouzas et al. (1988). In a group of 40 children the area of the carpus and metacarpal bones of the left hand were measured. The regressions of the area of each bone with age were calculated. Almost all regressions were highly significant (Figure 19.6). For the same children the EEG relative power in each band was calculated. The regression of relative power with age was not significant, as it was expected. From these results it was concluded that, although deviant patterns of EEG maturation exist, the biological development may be normal, which suggests that the brain is more sensitive to risk factors than other body tissues (Fernández-Bouzas et al., 1988).

As with our sample of children we were not able to reach any conclusion in relation to risk; because the effect was contaminated by the socioeconomic factor, a different sample of children was then analyzed. The sample was made up of American children recorded and analyzed at Brain Research Laboratories of New York University by Dr. E. Roy John and Dr. Leslie Prichep. All children had a high socioeconomic status. Seventy-three children with no history of risk were compared with 56 children with risk antecedents (low weight at birth, forceps, perinatal anoxia, head trauma with loss of consciousness, etc.). Students' t tests were calculated with Z scores using American norms. Higher significant values ($p<0.05$) were observed in the risk group for total absolute power in FZ, Fp1, Fp2, F3, F4, C3, C4, Cz, F7, T3, and T5 (Figure 19.7); for delta-absolute power in Fp2, Fz, F3, and F7; for theta-absolute power in Fz, F3, F4, Cz, C4, 02, F7, and T5; for alpha-absolute power in Fz, Fp2, F3, F4, C3, Cz, T3, T4, T5, and T6; and for beta-absolute power in Fz, F3, T5, and T6. It is possible to observe that the anterior region and the posterotemporal leads were consistently affected by risk.

These differences were very similar to those already mentioned in the Mexican group. However, the comparison of relative power between risk and no-risk groups gave differences in the opposite expected direction: children without risk

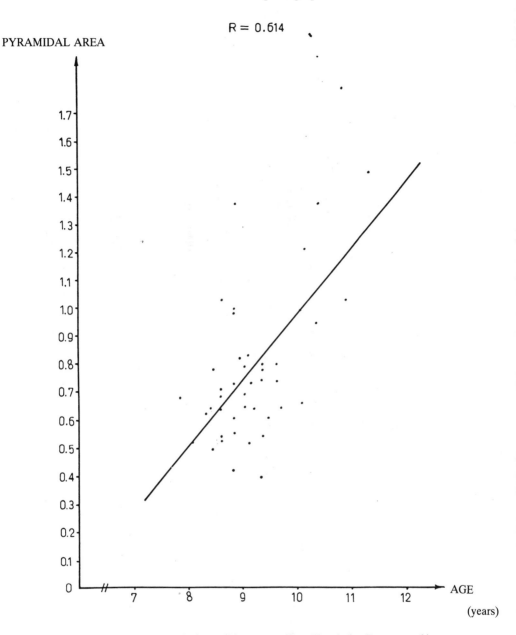

FIGURE 19.6. Age-regression of triquetral bone area (from Fernández-Bouzas et al.).

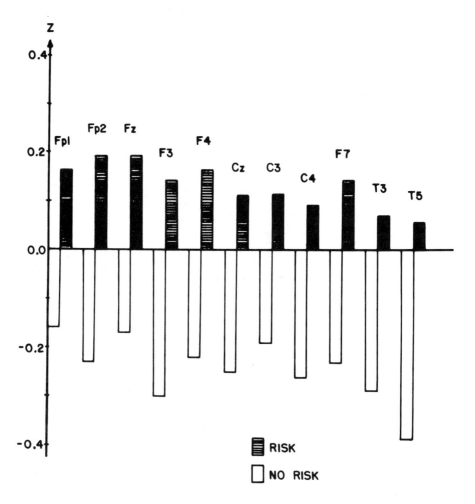

FIGURE   19.7. Z values of total power in American children with and without risk.

presented higher percentage values of delta in 01 and T5 and lower amount of alpha in T3, T4, and T5. Such differences are difficult to explain, but if it is taken into consideration that from 76 comparisons only four were significant, it is possible that they are only due to chance. This was not the case for absolute power where consistent differences were obtained. Therefore, it was possible to conclude that risk antecedents affect absolute but not relative power.

19.4 Conclusions

EEG matures at the same rate in children who grew up with adequate nutritional, sanitary, and environmental conditions independently of ethnic or cultural factors.

Sex differences were observed that are possibly related to a pubertal spurt: slopes of the regression equations in function of age were higher in girls than in boys. In rural populations of children, psychosocial handicaps were as important as biological risk factors in producing abnormalities in EEG maturation. In urban areas children with low SES had higher values of absolute power and a higher percentage of delta and lower of alpha than children with good SES, suggesting a maturational lag. Risk factors associated with perinatal problems were also associated with higher values of absolute power in frontal regions, even in samples of children with good SES.

Acknowledgments. For our study of the Mexican rural population (Section 19.2.1), we acknowledge the use of the facilities of the Brain Research Labs of New York University, the assistance of Drs. E. Roy John and Leslie Prichep, as well as the help provided by Alfredo Toro in making these analyses.

References

Ahn, H., Prichep, L., John, E.R., Baird, H., Trepetin, M., Kaye, H. (1980): Developmental equations reflect brain dysfunctions. *Nature* 210, 1259–1262

Díaz de León, A.E., Harmony, T., Marosi, E., Becker, J., Alvarez, A. (1988): Effect of different factors on EEG spectral parameters. *Int. J. Neurosci.* 43, 123–132

Fernández-Bouzas, A., Harmony, T., Marosi, E., Becker, J., Fernández, T., Méndez García, R., Perez, M.D., Mendieta, S. (1988): Estudio de la maduración ósea y de EEG en niños con antecedentes de riesgo. XXXI Congreso Nacional de Ciencias Fisiológicas, Querétaro

Gasser, T., Bächer, P., Möcks, L. (1982): Transformation toward the normal distribution of broad band spectral parameters of the EEG. *Electroenceph. Clin. Neurophysiol.* 53, 119–124

Harmony, T., Alvarez, A., Pascual, R., Ramos, A., Marosi, E., Díaz de León, A.E., Valdés, P., Becker, J. (1988): EEG maturation in children with different economic and psychosocial characteristics. *Int. J. Neurosci.* 41, 103–113

John, E.R., Ahn, H., Prichep L., Trepetin, M., Brown, D., Kaye, H. (1980): Developmental equations for the EEG. *Science* 210, 1255–1258

John, E.R., Prichep, L., Ahn, H., Easton, P., Fridman, J., Kaye, H. (1983): Neurometric evaluation of cognitive dysfunctions and neurological disorders in children. *Progr. Neurobiol.* 21, 239–290

Matousek, M., Petersén, I. (1973): Frequency analysis of the EEG in normal children and in normal adolescents. In: *Automation of clinical EEG.* Kellaway, P., Petersén, I. (eds.). New York: Raven Press, pp. 75–102

Morrison, D.F. (1976): *Multivariate statistical methods.* New York: McGraw-Hill

Yamane, T. (1976): *Statistics; an introductory analysis.* Habana: Edicion Revolucionaria

20

Brain Stimulation of Comatose Patients: A Chaos and Nonlinear Dynamics Approach

Robert W. Thatcher

20.1 Introduction

A recent and important contribution to the neurophysiological bases of human consciousness lies in the application of the mathematics of nonlinear dynamics or chaos theory to human electrophysiology. Recently, Freeman (1987a; 1987b) has used chaos theory to model state changes in the electroencephalogram (EEG), in which a relatively small number of parameters can determine brain state from seizure, deep anesthesia, sleep, awakeness, etc. One advantage of the nonlinear dynamic models of neural networks is simplification, whereby a small number of parameter changes can result in large-system state changes that may involve billions of individually interacting elements (Thompson and Stewart, 1986). Another advantage is the consistency of mathematics from the laws of physics to the laws of biology in which common, and often poorly understood, phenomena are explained by a single mathematical model (Peitgen and Richter, 1986; Swinney, 1983; Swinney and Gollub, 1981; Swinney and Roux, 1984; Thompson and Stewart, 1986).

In this context this chapter examines the application of nonlinear dynamics to the identification of EEG features that may predict changes in brain state following brain stimulation in comatose patients. This work arises from the early work of Hassler and colleagues (Hassler, 1979; Hassler et al., 1969) and more recent efforts by Francis Cohadon and colleagues in Lyons, France, and by Hosobuchi and Yingling at the University of California, San Francisco, to awaken patients who have been comatose for more than 2 years. Preliminary studies by Cohadon and colleagues have reported the successive awakening of eight of 13 patients stimulated with brainstem- and thalamic-implanted electrodes. Hosobuchi and Yingling (personal communication, 1988) are attempting to replicate and extend the Cohadon results while exploring different stimulus parameters and electrode locations. These clinical trials are not only of considerable clinical importance but they also represent a rare opportunity to investigate the genesis and maintenance of human consciousness. In recognition of work performed at the University of

Maryland and the presence of a large EEG database from noncomatose and comatose patients (Thatcher et al., 1983, 1987, 1989, 1990), Drs. Hosobuchi and Yingling are collaborating with Drs. Thatcher, Geisler, and Sestokas of the Applied Neuroscience Laboratory at Shock Trauma (University of Maryland Medical Systems) to evaluate features of the EEG that help identify levels of awareness and predict transitions from coma to awareness. Such information may be useful, especially when the patient is behaviorally comatose and unresponsive to conventional neurological examination. The study is limited to patients with minimal damage to cortical and supporting structures, such as the basal ganglia and thalamus, because such patients, after the risk and expense of surgical intervention, have a reasonable chance of leading normal and productive lives once they are awakened from their comatose states (Thatcher et al., 1990).

Three critical areas of experimental design were used in this collaboration: (1) the use of nonlinear dynamics to identify EEG state changes that may characterize the transition from coma to wakefulness, (2) the anatomical location of electrodes, and (3) parameters of electrical stimulation.

The clinical importance of the research into brain stimulation of comatose patients is illustrated by the fact that in 1975 the total number of individuals who suffered head injury in the United States, due to all causes, was estimated to be approximately 8 million (Wilder, 1976). Of these individuals approximately 70% suffered only mild head injury, defined by a Glasgow Coma Score (GCS) equal to or greater than 13 and either no loss of consciousness (e.g., mild concussion) or a loss of consciousness of no longer than 20 minutes (Langfitt and Gennarelli, 1982; Thatcher et al., 1989). The remainder, or approximately 2 million people, suffered moderate to severe brain damage defined by much longer periods of unconsciousness and greater depths of unconsciousness, e.g., GCS between 3 and 12 (Wilder, 1976). The state of being comatose is typically defined as a GCS less than 6 or a patient who does not respond to command and is not aware of his or her surroundings, e.g., GCS between 3 and 6. Although prolonged coma is all too common, the majority of the severely injured who survive the initial few weeks after admission to the hospital eventually recover consciousness to the extent that they are aware of their surroundings and gain willful control over themselves. However, the steps from deep coma to intermediate coma to the twilights of awareness require considerable time with very specific stages of improvement along the way. Figure 20.1 shows the various levels of the GCS from which comatose patients gradually climb, achieving more and more mental competencies.

20.2 Measuring EEG Transitions from Coma to Awakeness

Our initial design was to use scalp-recorded EEG from individual patients whose EEG had been repeatedly measured as they progressed from deep coma (GCS 3 to 6) to intermediate coma (GCS 7 to 12), to awakeness (GCS 13 to 15). In this way we could identify EEG features that characterize the transition from coma to

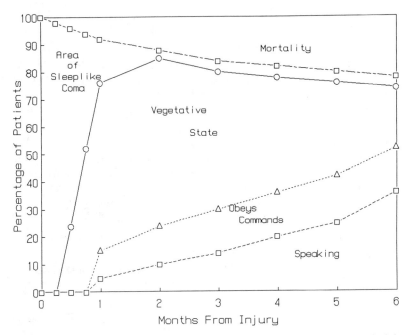

FIGURE 20.1. Time course of recovery of neurological functions in severely brain-injured patients. In general there is a shift from deep coma to a vegative state to a slow recovery of functions, such as obeying commands and speaking. The mortality rate tends to decrease as a function of time following injury.

awakeness and then empirically test nonlinear dynamical assumptions and models that may predict transitions of consciousness in comatose patients. We recognized that, even if a specific model were not derived, nonetheless, an empirical study of comatose patients' EEGs could help identify some of the practical steps necessary to eventually develop a nonlinear dynamical model. However, as of this date we have not completed the empirical aspect of this design. Therefore, this section only presents the general procedures and concepts that underlie nonlinear dynamical analyses of the human EEG. We consider these procedures as essential for the eventual empirical analyses.

20.2.1 EEG Synchrony versus Desynchrony

The human waking EEG is characterized by a narrow band spectrum with over 95% of the energy at frequencies ranging from near zero to approximately 30 Hz. For instance, in normal adult posterior cortex over 70% of the energy of the EEG is in the alpha band (e.g., 7 to 13 Hz), with a decreasing gradient extending anterior to about 45% at the frontal poles (John et al., 1980; Matousek and Petersen, 1973). The frequency spectrum of a normal adult typically exhibits a strong alpha peak (e.g., 10 Hz) with secondary peaks near the subharmonic and harmonic of the alpha frequency, e.g., near 5 Hz (theta band) and 20 Hz (beta band). The waking EEG is also characterized by intermediate bursts of synchronized rhythms embedded in a sea of desynchronized electrical activity with the interburst intervals occurring on a quasi-random basis (Freeman, 1987b). The EEG recorded during active tasks, for example, is characterized by a diminution in the number and amplitude of synchronized bursts and a general increase in desynchrony (Morrell, 1962). The spatial extent of the desynchronized EEG can also expand and contract depending upon the demands of the environment, such as in classic conditioning experiments (John, 1967; Morrell, 1962). Several studies have shown that diffuse and widespread EEG desynchronization observed during the initial stages of classic conditioning diminishes and contracts during the intermediate stages of conditioning to eventually become concentrated, like islands of desynchrony, in the neighborhood of the sensory and motor cortex involved in the mediation of the task. For example, persistent EEG desynchrony was confined either to the visual or auditory or somatosensory cortex depending upon whether the conditioned stimulus was, respectively, a visual or auditory or somatosensory stimulus (Gasteau et al., 1957).

In human EEG studies the power spectral measure called coherence is frequently used to measure the degree of desynchrony within and between regions of the human scalp. Coherence is mathematically analogous to a cross-correlation coefficient in the frequency domain (Bendat and Piersol, 1980) and provides information about the degree of coupling or shared activity between two spatially separate regions. There are two advantages of EEG coherence measures over other EEG measures: (1) coherence reflects the operation of the white matter of the neocortex through which cortico-cortical associations are made (Thatcher et al., 1986; Thatcher, 1990) and (2) coherence can be used to estimate such network properties as conduction velocity, frequency dispersion, direction vectors, and the magnitude of coupling (Bendat and Piersol, 1980). Consistent with the classic conditioning animal experiments, coherence measures of EEG synchrony in normal people have also shown certain systematic relationships. Figure 20.2 shows an example of the relationship between the magnitude of coherence and IQ in 199 children ranging in age from 5 to 16 years. This negative relationship between the magnitude of synchrony and intelligence was consistently observed in nearly all of the electrode pairs that were examined (Thatcher et al., 1983). Recently Rene Hernandes (1988) replicated this phenomenon in central, parietal, temporal, and occipital regions in an independent group of children. These findings were interpreted (Hernandes, 1988; Thatcher et al., 1983) in terms of information theory in which a desynchronized system represents a state of maximal readiness and maximal information capacity in an information theoretical sense (Shannon and Weaver, 1949).

20.3 Nonlinear Neural Network Models

Although the application of nonlinear dynamics to formulate models of human EEG is in a primitive stage of development, it may be helpful to briefly review

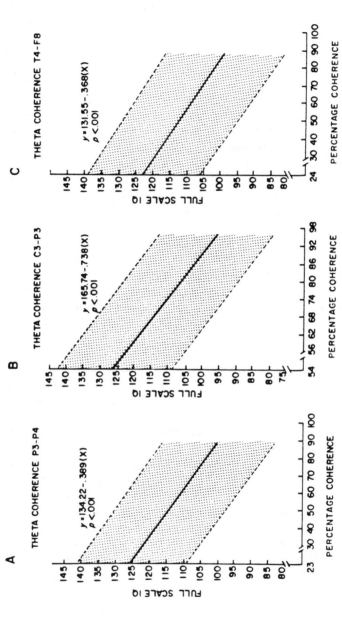

FIGURE 20.2. Polynomial regression analyses with full-scale IQ as the dependent variable and percentage coherence as the independent variable. A, Interhemispheric coherence between homologous parietal (P3 and P4) derivations. B, Left intrahemispheric coherence between left central (C3) and left parietal (P3) derivations. C, Right intrahemispheric coherence between the right anterior temporal (T4) and right frontal (F8) derivations. Dashed lines represent +– 1 S.E. Reprinted with permission from Thatcher et al. (1983): Hemispheric EEG asymmetries related to cognitive functioning in children. In: *Cognitive Processing in the Right Hemisphere*, A. Percuman, ed. New York: Academic Press.

some of the progress that has been made to date. The general feasibility of modeling EEG using ordinary differential equations (ODEs) was first demonstrated by Freeman (1975), who found that processes could be simulated that were statistically indistinguishable from real EEG. Nunez (1981) extended these efforts specifically to human EEG by taking into consideration variables such as skull volume, brain geometry, conduction velocities of cortical association fibers, and the distribution of short versus long distance cortico-cortical association fibers. Although both Freeman's and Nunez's models were nonlinear in nature, they were not specifically models of chaos.

The first neural network model of chaos was developed by Wilson and Cowen (1972, 1973). This is an important model because it is capable of exhibiting limit cycles that arise from a realistic neural network model. It has formed the foundation for a number of nonlinear dynamical neural models (see Kawahara, 1980; Skarda and Freeman, 1987) and is of considerable practical and historic relevance. The model contained two fundamental variables: the proportion of excitatory cells firing per unit time $E(t)$ and the proportion of inhibitory cells firing per unit time $T(t)$. We assume that E and I at time $(t+\tau)$ after a delay τ will be equal to the proportion of cells that are sensitive and also receive at least threshold excitation. Nonsensitive cells are those that, having recently fired, cannot fire again for their refractory period. Thus, if the absolute refractory period is r, the proportion of sensitive excitatory cells can be approximated:

$$E_s = 1 - r_e E$$

with a similar expression for I_s. The expected proportions of the subpopulations receiving at least threshold excitation per unit time will be a mathematical function of E and I, for which the proportion of excitatory cells is

$$\mathcal{L}(x) = \mathcal{L}\left[c_e E - g_e I + P(t)\right]$$

and for inhibitory cells is

$$\mathcal{L}(x) = \mathcal{L}\left[c_i E - g_i I + Q(t)\right]$$

Here, the coefficients are constants representing the average number of synapses per cell, and P(t) and Q(t) are external excitatory inputs.

After defining response functions (x), which depend on the probability distribution of neural thresholds (these are sigmoidal in shape), adjusting for coarse-graining assumptions and for a stable resting state condition, the following final differential equations were developed:

$$\tau_e \frac{dE}{dt} = -E + (k_e - r_e E)\mathcal{L}_e\left[c_e E - g_e I + P_{(t)}\right]$$

$$\tau_i \frac{dI}{dt} = -I + (k_i - r_i I)\mathcal{L}_i\left[c_i E - g_i I + Q_{(t)}\right]$$

With Q equal to zero and P equal to a certain constant value, the model can, with an appropriately chosen set of coefficients, exhibit a stable limit cycle as shown in Figure 20.3 (see Thompson and Stewart, 1986).

Recently, Kawahara (1980) extended this model to represent a system of mutually coupled Van der Pol equations. The issues of coupling coefficients and interconnectiveness within and between neural populations were addressed in this model. Thus, although a mature mathematic model has not yet been devised to describe the neurophysiological bases for the behavior of the EEG, considerable progress has nonetheless occurred (see also, Lopes de Silva et al., 1973).

Common features of all of these models are constants and coefficients that express the value of neural thresholds. Generally the thresholds depend on a probability distribution that has a sigmoidal shape, rising monotonically from zero and becoming asymptotic to a value equal to or near to unity. This aspect of the nonlinear modeling is relevant to the present discussion since it is known that the reticular activating system through its diffuse projection to the cortex and forebrain, can affect the neural thresholds in large populations of cells. Thus, electrical stimulation of the brainstem of comatose patients should have a profound effect on this important control parameter.

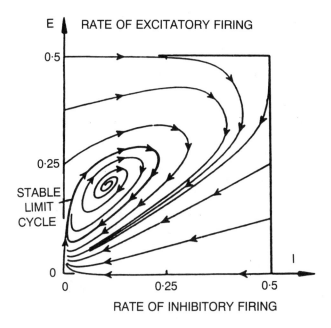

FIGURE 20.3. A stable limit cycle in the (E, I) phase space of a neural system. The limit cycle represents a steady oscillation in the two firing rates. Reprinted with permission from Thompson and Stewart (1986): *Nonlinear Dynamics and Chaos.* New York: John Wiley & Sons.

20.4 Nonlinear Dynamic Modeling of EEG

The role of EEG desynchrony in information processing has recently been mathematically modeled using nonlinear dynamics (Freeman, 1987a; Skarda and Freeman, 1987). Freeman and colleagues have shown that a bifurcation takes place when a neural system undergoes a major transition in its dynamics; for example, the transition from sleep to waking or from a normal waking state to seizure activity. These transitions are described by the language of nonlinear dynamics as a shift between two classes of attractors, e.g., from a limit cycle attractor to a chaotic attractor. The transition from a stable attractor to a chaotic attractor commonly involves a sequence of change from periodic motion to quasi-periodic motion to chaotic or nonperiodic motion, i.e., to a desynchronized state. According to this model, the state of chaotic EEG or desynchronized EEG is one of maximal information content, such as described in information theory (Shannon and Weaver, 1949). Skarda and Freeman (1987) argued that the state of chaos in neural networks constitutes a "ground state" or a condition of restless but bounded activity. This state resembles band-limited white noise; however, it is mathematically deterministic in the sense that it can be very simply and immediately turned on or off (Stewart and Thompson, 1986). In contrast to statistical randomness, which is determined by the central limit theorem, chaos is of a "low dimension" such that a small number of control parameters can significantly influence its presence or absence. For instance, Freeman (1987a) has estimated that the dimensionality of desynchronized EEG ranges from approximately 4 to 7. This low dimensional-control capability is ideal for any system that uses chaos in a functional sense frequently and with least effort. Changes in brain state from synchrony to desynchrony, for instance, are an example of where parametric control over chaos would be of biological importance.

Nonlinear dynamics can also be used to characterize EEG and single neuron behavior during the state transitions from sleep to dreaming to wakefulness, as well as during a pathological state, such as seizure or epilepsy. Figure 20.4 shows the behavior of pairs of individual neurons and the EEG during different stages of arousal in a rhesus monkey (Evarts, 1964), and Table 20.1 provides the nonlinear dynamical classifications. During waking a typically desynchronized EEG is present in Figure 20.4A with a seemingly random or unrelated time relationship between the spike discharges that emanated from the two neurons. Figure 20.4A shows that during waking the EEG is desynchronized, the interspike intervals for the two neurons are nonperiodic, and the cross-correlation between the two spike discharge time series is low. In terms of nonlinear dynamics the waking state is characterized by Chaotic-type behavior in all three times series measures. During slow wave sleep, in Figure 20.4B, the EEG exhibits increased synchrony, the interspike intervals show periodic burst patterns, and the cross-correlation between the two spike discharges is increased. In nonlinear dynamics terms the system is moving toward a limit cycle type of behavior in all three time series measures. The dreaming state is often referred to as the paradoxical sleep state since the EEG is desynchronized, the individual's eyes become extremely active, yet the threshold

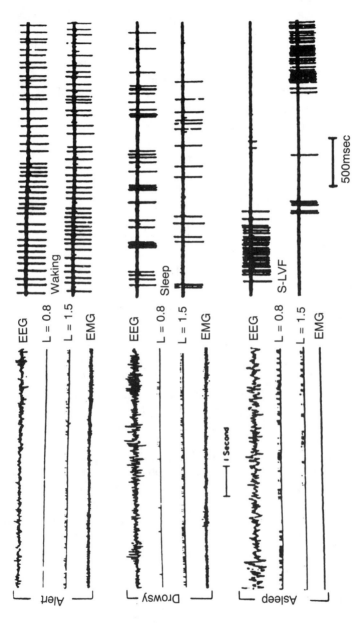

FIGURE 20.4. Patterns of unit discharge during waking, sleep (with EEG slow waves), and during paradoxical sleep (S-LVF). During waking, discharge tends to be regular with an absence of both short and long interspike intervals. During slow-wave sleep (*middle pair of traces*), there are bursts interspersed with periods of relative inactivity. With S-LVF, burst duration increases, intervening periods of inactivity become longer, and discharge frequency rises. The EEG is similar (low-voltage high-frequency activity) during waking and S-LVF. Note, however, that the pattern of discharge is markedly different during these two states. Reprinted with permission of the American Physiological Society from Evarts (1964): Temporal patterns of discharge of pyramidal tract neurons during sleep and waking in the monkey. *J. Neurophysiol.* 27: 152–171.

to awaken the individual is at its lowest. In Figure 20.4C, during dreaming or REM sleep the EEG is highly desynchronized; similar to the waking state, the interspike intervals are highly synchronized, even more than in the slow-wave sleep state, and the cross-correlation between the spike discharges is decreased. In nonlinear dynamics terms, during dreaming the EEG and the cross-correlation between spike trains are moving toward a Chaotic type of behavior while the interburst intervals move further toward the limit cycle type of behavior.

There are several generalizations that one may draw from these data. The first is that, because the cross-correlation between neurons (or the coherence) is the same in waking as it is in dreaming, one may conclude that critical parameters determining the transition from periodic-to-quasi-periodic-to-chaotic dynamics is the magnitude of coupling or forcing between neurons and the magnitude of coupling or forcing between neural networks. Therefore, such a coupling or forcing variable must be incorporated into any nonlinear dynamical model that purports to describe changes in brain state and in consciousness. The second is that the presence of the nonlinear transition sequences in brain state in Figure 20.4 and Table 20.1 helps narrow the search to identify the types of system behavior expected near the point of transition from periodic to chaotic. As Thompson and Stewart (1986) demonstrate mathematically, these behaviors can include intermittency, frequency locking, and period doubling. This indicates that one of the constraints on the development of a nonlinear model of consciousness would be that the model must account for the transition sequences actually observed in the human EEG as an individual changes brain state.

Table 20.1 Nonlinear Dynamical Associations During State Transitions[*]

Brain state	Nonlinear dynamic attractor type
Waking	EEG=quasi-periodic to chaotic
Waking	ISI=quasi-periodic to chaotic
Waking	Cross-correlation=low coupling
Slow-wave sleep	EEG=periodic (limit cycle)
Slow-wave sleep	ISI=periodic (limit cycle)
Slow-wave sleep	Cross-correlation=high coupling
Dreaming	EEG=quasi-periodic to chaotic
Dreaming	ISI=highly periodic (limit cycle)
Dreaming	Cross-correlation=low coupling
Seizure	EEG=highly periodic (limit cycle)
Seizure	ISI=highly periodic (limit cycle)
Seizure	Cross-correlation=high coupling

[*]Based on Freeman (1987a)
ISI=interspike interval.

20.5 How to Detect Chaos in the Human EEG

This section presents practical issues in the detection and analysis of nonlinear dynamical systems. Fortunately, high-speed computers and adaptable software have been developed to allow one to measure "chaos" in experimental data. An excellent software package for this purpose, which runs on an IBM-AT or compatible computer, was developed by Schaeffer et al. (1988). In order to facilitate the experimental application of the notions of nonlinear dynamics we draw explicitly on some of the features that have been made available by Schaeffer et al. (1988).

20.5.1 Power Spectral Analyses

An important practical question is how one goes about detecting chaos in the EEG and quantifying its transition states. The usual way of looking for order in a time series is to perform spectral analyses. Unfortunately, spectral analyses are not always a useful diagnostic tool for the detection of chaos. For example, Figure 20.5 shows power spectra for the Rossler attractor and the funnel (Schaeffer et al., 1988). Both of these attractors are chaotic and differ only in their parameter values, i.e., each has the same differential equations. However, in Figure 20.5A, we see very sharp spectral peaks, whereas in Figure 20.5B, there is a broad band and "featureless" spectrum. Farmer et al. (1983) have shown that the difference between these spectra is in the magnitude of "phase coherence." In Figure 20.5A, the cloud of initial points remains in phase with each other, even after many transits about the attractor. In contrast, in Figure 20.5B, the phase relationships of a set of initial points are rapidly destroyed, and the spectrum lacks prominent peaks. Thus, as Schaeffer et al. (1988) state, "The presence of sharp spectral peaks in a time series does not necessarily indicate a periodic attractor, nor does the absence of such peaks exclude the possibility of deterministic dynamics." In summary, caution must be exercised in relying solely upon power spectral analyses to analyze the nonlinear dynamics of the EEG.

20.5.2 The Phase Portrait

An especially useful diagnostic tool for analyzing nonlinear dynamical properties of a time series, such as an EEG, involves viewing the phase portrait. An important technical question is how one constructs phase portraits in cases where it is impossible to measure or know all of the state variables. As Schaeffer et al. (1988) point out, one can use Taken's method (Takens, 1980) of extracting the necessary information from a univariate time series. Takens (1980) showed that for any n-dimensional dynamical system for which measurements were made at t, $t+T$, $t+2T$, ... for an observable y, then for almost any choice of y and time lag T, the phase portrait can be obtained by plotting:

$$y(t) \; vs. \; y(t+T) \; vs \; y(t+2T) \; vs \ldots . \; y[t+(m-1)T]$$

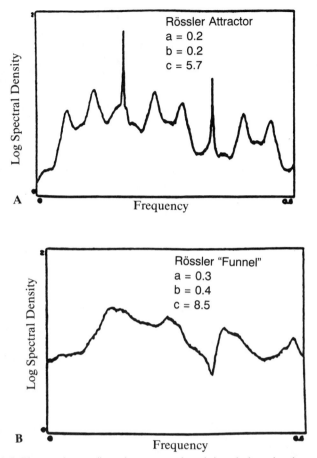

FIGURE 20.5. Phase-coherent (i.e., sharp spectral peaks) and phase-incoherent (i.e., weak or absent peaks) dynamics in Rossler's equations. A, The simple Rossler attractor. B, A variant on the funnel. Note the use of the log scale on the Y-axis. Reprinted with permission from Schaeffer et al., (1988): *Dynamical Software:* Vols I and II. Tucson: Dynamical Systems, Inc.

This is called "embedding" of the full *n*-dimensional system into a lower dimensional space. For practical purposes it is best to choose a time lag on the order of 10–30% of the period of an average orbital excursion. For spiky EEG data, a smaller lag may be necessary. It is important to emphasize, however, that the Takens method only provides a faithful reconstruction of the dynamical properties of the original time series. Thus, it is the dynamical properties, such as the fractal dimension, the Lyapunov exponents, etc., that must be emphasized. To be certain that the embedding process was conducted successfully, one should establish that the dynamical properties are indeed invariant with respect to the particular space in which the dynamics are represented or embedded.

20.5.3 Lyapunov Characteristic Exponents

The Lyapunov exponents measure the average rates at which nearby trajectories diverge or converge. This is an excellent method by which one examines the properties of an attractor. For example, for an attractor to be chaotic it must have at least one positive Lyapunov exponent. Mathematically, the Lyapunov exponents represent the rate and magnitude of deformation of a circle into an ellipsoid. For Lyapunov numbers greater than 1, the ellipsoid is expanding; numbers less than 1 imply contraction. For complete stability—the watershed between expansion and contraction—the number equals 0. Algorithms for computing Lyapunov exponents from experimental times series are given by Wolf et al. (1985) and Schaeffer et al. (1988). Figure 20.6 schematically shows the Wolf et al. (1985) method of computing the Lyapunov exponent.

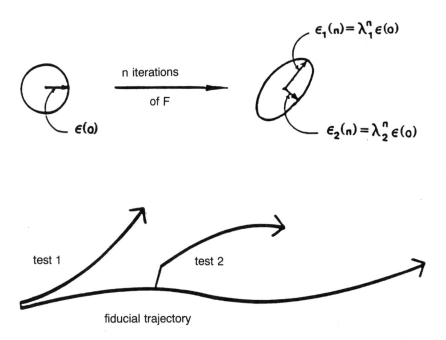

FIGURE 20.6. *Top,* Deformations of a circle of initial points into an ellipsoid. The Lyapunov numbers measure rates of expansion and contraction of the principal axes. For maps, Lyapunov numbers greater than 1 indicate expansion; numbers less than 1 imply contraction. For flows, the watershed between expansion and contraction is 0. In this case, one speaks of Lyapunov exponents. *Bottom,* Method used to estimate Lyapunov exponents. One begins by choosing a fiducial trajectory and a second nearby trajectory called the test trajectory. When the separation between the two trajectories becomes large, a new test trajectory is chosen at approximately the same distance and orientation as the first. When the equations of motion are known, this is accomplished by using (periodically renormalized) eigenvectors as test trajectories. For experimental data, one searches the data set for a suitable replacement. Reprinted with permission from Schaeffer et al. (1988): *Dynamical Software:* Vols I and II. Tucson: Dynamical Systems, Inc.

20.5.4 Kolmogorov-Sinai Metric Entropy

The metric entropy measures the average rate of information gained by measuring the neighborhood of a trajectory. The Kolmogorov-Sinai (K-S) metric is related to the Lyapunov exponent by equaling the sum of the positive Lyapunov exponents (Ruelle, 1983). In general, a positive K-S implies chaos. This follows from the fact that chaotic systems are sensitive to initial conditions, in which minute differences are eventually amplified to large-scale fluctuations. As a consequence, one can never predict the system's behavior into the indefinite future because one's predictions are degraded as one moves further into the future. The K-S metric measures the rate of information generation as the short-term predictability is transformed into long-term uncertainty. Based upon the methods of Schaeffer et al. (1988), the K-S entropy is computed by dividing the phase space into a minimal set of hypercubes sufficient to cover the attractor. Let p_i be the fraction of points that fall in the ith cube. Consider the quantity

$$I_{(e)} = -\sum_i P_i \log p_i$$

where e is the length of a side of the cube and $I(e)$ is information as defined by Shannon and Weaver (1949). From this one can define the information dimension as

$$d_i = \lim_{e \to 0} [I(e)/ln\,(1/e)]$$

To define the K-S entropy, space-time is divided into n bins of length e. Let p (i1, ... in) be the joint probability that x(t+dt) falls in cube i1, x(t+2dt) in cube i2, ..., x(t+n dt) in cube n. Then we can define

$$\tilde{I}(e) = -\sum_i p(i1, \ldots, in)\, ln\, p(i1, \ldots, in)$$

and the metric entropy as

$$K\text{-}S = \lim_{e \to 0}\ \lim_{n \to \infty}\ (1/n)\, \tilde{I}(e)$$

20.5.5 Fractal Dimension

A fractal dimension is where successive magnifications of a system reveal structure within structure, within structure ... ad infinitum. Whereas the Lyapunov exponents measure the separation of neighboring trajectories, the fractal dimension is defined with respect to particular portions of phase space. Similar to the K-S entropy, one computes the fractal dimension by dividing the phase space into hypercubes and retaining a minimal set that covers the attractor. However, caution should be exercised because chaotic systems do not always have a noninteger or fractal dimension (Grebogi et al., 1984).

There are a great many different dimension-like quantities published in the fractal literature (Farmer et al., 1983; Grassberger and Procaccia, 1983a and b; Mandelbrot, 1977; Schaeffer et al., 1988). In general, these dimensions fall into two classes: metric dimensions and probabilistic dimensions. Metric dimensions are independent of the frequency with which various regions of the attractor are visited, whereas probabilistic dimensions explicitly utilize the frequency distribution, sometimes called the natural measure of the attractor. If the data under study constitute a known set, such as the long-term dynamics of a system of equations, then the metric and probabilistic dimensions are comparable (Farmer et al., 1983; Schaeffer et al., 1988). However, for experimental data, such as the EEG, the probabilistic dimension is easier and more accurate to estimate (Schaeffer et al., 1988). An example of a probabilistic fractal dimension that is related to the Lyapunov exponent is the dimension-like quantity called the Lyapunov dimension, d, as defined by the equation:

$$d_L = j + \Sigma \, i/ \quad | \quad j + i \, |$$

where the λ's are the Lyapunov exponents arranged in order of decreasing magnitude and j is chosen so that

$$\sum_{}^{j-} \lambda_i > 0$$

$$\sum_{}^{j+1} \lambda_i < 0$$

If D is a intermediate value between integers (e.g., 2.47), then we can expect to observe fractal structure in lower-dimensional maps (Schaeffer et al., 1988).

Another and more computationally efficient method of estimating the fractal dimension of a time series is that proposed by Grassberger and Procaccia (1983a and b). In their algorithm one replaces cube counting with the measurement of distances by computing the distance between all points of a time series and then asking what fraction is less than a series of predetermined length scales. The average of these lengths is defined as the correlation integral for different length scales, g. For example

$$C(g) = \lim_{N \to \infty} \{1/ [N (N\text{-}1)] \} \sum_{i,j} H (g, D_{ij})$$

where D_{ij} is the distance between points X and X , and $H (g, D)$ is the Heaviside function

$$H (g,D) = \begin{cases} 0, g < D \\ 1, g > D \end{cases}$$

$C(g)$ increases from zero for g very small to 1 for g very large. Grassberger and Procaccia (1983a and b) show that for small g

$$C(g) \sim g^h$$

and the exponent, n, is a lower bound on the information dimension. In more detail, they propose that

$$n < d_I < d_H$$

where d_H is the Hausdorf dimension and the exponent, n, is referred to as the correlation dimension. Freeman (1987a) successfully used the Grassberger and Procaccia (1983a and b) method to measure the fractal or chaotic dimension of olfactory EEG yielding values between 4 and 7.

20.6 The Torus Route to Chaos in Periodically Forced Systems

As stated previously, it is presumptuous to assume that a mature nonlinear model of brain state control can be developed today. However, of the various nonlinear dynamical models available, the toroidal flow model represents a class of models that seem most appropriate for the brain stimulation experiments that are to be conducted on comatose patients. The torus is a reasonable model to start with because toroidal flow is produced when a system that oscillates on its own is subjected to periodic forcing at a different frequency. If we assume that comatose patients exhibit naturally oscillating neural systems (e.g., EEG), then brain stimulation at periodic intervals and/or periodic frequencies and/or at periodic intensities may produce dynamic toroidal behavior.

By toroidal behavior we mean the motion of trajectories on the surface of a torus, i.e., a doughnut. Toroidal motion contains two classes of dynamics: (1) the motion is periodic in which the orbit winds around the torus an integer number of times before repeating itself, and (2) the motion never repeats itself and in fact covers the entire surface of the torus. In the first case the motion is said to be phase-locked, and in the second it is said to be quasi-periodic. As shown by Thompson and Stewart (1986), a change in parameter value of a dynamical system often results in the following sequence:

$$\begin{array}{lcl}
\text{Limit Cycle} & \to & \text{Quasi-periodicity} \\
\text{Limit Cycle} & \to & \text{Phase Locking} \\
\text{Limit Cycle} & \to & \text{Chaos}
\end{array}$$

A frequently used mathematical model of toroidal flow is the following system of equations:

$$dx/dt = f(x, y, u)$$

$$dy/dt = g(x, y)$$

where f(x,y,u) and g(x,y) are nonlinear functions and u=u(t) is a periodic function of time with period T. In this model there are two frequencies of interest: the inherent frequency of the nonlinear oscillator and the forcing frequency. As mentioned above, the inherent frequency of the system can represent the background EEG and/or sleep-wakefulness cycle of the comatose patient, and the second forcing frequency can represent various stimulus parameters used to awake the patient.

Although there are many ways to analyze this system of equations, one of the more interesting is to plot x and y at intervals equal to the period, T, of forcing. The resulting construction is called a time-one map or a Poincare map. The time-one map will be a closed curve topologically equivalent to a circle. In order to represent the positions of successive points, Schaeffer et al. (1988) introduced a construction called a circle map, which gives the sequence of rotations on the torus. This map, which is shown in Figure 20.7, is produced as follows:

Impose coordinate axes at the "center of mass."

To each point, i, assign an angle, $y(i)$.

Plot the y's in temporal sequence, i.e., for all pairs of points, i and $i+1$, plot $y(i+1)$ vs. $y(i)$.

The rotations on the torus are defined for the ith rotation as

$$r(i) = y(i + 1) - y(i)$$

This allows us to compute the so-called rotation number r, which is simply the long-term average of the individual rotations:

$$r = \lim_{n \to \infty} \frac{1}{n} \sum_i \frac{r(i)}{2\pi}$$

Thus, if the average number of rotations is 360° then $r=1$.

20.7 Where Should the Electrodes Be Placed?

It is obvious that there is a limit to the number of brain-stimulating electrodes that can be placed in a comatose patient. A subset of candidate brain regions that are likely to have a low threshold for activation must be selected. The following sections review some of the literature concerning possible neuroanatomical control networks responsible for the sleep-wakefulness cycle and the known inhibitory and facilitatory brainstem and forebrain regions.

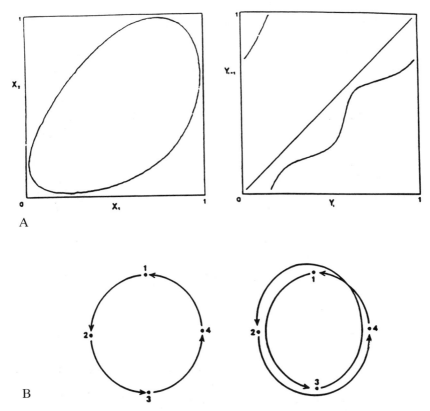

FIGURE 20.7. A, Invariant circles and circle maps. *Left,* For quasi-periodic motion, the time-one map is a closed curve topologically equivalent to a circle. The lines indicate the sequence of points. *Right,* A circle map is constructed by imposing coordinate axes at the center of mass (marked with a cross) of the invariant circle and assigning an angle, y(i), to each point. One then plots y(i+1) versus y(i) for all pairs, i and (i+1). B, For periodic orbits, the rotation number, r, is simply the number of rotations required to visit all points on the time-one map divided by the number of points. *Left,* r=1/4=0.25. *Right,* r=2/4=0.5. Reprinted with permission from Schaeffer et al. (1988): *Dynamical Software:* Vols I and II. Tucson: Dynamical Systems, Inc.

20.7.1 Sleep-Wakefulness Control Systems

An understanding of some of the anatomical systems involved in the regulation of sleep, wakefulness, and arousal states has come about through the use of three techniques: electrical brain stimulation, brain lesions, and neurochemical manipulations. For example, relatively low levels of high-frequency stimulation in the midline thalamus and reticular formation can transform a sleeping animal whose EEG is dominated by slow waves to one displaying an activated EEG and behavioral arousal. This effect is mediated largely by excitatory drives exerted on cortical, limbic, and basal ganglia structures (Thatcher and John, 1977). Lesions of the mesencephalic and pontine reticular formation result in a comatose state, characterized by a slow-wave EEG at intermediate levels of coma to a disorganized and low-amplitude EEG at deep coma (Bricolo et al., 1978). The symptoms of the sleeping disease—encephalitis lethargical (Economo, 1918)—are largely the result of lesions in the central gray of the brainstem. Similarly, tumors in this region produce prolonged states of somnolence (Fulton and Bailey, 1929).

These classic findings, which constituted an important contribution to the development of contemporary neuroscience, led to the formulation of the concept of an ascending reticular activating system involved in the control of sleep-wakefulness (Lindsley et al, 1949; Moruzzi and Magoun, 1949). Damage to the sensory-specific systems had little effect on sleep-wakefulness, whereas lesions located a few centimeters more medially would produce a depressed, somnolent state.

The possibility of reciprocal corticoreticular influences on the control of arousal was demonstrated through electrical stimulation studies by French, Hernandez-Peon, and Livingston in 1955. Lindsley (1950) was one of the first to emphasize the role of descending cortical influences in the control of arousal and attention, as well as in emotion and motivation. The recognition that the limbic system and cortex-made contributions to the reticular formation made it possible to conceive of how internal states, such as thoughts, worries, and apprehensions, might generate arousal activity in the reticular formation.

Further emphasis on reciprocal and dynamic control mechanisms in the regulation of sleep-wakefulness was provided by the finding that sleep can be produced not only by the reduction of activity in a tonic activating system but also by the influence of an active sleep-producing system. Hess (1954) and others (Akimoto et al., 1956; Hess et al., 1953) demonstrated that low-frequency stimulation of diencephalic structures would cause otherwise alert cats to select a likely place, after which they would curl up and then proceed to go to sleep much as a normal animal. In a study of the hypothalamic region of rats, Nauta (1946) found that lesions of the posterior thalamus produced prolonged somnolence, whereas lesions of more anterior hypothalamic regions produced rats incapable of sleeping. These results fit in with the findings of Economo (1918) who found, in some encephalitic patients, prolonged sleeplessness associated with lesions in the anterior hypothalamus.

More recent studies have mapped out the areas within the hypothalamus, limbic system, and reticular formation involved in the control of sleep and wakefulness (Hernandez-Peon and Chavez-Ibarra, 1963; Hernandez-Peon et al., 1963; Jouvet, 1969). Hypnogenic regions have been found throughout the pons, upper mesencephalon, diencephalon, and limbic system, as well as the orbital frontal cortex (Akert et al., 1952; Hess, 1954). A reciprocal activation-suppression relationship exists within the pontine reticular formation. Batini et al. (1959) demonstrated that lesions of the rostral pons gave rise to an EEG picture of arousal, whereas lesions a few millimeters lower, at the midpontine level, resulted in an EEG pattern typical of sleep. The inference drawn from these studies was that a mechanism present in the lower part of the reticular formation brings about sleep by inhibiting the upper reticular formation (Moruzzi, 1960). This inference was supported by experiments that blocked the upper and lower brainstem structures separately through the perfusion of barbiturates in the carotid and vertebral arteries, respectively (Magni et al., 1959). Perfusion of the caudal half of the brainstem through the vertebral arteries resulted in activation of the EEG, whereas barbiturate perfusion of the upper portion via the carotids resulted in a sleep-like EEG.

These data support a oscillatory model of brain-state control and regulation in which at least two systems are involved and distributed within both diencephalic-limbic and reticular formation structures, one a sleep-producing or hypnotic system, and the other an awakening or arousal system. These systems appear to be reciprocally connected so that activity in one will suppress the other and vice versa, thus giving rise to the sleep-wakefulness cycle (Hernandez-Peon and Chavez-Ibarra, 1963). This substrate of a reciprocally connected control circuit is typical of many neurophysiological systems (Thatcher and John, 1977).

20.7.2 Reticular and Diencephalic Facilatory-Inhibition Centers

Some of the dynamic reciprocal control centers in the diencephalon and brainstem are shown diagrammatically in Figure 20.8. This set of relations (pluses representing facilatory outflows and minuses representing inhibitory influences) was discovered by Magoun and collaborators (Bach & Magoun, 1947; Lindsley et al., 1949; Magoun & Rhines, 1948). These workers demonstrated that, when stimulated, the rostral portion of the reticular formation (involving mesencephalon and pontine regions labeled as structure 5 caused facilitation or enhancement of spinal reflexes and cortically initiated movement. This was also the region of lowest threshold for arousal and therefore should be a major candidate for deep brain stimulation in comatose patients. A second, more dorsal facilitatory region (labeled 6) is related to the vestibulospinal system. Stimulation of this system affects tonic facilitatory mechanisms involved in the control of posture. In contrast, stimulation of area 4 (caudal reticular formation) results in an inhibitory effect on spinal motor outflow and has been shown to reduce spasticity and hyperreflexia. Stimulation of the other inhibitory regions (1, 2, or 3) reinforces the suppressor or inhibitory role of the caudal reticular formation. Lesions of the inhibitory regions result in increased spasticity or hyperreflexia due to the dominance of the facilitatory reticular formation system. Thus, it appears that an appropriate balance between inhibition and facilitation can be maintained by such a mechanism, with the normal state of a waking animal favoring facilitation of spinal reflexes and motor outflows. In relation to drowsiness or sleep, there is a progressive tendency for the inhibitory system to dominate, resulting in sluggish reflexes and loss of postural control.

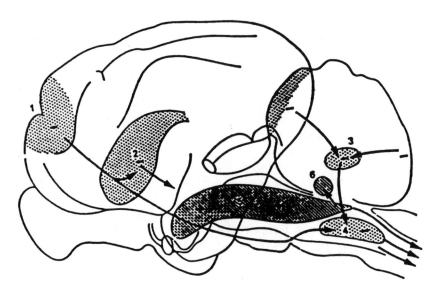

FIGURE 20.8. Excitatory and inhibitory influence of stimulation of the nuclei of the reticular formation: (5) and (6) facilatory, and (4) inhibitory zones of the brainstem reticular formation and connections running to it from the cortex (1), thalamus (2), and cerebellum (3). Reprinted with permission from Luria (1973): *The working brain.* Middlesex, England: Penguin Press.

This complex set of interrelationships reveals a system with enormous capabilities for the regulation of levels of arousal. First, the system possesses wide-ranging ascending influences so that its outflow can modulate the level of excitation of extensive regions of the cortex. In the case of absent or diminished outflow, the tonic level of cortical excitability will drop. Second, the inputs to this system from relay nuclei, limbic system, and cortex can act upon it so as to cause alterations in the level of arousal. Finally, outputs from this system to structures that are possible sources of such inputs provide the capability of fine adjustments, increasing the effectiveness of particular inputs so small changes in afferent intensity will cause major changes in arousal level, or decreasing the effectiveness of other inputs so they are temporarily excluded from acting upon the system.

20.8 The Parameters of Brain Electrical Stimulation

We come finally to the question of what are the best stimulus parameters, e.g., stimulus duration, stimulus intensity, on-off cycles, etc. It should now be clear that the critical observation concerns the neural dynamics during the transition from one brain state to another. Careful monitoring of the comatose patient's EEG is essential in order to describe the time series events preceding, during, and following changes in brain stimulation. If we follow a simple toroidal model of forced oscillations, then we should look for periodic, quasi-periodic, phase locking, and chaotic behaviors in the EEG, which are deterministically related to the stimulus parameters. Thus, it is desirable to record from as many electrodes as possible, including the stimulating electrodes (i.e., during the pre-and poststimulation periods to detect afterdischarges) with a high enough sampling rate (e.g., greater than 1000 Hz) to resolve spike discharges and rapid bifurcations.

Brain stimulation studies of comatose cats have shown that brief but daily stimulation of brainstem locations with low arousal thresholds was optimal in awakening the animal (see Chapter 3). Because these studies were conducted on surgically lesioned animals it was possible to determine the arousal thresholds before brain surgery. Obviously, this is not possible in brain-injured comatose patients. However, the power spectrum of the cortical EEG can reveal a circadian pattern or oscillatory pattern of changes in cortical excitability. The idea is to use these measure to deliver electrical stimulation when the patient is most aroused.

In the Cohadon preliminary studies, daily stimulation of the nucleus centrum median was used without systematic exploration of optimal stimulus parameters. However, Drs. Hosobuchi and Yingling used a 20-channel EEG and evoked potential system to record from scalp and deeply implanted electrodes. This system allows them to evaluate stimulus intensity thresholds for afterdischarge and maximum effect on the arousal pattern of the cortical EEG. That is, they can quantify the magnitude of change in EEG delta and alpha activity produced by different stimulus parameters, such as stimulus frequency, stimulus duration, and stimulus intensity, and then utilize the most effective stimulus settings.

20.9 Discussion

In theory it should be possible to obtain estimates of the dimensions of nonlinear dynamics of the cortical and subcortical EEG that are correlated with the recovery process in comatose patients and that characterize state changes that are parallel to or accompany changes in levels of awareness. Although this experimentation may not provide a mathematical model of consciousness, it nonetheless should add to our understanding of some of the dynamical neurophysiological processes that accompany the transition from coma to awakeness. An accurate nonlinear dynamical analysis may aid in the development of mathematical models of brain-state properties, such as stability and instability, expansion and contraction, excitation and inhibition, competition and cooperation, and the creation, transition, and destruction of brain states. It may provide clues to the critical dimensionalities that are relevant to the process of consciousness and point toward future experimentation to reveal more about the underlying neurophysiology. In a practical sense, a mature nonlinear dynamical model of brain-state changes may aid in the remediation of neurological disorders by providing reliable predictors in patients who are comatose and unable to respond to conventional neurological and neuropsychological evaluation. In the present context, such a model may guide experimental procedures to identify the optimal stimulus parameters and electrode locations used in the brain stimulation of comatose patients.

Acknowledgments. I would like to thank Dr. Anthony Sestokas for his interest and enthusiasm regarding collaboration in the study of brain stimulation of comatose patients and his approval and encouragement for the collaboration between the Shock Trauma Neurometrics Laboratory and Drs. Hosobuchi and Yingling at the University of California. The Shock Trauma patient EEG data that were used to begin the nonlinear dynamical analyses were collected while I was the Principal Investigator and/or the Scientific Director of the UMES-Shock Trauma Neurometric Clinical service. I also acknowledge Dr. Fred Geisler who gave permission for access to the neurotrauma EEG data as the physician of record of the neurotrauma patients.

References

Akert, K., Koella, W., Hess, R. (1952): Sleep produced by electrical stimulation of the thalamus. *Am. J. Physiol.* 168, 260–267

Akimoto, H., Yamaguchi, N., Okabe J., Nakagawa, T., Nakamura, I., Abe, K., Torri, H., Masahashi, K. (1956): On sleep induced through electrical stimulation of dog thalamus. *Folia. Psychiatry Neurol.* 10, 117–146

Bach, L., Magoun, H.W. (1947): The vestibular nuclei as an excitatory mechanism for the cord. *J. Neurophysiol.* 5, 331–337

Batini, C., Moruzzi, G., Palestini, M., Rossi, G., Zanchetti, A. (1959): Effects of complete pontine transections on the sleep-wakefulness rhythm: The midpontine pretrigeminal preparation. *Arch. Ital. Biol.* 97, 1–12

Bendat, J.S., Piersol, A.G. (1980): *Engineering applications of correlation and spectral analysis.* New York: John Wiley & Sons

Bricolo, A., Turazzi, S., Faccioli, F., Odorizzi, F., Sciarretta, G., Erculiani, P. (1978): Application of compressed spectral array in long-term EEG monitoring of comatose patients. *Electroenceph. Neurophysiol.* 45, 221–225

Economo, C. Von. (1918): *Die Encephalitis Lethargica.* Vienna: Deuticke. Cited in Ochs, S. (1965): *Elements of neurophysiology.* New York: John Wiley & Sons

Evarts, E.V. (1964): Temporal patterns of discharge of pyramidal tract neurons during sleep and waking in the monkey. *J. Neurophysiol.* 27, 152–171

Farmer, J.D., Ott, E., Yorke, J.E. (1983): The dimension of chaotic attractors. *Physica* 7D, 153–180

Freeman, W.J. (1975): *Mass action in the nervous system.* New York: Academic Press

Freeman, W.J. (1987a): Simulation of chaotic EEG patterns with a dynamic model of the olfactory system. *Biol. Cybern.* 56, 139–150

Freeman, W.J. (1987b): Analytic techniques used in the search for the physiological basis of the EEG. In: *Methods of analysis of brain electrical and magnetic signals. EEG handbook (revised series, vol. 1).* Gevins, A.S., Remond, A. (eds.). Amsterdam: Elsevier

Fulton, J.F., Bailey, P. (1929): Tumors in the region of the third ventricle: Their diagnosis and relation to pathological sleep. *J. Nerv. Ment. Dis.* 69, 1–25

Gasteau, H., Jus, A., Jus, C., Morrell, F., Storm Van Leewen, W., Dongier, S., Naquet, R., Regis, H., Roger, A., Bekkering, D., Kamp, A., Werre, J. (1957): Etude topographique des reactions electroencephalographiques conditioness chez l'homme. *Electroenceph. Clin. Neurophysiol.* 9, 1–14

Grassberger, P., Procaccia, I. (1983a): Characterization of strange attractors. *Phys. Rev. Lett.* 50, 346–249

Grassberger, P., Procaccia, I. (1983b): Measuring the strangeness of strange attractors. *Physica* 9D;, 189–208

Grebogi, C., Ott, E., Pelikan, S., Yorke, J.A. (1984): Strange attractors that are not chaotic. *Physica* 13D, 261–268

Hassler, R. (1979): Striatal regulation of adverting and attention directing induced by pallidal stimulation. *Appl. Neurophysiol.* 42, 98–102

Hassler, R., Dalle, G., Bricolo, A., Dieckmann, G., Dolce, G. (1969): Behavioral and EEG arousal induced by stimulation of unspecific projection systems in a patient with post-traumatic apallic syndrome. *Electroenceph. Clin. Neyurophysiol.* 27, 306–310

Hernandez, R.S. (1988): *The effects of task condition on the correlation of EEG coherence and full-scale IQ.* Doctoral dissertation, Maharishi International University, India

Hernandez-Peon, R., Chavez-Ibarra, G. (1963): Sleep induced by electrical or chemical stimulation of the forebrain. *Electroenceph. Clin. Neurophysiol.* 24 (Suppl.), 118–198

Hernandez-Peon, R., Chavez-Ibarra, G., Morgane, P.J., Timo-Iaria, C. (1963): Limbic cholinergic pathways involved in sleep and emotional behavior. *Exp. Neurol.* 8, 93–111

Hess, R., Koella, W.P., Akert, K. (1953): Cortical and subcortical recordings in natural and artificially induced sleep in cats. *Electroenceph. Clin. Neurophysiol.* 5, 75–90

Hess, W.R. (1954): *Diencephalon-autonomic and extra-pyramidal functions.* New York: Grune & Stratton

John, E.R. (1967): *Mechamisms of memory.* New York: Academic Press

John, E.R., Ahn, H., Prichep, L., Trepetin, M., Brown, D., Kaye, H. (1980): Developmental equations for the EEG. *Science* 210, 1255–1258

Jouvet, M. (1969): Biogenic amines and the states of sleep. *Science* 163, 32–41

Kawahara, T. (1980): Coupled Van der Pol oscillators: A model of excitatory and inhibitory neural interactions. *Biol. Cybernet.* 39, 37–43

Langfitt, T.W., Gennarelli, T.A. (1982): Can the outcome from head injury be improved? *J. Neurosurg.* 56, 19–25

Lindsley, D.B. (1950): Emotions and the electroencephalogram. In: *Feelings and emotions: The Mooseheart Symposium.* Reymert, M.L. (ed.). New York: McGraw-Hill

Lindsley, D.B., Schreiner, L., Magoun, H.W. (1949): An electromyographic study of spasticity. *J. Neurophysiol.* 12, 197–205

Lopes da Silva, F.H., Van Lierop, T.H., Schrijer, C.F., Storm van Leeuwen, W. (1974): Organization of thalamic and cortical alpha rhythms: Spectra and coherences. *Electroenceph. Clin. Neurophysiol.* 35, 627–639

Luria, A.R. (1973): *The working brain.* Middlesex, England: Penguin Press

Magni, F., Moruzzi, G., Rossi, G.F., Zanchetti, A. (1959): EEG arousal following in-activation of the lower brain stem by selective injection of barbiturate into lower brain stem circulation. *Arch. Ital. Biol.* 97, 33–46

Magoun, H.W., Rhines, R. (1948): *Spasticity: The stretch reflex and extrapyramidal systems.* Springfield, IL: Charles C. Thomas

Mandelbrot, B. (1977): *Fractals: Form, chance and dimension.* San Francisco: Freeman Publishing Co.

Matousek, M., Petersen, I. (1973): Frequency analysis of the EEG background activity by means of age dependent EEG quotients. In: *Automation of clinical electroencephalography.* Kellaway, P., Petersen, I. (eds.). New York: Raven Press

Morrell, F. (1962): Electrophysiological contributions to the neural basis of learning. *Physiol. Rev.* 41, 443–476

Moruzzi, G. (1960): Synchronizing influences of the brain stem and the inhibitory mechanisms underlying the production of sleep by sensory stimulation. *The Moscow Colloquim on Electroencephalography of Higher Nervous Activity, Electroencephalography and Clinical Neurophysiology* 13, 231–256

Moruzzi, G., Magoun, H.W. (1949): Brain stem reticular formation and activation of the EEG. *Electroenceph. Clin. Neurophysiol.* 1, 455–473

Nauta, W.J.H. (1946): Hypothalamic regulation of sleep in rats: An experimental study. *J. Neurophysiol.* 9, 285–316

Nunez, P. (1981): *Electric fields of the brain: The neurophysics of EEG.* New York: Oxford University Press

Peitgen, H.O., Richter, P.H. (1986): *The Beauty of fractals.* New York: Springer-Verlag

Ruelle, D. (1983): Five turbulent problems. *Physica* 7D, 40–44

Schaeffer, W.M. Truty, G.L., Fulmer, S. (1988): *Dynamical software: vols. I and II* Tucson: Dynamical Systems, Inc.

Shannon, C.E., Weaver, W. (1949): *The mathematical theory of communication.* Urbana, IL: University of Illinois Press

Skarda, C.A., Freeman, W.J. (1987): How brains make chaos in order to make sense of the world. *Behav. Brain Sci.* 10, 161–195

Swinney, H.L. (1983): Observations of order and chaos in nonlinear systems. *Physica* 7D, 3–15

Swinney, H.L., Gollub, J.P. (eds.). (1981): *Hydrodynamic instabilities and the transition to turbulence.* New York: Springer-Verlag

Swinney, H.L., Roux, J.C. (1984): Chemical chaos. In: *Nonequilibrium dynamics in chemical systems.* Vidal, C. (ed.). New York: Springer-Verlag

Takens, F. (1980): Detecting strange attractors in turbulence. In: *Dynamical systems and turbulence.* In: *Springer lecture notes in mathematics,* vol. 898. Rand, D.A., Young, L.S. (eds.). New York: Springer-Verlag, pp. 366–381

Thatcher, R.W., John, E.R. (1977): *Functional neuroscience: vol. I. Foundations of cognitive processes.* Hillsdale, NJ: Erlbaum Associates

Thatcher, R.W., McAlaster, R., Lester, M.L., Horst, R.L., Cantor, D.S. (1983): Hemispheric EEG asymmetries related to cognitive functioning in children. In: *Cognitive processing in the right hemisphere.* Perecuman, A. (ed.). New York: Academic Press

Thatcher, R.W., Krause, P., Hrybyk, M. (1986): Corticocortical association fibers and EEG coherence: A two compartmental model. *Electroenceph. Clin. Neurophysiol.* 64, 123–143

Thatcher, R.W., Walker, R.A., Giudice, S. (1987): Human cerebral hemispheres develop at different rates and ages. *Science* 236, 1110–1113

Thatcher, R.W., Walker, R.A., Gerson, I., Geisler, F.H. (1989): EEG discriminant analyses of mild head trauma. *Electroenceph. Clin. Neurophysiol.* 74, 94–106

Thatcher, R.W., Cantor, D.S., McAlaster, R., Geisler, F.H., Krause, P. (1990): Comprehensive prediction of outcome in head injured patients: Development of prognostic equations. *Ann. NY Acad. Sci.* (in press)

Thompson, J., Stewart, H. (1986): *Nonlinear dynamics and chaos.* New York: John Wiley & Sons

Wilder, C.S. (1976): Health Interview Survey. Rockville, MD: National Center for Health Statistics, Department of Health, Education and Welfare

Wilson, H.R., Cowen, J.D. (1972): Excitatory and inhibitory interactions in localized populations of model neurons. *J. Biophysics* 12, 1–24

Wilson, H.R., Cowen, J.D. (1973): A mathematical theory of the functional dynamics of cortical and thalamic nervous tissue. *Kybernetik* 13, 55–80

Wolf, A., Swift, J.B., Swinney, H.L., Vastano, J.A. (1985): Determining Lyapunov exponents from a time series. *Physica* 16D, 285–317

Part VII

Brain Imaging

21

The Statistical Analysis of Brain Images

Pedro A. Valdés-Sosa and Rolando Biscay Lirio

21.1 Introduction

Mental processes are the result of the spatially and temporally distributed activity of neural masses. Advances in neurosciences in recent years have generated new, formalized theories about the dynamics of these neural ensembles (Lopes da Silva et al., 1987, Chapters 2, 3, and 6) and also new techniques for obtaining either static *anatomical* images (CT, MRI) of the structural constraints in which they are embedded or time-varying *functional* images (PET, EEG, MEG) of their activity.

Deciphering the machinery of the mind requires a combination of theoretical work and empirical observation, the latter based on data gathered from both experimental neurosciences and clinical studies. However, for many purposes the verification of theory by experiment will ultimately be reduced to the analysis of brain images collected in an experimental design. The inherent noisiness of instrumental observation and biological fluctuation dictates the use of statistical criteria for such inferences.

This chapter deals with several important issues concerning the statistical analysis of brain imaging. Many of these are being attacked in an isolated fashion, others are yet unsolved. Their uniqueness stems from a series of particular problems that many times place their analysis beyond the pale of traditional statistics. These problems are caused by the following:

The sheer volume of data that is generated by imaging apparatus and processing makes megabyte to gigabyte files of voxels common for each subject in each session.

This information is not independent, but rather is highly structured. This spatio-temporal structure is imposed by the anatomofunctional constraints of the brain, as well as by the experimental design under which the data are collected. Multiple intrasubject recordings, sometimes of different types of raw data, further complicate the issue, introducing a repeated-measures situation.

The number of variables is usually much larger than the number of subjects, a situation for which traditional multivariate statistics was not conceived.

This chapter attempts to devise a general framework for describing brain images and their associated statistical problems. It then uses this framework to explore the nature and origin of the difficulties just mentioned. Some possible general solutions are sketched out. These ideas are briefly exemplified in specific experimental situations. A choice of examples, selective rather than exhaustive, are followed systematically through the text.

21.2 Brain Images and Their Statistical Problems

21.2.1 Images: Anatomical and Functional; Static and Dynamic

This section examines two different types of images. The first type are representations of geometrical objects, such as the head, the brain, or parts of other anatomical entities, i.e., the scalp or the meninges. These shall be termed *anatomical images*. Typical examples are images obtained by x-ray and magnetic resonance tomography or simply photographs or drawings of sliced brains in stereotaxic charts.

Physiologists are more interested in the activity of the cells that compose these anatomical objects. The distribution of activity over anatomical objects is found in *functional images*. Examples are metabolic three-dimensional maps obtained by 2-deoxyglucose labeling, positron emission tomography, or two-dimensional maps of the electrical or magnetic fields of the brain at different scalp locations.

Both types of images are fundamental in understanding mental activity and, ultimately, in aiding neuropsychiatric patients. As an example, a survey of imaging studies in psychiatry has been presented by Andreasen (1988), stressing the complementary knowledge offered by different techniques.

From the geometrical point of view, anatomical images may be considered representations of three-dimensional (3-D) objects, two-dimensional (2-D) surfaces, or one-dimensional (1-D) contours. A slightly more rigorous description would be that anatomical images are differentiable manifolds $\mathbf{A}$ defined by a mapping:

$$\mathbf{A}: U^k \succ\!\!\!-\!\!\!-\!\!\!- \mathbb{R}^p$$

$$u \;-\!\!\!-\!\!\!- \mathbf{A}(u)$$

[1]

Where U is a real interval that, without loss of generality, may be considered as $[0, 1]$; k is the intrinsic dimensionality of the anatomical object under consideration; and p the dimensionality of the space in which it is embedded ($p = 2$ corresponding to plane and $p = 3$ to solid geometry). Values of $k = 1, 2,$ and 3 correspond, respectively, to contours, surfaces, and 3-D objects. This is of course the usual parametric representation of such objects.

As concrete examples (Figure 21.1A), consider a brain slice from an atlas or a single CT scan that constitutes a plane region ($k = 2$) on a sheet $p = 2$. The edge

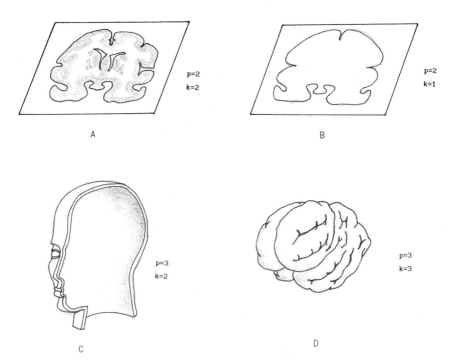

FIGURE 21.1. Types of anatomical images. Anatomical images are the representation of subsets of head structures. Two parameters—p, the embedding dimensionality, and k, the intrinsic dimensionality—serve to classify anatomical images. Some examples are given here. In A and B, the embedding dimensionality is a plane that contains cuts of the brain ($p = 2$). A, Slice of a whole brain ($k=2$) and B, its contour ($k=1$). C, Surface of the skin, a two-dimensional manifold ($k = 2$) embedded in space ($p = 3$), which may also contain a three-dimensional brain as in D.

of the brain slice or CT scan (Figure 21.1B) is a closed line, a contour ($k = 1$) on a sheet ($p = 2$). In three-dimensional space ($p = 3$) one may encounter surfaces ($k = 2$), such as the outer or inner faces of the skin (Figure 21.1C), skull, meninges, or brain (as represented in electrophysiological modeling, e.g., Hamalainen and Sarvas (1987) or fully three-dimensional objects ($k = 3$), such as the brain (Figure 21.1D) or head.

21.2.1.1 Example: Study of Cortical Atrophy in Psychiatry

A number of studies report cortical atrophy in psychiatric patients (see Andreasen, 1988 for a review). In particular Ishii (1983) has examined this issue for different alcoholic groups. CT scans level to the plane parallel to the orbitomeatal line were obtained for 75 alcoholic patients and 94 control subjects. Measurements were made of different anatomical structures, and several standard indices reflecting cortical atrophy were calculated. Differences between controls and alcoholic patients were substantiated, taking into consideration an important covariable: age.

From the point of view of our notation these slices constitute contours on a plane (Figure 21.1B) with the parameter **u** being arc length around the contour. The standard indices calculated by Ishii are summarizations of this basic information.

We may now formalize the concept of functional image **F** as a vector field defined upon an anatomical object by a mapping:

$$\mathbf{F}: \mathbf{A}(U^k) \succ\!\!\!-\!\!\!-\!\!\!- \mathbb{R}^q$$

$$\mathbf{A}(u) \text{ ———— } \mathbf{F}(\mathbf{A}(u))$$

[2]

The dimensionality q reflects the number of measures considered at each point of **A**. For $q = 1$ we have univariate functional images, such as the distribution of oxygen consumption, which takes values for each point of the brain in three dimensions ($k = 3$, $p = 3$).

21.2.1.2 Example: Mapping Memory

A study was carried out by John et al. (1986) with the purpose of determining which neural masses mediate memory. Labeled with two discriminable radio-active tracers (^{14}C and ^{18}F) 2-Flurodeoxyglucose (FDG), was injected into split-brain cats during the differential presentation of stimuli to each hemisphere. One side of the brain served to estimate the metabolic variability of nonspecific influences. The other side also included changes related to learning.

When one hemisphere received unspecific and the other learning-specific input, ^{14}C-labeled FDG was injected into the cat. Afterward ^{18}F-labeled FDG was administered when both hemispheres were receiving unspecific input. Serial sections of the cat's brain were then placed successively upon two photographic emulsions, the pattern of film exposure reflecting the amount of labeled substance trapped in the cells of each brain region. Because the half-life of ^{18}F is very short relative to that of ^{14}C, two sets of pictures could be obtained. These pictures were then digitized and transformed into quantities that reflect metabolic activity.

What has resulted, are two scalar ($q = 1$) functional images **F(A)** that reflect brain metabolism during the training period. **A** is the cat brain, and **F(A)** is brain metabolism as measured by the two tracers. A cross-section of **F** is shown in Figure 21.2 for ^{14}C and ^{18}F. Analyzing these images the authors conclude that 5–100 million neurons were involved in the formation of a memory trace.

21.2.1.3 Example: Cortical Anatomy of Word Processing

A similar example is given by PET images of humans during different behavioral conditions as obtained by Petersen et al. (1988) (Figure 21.3). In this case the images **F** localized brain activity via measurement of regional cerebral blood flow using ^{15}O-labeled water. The resulting measure is also a scalar ($q=1$). The behavioral conditions were increasingly complex; each intended to add a small set of information-processing operations to that of its subordinate (control) condition.

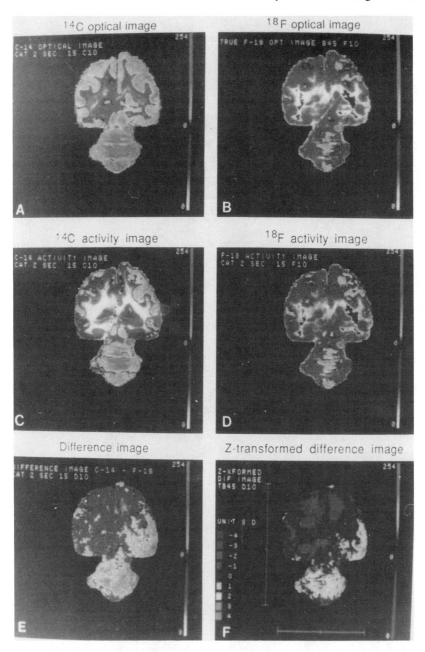

FIGURE 21.2. Mapping memory with metabolic tracers. A and B are the maps for a certain brain slice of ^{18}F and ^{14}C activity respectively. C, Difference map between A and B. D,Z-transformed map reflecting learning-specific metabolic activity. Reprinted with permission from John et al. (1986): Double-labeled metabolic maps of memory. *Science* 233:1167-1175. ©1986 AAAS.

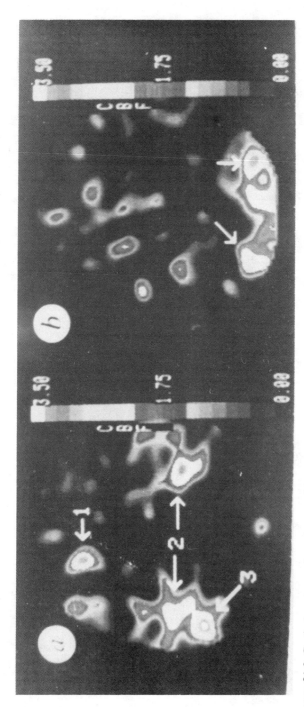

FIGURE 21.3. Cortical anatomy of word processing. Difference maps obtained from horizontal PET scan 1.6 cm. above AC-PC line. Both A and B are obtained by averaging difference images obtained during control and passive stimulation. A, Auditory stimulus and B, Visual Stimulus. Reprinted with permission of MacMillan Journals Ltd. from Petersen et al. (1988): Positron emission tomographic studies of cortical anatomy of single-word processing. *Nature* 331:585–589.

The behavioral conditions for the visual part of the experiment were (1) simple visual fixation without word presentation, (2) word presentation without speech, (3) uttering each word after presentation, and (4) saying a use for each word. Comparing pairs of successive images the authors found specific localized cortical involvement for each set of operations.

A different type of example is the distribution of the electromagnetic field in the head. If one measures *inside the head* and not only on the scalp, $\mathbf{A}$ is the head ($k = 3, p = 3$) and $\mathbf{F}$ is a four-vector (three components for the magnetic field and one for potential) with $q = 4$.

Time adds an additional classification of brain images. Structure and function vary in time, from the long-term scales related to evolution through the medium-term scales of individual maturation to the very short-term scales involved in mentation. *Static images* are those that do not depend upon the time parameter t. On the other hand the term *dynamic* is applied to an ordered set of related time-varying images. The time parameter t is specifically included when necessary as $\mathbf{A}(\mathbf{u};t)$ and $\mathbf{F}(\mathbf{A};t)$.

21.2.1.4 Example: Brain Electrical Maps and their Derivatives

Neurophysiologists usually measure time-varying potential values $\mathbf{F}$ ($q = 1$) on an $\mathbf{A}$ that is the surface of the scalp ($k = 2, p = 3$). These images may be subject to successive transformations. Because their domain of definition is a two-dimensional manifold they all receive the denomination of *maps*. In particular the following example is referred to repeatedly in this chapter.

21.2.1.5 Example: Neurometrics

John et al. (1988) have described procedures for obtaining maps of parameters obtained by means of computerized analysis of the EEG. These parameters have been demonstrated to reflected brain function and pathology and have been found to vary with location and age. John et al. (1988) created a normative database (training sample) with samples of normal subjects and of patients with neurological impairment, subtle cognitive dysfunctions, and psychiatric disorders (including dementia and primary depression).

Standardized neurometric maps may be obtained by transformation of the EEG parameters relative to the average values of the normal subjects in the normative data bases. An essential contribution made by neurometrics is the regular use of EEG developmental equations—regression equations of the EEG parameters that provide the age-dependent average and standard deviations for the normal population.

It was found that the neurometric maps of pathological subjects are different from those of normals and that different disorders are characterized by distinctive patterns. This opens the possibility of computerized differential classification of new (test) individuals with such disorders. Other types of neurometric maps may be obtained, depending on the parameter set used to describe the EEG (see Chapter 4). Also, neurometric maps may be obtained from voltage values or from current source density calculations (Pascual et al., 1988b).

21.2.1.6 Example: ERP Studies of Face Semantics

Event-related potentials (ERPs) provide a window with high time resolution into brain activity. In Chapter 13 Valdes and Bobes report on an experiment to study brain involvement in face recognition. Experimental subjects are presented with (a) familiar faces and (b) those same faces with some features pasted in from another face. Observers unfamiliar with these faces made up the control group.

ERPs were extracted from the EEG of these subjects for multiple channels. The set of values for several channels for all time instants after the stimulus constitute a dynamic ERP map. Examination of these maps identified a specific negative component associated with face feature mismatch similar to N400. Evidence was found for separate spatiotemporal patterns of activity.

In some functional images the anatomical object is considered to be a given, and it is convenient to omit explicit dependence upon it. In these cases $F(u)$ will be used in place of $F(A(u))$, for static images and similar $F(u;t)$ for dynamic ones. Thus, both types of maps may be similarly parameterized, and sometimes it is convenient to refer to an image without specifying whether it is anatomical or functional. In this case the symbol I will substitute for A and F.

This should not obscure the fundamental conceptual difference between both types of maps. An anatomical map is a set of points in space; a functional map is the mapping of values onto this set of points.

In practice everything is sampled, and actual anatomical or functional images are limited to measurements on a finite sample of u_i. For regular grids of u_i, scalar values of I take the particular names *voxel* ($p=3$) and *pixel* ($p=2$). We shall use the former term in general.

21.2.1.7 Example: The 10/20 System

Brain electrical maps are usually reported as values at a fixed parametric representation obtained as certain percentages of certain skull landmarks in the 10/20 system.

21.2.2 Matching of Anatomical Images

Even for the same type of anatomical image A there is considerable inter- and sometimes intraindividual variation. Normal head shapes vary with systematic variables, such as age, and are considerably different from person to person. In pathology the brain may be pushed aside by an expansive lesion, etc.

For many purposes this variability is ignored, and a "canonical" object A is considered as exemplified in stereotaxic atlases. Furthermore, in the study of functional images the explicit dependency upon A is sometimes obviated, as mentioned above, by a procedure based upon the assumption that the u_i are fixed and all anatomical objects are the same.

The assumption of fixed sampling points and/or fixed geometry is not always true, a difficulty to be overcome by MATCHING anatomical images. It is an important problem that must be solved not only to allow the statistical comparison of different individuals but also to align slices from the same brain into a 3-D image. Matching is also essential if different functional images (e.g., PET versus EEG) are to be compared or combined or if detailed physical or physiological theories are to be explored as in the testing of predictions of the electromagnetic fields produced in realistically shaped head models (see Chapter 22).

To date, matching methods have been based upon simple movements and scaling of objects. Hibbard et al. (1987) have introduced the principal axis transform and Fourier methods in order to align rat brain slices into a whole (Figure 21.4). Thus, slices are rotated, centered, and scaled to obtain a global match between slices.

Perhaps the notions of topological transformations of morphology introduced by D'Arcy Thompson (Figure 21.5) are more to the point. Continuous local deformations map one object to another. This goal has been pursued in multi-resolution elastic matching (Bajcsy and Kovacic, 1989). Yet, another way to deal

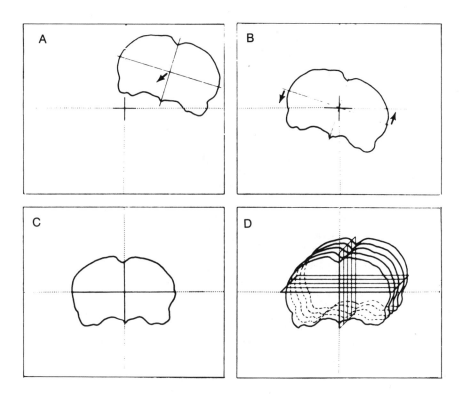

FIGURE 21.4. Principal axis transform for matching of anatomical images. Hibbard et al. (1987) have achieved intrasubject matching of images of brain slices using the principal axis transformation. Images are rotated and centered but not deformed. Reprinted with permission from Hibbard et al. (1987): Three-dimensional representation and analysis of brain energy metabolism. *Science* 236:1641-1646. ©1987 AAAS.

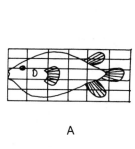

A

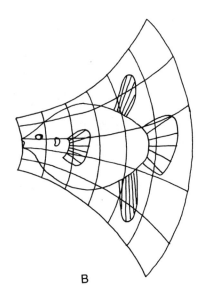

B

FIGURE 21.5. Topological transformation of two different types of images as an example of anatomical matching with image deformation.

with this problem is to consider anatomical variations from a statistical point of view (Biscay et al., 1988). This method is explored in section III.

21.2.3 Probabilistic Models for Brain Images

A feature common to all the aforementioned examples is the need to gather samples of images $I(u)$ collected according to certain experimental designs in which experimentally controlled covariables, such as time, age, sex, subject, experimental condition, etc., are either modified or recorded systematically. The set of covariables is denoted explicitly by means of the variable d and the associated image as $I(u;d)$. The joint variable $x = (u;d)$ permits the compact expression $I(x)$. The set of values of x is $X = U^k \times D$, where D is the set of values of the covariable d.

As mentioned in Section 21.1, variability enters the picture as an inherent characteristic of experimental data, be it anatomical or functional, thus making the use of statistical methods mandatory. These methods can be classified and understood (Valdes, 1984) in terms of three categories: (1) the statistical questions being asked, (2) the probabilistic mode on which they are based, and (3) the type of measurement being made, which is in this case images.

The statistical questions associated with many different experimental problems appear to be essentially invariant when cross-tabulated with the other two categories (see Valdes, 1984, for a number of examples) so we may turn our attention to the category of probability models. There is a great variety of probability models (Loeve, 1967). These models are continuously being enriched and applied to increasingly abstract sample spaces (Grenander, 1981). An exciting new development is their convergence with logical or syntactical formulations originating in artificial intelligence and pattern recognition (Fu, 1980; Grenander, 1983). Some of these may be essential for the future development of brain image analysis (see Winter, 1984).

For the purpose of this paper, however, the probabilistic models assumed are *Gaussian*. This is not such a restrictive decision since a great majority of practical applications are amenable to such models. They are more so if it is taken into account that suitable data transformations may ensure univariate and multivariate Gaussianity (Biscay et al., 1989a). The models presented may be considered as a first approximation to more sophisticated versions.

Assuming Gaussianity, the distribution of the sample of images $\mathbf{I}$ will be completely specified by its means and covariance functions:

$$\mu = \{\mu(\mathbf{x}): \mathbf{x} \in X\},$$

$$\Sigma = \{\sigma(\mathbf{x},\mathbf{x}'): \mathbf{x},\mathbf{x}' \in X\}$$

where $\mu(\mathbf{x}) = E\{\mathbf{I}(\mathbf{x})\}$ and $\sigma(\mathbf{x},\mathbf{x}') = \text{cov}\{\mathbf{I}(\mathbf{x}), \mathbf{I}(\mathbf{x}')\}$

Therefore an observed image may be modeled as

$$\mathbf{I}(\mathbf{x}) = \mu(\mathbf{x}) + \mathbf{e}(\mathbf{x}) \tag{3}$$

with $\mathbf{e}$ distributed as $N(0,\Sigma)$

That is to say, an observed image may be considered as a "true image" expressed by the mean, corrupted by noise. Therefore, most of the interesting statistical questions may be phrased as a hypothesis about μ. On the other hand, the structure of Σ determines the peculiar fashion in which the data, the questions, and their tests are represented.

21.2.4 Function Expansion Approach to Brain Image Analysis

This chapter proposes the following approach to brain image analysis: a choice will be made of Nd appropriate basis images $\mathbf{B}_j(\mathbf{u})$, valid for all design conditions, that will simplify the structure of Σ and reduce the problems to a succession of simpler ones in more traditional formats. Thus,

$$I(u;d) = \sum_{j=0}^{Nd-1} \mathbf{I}_j^*(d)\, B_j(u) \tag{4}$$

where $\mathbf{I}_j^*$ is a scalar that is the projection of the image $\mathbf{I}$ onto the corresponding jth basis function $\mathbf{B}_j(\mathbf{u})$.

Let $\mathbf{I}^*$ be the vector obtained by stacking the Nd scalars $\mathbf{I}_j^*$. Accordingly,

$$\mathbf{I}^*(d) = \mu^*(d) + e^*(d) \qquad [5]$$

which is a transformation of eq. 3 into the domain of the basis functions.

In the case of dynamic images the expansion is

$$\mathbf{I}(u;t,d) = \sum_{j=0}^{Nd-1} \mathbf{I}_j^*(d)B_j(u;t) \qquad [6]$$

and an expression similar to eq. 5 is obtained.

$e^*(\mathbf{d})$ is a random, multivariate, Gaussian vector with covariance Ψ. Usual basis functions have some sort of order that may be helpful. $\mathbf{B}_0$ is usually the constant image, representing a dc level that may be eliminated or not. Higher order $\mathbf{B}_j$ usually contains more "noise," and their elimination leads to smoothing and sometimes to drastic dimensionality reductions. Sometimes it is possible to use standard multivariate statistical methods for analyzing $\mathbf{I}^*$.

A particularly important foundation of statistical analysis is that of the Karhunen-Loeve expansion (known as principal component analysis), which diagonalizes Ψ. It is of particular importance since $\mathbf{I}^*$ will now have independent components and univariate statistics may be applied to each one. Global answers for the entire vector may be obtained by simple linear combinations of the univariate component statistics.

When the original Σ is structured, the Karhunen-Loeve expansion may be quite simple, and efficient computational algorithms are available. Such is the case for *stationary* images. For this type of image the covariance $\sigma(\mathbf{x},\mathbf{x}')$ depends only on the distance between $\mathbf{x}$ and $\mathbf{x}'$. The approximate basis functions are then products of complex exponentials, and the FFT (possibly multidimensional) is a convenient computational algorithm.

The price paid for simplicity and computational ease is the need for the use of complex random variates. As discussed in Brillinger (1975), the univariate and multivariate Gaussian distribution have their complex counterpart, as do most standard statistical real procedures.

In many cases it is convenient to use a Fourier basis even if the image is not stationary. This is not so restrictive as it may seem. Cramer introduced the notion of certain classes of nonstationary processes, that of harmonizable processes (Loéve, 1967). In this case Ψ is not diagonal, a component-by-component analysis is not possible, and multivariate methods are necessary. This added complexity is offset by the possibility of analyzing nonstationary spatiotemporal data. The methods may still be computationally effective due to the speed of FFT algorithms and the dimensionality reduction because of the elimination of "noise components" mentioned above.

21.3 Analysis of Anatomical Images

There is an emerging field of quantitative analysis of anatomical images that is commonly known as morphometrics. As an example, consider the work of Takeda and Matsuzawa (1984) in developing age-dependent norms from age 10 to 88 for a normative data base of 980 normal subjects. Ishii (1983) developed similar procedures and classified different groups of alcoholics. Striking parallelism exists between work in electrophysiology and x-ray computer-tomography. In fact the examples just given are isomorphic to those developed by neurometrics (John et al., 1988).

Most of these studies are based upon selected parameters that are extreme condensations of the original data. In the study of cortical atrophy the original data are contours of very irregular shapes. The standard practice is to make either linear measurements between some standard points or to measure the area enclosed by the contour.

Using the general approach outlined in Section 21.1 it can be demonstrated that it is possible to perform statistical analysis upon the *shapes* of the cortex in a fashion that brings neurometrics and morphometrics even closer together methodologically. The basis for contour expansion is selected according to geometrical considerations.

Each sample image is $I(u)$, a closed and smooth plane contour (e.g., $k = 1$, $p = 2$) as in Figure 21.1B. What is actually analyzed is the set of values that are the contour counterpart of pixels. These are obtained by uniform sampling around the contour in a counterclockwise orientation. The image is a set of N values for discrete values of the parameter u ($u = 0, \ldots N - 1$), starting at an arbitrary initial point $I(O)$. $I(O)$ will be equal to $I(N)$. Each point $I(u) \in \mathbb{R}^2$ may be represented alternatively as a complex number, $I(u) \in \mathbb{C}$.

The *shape* of a contour must be invariant to the following transformations:

Translation: moving a contour about in the complex plane is

$$I \longrightarrow I + c\,\mathbf{1}$$

where $\mathbf{1}$ is the vector $\mathbf{1} = (1, \ldots, 1)^t$ and c is a complex number.

Scale change:

$$I \longrightarrow \alpha\,I$$

where α is a positive real number.
Rotation:

$$I \longrightarrow \exp(i\theta)\,I$$

where θ is the angle through which the contour is rotated and $i=\sqrt{-1}$.
Changing the initial point of digitization:

$$(I(0), \ldots, I(N-1)) \longrightarrow (I(\phi), I(\phi+1), \ldots, I(N-1), I(0), \ldots, I(\phi-1))$$

where ϕ is the number of points by which the contour representation is shifted.

In consonance with the above, the *shape* of a contour may be mathematically defined as the equivalence relation on the N-dimensional complex linear space $\mathbb{C}^N$ that is induced by the group of transformations that consist of translations, scaling, rotations, and cyclic permutations of the components.

In order to model the stochastic variations around a morphometric pattern eq. 3 is adapted to the present situation as follows:

$$I(u) = c + \alpha \exp(i\theta) \, \Pi(u-\phi) + e(u) \qquad (u = 0 \ldots N-1) \qquad [7]$$

where $\Pi(u)$ is the population morphometric pattern, and C, α, θ, and ϕ are parameters corresponding to translation, scaling, rotation, and change of the initial point for sampling in the observed contour. Note that in comparison with eq. 3, in eq. 7 a structure has been given to the mean $\mu(u)$, depending on the set of (nuisance parameters) that reflect the geometrical invariance inherent to the concept of shape.

Eq. 7 expresses the idea that the observed contour is obtained by a transformation of the population pattern Π that does not change the shape, with the addition of a stochastic deformation e.

Alternatively this model may be expressed in the discrete Fourier basis:

$$B_\omega(u) = \exp(-i2\pi u\omega/N)$$

thereby transferring the model to the frequency domain:

$$I_\omega^* = \alpha \exp(i2\pi(\theta + \phi\omega)/N) \, \Pi_\omega^* + e_\omega^* \qquad [8]$$

Frequency ω will take the integer values in $[0 \ldots T-1]$. If the frequency $\omega = 0$ is eliminated the effect of translations is implicitly eliminated.

The I_ω^* are well known in the pattern recognition and image processing literature as the *Fourier shape descriptors* (FSD) (Dekking and Van Otterloo, 1986; Persoon and Fu, 1977) of the contour I. The representation of contours using this basis has the following advantages. In general, it results in a compact coding (low dimensionality). There is a simple parametric expression for the transformations for which shape is invariant. Usual geometric features (perimeter, area, curvature) may be efficiently calculated from the FSD.

Figure 21.6 illustrates the use of FSD in representing the contours of the cerebral hemispheres. In that figure, $S(\omega)$ is the absolute value of I_ω^* for a sample of a normal subject. It is also apparent that finer detail is conserved as one adds elements to the basis (5, 30, and 40). A balance must be struck between smoothing and retention of detail, which seems to be optimal for about 30 coefficients.

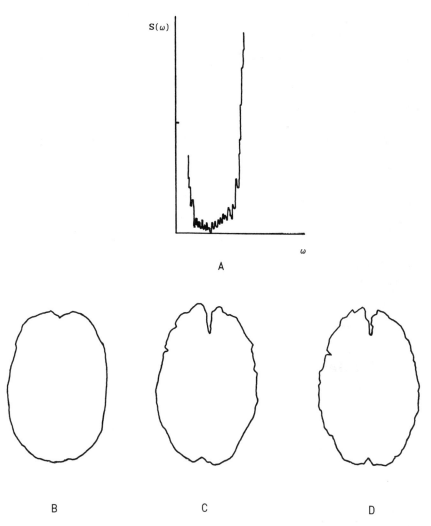

FIGURE 21.6. Representation of a brain contour by means of Fourier shape descriptors (FSD). $S(\omega)$ is the shape spectrum. Successively better reconstructions are obtained by means of 5, 30, and 40 Fourier coefficients.

It is important to consider that the same type of representation that describes cortical outlines also serves to describe the electrical activity it produces.

Eqs. 7 and 8 takes into consideration the geometric invariance inherent to the concept of shape (by means of the parameters C, α, θ, ϕ) inheriting the advantages of FSD, and the existence of stochastic additive deformations $\mathbf{e}(u)$.

In practice it is natural to assume that the deformation or "noise" $\mathbf{e}(u)$ is stationary. With this assumption $\mathbf{e}(u)$ is a circulant stationary stochastic process, and the Fourier basis is the exact Karhunen-Loeve transform. As mentioned in

Section I the coefficients e_ω^* are uncorrelated and their variances σ_ω^2 define the spectrum of the process e. Homogeneous markovian models (e.g., circular auto regressive, Kashyap and Chellappa, 1981) are particular cases of the present model when the spectrum is a rational function.

Eqs. 7 and 8 may be used as a basis for statistical inference about morphometric patterns (Biscay et al., 1988). Some notable examples are described below.

21.3.1 Estimation of a Population Morphometric Pattern Π

This estimation is made on the basis of a sample of contours $I_1, \ldots, I_M$ from the same population defined by eq. 7. The log-likelihood function in the frequency domain is (up to a constant)

$$L = N \sum_\omega \ln(\sigma_\omega^2) + \sum_m \sum_\omega |I_{\omega m}^* - \alpha_m \exp(i\theta_m + i\omega \phi_m) \Pi_\omega^*|^2/\sigma_\omega^2 \qquad [9]$$

where the parameters σ_m, θ_m, ϕ_m determine the translation, scale, rotation, and initial point change of the pattern Π associated to each observed I_m. The estimation of the pattern Π may be carried out by means of the maximum likelihood (ML) method, i.e., maximizing eq. 9 with respect to Π, α_n, θ_n, ϕ_n and σ. The parameters of the transformations are nuisance parameters.

21.3.2 Tolerance Regions

The ML estimation of the pattern Π and the spectrum σ on the basis of a sample permits the construction of the appropriate tolerance regions to decide about the hypothesis H: does a new sample contour I have the same morphometric pattern Π?

For this purpose the following Mahalanobis-type distance is defined between the shape of two contours: I and J:

$$M(I,J;\sigma) = \inf \sum_\omega |I_\omega^* - \alpha \exp(i\omega\phi + \theta) J_\omega^*|^2/\sigma_\omega^2 \qquad [10]$$

where the infimum is with respect to the nuisance parameters θ, α, ϕ. This operation corresponds to the set of geometric transformations that achieve the best ML match between both images.

If I follows eq. 7, then $M(I,\Pi;\sigma)$ follows an approximate χ^2 distribution with $2N$ degrees of freedom. This is the statistic that is obtained by a likelihood ratio for hypothesis H. Therefore, an approximate tolerance region at the $1 - \alpha$ level for I is given by $M(I,\Pi,\sigma) \leq \chi^2(1 - \alpha;2N)$.

Alternatively, one may be interested in variation in size. The problem here is to find an optimal match between a sample and the pattern shape and to evaluate the fluctuations in size.

Figure 21.7 illustrates these concepts. Figure 21.7B shows a sample of "noisy" contours of brain hemispheres. From these the estimated morphometric pattern

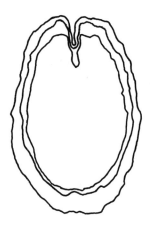 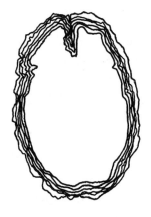

A   B

FIGURE 21.7. Confidence intervals for brain contours obtained by means of Fourier shape descriptors. A, 0.01 level confidence intervals of brain image. B, Superposition of sample contours.

was extracted. Considering the fluctuations in shape, the 0.01 tolerance region was obtained as shown in Figure 21.7A. Brain hemispheres with "normal" size and shape brain hemispheres fall within it.

When **I** falls out of the limits of this region there is evidence for significant deviations in *shape* (or size) from the population (normal) pattern. The contour would be considered pathological. It is suggested that studies of cortical atrophy in psychiatric patients explore this procedure.

21.3.3 Comparison of Shapes

A frequent problem in morphometrics is the comparison of the shapes of two samples of contours—$\mathbf{I}_1, \ldots, \mathbf{I}_M$ and $\mathbf{J}_1, \ldots, \mathbf{J}_R$— corresponding to two populations of interest, e.g., schizophrenics and normals. The comparison may be carried out by means of the distance of eq.10 with estimated morphometric patterns $\hat{\Pi}_1$ and $\hat{\Pi}_2$; that is, by means of:

$$T = M(\hat{\Pi}_1, \ \hat{\Pi}_2 \ ; \ \hat{\sigma})$$

where $\hat{\sigma}$ is the estimation of the spectrum of the deformations (assuming σ is the same for both populations). This is the statistic obtained by likelihood ratios for this problem. The distribution under the null hypothesis is proportional to a F distribution.

21.3.4 Contour Filtering

Given a contour $\mathbf{I}$ observed in eq. 7, in order to obtain the component $\hat{\mathbf{I}}(u) = \alpha \exp(\theta iu\phi)\Pi(u)$ it is necessary to eliminate the deformation $\mathbf{e}$ from $\mathbf{I}$. This operation of "morphometric Wiener filtering" may be carried out optimally (in mean square sense) by means of

$$\hat{\mathbf{I}}_\omega^* = H_\omega \ \mathbf{I}_\omega^*$$

where $H_\omega = |\Pi_\omega^*|^2/[|\Pi_\omega^*|^2 + \sigma_\omega^2]$

21.3.5 Classification of Shapes

Suppose that a contour $\mathbf{I}$ is to be classified into one of two populations with patterns Π_1 and Π_2 (and the same σ). The morphometric Mahalanobis distance (eq. 10) offers a statistical metric for this problem: classify $\mathbf{I}$ in population Π_1 if

$$M(Z,\Pi_1;\sigma) < M(Z,\Pi_2;\sigma)$$

This is the ML classification rule applied to this situation.

21.3.6 Regression of Shapes Upon Age

Consider in eq. 7 that the images depend on age either in a linear or nonlinear fashion. Then age-regression equations in the spirit of neurometrics may be obtained for shape. Additionally the size parameter α may be also age regressed. Consider the example of *study of cortical atrophy* (Section 21.2.1.1). The sample images are $\mathbf{I}(u;g,i,a_i)$, where $d = (g,i,a_i)$. i index the ith individual in group g (alcoholic, normal) with age a_i. Sections 21.3.3 and 21.3.4 permit us to analyze this example based upon eq. 8 with age-dependent parameters:

$$\mathbf{I}(u;g,i,a_i) = C_i + \alpha(g,a_i) \exp(i\theta) \Pi(u - \phi; \ g,a_i) + \mathbf{e}(u;g,i)$$

The Mahalanobis-type distance introduced with eq. 10 may be used to answer the question of equality of alcoholic and normal shapes.

A set of three-dimensional Fourier descriptors has been developed by Park and Lee (1987). Work is in progress to extend the methodology outlined here to three-dimensional surfaces.

21.4 Analysis of Functional Images

In this section the discussion of the examples of functional images introduced in Section 21.2 is concluded. By no means is this discussion exhaustive. Rather, the function expansion approach outlined above and discussed in detail for anatomical images is also valid for functional images, applied with differing levels of complexity.

21.4.1 Mapping Memory

The fundamental question here is how to identify neural masses specifically related to memory. The sample images are $I(u;h,l)$, where h is hemisphere (left, right) and l the label ($^{14}C, ^{18}F$). The implicit model is, in our notation,

$$I(u;h,l) = \alpha(u;h) + \beta(u;h,l) + e(u;h,l) \qquad [11]$$

That is, it is an image version of a two-way nested ANOVA. In eq. 11 $\alpha(u;h)$ is a nonspecific hemispheric effect, and $\beta(u;h,l)$ is a specific effect that by design could only be nonzero for h (right) and $l = {}^{14}F$. Forming the image

$$\Delta I(u;h) = I(u;h,{}^{14}C) - I(u;h,{}^{18}F) = \Delta\beta(u;h) + e'(u,h)$$

will eliminate the unspecific factors α from both hemispheres, where $e'(u,h)$ is the noise of the difference. It therefore remains to test whether $\beta(u;right,{}^{14}C)$ is different from zero.

This is a complex problem. The answer may be with respect to the whole image, to subregions, or to individual voxels. As an example of an analysis at the latter level, John et al. (1986) employed the following procedure. For each individual voxel of the right hemisphere the Z-transform was performed:

$$Z(u;h) = (\Delta I(u;h) - mean)/s.d$$

where the mean and standard deviation (s.d.) are calculated over $\Delta I(u;left)$. A color-coded map of Z (Figure 21.2) is the final result. Other scaling methods are possible (John et al., 1986); all are based upon introducing a common metric for deviation from the null hypothesis, thus rendering detailed knowledge of basal regional metabolism superfluous.

This is a simple application of Gaussian univariate statistics. On the left side very little activity is evident. On the right, specific learning activity in certain anatomical structures is clearly evident. As an exploratory tool the Z-transform has proved to be extremely useful in many fields (John et al., 1987).

However, the use of univariate statistics in image statistics introduces a number of problems that traditional statistics has not solved satisfactorily.

As mentioned in the introduction, the anatomical, physiological, (and sometimes even experimental) constraints of brains introduce a high degree of correlation between voxel values. Σ is certainly not diagonal. Use of univariate statistics inflates the type I error and makes the assignment of probability levels difficult if not impossible. This point is discussed in detail in Valdes (1984) and demonstrated for neurometric data in John et al. (1988), and is caused by the fact that implicitly Z maps impose rectangular confidence intervals instead of the el-lipsoidal ones that would result from taking into consideration the correlations between voxels. The practical consequence is the detection of nonactive areas as active.

A gigantic number of multiple tests are being performed. Statistically orthodox simultaneous confidence intervals are extremely wide. This leads to the next problem.

Even if the type I error could be controlled by some multiple comparison method (which is not so easy!), the high error variance and relatively low sample sizes would inflate the type II errors to an unknown degree. The practical conse-quence here is that, in trying to avoid the type I error, active sites may be missed by being over cautious.

These shortcomings of traditional univariate statistics indicate the need for caution in the interpretation of Z-transformed maps. They are important tools in exploratory data analysis, but present difficulties for confirmatory statistics. The consistency of new experimental results is desirable.

The alternative function expansion procedure would involve (1) expressing the images in terms of a Karhunen-Loeve expansion, (2) carrying out global multi-variate tests in the reduced representation space, and (3) identifying by external criteria regions of interest (ROI). Tests that are restricted to these regions may be obtained by means of the union intersection (UI) principle of S.N. Roy (Mardia et al., 1979). The best multiple confidence intervals for individual voxels would also be obtainable from UI theory.

21.4.2 Cortical Anatomy of Word Processing

The sample images are $\mathbf{I}(\mathbf{u};b,i)$, where b is the behavioral condition and i the subject. The statistical problems encountered here are similar to those of the preceding example. In an attempt to control type II errors, images from different subjects are pooled.

This is a *repeated-measures* situation. The complications are due to the fact that voxels from the same subject may have *further* correlation patterns than those mentioned above. This problem may be solved sometimes by using differences between images (but see Bock, 1975, for more complicated situations and solu-tions).

In the original paper active voxels were identified by using an outlier detection test and then a Z transform. The same comments discussed above apply here. A formal model for this situation is

$$\mathbf{I}(\mathbf{u};b,i) = \alpha(\mathbf{u};i) + \beta(\mathbf{u};b) + \mathbf{e}(\mathbf{u};b,i)$$

where the α is a random effect due to each subject and the effects of each behavioral condition β are considered fixed. In order to eliminate the repeated measure effect due to α, Petersen et al. (1988) formed the images

$$\Delta I(\mathbf{u}; b, i) = \mathbf{I}(\mathbf{u}; b, i) - \mathbf{I}(\mathbf{u}; b\text{-}1, i) = \Delta\beta(\mathbf{u}; b) + \mathbf{e}'(\mathbf{u}; b, i) \quad (b=2, 3, 4)$$

to which univariate and function expansion methods may be applied.

21.4.3 Neurometrics

The sample images are $\mathbf{I}(\mathbf{u}; s, g, i, a_i)$, where s is the type of set (training, test), g is the diagnostic category, i the individual, and a_i his or her age. Here the model is

$$\mathbf{I}(\mathbf{u}; s, g, i, a_i) = \mu(\mathbf{u}; g, a_i) + \mathbf{e}(\mathbf{u}; g, i) \qquad [12]$$

That is, there is a "pattern" for each diagnostic group, depending on age and corrupted by noise that is group independent with covariance Σ. The treatment of this problem entails obtaining the regression equations of **I** with respect to age, (John et al., 1987) and performing discriminant analysis with group means that depend on age. Discriminant analysis is based upon the use of Mahalanobis distance, a rescaled metric that is the multivariate counterpart to the Z-transform.

However, this example highlights the need for an adequate functional representation for images. When the number of variables (voxels and types of neurometric parameters) is small, simple multivariate statistical methods, as available in standard packages, usually provide adequate results.

Even so, with as few as two variables, the situation might not be so simple. In Chapter 4 Alvarez et al. report that the values of peak α frequency for 192 subjects (ages 5–12) for corresponding right-left homologous voxels were very highly correlated, a not very surprising fact if one considers the presence of massive transcallosal connections. The covariance matrix of just these two variables is nearly singular. An increasing number of variables only aggravates the presence of multicolinearities.

This magnitude of the problem is reflected quantitatively by the Karhunen-Loeve expansion of all variables of a complete neurometric model (ξα). Figure 21.8 shows the distribution of the eigenvalue of the sample covariance matrix, which appears to decrease exponentially, rapidly approaching zero. Singularity or near singularity of Σ is certainly a nonstandard situation for statistical theory. Though the singular multinormal distribution is well studied (Rao and Mitra, 1971), the state of procedures to decide on the number of components and to perform statistical inference is far from finished.

John et al. (1987) have described the increasing unreliability of Mahalanobis distances with an increasing number of variables. Setting the smaller correlations to zero improved the situation somewhat, but is an ad hoc procedure.

Valdes et al. (1985) proposed partitioning the sample space into "normal" and

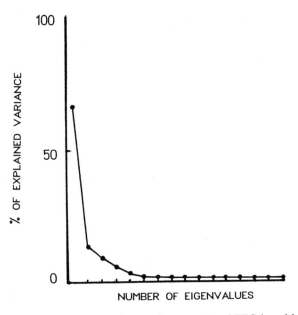

FIGURE 21.8. Eigenvalues of the sample covariance matrix of EEG broad band spectral measures. The rapid decrease of the eigenvalues reflects the high degree of redundancy present in the data.

"pathological" subspaces, the former reflecting the constraints of normal brain functioning. Biscay et al. (1989b) have generalized and given statistical justification to this proposal by defining

$$M(\mathbf{I^*},\mu^*) = M_N(\mathbf{I^*},\mu^*) + C \, \|\mathbf{I^*} - \mu^*\|_P^2 \qquad [13]$$

where $M_N(\mathbf{I^*},\mu^*)$ is the Mahalanobis distance in the space E_N spanned by the first K principal components, $\|\mathbf{I^*} - \mu^*\|_P$ is the usual Euclidean distance in the space E_P orthogonal to E_N, and C is a constant defined to guarantee that eq. 3 has an approximate χ^2 distribution.

The Mahalanobis distance in E_N is defined by

$$M_N(\mathbf{I^*},\mu^*) = \sum_{j}^{K} (\mathbf{I}_j^* - \mu_j^*)^2/\lambda_j$$

and the Euclidean distance in E_P as

$$\|\mathbf{I^*} - \mu^*\|_P^2 = \|\mathbf{I}_P^* - \mu_P^*\|^2$$

where $\mathbf{I}_P^*$ and μ_P^* are the projections of E_P of $\mathbf{I^*}$ and μ^*, respectively. This solution apparently retains both the advantage of reduced dimensionality (using the first KPC) and keeping the information available in the last p-K. Biscay et al. (1989b) demonstrated, both theoretically and with cross-validation of simulation experiments, that there are substantial gains in classification rates for linear and nonlinear discriminants.

21.4.4 ERP Studies of Face Semantics

The sample images are $I(\mathbf{u};t,c,i)$, where c is the experimental condition (congruent face, incongruent face) and i is the individual. The model is

$$I(\mathbf{u};t,c,i) = \alpha(\mathbf{u};t,i) + \beta(\mathbf{u};t,c) + e(\mathbf{u};t,c,i) \qquad [14]$$

which is similar to that of eq. 11. Pooling subjects leads to a repeated-measures situation modeled by the random effects images α. The following difference images are defined as

$$\Delta I(\mathbf{u};t,i) = I(\mathbf{u};t, \text{ congruent}, i) - I(\mathbf{u};t, \text{ incongruent}, i) \qquad [15]$$
$$= \Delta\beta(\mathbf{u};t) + e'(\mathbf{u};t,i)$$

The hypothesis of interest is that $\Delta\beta(\mathbf{u},t) \neq 0$. Differences with eq. 11 are that in eq. 13 the image is limited to the scalp and is dynamic; hence, not only the question of *where* activity is occurring but also *when* is of importance. This illustrates the appearance of an "image version" of time series analysis collected in an experimental design.

In dynamic images the correlation structures and dimensionality problems are now in two different interrelated domains: space and time. In order to answer questions about eq. 13 a number of solutions have emerged, none of which is conclusive.

21.4.4.1 Univariate Solutions Based upon a Voxel-by-Voxel Basis

These solutions include the following:

Usual multivariate statistics is applied by stacking values at time instants $t_1, \dots, t_T$. If T is large, singularity of temporal covariances occurs.

Assuming stationarity, Fourier analysis makes each frequency asymptotically independent. Complex univariate statistics may be applied to each frequency. As applied to eq. 14 the definition of a "difference spectrum" as the analog to the t test has been defined in Valdes et al. (1988) and Carballo et al. (1989). Complex ANOVA models for ERP studies have been discussed in Brillinger (1981).

More specific parameterizations of stationary Σ are possible. Linear (ARMA) models for the EEG have also been applied (Gersch, 1987), which are representations in terms of rational polynomials of the spectrum. Nonlinear models for the EEG spectrum have also been explored (Chapter 4).

The restriction to stationary background processes may be relaxed by use of the theory of harmonizable processes (Loeve, 1967). This approach has been propounded by Valdes et al. (1988) and Carballo et al. (1989).

21.4.4.2 Multivariate Time Series Analysis

The analysis of the variation of single voxels over time ignores correlations between different voxels and spatial correlations. One solution to this problem is the application of multivariate time series analysis. In this type of analysis the voxels of an image $I(u;t_0,d)$ for a particular time t_0 constitute a vector. The resulting set of dynamic vectors $I(,d)$ possess a mean $\mu(t,d)$ and covariance matrix $\Sigma(t)$.

If the components of $I(t;d)$ and their covariances are stationary, then a Fourier transformation of each component will yield observations $I_\omega^*(d)$ that are approximately independent. Each frequency may be then treated using complex multivariate statistical methods, the variance and covariance information residing in a frequency-dependent hermitian matrix $\Sigma^*(\omega)$ known as the cross-spectral matrix.

Such an approach has been developed by Brillinger (1975) who has extended nearly all classical multivariate analysis to multiple time series, and been applied to ERPs by Rawlings et al. (1984, 1986) (see Chapter 4). Rational and nonlinear representations for the cross-spectral matrix have been calculated by Gersch (1987), Pascual et al. (1988a), and Alvarez et al. in Chapter 4.

In this context the answer to the significance of $\Delta\beta(u;t)$ may be sought for different regions of space and time by means of the UI principle. In order to obtain an answer, a number of assumptions (stationary, homogeneity of variance) have to be adopted. They should be checked, or in the worst case, one should consider the results as exploratory.

This small review of examples may indicate the utilization of the functional expansion approach indicated in Section 21.2. Adequate a priori knowledge should be incorporated into the choice of the basis function and the parameterization of the mean. It would be ideal if the basis itself would be suggested by the class of solutions of specific neurophysical models of the brain. Without a greater emphasis on fundamental modeling the statistics of brain images will continue to be mostly exploratory in nature.

21.5 Combining Information from Different Images

Many groups are now engaged in trying to combine information from different types of images. For example, the results of PET, CT, and MRI scans are being correlated (Andreasen, 1988; Kling, 1986). This objective brings to mind procedures already developed in multivariate statistics in which two or more sets of different types of variables are correlated, a prime example being canonical correlations. The fully developed image analogs are yet to be developed.

In terms of the general notation introduced in Section 21.2, one could conceive of the joint study of various anatomical images, such as CT and MRI scans, that reflect different physical information. This could be attacked by means of the morphometric methods sketched in Section 21.3.

The joint study of functional images—for example, correlating PET and EEG studies—could be addressed by the methods described in Section 21.4. Instead of considering *scalar* fields over the head these would be *vector fields* with a vector formed by the pair (EEG, PET) measurement at each point. In this instance using the notation devised in Section 21.1, q is greater than 1.

On the other hand, the comparative study of anatomical and functional images is also of interest; some examples of such study are presented in this section. The first concerns modeling the origin of the electrical and magnetic fields of the brain and their reflection as measurable time-varying images. As discussed in the section on functional images, brain electrical maps (Duffy, 1986; John et al., 1988) provide useful insight into mechanisms of brain functioning. These functional images obtained from voltage values are mapped onto a standard head via the 10/20 system. The newer field of magnetic recordings (Williamson and Kaufman, 1987) uses the same approach.

For some years, neural models of the origin of the EEG (and consequentially the MEG) have been formulated (Lopes da Silva, 1987; Nunez, 1981). The field is ripe for precise numerical predictions that can be checked by experiments. The usefulness of these predictions is illustrated by combined CT and EEG studies of schizophrenics (Morisha, 1985) in which the appearance of EEG "hypofrontality" was dependent on concomitant cortical atrophy. A combination of morphometrics with neurometrics seems in order.

For these purposes an initial approximation may be a standard head modeled as a sphere (Scherg and Von Cramon, 1985; Chapter 22). However, there is a trend toward modeling in realistic geometry (Barnard et al., 1967; Hamalainen and Sarvas, 1987). The solution of the so-called direct problem and inverse problem calls for the connection, provided by physical equations, of anatomical and functional images.

The model used for this purpose is a piecewise homogeneous conductor: brain,

16 Slices   31 Slices

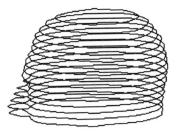   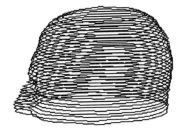

FIGURE 21.9. Three-dimensional Fourier description of skin surface reconstructed from CT scans. Such descriptions for brain, skull, and skin are used in physical modeling of the head to combine information provided by anatomical and physiological images.

skull, and skin, for example. A system of integral equations involving surface integrals over the boundaries of the regions must be solved numerically.

In recent work (Gonzalez et al., 1989) a connection has been established between the morphometric harmonic representation described in Section 21.3 and the solution to these problems. A three-dimensional Fourier description (FD3) is obtained for the boundaries between each region. Figure 21.9 illustrates such a representation of the surface of the skin, obtained from a series of CT contours. A projective method (collocation) is implemented based upon the FD3 representation just mentioned and a similar expansion of the functional image.

Sometimes it is of particular importance to assess the correspondence between information given by functional images and that given by anatomical images. A study in this direction is the evaluation of the accuracy with which electrical functional images actually localize perturbations of neural activity at specific sites as identified by localized alterations of anatomical images.

Unfortunately, this problem, to our knowledge, has not been studied statistically, with the exception of work carried out to assess the error rates of different types of diagnostic imaging procedures (Swets, 1988). In that context some extensions of the traditional ROC methodology are being implemented for the evaluation of radiological procedures (Starr et al., 1975).

In this section, we introduce a conceptual framework for the evaluation of localization procedures and some illustration with clinical material. To perform this task, measures must be designed that fulfill certain requirements. They must reflect the discrepancy between the proposed localization region D of the lesion and the actual one T. The measure of discrepancy E will reflect the difference between the two sets, D and T. E must fulfill the additional requirements. It must be invariant to image translations and rotations and to scale changes. A false-positive proposed localization should be given greater weight the farther away it is from the actual localization, and a false-negative localization should also be given weight, this time in correspondence with the proposed localization. E is invariant to image translations and rotations if the measure is constructed from the Euclidean distance. Invariance to scale changes is guaranteed by scale standardization.

The measure proposed for quantifying localization errors is

$$E(D, T) = FPE(D, T) + FNE(D, T)$$

where

$$FPE(D, T) = \int_D d(a, T)da$$

$$FNE(D, T) = \int_D d(a, D)da$$

and $d(a,A)$ is the Euclidean distance between point a and the set A.

FPE (false-positive error) is a set measure of how far each point of the proposed localization actually is from the totality of the lesion. The FNE (false-negative error) is a set measure of how far each point of the actual lesion is from the totality of the proposed localization. We adopt the convention, $d(x,\emptyset)=1$, which is the maximum distance possible ($\emptyset$ is the null set). Note that in the absence of lesion the error reduces to

$$E(D, \phi) = \int_D dx = area(D)$$

When there is a lesion but no detection is proposed the measure reduces to

$$E(\phi, T) = \int_T dx = area(T)$$

The specific issue chosen as an example is an assessment of how much hard evidence there is for the assertion that maps obtained with the Laplacean montage provide superior localization of brain lesions than those originating from ordinary voltage montages. In this case the region D was marked upon a 3-D reconstruction of the head by a panel of observers, and the region T was derived by independent evaluation of clinical, tomographic, and necrological evidence (Table 21.1).

Another panel of six neurometric experts (from two countries) evaluated a total of 30 maps with lesions and 60 without, each repeated as a voltage map and a Laplacean one. The maps were presented in a double-blind situation, completely at random by the computer.

The measures proposed above were calculated for each map in its voltage and Laplacean version. Because the same map was viewed once as a Laplacean map and once as a voltage map, a paired T-test was used to assess the significance of the differences. The results are presented in Table 21.1. Numerically, the Laplacean version is better in an overall sense, with a higher tendency to detect nonexisting lesions. Voltage measures tend to err by omission. The higher false-positive rates for localization by the Laplacean may have to do with the amplification of spatial noise discussed in Chapter 22. However, none of these tendencies is significant at the 0.05 level.

These results indicate that in spite of subjective clinical consensus in favor of the Laplacean montage and substantial theoretical reasons (Pascual et al., 1988) for its adoption, there is no overwhelming evidence in its favor. However, the differences observed are in the theoretically expected direction, and further research on this problem is certainly warranted.

This discrepancy between theory and reality underscores the clear need for experimental and theoretical work on the localization of the complex and some-

TABLE 21.1. Paired T-test: voltage-Laplacean montages (d.f.=267)

Measure	T-test	Problem
EFP	−0.1757	0.4303
EFN	1.3802	0.0843
E	0.6441	0.2600

times diffuse systems that mediate behavior. The inconclusive, yet promising results obtained with well-defined anatomical lesions provide matter for thought when trying to pinpoint the more elusive correlates of cognitive function. Pinpointing those correlates will undoubtedly be the subject of future work.

References

Andreasen, N.C. (1988): Brain imaging: Applications in psychiatry. *Science* 239, 1381-1388

Bajcsy, R., Kovacic, S. (1989): Multiresolution elastic matching. *Comput. Vis. Graphics Image Processing* 46, 1-21

Barnard, A.C.L., Duck, I.M., Lynn, M.S. (1967): The application of electromagnetic theory to electrocardiology, I. *Biophys. J.* 7, 443-462

Biscay, R.J., Valdes, P., Aubert, E. (1988): Morphometric analysis of contours and surfaces. *Revista CENIC Ciencias Biologicas* 19, 3

Biscay, R.J., Valdes, P., Pascual, R., Jimenez, J.C., Alvarez, A., Galan, L. (1989a): Multivariate Box-Cox transformations with applications to neurometric data. *Comput. Biol. Med.* 19, 4

Bock, R.D. (1975): *Multivariate statistical methods in behavioral research.* New York: McGraw-Hill

Brillinger, D.R. (1975): *Time series: Data analysis and theory.* New York: MRW-Inc.

Brillinger, D.R. (1981): The general linear model in the design and analysis of evoked response experiments. *J. Theoret. Neurobiol.* 1, 105-119

Carballo, J.A., Valdes, P., Valdes, M. (1989): The detection of event related potentials. *Int. J. Neurosci.*

Dekking, F.M., Van Otterloo, P.J. (1986): *Fourier coding and reconstruction of complicated contours. IEEE Trans. Systems Man Cybern.* SMC-16, 395-404

Duffy, F.H. (1986): *Topographic mapping of electrical brain activity.* Boston: Butterworths

Fu, K.S. (1980): Syntactic image modeling using stochastic tree grammars. *Comput. Graphics Image Processing* 12, 136-152

Gersch, W. (1987): Non-stationary multi channel time series analysis. In: *Handbook of electroencephalography and clinical neurophysiology:* vol. 1. Gevins, A.S., Rémond, A. (eds.). Amsterdam: Elsevier

Gonzalez, S., Grave, R., Biscay, R., Jimenez, J.C., Pascual, R., Lemagne, J., Valdes, P. (1989): Projective methods for the magnetic direct problem. In: *Advances of biomagnetism.* Williamson, S., Kaufman, L. (eds.). New York: Plenum Press

Grenander, U. (1981): *Abstract inference.* New York: Wiley

Grenander, U. (1983): *Tutorial in pattern theory.* Providence, RI: Division Applied Mathematics, Brown University

Hamalainen, M.S., Sarvas, J. (1987): Realistic conductivity geometry model of the human head for interpretation of neuromagnetic data. *Technical Report TKK-F-A614.* Helsinki: Helsinki University of Technology

Hibbard, S.L., McGlone, J.S., Davis, D.W., Hawkins, R.A. (1987): Three-dimensional representation and analysis of brain energy metabolism. *Science* 236, 1641-1646

Ishii, T. (1983): A comparison of cerebral atrophy in CT scan findings among alcoholic groups. *Acta Psychiatr. Scand.* 68, (Suppl.), 309

John, E.R., Tang, Y., Young, R., Brill, A.B., Young, R., Ono, K. (1986): Double-labeled metabolic maps of memory. *Science* 233, 1167-1175

John, E.R., Prichep, L.S., Easton, P. (1987): Normative data banks and neurometrics: Basic concepts, methods and results of norm construction. In: *Handbook of electroencephalography and clinical neurophysiology: vol 1*. Gevins, A.S., Rémond, A. (eds.). Amsterdam: Elsevier

John, E.R., Prichep, J.F., Easton, P. (1988): Neurometrics: Computer-assisted differential diagnosis of brain functions. *Science* 239, 162-169

Kashyap, R.L., Chellappa, R. (1981): Stochastic models for closed boundary analysis: Representation and reconstruction. *IEEE Trans. Information Theory* IT-27, 627-637

Kling, A.S. (1986): Comparison of PET measurement of local brain glucose metabolism and CT measurement of brain atrophy in chronic schizophrenia and depression. *Am. J. Psychiatry* 143, 175-180

Loéve, M. (1967): *Probability theory*. New York: Springer-Verlag

Lopes da Silva, F.H. (1987): Dynamics of generation of EEGs as signals of neural populations: models and theoretical considerations. In: *Electroencephalography: Basic principles, clinical applications and related fields*. Niedermeyer, E., Lopes da Silva, F. (eds.). Baltimore: Urban and Schwarzenberg

Mardia, K.V., Kent, J. T., Bibby, J.M. (1979): *Multivariate analysis*. London: Academic Press

Morihisa, J.M. (1985): Structure and function: Brain electrical activity mapping and computed tomography in schizophrenia. *Biol. Psychiatry* 20,3-19

Nunez, P.L. (1981): *Electric fields of the brain. The neurophysics of the EEG*. New York: Oxford University

Park, K.S., Lee, N.S. (1987): A three-dimensional Fourier descriptor for human body representation/reconstruction. *Comput. Biomed. Res.* 20, 125-140

Pascual, R., Valdes, P., Alvarez, A. (1988a): A parametric model for multi-channel EEG spectra. *Int. J. Neurosci.*

Pascual, R., Gonzalez, S., Valdes, P., Biscay, R. (1988b): Current source density estimation and interpolation based on the spherical harmonic Fourier expansion. *Int. J. Neurosci.*

Persoon, E., Fu, K.S. (1977): Shape discrimination using Fourier descriptors. *IEEE Trans. Systems Man Cybern.* SMC-7, 170-179

Petersen, S.E., Fox, P.T., Posner, M.I., Mintun, M., Raichle, M.E. (1988): Positron emission tomographic studies of cortical anatomy of single-word processing. *Nature* 331, 585-589

Rao, C.R., Mitra, S.K. (1971): *Generalized inverse of matrices and its applications*. New York: John Wiley and Sons

Rawlings, R.R., Eckardt, M.J., Begleiter, H. (1984): Multivariate time series discrimination in the spectral domain. *Comput. Biomed. Res.* 17, 352-361

Rawlings, R.R., Rohrbaugh, J.W., Begleiter, H., and Eckhardt, M.J. (1986): Spectral methods of principal components analysis of event-related potentials. *Comput. Biomed. Res.* 19, 497-507

Scherg, M., Von Cramon, D. (1985): Two bilateral sources of the late AEP as identified by spatio-temporal dipole model. *Electroencephalogr. Clin. Neurophysiol.* 82, 32-44

Starr, S.J., Metz, E., Lasted, L.B., Goodenough, D.J. (1975): Visual detection and localization of radiographic images. *Diagn. Radiol.* 116, 533-538

Swets, J.A. (1988): Measuring the accuracy of diagnostic systems. *Science* 240, 1285-1293

Takeda, S., Matzuzawa, T. (1984): Brain atrophy during aging: a quantitative study using computed tomography. *J Am. Geratr. Soc.* 32, 520-524

Tomoyuki, I. (1983): A comparison of cerebral atrophy in CT scan findings among alcoholic groups. *Acta Psychiatr. Scand.* [Suppl.] 68

Valdes, M., Bobes, M.A., Perez, M.C., Perera, M., Carballo, J.A., Valdes, P. (1987):

Comparison of auditory evoked potential detection methods using signal detection theory. *Audiology* 26, 166-178

Valdes, M., Bobes, M.A. (1989): Making sense out of faces: ERP evidence for multiple memory systems. In: *Machinery of the Mind* John, E.R., Harmony, T., Prichep, L.S., Valdes-Sosa, M., and Valdes-Sosa, P. (eds.). New York: Raven Press, 1991

Valdes, P. (1984): Statistical basis. In: *Functional Neurosci. vol III*. Harmony T. (ed.). New Jersey: Erlbaum Associates

Valdes, P., Biscay, R.J., Pascual, R., Jimenez, J.C., Alvarez, A. (1985): A quantitative description of development of neurometric parameters. *Technical report* NC-003, Havana, Cuba: National Center for Scientific Research

Valdes, P., Pascual, R., Jimenez, J.C., Carballo, J.A., Biscay, R.J., Gonzalez, S. (1988): Functionally based statistical methods for the analysis of the EEG and Event Related Potentials. In: *Progress in computer-assisted function analysis*. Willems, L., van Bemmel, J.H., Michel, J. (eds.). Berlin: North-Holland

Williamson, S.J., Kaufman, L. (1987): Analysis of neuromagnetic signals. In: *Handbook of electroencephalography and clinical neurophysiology,* vol 1 Gevins, A.S., Remond, A. (eds.). Amsterdam: Elsevier Science Publishers B.V.

Winter, J. (1984): Automated computer tomography image analysis using contour map topology. *IEEE Trans. on Medical Imaging* 3, 163-169

22

The Physical Basis of Electrophysiological Brain Imaging: Exploratory Techniques for Source Localization and Waveshape Analysis of Functional Components of Electrical Brain Activity

ROBERTO D. PASCUAL-MARQUI, ROLANDO BISCAY LIRIO, AND
PEDRO A. VALDÉS-SOSA

22.1 Introduction

One of the fundamental problems of electrophysiology is the determination of the neuronal generators corresponding to the functional components of brain electrical activity, based on measurements of scalp voltage differences. The solution to this problem (inverse problem of electrophysiology) would be a new type of tomography producing three-dimensional (3D) images containing functional information about the brain.

Unfortunately, it is well known that the scalp potential field does not determine uniquely the generators inside the head, due to the existence of an infinite number of distinct solutions to this problem (Katznelson, 1981). The main difficulty lies in the fact that there does not exist any known method for selecting the actual solution.

Recently, magnetic field measurements, derived from the same current sources in the brain that produce scalp voltage differences, are being used. Although these measurements provide useful complementary information, they do not solve the problem of not being unique (Sarvas, 1987). This chapter focuses on the electrical inverse problem.

In the literature, the most widely used and most popular solution is based on the assumption that the neuronal generators can be represented by one or several current dipoles, with time-varying moments and with fixed or time-varying locations (for a review, see Fender, 1987). In a limited number of cases under certain physiological conditions, this type of model has been validated to a satisfactory degree of approximation, However, there is no a priori reason why this particular

435

solution should approximate in any form the true one. Furthermore, it should be used with caution since it may produce meaningless source localizations.

Due to the very high time resolution of current digital recording equipment (e.g., 19 scalp potential differences every 10 msec, instantaneous inverse solutions of a typical 2.56-sec segment would produce 256 images. Such a large amount of information is extremely difficult to analyze by simple visual inspection. This situation can be improved by first estimating the functional components followed by their inverse solutions, thereby producing more meaningful results. The importance of this methodology is based on the hypothesis that each functional state of the brain produces a unique physiological component in electrical activity (Lehmann et al., 1987). Different functional states can correspond to different forms and levels of information processing, to different states of consciousness, to different pathologies, and the like. However, component estimation is also an ill-posed problem (Mocks, 1987).

This chapter deals with difficulties and solutions of the inverse problem of electrophysiology. Classical techniques of analysis are reviewed, their flaws exposed, and, when possible, remedies are proposed. Incorrect methods of widespread use in the literature are also exposed, and their correct solution is given. New inverse solutions, other than dipole localization methods, are developed for use as complementary exploratory tools for source localization. New component estimation algorithms are presented.

Section 22.2 describes the physical model of the head and the corresponding equations for the direct and inverse problems, together with the definition of a component model. In Section 22.3, exploratory methods for two-dimensional (2D) localization and for waveshape identification are reviewed. Three important problems are discussed: the time domain DC level, the reference electrode, and the 2-D resolution of voltage data and surface current source density (SCSD) data (in this chapter, SCSD is used to designate the more familiar terms "source derivations," "Laplacean derivations," etc.). In Section 22.4, the inverse problem is discussed, and new solutions are presented. Section 22.5 briefly reviews component models. New methods for estimating and identifying the number of components, and for segmentation of brain electrical activity are presented.

22.2 Basic Models and Methods

The physical model of the head consists of a piecewise homogeneous conductor surrounded by air. A concise formulation of the corresponding equations for the electrical and magnetic fields under the quasi-static approximation (introduced by Plonsey, 1969) can be found in Sarvas (1987).

The partial differential equation relating electrical potential (Ψ) and volume current source density (VCSD) is

$$\nabla^2 \Psi = I/\sigma \qquad [1]$$

where σ is conductivity, I is the VCSD defined as $I = \nabla J$, and $\mathbf{J}$ is the impressed current density due to the electromotive force arising from biochemical reactions at the microscopic level. It is important to emphasize that the function $I(v)$, where v denotes volume coordinates, completely characterizes the spatial distribution of the neuronal generators that produce the scalp potential field $\Psi(s)$, where s denotes surface coordinates.

Eq. 1 corresponds to Poisson's equation for regions where sources exist. In regions where there are no sources, Laplace's equation holds:

$$\nabla^2 \Psi = 0 \qquad\qquad [2]$$

In the direct problem (for known source I and unknown electrical potential), these equations must be solved under the following boundary conditions at each interface separating two different conductors σ' and σ'':

$$\Psi' = \Psi'' \qquad\qquad [3]$$

$$\sigma' \delta\Psi'/\delta\eta = \sigma'' \delta\Psi''/\delta\eta \qquad\qquad [4]$$

Eqs. 3 and 4 express the continuity of the electrical potential and of the normal current density component across the interface, respectively.

The solution to the direct problem for the scalp potential may be written as a functional equation:

$$\Psi(s) = \mathbf{A} \{I\} + c \qquad\qquad [5]$$

where $\mathbf{A}$ represents the functional operator ($\mathbb{R}^3 - \mathbb{R}^2$) and c is the reference potential constant. In the direct problem there are no theoretical difficulties, meaning that there exists a unique solution for the scalp electrical potential corresponding to a known VCSD, except for the additive constant determined by the reference potential.

In the inverse problem, roles are reversed: only $\Psi(s)$ is known, and $I(v)$ is unknown. For this case, the inverse operator does not exist, and as a consequence, an infinite number of distinct VCSDs satisfies eq. 5 for any given $\Psi(s)$. In other words, knowledge of the electrical potential on every point of the scalp is insufficient information for determining the neuronal generators.

One general methodology for obtaining particular inverse solutions is based on restricting the VCSD to be given as a parametric family of functions, denoted as $I(v;\vartheta)$, where the functional form is completely known while only the vector of parameters ϑ is undetermined. The method of solution in this case is as follows. (1) Substitute $I(v;\vartheta)$ in eq. 5 and solve. This gives the theoretical scalp voltage dependent on the unknown ϑ, denoted as $\Psi(s;\vartheta)$. (2) Minimize, with respect to ϑ and c, some measure of distance between the experimental data (denoted as $\hat{\Psi}[s]$) and the theoretical voltage with an additive unknown constant ($\Psi(s;\vartheta)+c$). It is important to point out that in general, this method does not guarantee unique estimators for the parameters. Other forms of mathematical restrictions for the VCSD are presented in Section 22.4.

Special emphasis is given throughout this chapter to a simple head model consisting of a homogeneous conducting unit radius sphere surrounded by air, which seems to constitute a fairly acceptable approximation (Henderson et al., 1975). In principle, all the methodology developed here can be extended to a more realistic model (inhomogeneous conductor with complicated geometry). However, such a task would entail great mathematical difficulties, obscuring the discussion of the inverse problem, and would not allow the development of practical inverse solutions.

Current numerical solution methods for the direct problem, which perhaps would work for realistic models, have only been tested in simulation experiments for spherical shell models with one dipole as the source. Despite the great effort spent in developing more efficient methods (Gonzalez et al., 1989a; Hamalainen and Sarvas, 1987; Meijs et al., 1987), they do not seem to be ripe enough yet for their use in obtaining inverse solutions. One of the main concerns of this chapter is the critical analysis of current inverse solutions used in the literature and the development of new practical (albeit approximate) ones.

Under spherical symmetry, a parametric orthogonal expansion for the scalp voltage can be given in terms of spherical harmonic functions:

$$\Psi(\theta,\phi) = \sum_{l=1}^{n} \sum_{m=-1}^{l} A_{lm} Y_{lm}(\theta,\phi) + A_{00} Y_{00} \qquad [6]$$

where (θ,ϕ) represents the usual surface spherical coordinates. The second term on the right-hand side contains the constant spherical harmonic function Y_{00} and the only reference potential dependent coefficient A_{00}. This type of expansion has a very interesting physical interpretation, corresponding to the well-known multipole expansion (Jackson, 1975). Also, eq. 6 allows a straightforward calculation of the surface Laplacean (SCSD):

$$I_s(\theta,\phi)/\sigma = -\nabla_s^2 \Psi(\theta,\phi) = \sum_{l=1}^{n} \sum_{m=-1}^{l} l(l + 1) A_{lm} Y_{lm}(\theta,\phi) \qquad [7]$$

where σ is the conductivity at the surface of the scalp, I_s is the SCSD, and ∇_s^2 is the surface Laplacean operator in spherical coordinates. As can be seen from this expression, the Laplacean is an spatial filtered version of the scalp potential.

The expansion given by eq. 6 is very useful in many theoretical and practical problems, as is shown throughout this chapter. The estimation of the spherical harmonic coefficients based on a finite number of scalp voltage measurements is given in Appendix 22.A.

The basic equation for the simple head model relating scalp voltage and VCSD can be obtained by solving eq. 1 with boundary condition $\partial\Psi/\partial\eta = 0$ (eq. 4) at the scalp/air interface. The solution gives the following integral operator for eq. 5:

$$A = 4\pi \int_0^{2\pi} \int_0^{\pi} \int_0^1 \sum_{l=1}^{\infty} \sum_{m=-1}^{l} \frac{r^l}{1} Y^*_{lm}(\theta,\phi) Y_{lm}(\theta s, \phi s) r^2 \sin\theta \, dr \, d\theta \, d\phi \qquad [8]$$

where (r,θ,ϕ) are the volume coordinates of the VCSD and $(\theta s, \phi s)$ represent the scalp coordinates of the voltage (all in terms of classical spherical coordinates). The inverse problem (eq. 5 with **A** given by eq. 8) corresponds to a Fredholm integral equation of the first kind (Baker, 1977; Goldberg, 1979).

The presentation of the direct and the inverse problems up to this point was restricted to instantaneous scalp potential field and VCSD. Time-varying sources were not considered. One approach that takes this into account is the formulation of a time-varying component model for the sources:

$$I(t,v) = \sum_{k=1}^{Nk} f_k(t) \; I_k(v) \tag{9}$$

where N_k is the number of components, and for the k-th component, $I_k(v)$ represents the VCSD and $f_k(t)$ is its time-dependent intensity function. Note that $I_k(v)$ completely characterizes the spatial distribution of the brain structures corresponding to the k-th functional state. A more general component model would include the possibility of "wave" type components, where time and space coordinates are mixed. The component model considered here is limited to the case of separated time and spatial functional dependencies (eq. 9).

Based on eqs. 5 and 9, the scalp voltage for the component model is

$$\Psi(t,s) = \sum_{k=1}^{Nk} f_k(t) \; \psi_k(s) + c(t) \tag{10}$$

where

$$\Psi_k(s) = A\{I_k\} + c_k \tag{11}$$

and $c(t)$ represents the time variations of the reference potential. For each component, $f_k(t)$ is the mathematical representation of what is known as the "wave-shape," and $\Psi_k(s)$ is the topographical distribution of the potential field determined by the underlying component generators. The component model given by eq. 10, without the reference electrode constant, has been studied by Mocks (1987).

Basically, the solution to the functional inverse problem for the component model may be carried out in two steps; first by estimating the scalp potentials $\Psi_k(s)$ (eq. 10), followed by solving the inverse problem for each component (eq. 11). Alternatively, if the source components in eq. 9 are known parametric functions $I_k(v,\vartheta_k)$ with unknown vectors ϑ_k, then the voltage component model is

$$\Psi(t,s) = \sum_{k=1}^{Nk} f_k(t) \; \Psi_k(s;\vartheta_k) + c(t) \tag{12}$$

In this case, estimation of the waveshapes and the parameters constitutes the solution.

22.3 Exploratory Methods

The two most widely used exploratory tools for approximate waveshape identification and 2D localization are the visual inspection of multichannel time
recordings and of topographic maps of brain electrical activity. The term "2D
localization" refers to the determination of the scalp surface coordinates nearest to
the underlying neuronal generators.

Visual inspection of multichannel time recordings may give some approximate
information on the waveshapes ($f_k(t)$ in eq. 10) of the components of brain
electrical activity and their approximate 2D localization, e.g., the N400 component
is strongest at the Cz electrode. Similarly, when analyzing spontaneous multichannel electroencephalogram (EEG) activity, visual inspection of the frequency
spectra may give insight into the approximate localization of rhythmic activity
components, e.g., alpha generators located somewhere in the occipital area are
seen as large spectral peaks at 01 and 02 electrodes in the 9–10 Hz frequency
range.

Topographic maps are now in common use in the literature (Duffy, 1986). They
are usually analyzed by simple visual inspection of single instantaneous maps or
sequences (motion pictures). They may correspond to voltage data (with possibly
different reference electrodes), to SCSD data, or to other derived measures of
electrical activity. The presence of sharp local maxima or minima in a map usually
corresponds to underlying prominent sources, thus providing approximate information for 2D localization (see 22.3.3). An interpolation method for the construction of topographic maps can be based on the spline function defined for a
spherical surface, as described in Appendix 22.A (see also Pascual et al., 1989).

The examination of many instantaneous topographic maps (256 in a 2.56-sec
EEG segment, sampled at 100 Hz) is possibly not the best procedure for determining candidate generators. Lehmann and Skrandies (1980) propose a method for
reducing the number of maps to be analyzed by selecting the time instants of
maximum global field power, corresponding to a high signal-to-noise ratio for the
spatial information. In simple cases, minima may correspond either to the polarity
inversion of one component or to a shift from one component to another. The
theoretical global field power, in the case of spherical symmetry, has a simple
expression in terms of the spherical harmonic coefficients, corresponding to the
instantaneous spatial power of the signal:

$$\int_0^{2\pi} \int_0^{\pi} [\Psi(\theta,\phi) - \bar{\Psi}]^2 \ \sin\theta \ d\theta \ d\phi = \sum_{l=1}^{n} \sum_{m=-l}^{l} |A_{lm}|^2 \qquad [13]$$

where Ψ is the mean potential value over the surface.

Sections 22.3.1 and 22.3.2 analyze two important sources of errors that may
produce misleading results when using these exploratory tools. Section 22.3.3
shows how to enhance the resolution of 2D localization in topographic maps.

22.3.1 Time Domain DC Level Problem

The built-in high-pass filter of current amplifier equipment distorts the waveshape of measured signals and furthermore introduces an unremovable uncertainty in their DC levels (Barr and Chan, 1986). Deconvolution techniques may be used for restoring the original signal, except for the actual DC level (Oppenheim and Schafer, 1975). Two examples illustrate this problem.

Figure 22.1A corresponds to scalp voltage versus time produced by a hypothetical component due to a cortical generator. At the time of the maximum, most of the pyramidal neurons undergo EPSPs at distal dendrites, which makes each neuron act as an equivalent dipole that does not invert polarity. However, the measured voltage would correspond to that shown in Figure 22.1B. The waveshape is distorted and would lead to the incorrect conclusion that the generator inverts its polarity.

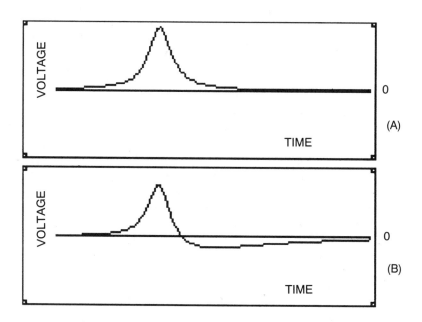

FIGURE 22.1A and 22.1B. Theoretical scalp voltage (average reference) versus time produced by a hypothetical component, the source of which is an equivalent dipole located under the measurement point. The curves are proportional to the theoretical component waveshapes [f(t), t in seconds]: 1A: $f(t) = 1/[1 + 4(t - 0.5)^2]$; 1B: $f(t) = 1 + \sin(20\pi t)$. Each curve consists of 128 measurements sampled every 10 msec.

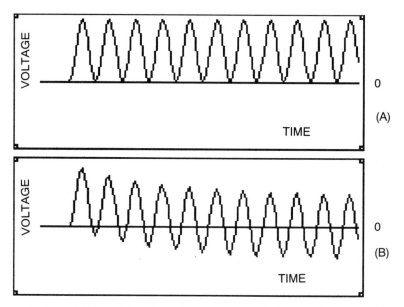

FIGURE 22.2A and 22.2B. Distortion of the waveshapes in Figures 22.1A and 22.1B, respectively, due to the high-pass filter in the recording amplifier. The curves were digitally high-pass filtered with a first-order Butterworth filter, cut-off at 0.5 Hz. 1B, Note a significant polarity inversion of the source immediately after the maximum. 2B, Note the shift in DC level toward zero, introducing a polarity inversion as well.

Figure 22.2A corresponds to a hypothetical alpha rhythm that may be produced by a model, such as the one described by Lopes da Silva et al. (1974) applied to the cortex, where the pyramidal neurons undergo EPSPs at distal dendrites and IPSPs at basal dendrites. In this case, the equivalent dipole field of each neuron does not invert in polarity. Figure 22.2B illustrates the measured voltage with conventional amplifiers, where the distortion effect on the periodic waveshape is seen clearly. These results would also lead to an incorrect interpretation.

The data presented in Figures 1 and 2 were simulated with a 100-Hz sampling rate, 1.28-sec segments, and a conventional high-pass filter (0.5 Hz cut-off, first-order Butterworth filter).

22.3.2 Reference Electrode (Spatial DC) Problem

Another source of error in exploratory waveshape estimation and 2D localization is related to the reference electrode (Lehmann, 1987). Due to well-established physical laws, there does not exist an "ideal inactive" reference electrode in a finite volume conductor (Katznelson, 1981).

The assumption that some electrode sites should be more inactive than others has misled some researchers to engage in a futile search for a "moderately" inactive reference electrode. Obviously, this problem cannot be solved by looking at the data with several different references, as suggested by Van Petten and Kutas (1988). The activity or inactivity of a point on the scalp cannot be determined by its voltage difference with respect to other points with equally unknown activity. Rather, it is related to the amount of current emerging (or entering) from that point, which is determined by the unknown spatial distribution of the neuronal generators. Therefore, in disagreement with Van Petten and Kutas (1988), methods for the quantitative analysis of brain electrical activity should be invariant to the reference electrode: experimental results should not ultimately depend on the reference.

22.3.2.1 Linked Earlobes Reference

One particular type of reference electrode, linked earlobes, has been the subject of some debate in the literature. Katznelson (1981) has cautioned against its use because of the scalp potential field distortion produced by current flow between the earlobes. In agreement with this hypothesis, Fender (1987) claims that all dipole localization methods based on linked earlobes reference recordings are of dubious validity. Van Petten and Kutas (1988) argue that naturally occurring interhemispheric asymmetries would tend to appear more symmetrical.

However, experimental evidence and theoretical arguments (Gonzalez et al. 1989b) point to the fact that no potential field distortion occurs, since the skin/electrode interface has too large an impedance to allow current to flow between the earlobes. Nevertheless, even though linked earlobes are just as good a reference as any other, it does not solve the main problem.

22.3.2.2 Average Reference

The average reference was originally proposed by Offner (1950). It has the advantage of providing reference-independent measurements. The question is if this new reference produces meaningful electrical activity.

In the case of spherical symmetry, according to eq. 6, the mean voltage over the scalp is equal to A_{00}. This coefficient is the only one that is reference dependent. Setting A_{00} to zero will not change the remaining coefficients, thus producing meaningful average reference scalp voltage data independent of the reference used. Furthermore, in terms of the multipole expansion, A_{00} corresponds to the mean value of the sources inside the head, which should be zero due to electroneutrality. This is another reason that supports the use of the average reference. It is important to emphasize that the arguments given above for spherical symmetry hold true for any given geometry of the scalp surface.

Average reference voltage is determined uniquely by the scalp geometry in the theoretical case, when scalp voltage is known over the whole surface. However, estimation based on a finite number of measurements is not unique. The classical estimator (Offner, 1950) consists of subtracting from each electrode the average voltage over all electrodes at each time instant. A different estimator is given in

Appendix 22.A, based on the spherical harmonic expansion for finite dimensional data.

22.3.2.3 SCSD

The SCSD does not depend on the reference electrode (Katznelson, 1981), and furthermore, it has two very important advantages over voltage data: (1) its physical interpretation corresponds to the amount of current emerging (or entering) from each point on the scalp due to underlying sources (the neuronal generators) (Katznelson, 1981) and (2) it has higher spatial resolution in the localization of underlying sources (see Section 22.3.3).

One method for calculating the SCSD is based on eq. 7, in which the spherical harmonic coefficients can be estimated as outlined in Appendix 22.A. For a brief review of other estimation procedures see Pascual et al. (1989).

22.3.3 Voltage Data versus SCSD Data: 2D Resolution of Topographic Maps

The spherical head model was used in simulation experiments for comparing the resolution capabilities of different types of data. The equation used for the theoretical average reference scalp voltage due to a current dipole inside the head (derived from eqs. 5 and 8) was

$$\Psi(\mathbf{Y}) = \mathbf{Q} \mathrm{o} \sum_{l=1}^{\infty} b_l \, r^{l-2} \, \{l \, P_l \, (g) \, \mathbf{X} + P_l{}'(g) \, [r \, \mathbf{Y} - g \, \mathbf{X}]\} \qquad [14]$$

where $\mathbf{Y}$ and $\mathbf{X}$ are position vectors of the measurement point and the dipole location, respectively, $\mathbf{Q}$ is the dipole moment; P_l' is the derivative function of the Legendre polynomial P_l, "o" denotes scalar product; $r = (\mathbf{X} \mathrm{o} \mathbf{X})^{1/2}$; $g = \mathbf{X} \mathrm{o} \mathbf{Y}/r$; and $b_l = (2l + 1)/(\sigma 1)$. If the dipole is located at the center of the head ($r = 0$), then

$$\Psi(\mathbf{Y}) = b_l \, \mathbf{Q} \mathrm{o} \mathbf{Y} \qquad [15]$$

The SCSD is obtained by setting $b_l = (2l + 1)(l + 1)/\sigma$ in eqs. 14 and 15. The fourth iterated surface Laplacean corresponds to

$$b_l = (2l + 1) \, l^3 \, (l + 1)^4/\sigma$$

Figure 22.3 shows the scalp voltage (a,b,c) and the SCSD (d,e,f) from two radial dipole sources that are both located at the same distance from the center and in all cases separated by an angle of 57.3°. There are two distinct maxima for measurements directly over the sources when the dipoles are nearest to the surface (3a and 3d, r = 0.85). Slightly nearer the center, at r = 0.6 (3b and 3e), the two voltage field maxima overlap to a much greater extent than for SCSD. Finally, for deep dipoles at r = 0.3 (3c and 3f), the voltage field is incapable of distinguishing the existence

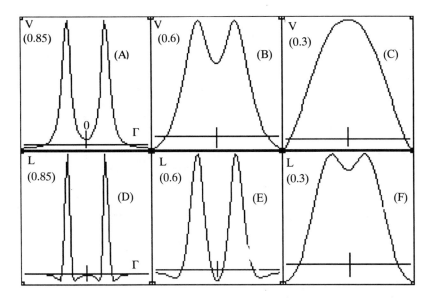

FIGURE 22.3. The curves correspond to voltage (A, B, C) and to SCSD (D, E, F) along an arc over the surface of an ideal head (homogeneous conducting unit radius sphere of air). The sources are two radial dipoles, both at equal distances from the center. The dipoles are separated by an angle of 57.3° in all cases. Voltage and SCSD were evaluated along the arc defined by the intersection of the head surface and the plane passing through the center and containing the two dipoles. The Abscissa is the angle between the position vector of the measurement point and the midpoint between the dipoles and varied in the range −90° to +90°, corresponding to the upper hemisphere. Each radial dipole produces a voltage and a SCSD peak on the surface. The peaks get broader as the dipole approaches the center. The results show that the voltage peaks are always considerably broader than those for the SCSD. For the deep dipoles (r=0.3, 3c and 3f), there appears one very broad voltage peak indicating only one source, whereas the SCSD can still discriminate unambiguously the existence of both sources. In this simulation experiment, SCSD has higher 2D resolution than voltage data.

of two sources (there appears only one broad maximum), whereas the SCSD can still discriminate unambiguously.

These results prove that the SCSD has higher 2D resolution than voltage measurements, i.e., SCSD provides much more accurate information on the scalp coordinates nearest to the underlying sources than voltage field data. It is our belief that these results would not change substantially if a more realistic head model were used. The conclusions of Perrin et al. (1987) are in complete disagreement with the results presented here, possibly the result of their use of an incorrect definition for 2D resolution.

Figure 22.4 compares global field power for actual evoked response (ER) data based on average reference voltage (SE[ARV] in Figure 22.4A) and on SCSD (SE[L] in Figure 22.4B). Both curves appear to be very similar. However, there

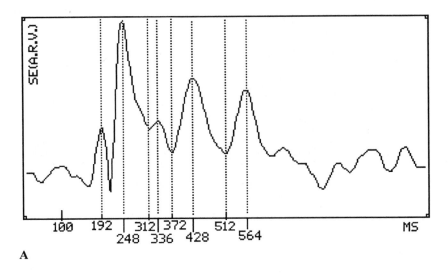

A

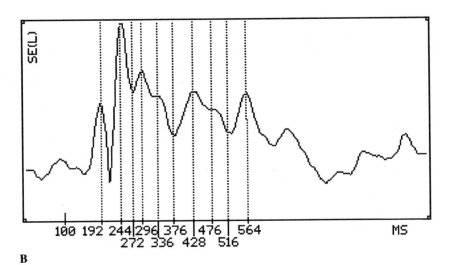

B

FIGURE 22.4A. Global field power based on average reference voltage [SE(ARV) in A] and on SCSD [SE(L) in B] for evoked response data is compared. Measurements were made with 19 electrodes (10/20 international system), a recording time of 1.024 sec, and a sampling rate of 250 Hz. The stimulus consisted of a 1-sec visual presentation of a word on the computer screen of the MEDICID 03M Neurometric System, after which the subject was asked to pronounce a semantically associated word. In the figure, stimulus onset (appearance of the word) occurred at t=100msec. The data correspond to the grand average for 15 subjects, 60 trials each. The interesting feature here is that SCSD global field power contains new maxima not present for voltage at 296 msec. and 476 msec. This may indicate the existence of two components not detected by voltage.

are two new maxima in the SCSD global field power at 296 msec and 476 msec, possibly indicating the existence of two components not detected by voltage. Although further experimental confirmation is needed in this particular ER experiment, these results indicate that SCSD global field power has higher resolution for distinguishing different components, which corresponds to its higher 2D resolution.

All these results suggest the possibility that higher-order spatial derivatives may provide higher 2D resolution. Indeed, this is the case, as illustrated in Figure 22.5, corresponding to the same two deep dipoles used in Figures 22.3C and 22.3F. A comparison is now made between voltage (5a), SCSD (5b), and the fourth iterated surface Laplacean (denoted as HR in 5c). These results show that 2D resolution can be considerably enhanced by this method, as seen in Figure 22.5C, where there is a complete separation (no overlap) between the two maxima.

Figure 22.6 corresponds to a single deep tangential dipole (r=0.3) in which voltage (6a), SCSD (6b), and HR (6c) are compared. The measure of resolution in this case is given by the arc distance between the maximum and minimum (smaller distance corresponds to higher resolution). As clearly seen and in agreement with the previous results, voltage has lower resolution, followed by SCSD, and finally by HR, which is significantly higher than the latter two.

These theoretical results are very encouraging. Care must be taken, however, when dealing with a finite set of scalp voltage measurements (typically only 19 electrodes for the 10/20 international system) contaminated with noise. It is a well-known fact that high-pass filtering increases noise considerably. One solution to this problem is to use smooth estimators in which the degree of smoothness is determined by the measured data (cross-validation), as described in Pascual et al. (1989) for SCSD. As long as the number of measurements is sufficiently large, these smooth estimators are reliable despite noise amplification. Therefore, the

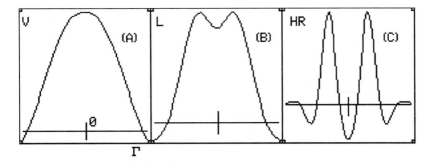

FIGURE 22.5. The same type of simulation experiment as used in Figures 22.3C and 22.3F was performed here. In this case, voltage (A), SCSD (B), and the fourth iterated surface Laplacean (denoted as HR in C) are compared. Higher-order surface derivatives have a significantly higher 2D resolution, as can be seen in C, where the two peaks are separated completely (compare with A and B.).

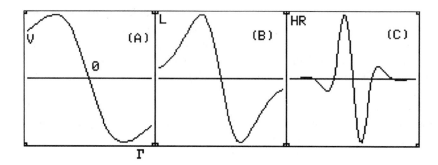

FIGURE 22.6. The simulation experiment here is similar to that of Figures 22.3 and 22.5. In this case the source was a single deep tangential dipole at r=0.3. Voltage (A), SCSD (B), and the fourth iterated surface Laplacean (denoted as HR in C) were evaluated along the arc defined by the intersection of the head surface with the plane passing through the center, containing the dipole and parallel to its moment. High 2D resolution corresponds to small arc distance between the maximum and the minimum. In agreement with the results in Figure 22.5, resolution is lowest for voltage, followed by SCSD, and significantly higher for HR.

price to be paid for higher resolution is to use more electrodes, which may be worthwhile for ensuring accurate 2D localization.

22.4 The Inverse Problem

In practical situations, a new difficulty is added to the notorious nonuniqueness problem: the amount of useful information is severely reduced and distorted due to the limited number of scalp voltage measurements contaminated with noise.

Despite these problems, particular solutions may be derived under suitable restrictions for the VCSD. Different types of solutions (restrictions) are considered, and their physical and physiological meanings are mentioned briefly in this section. All the solutions developed here can be used for instantaneous and for component (Section 22.5) scalp voltage distributions.

22.4.1 Dipole Localization Methods

The dipole model is an adequate approximation as long as the generator corresponds to a compact population of approximately parallel neurons, occupying only a few cubic millimeters (Fender, 1987). In this case the fitted dipole represents the sum of all microscopic dipole fields generated by such neurons.

Let $\Psi(s;\vartheta)$ represent the theoretical scalp voltage due to a single dipole (corresponding to eqs. 14 and 15) where there are six undetermined parameters in ϑ (three for the dipole moment and three for its location in the head). Nunez (1981) and Fender (1987) state that a minimum of six common reference measurements over the scalp is sufficient for estimating the six parameters. This is true only if a cephalic reference electrode is used, with location s_0. In such a case, estimation should be based on the model

$$\hat{\Psi}(s_i) = \Psi(s_i;\vartheta) - \Psi(s_0;\vartheta) + \varepsilon_i \qquad [16]$$

where s_i ($i = 1 \ldots$ Ne) are the scalp coordinates of the Ne electrodes, $\hat{\Psi}(s_i)$ are the actual measurements, and ε_i represent the errors. However, this is not the method used by Fender (1987).

In general, it is not true that six measurements are sufficient for determining the dipole. The incorrect methods used throughout the literature are based on the following models:

$$\hat{\Psi}(s_i) = \Psi(s_i;\vartheta) + \varepsilon_i \qquad [17]$$

and

$$\hat{\Psi}_{ar}(s_i) = \Psi(s_i;\vartheta) + \varepsilon_i \qquad [18]$$

where $\hat{\Psi}_{ar}(s_i)$ represents average reference data. In the reported literature, they have been used at least by Kavanagh et al. (1978), Sencaj and Aunon (1982), Sherg and Von Cramon (1985a, 1985b, and 1986), Fender (1987), and Stok (1987). Obviously, any results based on these models are of questionable validity.

The correct model for common reference measurements (see Section 22.2) is

$$\hat{\Psi}(s_i) = \Psi(s_i;\vartheta) + c + \varepsilon_i \qquad [19]$$

At least seven electrodes are necessary for estimating the seven unknown parameters: ϑ and c. The least squares estimator for the reference electrode constant is $[\bar{\Psi} - \bar{\Psi}(\vartheta)]$, in which both quantities are voltage averages over the scalp electrodes where $\bar{\Psi}$ is based on the measured voltages and $\bar{\Psi}(\vartheta)$ on the theoretical voltages. Substituting this result in eq. 19 gives

$$\hat{\Psi}_{ar}(s_i) = \Psi(s_i;\vartheta) - \bar{\Psi}(\vartheta) + \varepsilon_i \qquad [20]$$

Even this expression without the parameter c needs at least seven electrodes for the estimation of ϑ. An important fact that has been overlooked in the literature is that, if average reference data are used, then the average reference theoretical model must also be used, as shown in eq. 20. Furthermore, the least squares method based on any one of eqs. 16, 19, or 20 is slightly more complicated than that based on eqs. 17 or 18.

Note that eq. 17 would be correct only if the reference electrode were ideally inactive (eq. 19 with c = 0). On the other hand, eq. 18 would be correct only if the theoretical voltage average over the scalp electrodes were zero. However, these requirements cannot be guaranteed beforehand.

The estimation of any parametric model in general should be based on eqs. 16, 19, or 20, which are not limited to the case of single dipole sources.

22.4.2 Grid Solutions

In this method the brain is partitioned into disjunct volume elements (voxels), each containing a simple source, such as a dipole, with unknown moment and known location. This type of model has been studied with simulated MEG data (Jeffs et al., 1987).

From a theoretical point of view, this approach is very similar to that used in tomographic image reconstructions for MRI and CT scans (Censor, 1983). It would be expected that the dipoles represent the mean dipole moment inside each voxel. Unfortunately, this is not so, due again to the nonuniqueness problem. For example, consider two completely different neuronal generators producing exactly the same scalp electrical potential. To which one of these two different VCSDs would the grid solution approximate?

There are other difficulties as well (Jeffs et al., 1987), such as the insufficient number of scalp measurements, which limits severely the number of voxels, and the ill-conditioned nature of the set of equations to be solved in the estimation. However, neuroanatomical and neurophysiological knowledge can help set further constraints, possibly producing more meaningful solutions. Nevertheless, grid solutions need to be further developed for their use as exploratory tools in source localization.

22.4.3 Other Mathematical Restrictions and Solutions

The interpolation of a function, sampled at a finite set of points, is an ill-posed problem with an infinite number of solutions. A reasonable common-sense solution is to select the smoothest possible function, which is the fundamental basis of spline theory (Prenter, 1975). The same assumption is used for deriving a particular solution to the inverse problem. In this case, the statement of the problem is as follows: obtain the smoothest VCSD subject to the equality constraint given by eqs. 5 and 8. In mathematical terms, the smoothest condition can be ensured by minimizing the total square curvature of the VCSD in the head:

$$\int_0^{2\pi} \int_0^{\pi} \int_0^1 (\nabla^2 I)^2 \, r^2 \, \sin\theta \, dr \, d\theta \, d\phi \qquad [21]$$

The derived smooth solution is

$$I(r,\theta,\phi) = \frac{1}{4\pi} \sum_{l=1}^{\infty} \sum_{m=-l}^{l} l(3 + 2l) \, A_{lm} \, r^l \, Y_{lm}(\theta,\phi) \qquad [22]$$

where A_{lm} are the expansion coefficients for the scalp electrical potential (see eq. 6 and Appendix 22.A for estimation procedures). This solution corresponds to smoothly distributed neuronal generators throughout the brain, which is the opposite case of the (single or grid) dipole source model.

Another solution for the inverse problem may be obtained in the form of a VCSD confined to a spherical surface at $r_0 < 1$:

$$I(r,\theta,\phi) = I(\theta,\phi)\frac{1}{r^2}\,\delta(r - r_0) \qquad [23]$$

where the Dirac δ function is used, which gives the following solution:

$$I(\theta,\phi) = \frac{1}{4\pi}\sum_{l=1}^{\infty}\sum_{m=-1}^{l} l\, A_{lm}Y_{lm}(\theta,\phi)/r_0^l \qquad [24]$$

Similarly, the solution for a dipolar spherical surface at $r_0 < 2$ is

$$D(\theta,\phi) = \frac{1}{4\pi}\sum_{l=1}^{\infty}\sum_{m=-1}^{l} r_0\, A_{lm}Y_{lm}(\theta,\phi)/r_0^l \qquad [25]$$

These solutions (eqs. 24 and 25) correspond to sources lying on a spherical cortical surface at a distance r_0 from the center.

22.4.4 General Remarks on Inverse Solutions

The different solutions presented throughout Section 22.4 correspond to very different physiological conditions. In the general case there is no a priori information about which of these conditions, if any, actually holds. Therefore, in practice, the suggested procedure is to look at all inverse solutions, using them as complementary exploratory tools and exercising extreme caution in their interpretation. Candidate generator structures arrived at by these techniques need independent experimental confirmation.

22.5 Component Models

The source component model given by eq. 9 is based on the hypothesis that brain electrical activity is composed of several distinct subprocesses. Each component corresponds to the activation of some distinct brain structure or set of functionally related structures, characterized by the time course of the activation. The derived voltage component model (eq. 10) has a mathematical form that has been widely used in the literature (see review by Mocks, 1987), mostly as a principal components analysis model. Moreover, its magnetic field equivalent has been used by Ilmoniemi et al. (1987) in the definition of "alphons" as the physiological components of the MEG alpha rhythm.

As in the inverse problem, this type of model (eq. 10) has the difficulty that the estimators for the waveshapes $[f_k(t)]$ and voltage distributions $[\Psi_k(s)]$ are not unique. Particular solutions can only be obtained by introducing constraints (Mocks, 1987). An additional problem is the estimation of the number of components (N_k), which is usually solved by fitting the constrained model to the data sequentially, first with one component, then with two, etc., until there is a "good fit."

An example of a parametric source model, based on dipole components, was developed by Scherg and Von Cramon (1985a, 1985b). Its mathematical representation in our notation is:

$$\sum_{k=1}^{Nk} f_k(t)\Psi(s;\vartheta_k) + c(t)$$

where $\Psi(s;\vartheta_k)$ is the voltage due to the k-th dipole with unknown ϑ_k (location and moment), $f_k(t)$ is the unknown waveshape of the dipole activation (restricted to be a smooth, at most triphasic function), and $c(t)$ is an unknown function that accounts for the effect of the reference electrode. The imposed constraints ensure unique estimators when there are more electrodes than unknown dipole parameters.

In this section nonparametric models are considered. They have an important advantage over parametric models, provided that all $\Psi_k(s)$ are estimated correctly. On the one hand, exploratory methods for 2D localization can be used on each $\Psi_k(s)$. On the other hand, all available kinds of inverse solutions can be obtained from the scalp voltage components $\Psi_k(s)$, including any parametric inverse solution.

In Appendix 22.B, a new objective method for estimating the actual number of components is presented. It has the advantage of not requiring any assumption about the form of the component VCSDs $I_k(v)$, as is the case with parametric models. This is an important problem since different parametric models may produce different number of components. In such a case, doubt would remain as to which model gives the correct answer to how many different physiological states of the brain actually exist in an experiment. Furthermore, once the actual N_k is estimated, it can be used in the estimation of any type of component model. For example, if a parametric model, such as the dipole component model, does not give an adequate fit with the actual N_k, then the model is probably inadequate (although if it has a good fit does not necessarily mean that it is correct).

The basic assumptions underlying many segmentation methods of brain electrical activity are (1) distinct functional states do not overlap in time and (2) a certain amount of time is spent in each state, not changing randomly from one time instant to the next. For a brief review of these assumptions, see Lehmann et al. (1987). Traditional segmentation methods are based on the detection of changes in the stochastic properties of the signal, using time series analysis techniques. More recently, segmentation methods based on the detection of changes in the topographical distribution of the EEG (Lehmann et al., 1987) and the MEG (Ilmoniemi et al., 1987) have been developed. This new approach emphasizes the fact that different functional states correspond to different neuronal generators (with different spatial distributions), which is the basis of the component model proposed in this chapter (eqs. 9 and 10). In Appendix 22.C, an estimation-segmentation method is described.

The estimation procedures developed in Appendices 22.B and 22.C are based on the component model (eq. 10) for finite dimensional data, i.e., for brain electrical activity recorded from electrodes at location s ($i = 1 \ldots$ Ne), at time instants $\tau = 1, \ldots N_\tau$. In matrix notation the model is

$$\psi_\tau + \Phi \, F_\tau + c_\tau l \qquad\qquad [26]$$

where ψ_τ, F_τ and l are column vectors of sizes Ne, N_k and Ne, with elements $\Psi(\tau, s_i)$, $f_k(\tau)$ and ones, respectively; Φ is an (Ne × N_k) matrix with columns being the topographical voltage components $\Psi_k(s_i)$; and c_τ is the reference electrode time dependent scalar. The recorded brain electrical activity considered here may correspond to a spontaneous EEG, to a single trial ER, to an average ER, etc.

These estimation methods were developed very recently, and up to now, they have only been tested in simulation experiments using the component model with noise. All preliminary results are very encouraging.

Appendix 22.A
Estimation of the Spherical Harmonic Coefficients

A review of the methods presented in this appendix can be found in Besse and Ramsay (1986). Let $\Psi(\theta_i, \phi_i)$ represent the instantaneous measured scalp potentials (arbitrary common reference), $i = 1, 2, \ldots,$ Ne, at Ne electrodes. An interpolating spline can be defined as a function that passes through the measured values $\Psi(\theta_i, \phi_i)$ and that minimizes some measure of curvature (in order to guarantee smoothness), such as the total square curvature (TSC):

$$\text{TSC} = \int_0^{2\pi} \int_0^{\pi} (\nabla_s^2 \, \Psi)^2 \sin\theta \; d\theta \; d\phi \qquad\qquad [a1]$$

Due to the fact that the spherical harmonic functions are eigenfunctions of the surface Laplacean operator,

$$\nabla_s^2 \, Y_{lm} \, (\theta, \phi) = -l(l + 1) \, Y_{lm} \, (\theta, \phi) \qquad\qquad [a2]$$

the reproducing kernel is

$$K(\theta, \phi; \theta', \phi') = \sum_{l=1}^{n} \sum_{m=-1}^{l} \, Y_{lm} \, (\theta, \phi) \, Y^*_{lm} \, (\theta', \phi') / [l(l + 1)]^2 \qquad\qquad [a3]$$

where * denotes complex conjugate. The interpolating spline function then is

$$\Psi(\theta, \phi) = \sum_{i=1}^{Ne} C_i \, K(\theta_i, \phi_i; \theta, \phi) + \alpha \qquad\qquad [a4]$$

where α accounts for the arbitrary reference electrode and the C_i are the unknown coefficients of the expansion. Eqs. a3 and a4 define a spherical harmonic expansion as in eq. 6, where

$$A_{lm} = \sum_{i=1}^{N_e} C_i \, Y^*_{lm} \, (\theta_i,\phi_i)/[l(l+1)]^2 \qquad \text{[a5]}$$

It is worth noting that all the spherical harmonic coefficients A_{lm} are determined by only N_e parameters.

For fixed α, the C_i are obtained by solving the system (eq. a4):

$$\Psi(\theta_j,\phi_j) = \sum_{i=1}^{N_e} C_i \, K(\theta_i,\phi_i;\theta_j,\phi_j) + \alpha \qquad \text{[a6]}$$

for $j = 1 \ldots N_e$, which gives (in matrix notation):

$$C = K^{-1} \, (\psi - \alpha \, 1) \qquad \text{[a7]}$$

where C, ψ, and 1 are column vectors with elements C_i $\Psi(\theta_i,\phi_i)$, and ones ($i = 1 \ldots N_e$), respectively, the ($N_e \times N_e$) matrix K is

$$K_{ij} = \sum_{l=1}^{n} (2l + 1) \, P_l \, (\cos\partial_{ij}) \, /\{4\pi[l(l+1)]^2\} \qquad \text{[a8]}$$

P_l are Legendre polynomials and

$$\cos\gamma_{ij} = \cos\theta_i \, \cos\theta_j + \sin\theta_i \, \sin\theta_j \, \cos(\phi_i - \phi_j) \qquad \text{[a9]}$$

The TSC for equation a4 is

$$\text{TSC} = C^t K C \qquad \text{[a10]}$$

where C^t denotes transposition. Substituting eq. a7 in a10 and minimizing with respect to α gives

$$\alpha = 1^t \, K^{-1} \, \psi/1^t \, K^{-1} \, 1 \qquad \text{[a11]}$$

Substituting eq. a11 in a7 finally gives

$$C = K^{-1} \, (I \, 025 \, 11^t \, K^{-1}/1^t \, K^{-1}1) \, \psi \qquad \text{[a12]}$$

where I is the identity matrix.

The procedure described here for estimating the harmonic coefficients is equivalent to that described by Pascual et al. (1989). Smoothing splines were also considered in that paper for dealing with noisy data (measurement errors).

Average reference measurements (ψ_{ar}) under this model are obtained by substituting eq. a12 in system a6 and setting $\alpha = O$:

$$\psi_{ar} = H\psi \qquad \text{[a13]}$$

where

$$H = (I - 11^t K^{-1}/1^t K^{-1}1) \qquad [a14]$$

The classical method (Offner, 1950) in matrix notation corresponds to $K = I$ in eqs. a13 and a14. This points to the important fact that there does not exist a unique average reference; at least this is true in the case of a finite number of electrodes. In general, any arbitrary positive definite matrix K defines reference-independent measurements.

Appendix 22.B
Number of Components Estimation

The set of measurements ψ_τ (eq. 26) may be viewed as a set of points in $\mathbb{R}^{Ne}$, where each cartesian axis corresponds to the voltage measurement at the i-th electrode ($i = 1 \ldots Ne$). For reference-independent measurements, the component model (26) is

$$\psi_\tau^{ri} = \Phi^{ri} F_\tau \qquad [b1]$$

where $\psi_\tau = H\psi_\tau$, $\Phi^{ri} = H\Phi$, and H is given by eq. a14 for any arbitrary positive definite matrix K (see Appendix 22.A). It is worth noting that the reference-independent measurements ψ_τ^{ri} now lie in the subplace R^{Ne-1}. In what follows, for the sake of clarity in the equations, the superscript "ri" is dropped. If the number of electrodes is sufficiently large compared to the number of components ($N_k < Ne-1$), then the points lie on the smaller subspace R^{Nk}. Thus, the problem is reduced to the determination of the actual dimensionality of the set of reference-independent measurements, which is given by the number of nonzero eigenvalues of the matrix:

$$S = \sum_{\tau=1}^{N\tau} \psi_\tau \psi_\tau^t \qquad [b2]$$

In practice, however, due to measurement errors (e_τ), the estimation of N_k is made very difficult because the eigenvalues do not jump to zero in a clear-cut way. One objective solution to this problem is to use cross-validation (Biscay and Pascual, unpublished data). Let $S^{(\tau 0)}$ denote the sum of squared products (eq. b2), where the vector $\psi_{\tau 0}$ has been excluded from the summation. Let $\Gamma_k^{(\tau 0)}$ ($k = 1 \ldots Ne - 1$) denote the eigenvectors of $S^{(\tau 0)}$, corresponding to the eigenvalues in descending order ($\lambda_1^{(\tau 0)} \geq \lambda_2^{(\tau 0)} \geq \ldots$). The linear regression expressed by equation b1 can be equivalently written, after appropriate rotations, as

$$\psi_{\tau 0} = \sum_{k=1}^{\kappa} \alpha_k \Gamma_k^{(\tau 0)} + e_{\tau 0} \qquad [b3]$$

where α_k are unknown parameters and κ is the number of components in model b3. Exclusion of the i-th electrode in this model is denoted as

$$\psi_{\tau 0}^{(i)} = \sum_{k=1}^{K} \alpha_k \, \Gamma_k^{(\tau 0)(i)} + e_{\tau 0}^{(i)} \qquad [b4]$$

where the vectors $\psi_{\tau 0}^{(i)}$, $\Gamma_k^{(\tau 0)(i)}$, and $e_{\tau 0}^{(i)}$ have Ne $-$ 1 elements. Let $\hat{\alpha}_k^{(\tau 0)(i)}(\kappa)$ be the least squares estimator of α_k in model b4, and let $\hat{\psi}_{\tau 0}(\kappa, i)$ be the predicted scalar voltage at electrode i, given by substituting $\hat{\alpha}_k^{(\tau 0)(i)}(\kappa)$ in eq. b3. The cross-validation error is

$$CVE(\kappa) = \sum_{\tau 0 = 1}^{N\tau} \sum_{i=1}^{Ne} [\psi_{\tau 0}(i) - \hat{\psi}_{\tau 0}(\kappa, i)]^2 \qquad [b5]$$

where $\psi_{\tau 0}(i)$ is the measured scalar voltage (reference independent) at electrode i. The estimator for N_k is taken as the value of κ, which minimizes the cross-validation error.

Appendix 22.C
Components Estimation and Segmentation

Nonoverlapping states in the component model (eq. b1 for reference-independent data) correspond to vectors F_τ with only one nonzero element. Let L_τ denote the functional state (labeled with integers 1, 2, ... N_k) at time instant τ. Note that L_τ indicates the nonzero element of F_τ. Given an initial (possibly random) labeling L_τ, the proposed procedure for estimating the components (scalp voltage distribution vectors ψ_k and waveshapes $f_k(\tau)$) is the following:

1. For each state k = 1 ... N_k, and all τ such that $L_\tau = k$, minimize

$$\sigma_k^2 = \sum_\tau | \psi_\tau - f_k(\tau)\, \psi_k |^2 / \upsilon_k \qquad [c1]$$

with respect to $f_k(\tau)$ and ψ_k, where υ_k is the number of times instants in state k and V represents the Euclidean norm of vector V. The iterative minimization procedure is:

1a. Set $f_k(\tau) = 1$.
1b. Compute $\psi_k = \sum_\tau f_k(\tau)\, \psi_\tau / \sum_\tau [f_k(\tau)]^2$.
1c. Normalize ψ_k as $\psi_k / |\psi_k|$.
1d. Compute $f_k(\tau) = \psi_\tau^t \psi_k$.
1e. Go to step 1b and repeat until convergence, i.e., until the estimated ψ_k and $f_k(\tau)$ do not change significantly from one iteration to the next.
2. This step consists of updating the labels, where for each τ, L_τ will equal the value k, which minimizes

$$| \psi_\tau - \psi_k(\psi_k^t \, \psi_\tau) |^2 / (2\sigma_k^2) - \beta \, \mu_k \qquad [c2]$$

where $\beta = 1.5$ is an empirical constant, and μ_k is the current number of neighbors $\{L_{\tau-4}, L_{\tau-3}, L_{\tau-2}, L_{\tau-1}\}$ having label k. Return to step 1 and repeat until convergence, i.e., until the labels L_τ do not change from one iteration to the next.

The algorithm outlined in step 1 corresponds to the simple least squares estimation of $f_k(\tau)$ and ψ_k for given labels, based on the assumption of nonoverlapping states. In step 2, each measurement ψ_τ is reclassified (assigned to a possibly new state). Note that the classification rule combines the distance between ψ_τ and its projection onto the k-th scalp voltage distribution ψ_k, together with the number of neighbors in state k. This second factor ensures a certain degree of time continuity of each state, expressing that time instants close together tend to be in the same state. A more general discussion of this type of model, used in the analysis of images, can be found in Besag (1986).

Acknowledgments. The authors are grateful to Rolando Grave de Peralta for programming the component estimation and segmentation algorithm described in Appendix 22.C, together with the simulation experiments.

References

Baker, C.T.H. (1977): *The numerical treatment of integral equations.* Oxford: Cambridge University Press

Barr, R.E., Chan, E.K.Y. (1986): Design and implementation of digital filters for biomedical signal processing. *J. Electrophysiol. Tech.* 13, 73-91

Besag, J. (1986): On the statistical analysis of dirty pictures (with discussion). *J. Roy. Statist. Soc.* B48, 259-302

Besse, P., Ramsay, J.O. (1986): Principal components analysis of sampled functions. *Psychometrika* 51, 285-311

Biscay-Lirio, R., Pascual-Margui, R.D. (1989): Crossvalidation techniques for estimating dimensionality in general component models. (in preparation)

Censor, Y. (1983): Finite series-expansion reconstruction methods. *Proc. IEEE* 71, 409-418

Duffy, F.H. (1986): *Topographic mapping of electrical brain activity.* Boston: Butterworths

Fender, D.H. (1987): Source localization of brain electrical activity. In: *Handbook of electroencephalography and clinical neurophysiology, revised series: vol. 1. Methods of analysis of brain electrical and magnetic signals.* Gevins, A.S., Redmond, A. (eds.). Amsterdam: Elsevier, pp. 355-403

Goldberg, M.A. (1979): *Solution methods for integral equations: Theory and applications.* New York: Plenum Press

Gonzalez, S., Grave de Peralta, R., Biscay, R., Jimenez, J.C., Pascual, R.D., Lemagne, J., Valdes, P.A. (1989a): Projective methods for the magnetic direct problem. In: *Advances in biomagnetism,* New York: Plenum. (submitted)

Gonzalez, S., Pascual, R.D., Valdes, P.A., Biscay, R., Machado, C., Diaz, G., Figueredo, P., Castro, C. (1989b): Brain electrical field measurements unaffected by linked earlobes reference. *Electroenceph. Clin. Neurophysiol.* (in press)

Hamalainen, M.S., Sarvas, J. (1987): Realistic conductivity geometry model of the human head for interpretation of neuromagnetic data. Report TKK-F-A614. Helsinki: Helsinki University of Technology

Henderson, C.J., Butler, S.R., Glass, A. (1975): The localization of equivalent dipoles of EEG sources by the application of electrical field theory. *Electroenceph. Clin. Neurophysiol.* 39, 117-130

Ilmoniemi, R.J., Williamson, S.J., Hostetler, W.E. (1987): New method for the study of spontaneous brain activity. In: *Biomagnetism '87 - 6th International Congress on Biomagnetism - Tokyo.* Atsumi, K., Kotani, M., Ueno, S., Katila, T., Williamson, S.J. (eds.). Tokyo: Tokyo Denki University Press, pp. 182-185

Jackson, J.D. (1975): *Classical electrodynamics.* New York: John Wiley and Sons

Jeffs, B., Leahy, R., Singh, M. (1987): An evaluation of methods for neuromagnetic image reconstruction. *IEEE Trans. Biomed. Eng.* BME34, 713-723

Katznelson, R.D. (1981): EEG recording, electrode placement and aspects of generator localization. In: *Electrical fields of the brain.* Nunez, P. (ed.). New York: Oxford University Press, pp. 176-213

Kavanagh, R.N., Darcey, T.M., Lehmann, D., Fender, D.H. (1978): Evaluation of methods for three-dimensional localization of electrical sources in the human brain. *IEEE Trans. Biomed. Eng.* BME5, 421-429

Lehmann, D. (1987): Principles of spatial analysis. In: *Handbook of electroencephalography and clinical neurophysiology, revised series: vol 1. Methods of analysis of brain electrical and magnetic signals.* Gevins, A.S., Remond, A. (eds.). Amsterdam: Elsevier, pp. 309-354

Lehmann, D., Skrandies, W. (1980): Reference-free identification of components of checkerboard-evoked multichannel potential fields. *Electroenceph. Clin. Neurophysiol.* 48, 609-621

Lehmann, D., Ozaki, H., Pal, I. (1987): EEG alpha map series: brain micro-states by space-oriented adaptive segmentation. *Electroenceph. Clin. Neurophysiol.* 67, 271-288

Lopes Da Silva, F., Hoeks, A., Zetterberg, L. (1974): Model of brain rhythmic activity: The alpha rhythm of the thalamus. *Kybernetik* 15, 27-37

Meijs, J.W.H., Boom, H.B.K., Peters, M.J., Van Oosterom, A. (1987): Application of the Richardson extrapolation in simulation studies of EEGs. *Med. Biol. Eng. Comput.* 25, 222-226

Mocks, J. (1987): Decomposing event related potentials: A new topographic components model. Presented at Fourth International Conference on Cognitive Neurosciences, Dourdan, France

Nunez, P. (1981): *Electrical fields of the brain.* New York: Oxford University Press, p. 150

Offner, F.F. (1950): The EEG as potential mapping: The value of the average monopolar reference. *Electroenceph. Clin. Neurophysiol.* 2, 215-216

Oppenheim, A.V., Schafer, R.W. (1975): *Digital signal processing,* Englewood Cliffs, NJ: Prentice-Hall

Pascual-Marqui, R.D., Gonzalez-Andino, S.L., Valdes-Sosa, P.A., Biscay-Lirio, R. (1989): Current source density estimation and interpolation based on the spherical harmonic Fourier expansion. *Int. J. Neurosci.* (in press)

Perrin, F., Bertrand, O., Pernier, J. (1987): Scalp current density mapping: Value and estimation from potential data. *IEEE Trans. Biomed. Eng.* BME34, 283-288

Plonsey, R. (1969): *Biomagnetic phenomena.* New York: McGraw-Hill, p. 203

Prenter, P.M. (1975): *Splines and variational methods.* New York: John Wiley and Sons

Sarvas, J. (1987): Basic mathematical and electromagnetic concepts of the biomagnetic inverse problem. *Phys. Med. Biol.* 32, 11-22

Scherg, M., Von Cramon, D. (1985a): Two bilateral sources of the late AEP as identified by a spatio-temporal dipole model. *Electroenceph. Clin. Neurophysiol.* 62, 32-44

Scherg, M., Von Cramon, D. (1985b): A new interpretation of the generators of BAEP waves I-V: Results of a spatio-temporal dipole model. *Electroenceph. Clin. Neurophysiol.* 62, 290-299

Scherg, M., Von Cramon, D. (1986): Evoked dipole source potentials of the human auditory cortex. *Electroenceph. Clin. Neurophysiol.* 65, 344-360

Sencaj, R.W., Aunon, J.I. (1982): Dipole localization of average and single visual evoked potentials. *IEEE Trans. Biomed. Eng.* BME29, 26-33

Stok, C.J. (1987): The influence of model parameters on EEG/MEG single dipole source estimation. *IEEE Trans. Biomed. Eng.* BME34, 289-296

Van Petten, C., Kutas, M. (1988): The use of event-related potentials in the study of brain asymmetries. *Int. J. Neurosci.* 39, 91-99

23

Neurometric Functional Imaging: I. Subtyping of Schizophrenia

LESLIE S. PRICHEP, E. ROY JOHN AND F. MAS

23.1 Introduction

A vast literature, based upon conventional qualitative visual evaluations, suggests that a substantial proportion of psychiatric patients display EEG abnormalities. With the advent of computerized analysis of EEG data, these earlier impressions were substantiated by quantitative methods. An extensive review by Shagass (1975) not only confirms the existence of differences between normal subjects and patients with psychiatric disorders but also provides replicated reports of abnormal profiles distinctive for each disorder.

Reviewing more than 50 recent papers on the electrophysiology of schizophrenia, Grebb et al. (1986) concluded that it was now "well-established" that there was a high incidence of abnormal findings among schizophrenic patients. Among the consistently replicated findings were increased delta and theta but decreased alpha activity, and bilateral asymmetries. Summarizing a somewhat different selection of over 50 papers, Small (1983) concluded that schizophrenics were consistently reported to have increased delta and beta but decreased alpha activity, as well as marked bilateral asymmetries. Ford et al. (1986) concurred that frequency abnormalities, especially increased beta activity, were commonly encountered among schizophrenics, but also emphasized the prevalence of increased interhemispheric coherence, in agreement with Merrin et al. (1989) who cite three other reports with similar findings.

A very different picture emerges from studies of affective disorders. Summarizing over 100 articles on depression, Perris (1980) concluded that interhemispheric asymmetry, mean integrated amplitude, and its variance are the most consistently confirmed abnormal EEG measures in depressed patients.

Giannitrapani and Collins (1988), Soininen and Partaven (1988), Giaquinto and Nolfe (1988), and Goodin and Aminoff (1986) agree that a high proportion of patients with senile dementia display an abnormal electrophysiological profile, but one that is very different from that seen in schizophrenia or depression. Reviewing well over 100 papers, these workers agree that dementia patients are characterized by generalized increased delta and theta activity together with decreased alpha and beta activity.

In view of this abundant evidence that patients with different psychiatric disorders display abnormal electrophysiological profiles that are distinctively different from each other, it is not surprising that such patients can be differentially classified utilizing electrophysiological variables. Discrimination has been accomplished between dementia and normals by Giaquinto and Nolfe (1988), between Alzheimer's and non-Alzheimer's dementia and normals by Goodin and Aminoff (1986) and Giannitrapani and Collins (1988), and between Alzheimer's and multi-infarct dementia and normals by Leuchter et al. (1987). Discrimination between dementia and depressed patients and normal controls has been accomplished successfully by Brenner et al. (1986) and between Alzheimer's and vascular dementia and depressed by O'Connor et al. (1979). Abrams and Taylor (1979) described differential EEG profiles that discriminate among affective disorders, schizophrenics, and normals, as have Shagass et al. (1984), Merrin et al. (1989). Ford et al. (1986), separated schizophrenic from depressed from dementia patients, finding coherence variables to contribute most to the discrimination. These latter authors cited 17 papers that reported successful discrimination between schizophrenic patients and normals and 13 that reported discrimination between depressed patients and normals. Flor-Henry et al. (1983) performed principal components analysis of frequency and coherence variables in depressed, manic, and schizophrenic patients and normal controls, finding that overall coherence contributed most to the discrimination.

Our laboratory has previously reported marked electrophysiological differences between patients with a variety of psychiatric disorders and normal subjects, showing that patients with different disorders (including dementia, major affective disorders, and alcoholism) could be discriminated reliably from normal, as well as from one another (John et al., 1988b; Prichep, 1987) and that different subtypes of depressed patients could be identified with high accuracy (Prichep and John, in press). Our results were unusually robust, but otherwise compatible with the evidence cited above. It is of note that we have found that comparable discrimination can be achieved using different subsets of neurometric features. These subsets include features of frequency, coherence, absolute power, and asymmetry. This finding reflects substantial redundancy in the EEG measure set and may reconcile the different measure subsets found to be discriminating in the studies cited.

In this chapter, we focus on the utility of quantitative EEG for subtyping within disorders. Although such a capability would be of specific relevance to more effective treatment of patients, few papers other than our own address this issue. Our results were obtained using the neurometric methodology. Neurometrics uses a standardized procedure in which 19 channels of monopolar EEG and EP data are collected with on-line artifact rejection. Quantitative features are extracted, log transformed to obtain Gaussianity, age-regressed, and Z-transformed relative to population norms. Z-values or standard scores for these features (proportional to probabilities) are then used for all further analyses. The importance of each of these steps in enhancing the clinical utility of electrophysiological data has been discussed in detail elsewhere (John et al., 1987, 1988a, 1988b). In studies of large populations of children and adults, significant deviations from normal neurometric values were rarely found in normally functioning persons, independent of cultural or ethnic backgrounds (Ahn et al., 1980; Alvarez et al., 1987; John et al., 1980). However, significant departures from the normal range occurred frequently among patients with psychiatric disorders. Further, patients with different disorders showed distinctive neurometric profiles of electrophysiological dysfunction as demonstrated in the successful discriminants noted above.

In this chapter we present pilot data from 34 medication-free schizophrenic patients in whom a cluster analysis of neurometric EEG features was used to determine the existence of subtypes within the schizophrenias and to examine the relationship between cluster membership and response to drug treatment.

23.2 Subjects

Forty-six male patients were studied, all of whom met DSM-III criteria for schizophrenia. All received extensive clinical evaluations by multiple raters. These patients were from two different hospital sites: (1) Manhattan Psychiatric Center (MPC)[1], and (2) Manhattan Veteran's Administration Hospital (MVA)[2]. Thirty-four of the 46 patients were evaluated neurometrically both on- and off-medication. Off-medication periods were at least 7 days. Thirty-one patients were treated with Haldol and three with Mellaril.

23.3 Method

Each patient received a neurometric evaluation consisting of 2 minutes of artifact-free eyes closed resting EEG and a battery of four evoked potential (EP) conditions. Only the EEG results are discussed here. Monopolar recordings were obtained using the 19 electrodes of the 10/20 system (Jasper, 1958), referred to linked earlobes. The patients were seated comfortably in a dimly lit, sound-attenuated, shielded testing chamber.

All patients were evaluated using the Brief Psychiatric Rating Scale (BPRS) and were tested neurometrically within 24 hours of their BPRS ratings for the off- and on-medication condition.

---

[1] This work was done in collaboration with Dr. J. Volavka at Manhattan Psychiatric Center, N.Y.

[2] This work was done in collaboration with Drs. B. Angrist, J. Rotrosen, and F. Baruch at the Manhattan Veterans Administration Hospital, N.Y.

23.4 Data Analysis

23.4.1 EEG Feature Extraction

The feature extraction methods have been described in detail previously (John et al., 1987). Univariate and multivariate features were computed for absolute and relative power, coherence, and asymmetry in the four frequency bands for the 19 monopolar derivations, as well as for eight bipolar derivations.

Since Z-scores express the deviation of the disparate neurometric features from the predicted normative values in the common metric of relative probability, multivariate or composite features can be computed. Multivariates of two sorts were computed: (1) within each derivation across frequency bands for absolute power, relative power, coherence, or asymmetry and (2) across derivations in the anterior or posterior regions of each hemisphere, across the whole left and right hemispheres, and across the whole brain for every feature. Correction for inter-correlations among the features combined in each composite was accomplished by computing the Mahalanobis distance across the set of features. By procedures analogous to those used for univariate features, normative data were used to permit Z-transformation of these new composite features.

23.4.2 Cluster Analysis

The main purpose of this study was to identify subtypes within a schizophrenic population. The statistical procedure used was K-means clustering (BMDP KM). This method measures the Euclidean distance from each case to the center of each cluster. Cases are iteratively reallocated into the cluster whose center is closest. Analysis of variance criteria is applied to the variance within clusters and between clusters to seek the clearest structure (best separation) as the number of clusters is varied.

23.5 Results and Discussion

Figure 23.1 shows the group average topographic maps for Z relative power in the MPC, MVA, and combined schizophrenic population (MPC & MVA) in the off- and on-medication conditions. As can be seen in this figure, these two schizophrenic populations (MPC and MVA) showed very different neurometric profiles off drugs. Further, when combined, the total groups showed a neurometric profile that was different from either of the two subtypes of which it was comprised. The profiles on medication were more similar to each other, probably generally reflecting Haldol's known effect on the EEG of increasing theta and decreasing beta (Hermann & Schaerer, 1986). The gross discrepancies between the off-medication group averages suggested the presence of heterogeneity and the desirability of cluster analysis.

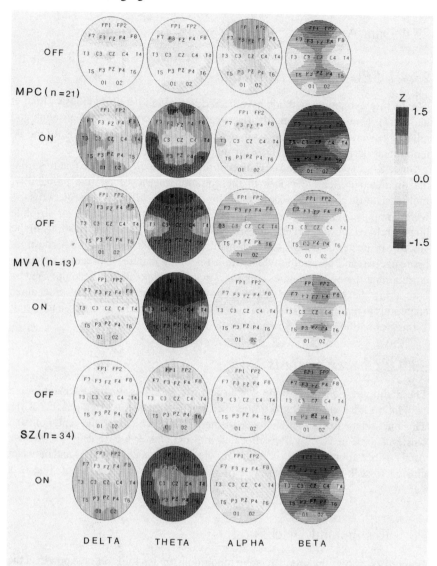

FIGURE 23.1. Group average topographic head maps for Z-relative power in the delta, theta, alpha, and beta frequency bands computed across groups of schizophrenic patients, off and on medication, referred from Manhattan Psychiatric Center *(top pair of rows)*, the Manhattan Veteran's Administration Hospital *(middle pair of rows)*, and the combined population from these two referral sources *(bottom pair of rows)*. These maps represent the mean relative power difference between each grouping and the normal reference group, expressed in standard derivation of the reference group. Density coding is proportional to the mean Z score for each group, in steps corresponding to those shown on the Z scale. The significance of Z scale values for group data can be estimated by multiplying the Z value by the square root of the group size. For example, to estimate the significance of a Z score of 0.5 for a group with 21 patients, the Z score would be multiplied by 4.6 giving a Z of 2.3 ($p<0.05$) for the group data.

Cluster analysis was performed using a subset of neurometric frequency variables from the off-medication condition and resulted in the identification of five clusters. Figure 23.2 shows the group average topographic maps for each of the five clusters off and on medication. It is of note that a similar structure was obtained from a cluster analysis based on coherence and asymmetry features. Vast differences between neurometric profiles can be seen in the five clusters. Particularly striking are the theta and alpha differences. Cluster I and V show excess theta, Cluster II and IV show deficit of theta, and Cluster III shows normal theta. Cluster III and IV show excess alpha, whereas the remaining clusters show deficits of alpha. Medication effects show the expected increase in theta and decrease in beta in all clusters.

Because the Veteran's Hospitals tend to see more chronic, less acute patients, we were concerned that the clusters might reflect "chronicity." Table 23.1 shows the distribution of cluster members by referral hospital and mean age of each cluster. It can be seen that all but one cluster has a mixture of patients from MVA and MPC. Cluster IV is composed exclusively of MPC patients and shows the lowest mean age. Also in relation to this point the theta excess seen in Figure 23.1 in the MVA group might have reflected a residual medication effect in a more chronic population. However, reviewing the topographic maps in a small group of "first-break" schizophrenic patients (n=6), patients with excess theta were identified, suggesting that this profile is related to a subtype of schizophrenia and not medication.

Table 23.1 also shows the mean total BPRS scores for each of the neurometric clusters of schizophrenic patients off-medication, on-medication, and the mean percentage change in total BPRS (Change/initial score × 100). Using a criterion of 25% change for defining a treatment responder, only Cluster I shows clinical improvement with medication. It is noted that the two highest mean total BPRS scores were obtained in Cluster I, comprised predominantly of patients referred from MVA, and in Cluster IV, comprised exclusively of patients referred from MPC.

In order to better define the clinical profile of the patients, a factor analysis of the BPRS items was computed (BMDP4M) based on the off-medication ratings. Five factors accounted for 70% of the variance and were easily interpretable. The symptoms loading highest in each of the first four factors obtained by principal components analysis with varimax rotation are shown in Table 23.2. For the convenience of discussion we have named the BPRS factors to reflect the symptom constellation they contain. These include active behavioral symptoms (Factor I), mood-related symptoms (Factor II), passive behavior symptoms (Factor III), and reality distortion (Factor IV).

Figure 23.3 shows the mean BPRS factor scores off-and on-medication for each of the neurometric clusters. Note that the initial BPRS factor profile is quite different for each of the five neurometric clusters. Response to treatment is also quite different. Several other important points are demonstrated in this figure. The concept of "clinical improvement" as reflected in changes in total scores of clinical scales is inadequate. Although all five clusters showed a decrease in the factor

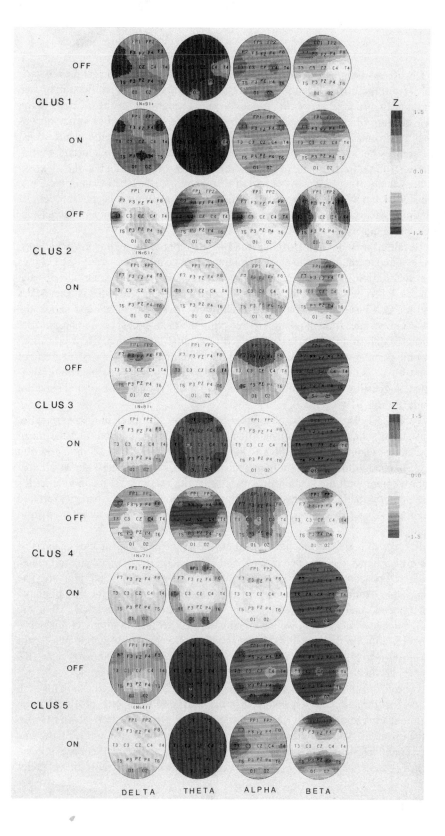

Table 23.1. Changes in total BPRS with medication for neurometric clusters of schizophrenic patients

Cluster	n	(MVA/MPC)	Mean age	Off med. total BPRS	On med. total BPRS	% change in total BPRS
I	9	6/3	43.7	61.8	45.1	27
II	6	3/3	44.2	54.8	48.2	12
III	8	1/7	36.4	53.4	45.7	14
IV	7	0/7	32.7	65.1	61.2	6
V	4	3/1	50.9	50.9	47.4	7

Table 23.2. BPRS factor analysis results

FACTOR I (Active Behavioral Symptoms)	FACTOR II (Mood-Related Symptoms)
tension hostility uncooperative excitement	anxiety guilt depressed mood

FACTOR III (Passive Behavioral Symptoms)	FACTOR IV (Evidence of Reality Distortion)
emotional withdrawal posturing blunt affect	concept disorganization hallucinations unusual thought content suspiciousness

FIGURE 23.2. Group average topographic maps for Z relative power in the delta, theta, alpha, and beta frequency bands computed for each of the five clusters of schizophrenic patients identified in the schizophrenic population, off medication *(first row of each pair)* and on medication *(second row of each pair)*. The number of patients in clusters 1 through 5 were 9, 6, 8, 7, and 4, respectively. These maps represent the mean relative power difference between each cluster and the normal reference group, expressed in standard deviations of the reference group. Density coding is proportional to the mean Z score for each cluster, in steps corresponding to those shown on the Z scale. The significance of Z scale values for group data can be estimated by multiplying the Z value by the square root of the group size.

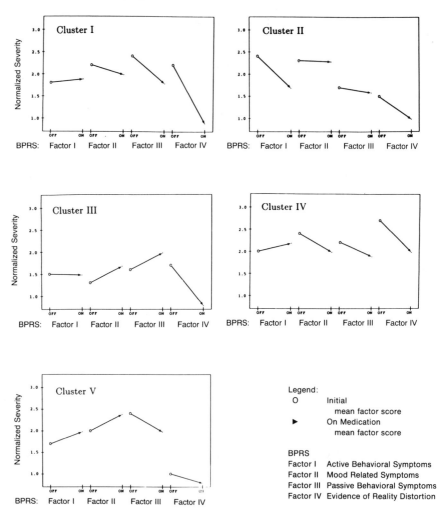

FIGURE 23.3. For each of the five neurometric clusters, the mean BPRS factor scores are shown for the initial evaluation (off medication) and the subsequent evaluation (on medication).

reflecting "evidence of reality distortion," quite varied effects are seen in the other factors. Cluster I and II showed the most consistent changes in the positive direction across factors. It is interesting that Cluster II in fact shows the most normalization neurometrically as seen in Figure 23.3.

This pilot study demonstrates clear pathophysiological differences among subtypes within the schizophrenic population. Studying the neurometric profiles for each cluster suggests that it would be unlikely that they would all respond to the same medication. Further, the relationship between neurometric profile, clinical profile, and response to treatment shown in the final figure suggests the importance of subtyping for more optimal assessments and treatment of the patients.

Acknowledgments. The authors acknowledge the work of Thomas Essig-Peppard and Mary Almas in the statistical analysis of the data and Kenneth Alper for help with the literature review. This work was supported in part by Cadwell Laboratories, Inc., Kennewick, Washington, USA.

References

Abrams, R., Taylor, M.A. (1979): Differential EEG patterns in affective disorders and schizophrenia. *Arch. Gen. Psychiatry* 36, 1355

Ahn, H., Prichep, L., John, E.R., Baird, H., Trepetin, M., Kaye, H. (1980): Developmental equations reflect brain dysfunction. *Science* 210, 1259–1262

Alvarez, A., Pascual, R., Valdez, P. (1987): U.S. EEG developmental equations confirmed for Cuban schoolchildren. *Electroenceph. Clin. Neurophysiol.* 67, 330–332

Brenner, R.P., Ulrich, R.F., Spiker, D.G., Sclabassi, R.J., Reynolds, C.F., Marin, R.S., Boller, F. (1968): Computerized EEG spectral analysis in elderly normal, demented and depressed subjects. *Electroenceph. Clin. Neurophysiol.* 64, 483–492

Flor-Henry, P., Koles, Z.J., Sussman, P.S. (1983): Multivariate EEG analysis of the endogenous psychoses. *Adv. Biol. Psychiatry* 13, 196–210

Ford, M.R., Goethe, J.W., Dekker, D.K. (1986): EEG coherence and power in the discrimination of psychiatric disorders and medication effects. *Biol. Psychiatry* 21, 1175–1188

Giannitrapani, D., Collins, J. (1988): EEG differentiation between Alzheimer's and non-Alzheimer's dementias. In: *The EEG of mental activities.* Giannitrapani, D., Murri, L. (eds.). New York: Karger, pp. 26–41

Giaquinto, S., Nolfe, G. (1988): The electroencephalogram in the elderly: Discrimination from demented patients and correlation with CT scan and neuropsychological data. In: *The EEG of mental activities.* Giannitrapani, D., Murri, L. (eds.). New York: Karger, pp. 55–60

Goodin, D.S., Aminoff, M.J. (1986): Electrophysiological differences between subtypes of dementia. *Brain* 109, 1103–1113

Grebb, J.A., Weinberger, D.R., Morihias, J.M. (1986): Encephalogram and evoked potential studies of schizophrenia. In: *Handbook of schizophrenia. The neurology of schizophrenia.* Nasrallah, H.A., Weinberger, D.H. (eds.). Amsterdam: Elsevier, pp. 121–140

Hermann, W.M., Scharer, E. (1986): Pharmaco-EEG: Computer EEG analysis to describe the projection of drug effects on a functional cerebral level in humans. In: *Handbook of electroencephalography and clinical neurophysiology.* (revised): vol 2.Lopes da Silva, F.H., Storm van Leeuwen, W., Remond, A. (eds.). Amsterdam: Elsevier

Jasper, H.H. (1958): The ten-twenty electrode system of the International Federation. *Electroenceph. Clin. Neurophysiol.* 10, 371–375

John, E.R., Ahn, H., Prichep, L., Trepetin, M., Brown, D., Kaye, H. (1980): Developmental EEG equations for the electroencephalogram. *Science* 210, 1255–1258

John, E.R., Prichep, L.S., Easton, P. (1987): Normative data banks and neurometrics: Basic concepts, methods and results of norms constructions. In: *Handbook of electroencephalography and clinical neurophysiology: vol. 3, Computer analysis of the EEG and other neurophysiological signals.* Remond, A. (ed.). Amsterdam: Elsevier, pp. 449–495

John, E.R., Prichep, L.S., Easton, P. (1988a): Neurometrics: Computer assisted differential diagnosis of brain dysfunctions. *Science* 239, 162–169

John, E.R., Prichep, L.S., Friedman, J., Essig-Peppard, T. (1988b): Neurometric classification of patients with different psychiatric disorders. In: *Statistics and topography in quantitative EEG*. Samson-Dollfus, D. (ed.). Paris: Elsevier, pp. 88–95

Leuchter, A.F., Spar, J.E., Walter, D.O., Weiner, H. (1987): Electroencephalographic spectra and coherence in the diagnosis of Alzheimer's-type and multi-infarct dementia. *Arch. Gen. Psychiatry* 44, 993–998

Merrin, E.L., Floyd, T.C.,Fein, G. (1989): EEG coherence in unmedicated schizophrenic patients. *Biol. Psychiatry* 25, 60–66

O'Connor, K.O., Shaw, J.C., Ongley, C.O. (1979): The EEG and differential diagnosis in psychogeriatrics. *Brit. J. Psychiatry* 135, 156–162

Perris, C. (1980): Central measures of depression. In: *Handbook of biological psychiatry*. van Praag, H. (ed.). New York: Marcel Dekker, pp. 183–225

Prichep, L.S., John, E.R. (in press): Neurometric characteristics of depressive disorders. In: *Plasticity and morphology of the central nervous system*. Cazzullo, C.L., Invernizzi, G., Sacchetti, E., Vitta, A. (eds.). M.T.P. Press

Prichep, L.S. (1987): Neurometric quantitative EEG measures of depression. In: *Cerebral dynamics, laterality and psychopathology*. Takahashi, R., Flor-Henry, P., Gruzelier, J., Niwa, S. (eds.). Amsterdam: Elsevier, pp. 55–69

Shagass, C. (1975): EEG and evoked potentials in the psychoses. In: *Biology of the major psychoses, research and public association of research in nervous and mental disease:* vol. 54. Freedman, D.X. (ed.). New York: Raven Press, pp. 101–127

Shagass, C., Roemer, R.A., Straumanis, J.J., Josiassen, R.C. (1984): Psychiatric diagnostic discriminations with combinations of quantitative EEG variables. *Brit. J. Psychiatry* 144, 581–592

Small, J.G. (1983): EEG in schizophrenia and affective disorder. In: *EEG and evoked potentials in psychiatry and behavioral neurology*. Hughes, R.R., Wilson, W.P. (eds.). Boston: Butterworths, pp. 25–44

Soininen, H., Partanen, J.W. (1988): Quantitative EEG in the diagnosis and followup of Alzheimer's disease. In: *The EEG of mental activities*. Giannitrapani, D., Murri, L. (eds.). New York: Karger, pp. 42–49

24

Neurometric Functional Imaging: II. Cross-Spectral Coherence at Rest and During Mental Activity

Leslie S. Prichep, E. Roy John, P. Easton, and R. Chabot

24.1 Introduction

Cross-spectral coherence is one way to quantify the dynamic interrelationships among brain regions, possibly reflecting the neurofunctional organization of the brain. The data reported in this chapter were from a pilot study performed in normal adults for the purpose of exploring whether mental activity would be reflected as different changes in cross-spectral coherence of the EEG (John et al., 1989).

The average cross-spectral coherence was computed for each region versus every other region, separately in each frequency band, and for the total spectrum at 0.5-Hz intervals. The 2-min EEG record was divided into samples of 15 sec each for the purpose of calculating means and standard deviations for the cross-spectral coherence values. All data reviewed here were computed as t-tests comparing the cross-spectral coherence of the appropriate baseline to the cross-spectral coherence obtained from the EEG samples recorded during each mental task, or t-tasks comparing the cross-spectral coherence between mental tasks.

24.2 Method

A baseline recording of 2 minutes of artifact-free EEG was collected from the 19 leads of the international 10/20 system referred to linked ears. The subject was seated comfortably in a dimly lit, sound-attenuated chamber. The EEG was then recorded at rest and during the performance of cognitive tasks. The cognitive tasks were the following.

24.2.1 Match/Mismatch

Word semantic (WS): In this task pairs of words were rear projected on a screen that was approximately 1 m from the subject, at a rate of one every 2 sec. The subject's task was to decide whether the two objects in the pair come from the same or different semantic categories. For example, CAT-DOG belong to the same category, whereas CAR-DOG belong to different categories.

Picture Semantic (PS): This task was the same as the above (WS), except the pairs of stimuli were line drawings of the objects.

Auditory Semantic (AS): This task was the same as WS, except the word pairs were presented through headphones.

Abstract Form (AF): In this task, pairs of abstract forms (rated low for eliciting names) were presented on the rear projection screen. The subject's task was to respond "same" or "different" abstract form. An example is shown in Figure 24.1.

The BL and subsequent EEGs for the match/mismatch tasks were recorded with eyes open.

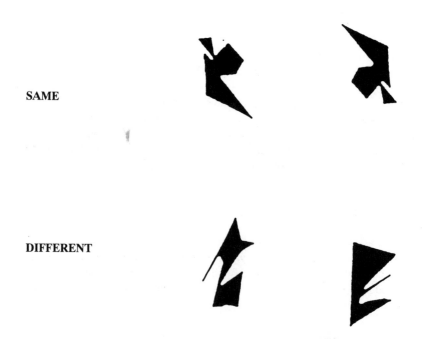

SAME

DIFFERENT

FIGURE 24.1. Example of two pairs of abstract forms, top pair are the same and the bottom pair are different.

24.2.2 Memory Tasks

"Short-term Memory" (STM): Eight two-digit numbers were read to the subject who was brought to 100% criterion. Each subject was then told to silently rehearse the numbers that were to be repeated at the end of the EEG recording.

"Long-term Memory" (LTM): This was a mental imagery task in which the subject was instructed to start with the letter A and recall an acquaintance whose name began with that letter in as much detail as he or she could, and then to continue with the letter B and so on until the end of the EEG period. This is a very poorly defined task, in which old memories with unspecified personal affective content are retrieved in a manner that cannot be confirmed. We rely upon the assurances given by our subjects that they actually perform the task. Most of them report it as extremely engrossing and pleasant.

The BL and subsequent EEGs for the memory tasks were recorded with eyes closed.

24.3 Results and Discussion

Topographic maps of cross-spectral coherence were constructed. At each scalp location a rosette of 19 leads is color coded to reflect the coherence between the fiducial electrode (shown as a brightened yellow spot) and each of the other 18 positions of the 10/20 system. This coherence is shown separately for each band and for the total spectrum. Maps can likewise be constructed for the significance of the differences in cross-spectral coherence.

Figure 24.2 shows differences in cross-spectral coherence between mental tasks for the Beta frequency band. Widespread differences are seen within the same class of tasks (LTM-STM and PS-WS), as well as across classes (PS-LTM). It is noted that coherence between anterior leads is significantly increased during the LTM as compared with STM task, whereas coherence between posterior leads and the rest of the head is increased in the STM task, possibly reflecting visual rehearsal used in the STM task. Also in the beta band, PS-WS shows little significant difference except in the right frontal regions where local coherence is increased in the PS task as compared to the WS task versus the affective content of the LTM task. This may reflect right anterior hemispheric involvement in the processing of picture/form information. Lastly, when comparison is made across class of task (PS-LTM) in the beta band, much lower coherence is seen in anterior and central regions in the PS compared with LTM tasks, whereas all posterior regions show increased coherence. This may reflect the visual nature of the PS task compared with the more frontal functioning involved in the LTM task.

Figure 24.3 shows a set of cross-spectral coherence difference maps (task minus BL) for a typical subject, observed in the beta band during the AF, WS, STM, and

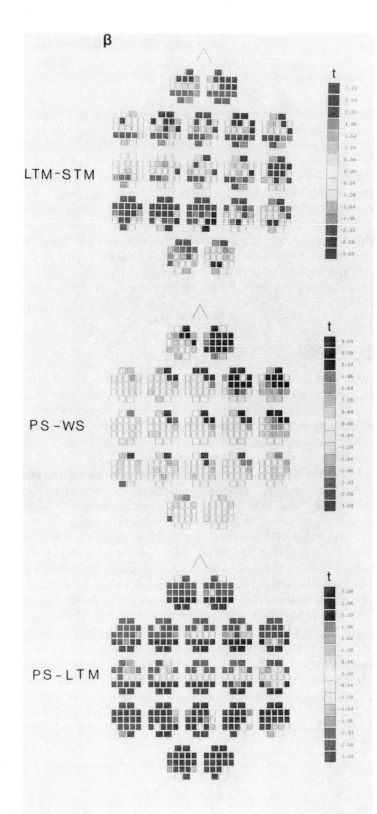

β

LTM-STM

PS-WS

PS-LTM

(1460)

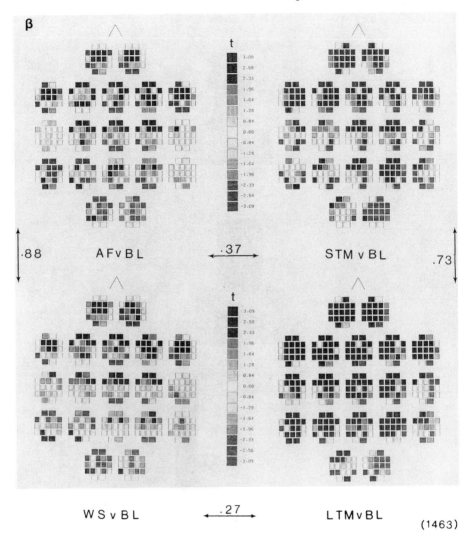

FIGURE 24.3. Topographic head maps of *t*-tests comparing the cross-spectral coherence values within the beta frequency range for a single subject for the abstract form (AF), word semantic (WS), short-term memory (STM), and long-term memory (LTM) cognitive tasks vs the appropriate baseline condition. Maps are density coded for the *t* value according to the *T* scale shown. Overall correlations (r) are also shown within and across types of cognitive task.

FIGURE 24.2. Topographic head maps of *t*-tests comparing the cross-spectral coherence values within the beta frequency range for a single subject for the long-term memory task vs the short-term memory task (LTM-STM, *top panel*), picture semantic vs word semantic task (PS-WS, *middle panel*), and the picture semantic vs long-term memory task (PS-LTM, *bottom panel*). Maps are density coded for the *t* value according to the *t* scale shown.

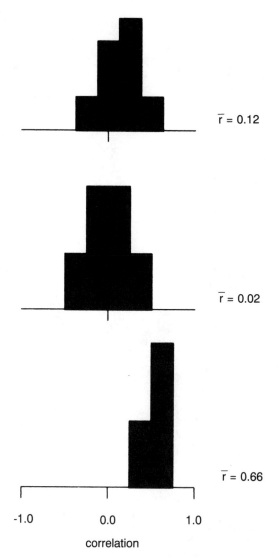

FIGURE 24.4. Histograms showing overall mean correlations (r) of cross-spectral coherence changes from baseline for total power for the same type of cognitive task between subjects (*top panel*), for different types of cognitive tasks within subjects (*middle panel*), and for the same type of cognitive tasks within subjects (*bottom panel*), (John et al., 1989d).

LTM tasks. The correlations between the matrix of difference Z scores for the two tasks in the same class are shown between the vertical arrows next to the topographical maps. Note that the changes in relationships among brain regions that occur during performance of these mental tasks involve widespread regions of the cortex. Different tasks in the same class cause similar patterns of new relationships to be established ($r=0.88$, AF vs WS; $r=0.73$, STM vs LTM), but different classes of mental activity appear to be mediated by different organization of brain relationships, reflected by the low correlations shown between the horizontal arrows ($r=0.37$, AF vs STM; $r=0.27$, WS vs LTM).

Correlations between the changed patterns of coherence were computed for the changes in cross-spectral coherence in the same type of tasks between subjects, different types of tasks within subjects, and the same tasks within subjects. Figure 24.4 shows the histograms of the correlation value in six normal subjects for total power. The bottom histogram shows that each subject displayed consistent patterns of organization of brain relationships within each class of tasks. When two tasks in the same class were compared within the same subject, the mean correlation coefficient, r, was equal to 0.66. Each different class of task within each subject was subserved by a different pattern of brain organization. The middle histogram shows that, when two tasks in different classes were compared within the same subject, the mean correlation coefficient, r, was equal to 0.02. Different subjects appeared to use distinctive styles. The mean correlation, r, between different subjects performing the same task was equal to 0.12. By chi-squared, the differences between these distributions were highly significant, ($p \leq 0.05$ for all comparisons). Quite similar results were obtained for the alpha and beta bands.

Within a subject, different mental tasks within a particular class appear to be mediated by brain activity organized in a characteristic way. Mental tasks in different classes are mediated by different organizations of brain activity. It appears that different individuals have different physiological cognitive styles. Finally, mental activity is not localized. Both cortical and subcortical structures participate in each of these mental tasks, and widespread brain regions appear to be engaged by each mental task.

These findings in humans are quite compatible with the findings from EEG, EP, and unit recordings, as well as from double-labeled 2-deoxyglucose metabolic maps in experimental animals, revealing the engagement of widespread brain regions in cognitive behavior (John, 1972; John et al., 1986, John, 1987, 1988).

Acknowledgments. This work was supported in part by a reasearch grant from Cadwell Laboratories, Inc., Kennewick, Washington, USA.

References

John, E.R. (1972): Switchboard versus statistical theories of learning and memory. *Science* 177, 850–864.

John, E.R. (1987): Distributed learning on vertebrates, electrophysiological studies. In: *Encyclopedia of Nueroscience: vol 1.* Adelman, G. (ed.). Boston: Birkhauser

John, E.R. (1988): Resonating fields in the brain: The Hyperneuron. In: *Springer Series on Brain Dynamics I.* Basar, E. (eds.). Hiedelberg: Springer-Verlag

John, E.R., Tang, Y., Brill, A.B., Young, R., Ono, K. (1986): Double-labeled metabolic maps of memory. *Science* 233, 1167–1175.

John, E.R., Prichep, L.S., Chabot, R.J. (in press): Quantitative electrophysiological maps of mental activity. In: *Dynamics of sensory and cognitive processing by the brain.* Basar, E., Bullock (eds.). Heidelberg: Springer-Verlag

25

Magnetic Fields of the Brain Resulting from Normal and Pathological Function

HAROLD WEINBERG, A. W. ROBERTSON, D. CRISP, AND B. JOHNSON

25.1 Introduction

Theoretical physics predicts that currents in biological media result in magnetic fields; however, the predicted fields are so small that they were measurable only recently. One of the first biomagnetic measurements was of fields associated with heart function, measured by Cohen et al. in 1970. Coil magnetometers of the kind they used are usually not sensitive enough for the detection of brain function, which is an order of 10^{-4} smaller than fields produced by the heart. Consequently it was not until a Josephson junction was incorporated into a superconductive quantum interference device (SQUID) that magnetometers with the required high sensitivity were available for the measurement of brain function. The SQUID is used as ultrasensitive magnetic flux detector. The problem with SQUIDs is that, in order to maintain their superconductivity, the sensor has to be cooled to the temperature of liquid helium ($4.2°$ K). In order to accomplish these low temperatures the SQUID is immersed in liquid helium inside a helium dewar. The rf-SQUID is a superconducting ring with one Josephson junction (weak link) in it. The dc-SQUID has two weak links in the ring. Flux transformers transfer flux from a sensing coil to the SQUID. For example, a closed loop of superconducting wire maintains the total magnetic flux inside the loop. If this loop contains two coils, coupled in series, a change of the magnetic flux through one of the coils causes a change in the magnetic flux in the other coil. Thus, magnetic flux is transferred from the sensing coil L1 to the signal coil Ls inside the SQUID (Figure 25.1). In order to increase the signal-to-noise ratio, differential magnetometers, referred to as gradiometers, are utilized in preference to the simple magnetometer (Figure 25.2). The first-order gradiometer has two sensing coils, L1 and L2. The gradiometers shown in Figure 25.2 are asymmetrical; the diameter of the main sensing coil L1 is smaller than that of the compensating coil L2. This two-coil system is insensitive to external homogeneous fields if the inductances are equal. The sensing coils, L1 and L2, are oppositely wound. The magnetic flux transferred into the SQUID (and thus also the output signal) is smaller than that for a

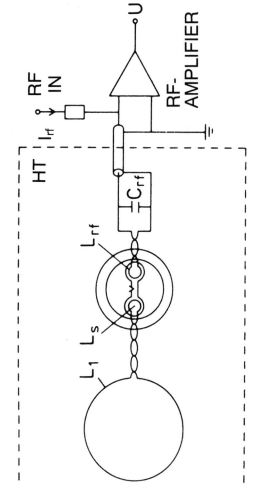

FIGURE 25.1 Flux transformer. The coupling of coils of a simple squid magnetometer. The flux transformer consists of two superconducting coils; L_1, L_s. The superconducting flux transformer usually consists of several ending coils and a signal coil.

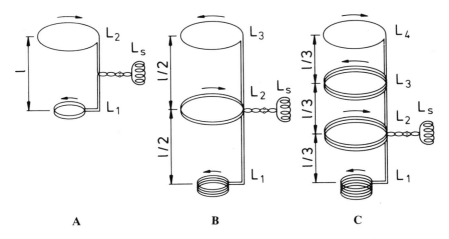

FIGURE 25.2. Gradiometer. (A), first-, (B) second-, and (C) third-order gradiometric (asymmetrical) flux transformers. L_1 is the main sensing coil; L_2, L_3, and L_4 are compensating coils.

magnetometer. For best discrimination of signal from noise the distance between the sensing coils, L1 and L2, should depend on the source (signal) location, i.e., the distance between the source and the coil L1. In Figure 25.2, second-and third-order gradiometric flux transformers are also shown (Vrba et al., 1982; Weinberg et al., 1984). A second-order transformer is insensitive not only to homogeneous external magnetic fields but also to the first gradients.

In the ideal case, the third-order gradiometer gives no output signal for homogenous magnetic fields or first- or second-order field gradients. A typical coil diameter for brain studies is between 1 and 2 cm. The gradiometers depicted in Figure 25.2 measure one component of the field (or gradient) only. However, the magnetic field is a vector field with three components. It is possible to measure all the three components by building an instrument with three orthogonal sensing coils or three orthogonal gradiometers, the centers of which coincide. The measurement of three components requires the use of three separate SQUIDs as sensors or utilization of a multiplexed system. In order to measure the distribution of magnetic fields over the head with a one-channel instrument, measurements have to be repeated several times. The sequential measurement at perhaps several tens of locations is not only time consuming but also constrains experimental paradigms to those that assume that the brain is processing the same inputs in the same way during successive measurements. Multichannel instruments (Figures 25.3 and 25.4) of up to seven channels have been recently developed and there are now under construction 60-channel instruments. Because SQUID magnetometers must be immersed in liquid helium the physical configuration and size of the measurement instruments make the typical system bulky and unwieldy. Typically, the diameter of the dewar tail is between 5 and 10 cm and the height of the dewar at least 50 cm. Due to the liquid helium container the instrument cannot be freely

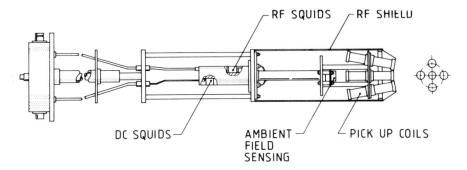

FIGURE 25.3 B.T.I. five-channel squid gradiometer. The dewar is not shown in the figure. From Williamson et al. 1985. The seven-channel system is similar in design.

tilted, and it weighs at least 1 kg. A typical distance between the main sensing coil and the skin is of the order of 1 cm. The instrument is sensitive to vibrations and to movements, and it has to be well supported. Depending on the size, the dewars stay cold between 1 day and about 1 week.

Magnetic fields that were clearly the result of electrical activity in the brain were first recorded by Cohen (1972). He observed spontaneous rhythms in the MEG that were in the alpha frequency range (8–12 Hz). Following those initial observations the field of magnetoencephalography (MEG) has grown rapidly, and

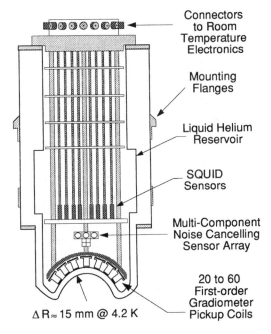

FIGURE 25.4. C.T.F. 20-channel system.

several recent reviews (romani et al., 1983, 1985; Weinberg et al., 1985b; Williamson et al., 1983) document the relevant research and theoretical issues. The use of MEG to study brain function has exclusively focused on methods of localizing electrical sources responsible for frequencies of the spontaneous EEG and sources responsible for "components" of evoked potentials (EP).

A component of an EP or an evoked field (EF) is a systematic fluctuation in the field (usually but not always a reversal of polarity) defined by its latency after sensory stimulation or its (negative) latency preceding muscle movement. When different stimuli or responses result in components and those stimuli and responses temporally overlap, components are sometimes said to overlap. However, it must be shown that the information content of different stimuli, presented simultaneously, differentially modify the component separately. Two broad categories of components are discussed in the literature, those that reflect properties of stimulus energy, called sensory evoked potentials, and those that index the nature of information processing within the brain, either with respect to stimulus input or response output. Those in the latter category are called event-related potentials (ERP). It is generally assumed that components of EPs can be localized to a much greater extent than components of ERPs.

25.2 The Inverse Solution

The accurate location of sources is theoretically made possible because magnetic fields are not attenuated or distorted by tissues of the head (Geselowitz, 1967). Methods of estimating source locations make use of the simplifying assumption that neural generators can be modeled as current dipoles and that the head can be approximated as a sphere. The implication of this assumption is that the component of the magnetic field that is radial to the surface of the sphere (i.e., the skull) is due to the primary current flow of a dipole tangential to the skull (Williamson et al., 1983). The radial components of the primary current do not contribute to the tangential (or radial) component of the magnetic field observed outside the head. Theoretically, this is because the field attributable to a radial dipole is canceled by the field due to the secondary sources, i.e., volume currents. The current dipole that best accounts for the observed data can be computed (Cuffin, 1982). However, given that it is impossible to know the total pattern of current flow in the intact, functioning brain at any instant, it is also impossible to determine the amount of error in the location of sources. Several attempts have been made to use computer modeling, but the accuracy of estimates of dipole locations in these studies is influenced by the assumptions used to develop the models (Okada, 1985; Romani and Williamson, 1983; Williamson et al., 1983). An important question arising when estimating the location of current sources is how to accommodate the actual shape of the head. The spherical approximation has in some cases caused significant errors. Physical models have been used by Weinberg et al. (1986) to study the problem of source location in which current dipoles were implanted in known positions within a human skull in vitro. They used a least-squared iterative method to find the parameters of a dipole such that the sum of squared differences between the recorded and predicted data is minimized; the smallest three-dimensional location error in the location of single dipoles were shown to be 3.5 mm using this procedure (Figure 25.5).

The concept of a source as an aggregate of distributed activity is one that has not been sufficiently appreciated, but is pivotal for any interpretation of how observed magnetic and electrical activity of the brain are related to sensory processing and cognitive function. When considering what is meant by a source, it is important to distinguish between sources associated with cognitive function (i.e., information processing that gains access to memory systems) and sources associated with those events of the brain that are the first stages of sensory processing (and indeed also the last stages of motor output). The distinction between sensory and cognitive processes implies a differentiation between those brain functions that code the so-called physical characteristics of input (i.e., stimuli) and those processes that extract the meaning of stimuli. The meaning of stimulus is defined by the demand characteristics of the stimuli and of the required behavior. For example, an auditory stimulus that changes in frequency or intensity produces changes in the brain attributable to registration processes. That same stimulus has cognitive value when it is given demand characteristics that require discrimination for the purpose of differential response output.

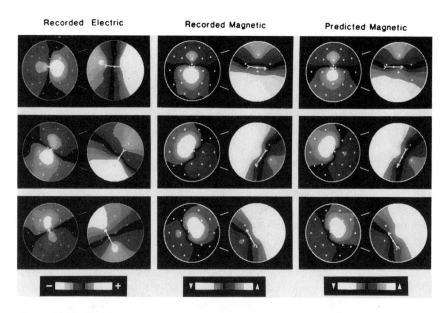

FIGURE 25.5A. Weinberg study of skull model. Isocontour maps for dipoles I, II, and III (top to bottom). For each pair of maps, the whole head view is shown on the left and the 5 times enlargement is on the right. For the maps of recorded electric and magnetic data, the arrow indicates the actual dipole location and orientation, whereas for the predicted magnetic maps the arrow shows the result of the dipole fitting procedure. A dot indicates the dipole center; the length of the arrow represents the actual length of the wires in the skull.

Recorded Electric   Recorded Magnetic   Predicted Magnetic

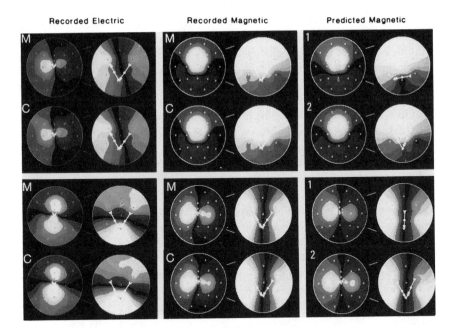

FIGURE 25.5B. Maps for two dipole combinations II+III (top) and II-III (bottom). For maps of recorded data the label "M" signifies that the data were measured by recording with both dipoles simultaneously energized (for II-III the direction of current flow was reversed in dipole III); "C" indicates that the map was calculated from the data for dipoles II and III shown in Figure 25.5A by adding or subtracting the two values at each position. Predicted maps are labeled with a number to indicate either a 1- or 2-dipole solution.

This basic distinction, arguable though it may be, is responsible for the de-facto differentiation between evoked and event-related potentials and also for the distinction between the so-called exogenous and endogenous components of ERPs and event-related (magnetic) fields (ERFs). The concept of a component in an event-related potential depends on the assumption that different types of processing may be occurring simultaneously, or successively, and that the electrical potentials occurring during this processing index the "algorithms" of that processing. Furthermore, the definition of a component is posthoc. If two voltage changes, in two separate intervals of time after input, can be made to vary independently, they are defined as components, or if components are said to overlap in time it means that different demand characteristics of the input modify the pattern of voltage changes in that interval. When processing is occurring simultaneously in different parts of the brain, the electrical changes associated with that processing may summate both functionally and electrotonically, i.e., through volume conduction.

Many of the studies of ERPs and their components utilize measures of component amplitude, latency, and distribution as indices of the neurophysiological strategy in use during information processing. These measures are said to index the nature of the programs and subroutines being developed and used by the brain. The description of neurophysiological strategies (i.e., a model of the brain's software) does not usually include statements about anatomical structure or biochemical processes. The approach of cognitive psychophysiology has been to use the EEG to model the strategies used in processing complex information. Those using MEG to study information processing have started from a very different approach—to locate sources in the brain. The assumption adopted by those using MEG is that an understanding of how the brain processes information must begin with a knowledge of the anatomical systems involved and the spatial and temporal patterning of excitation and inhibition within that system.

It is frequently not clear what is meant by a source or what is meant by localization or location of a source. Localization and location have two different implications. Localization usually implies that the source is localized within some small site or space, as distinct from distributed. The term "location" does not prejudge the data; the source could be local or distributed. A distributed system may be usefully described as a constellation of current dipoles distributed in some three-dimensional space, each having different vectors and each resulting from convergence and divergence of excitatory and inhibitory processes. There is not a single cell or collection of cells in the brain that is not subject to spatial summation. The place of convergence, however, is not necessarily a source.

Nunez (1988) has argued that there are four types of sources that must be distinguished: (1) localized and stationary, (2) localized and nonstationary, (3) distributed and stationary, and (4) distributed and nonstationary. By localized Nunez means that the dipole can be modeled as a single dipole. When a dipole is stationary it does not change its location, given that it is measured at the same time relative to successive (stimulus) inputs. Most approaches to the analysis of sources in MEG (and EEG) assume a stationary source for a component of evoked potentials. However, local (single) sources may be in different locations during the interval of a component, i.e., the (single) source(s) for that component may actually be moving systematically with respect to the interval of time over which the component is defined. Distributed sources are those located in different places and are simultaneously active. As (Nunez, 1988; Weinberg et al., 1988) pointed out, components of ERPs are almost certainly attributable to distributed, nonstationary sources. If a component of an ERP is one that develops over a period of as much as 50 msec, it is quite likely that, for an interval of that time, multiple sources are involved that are spatially distributed, simultaneously active, and changing over the interval of the defined component, i.e., they are nonstationary. The same analysis can be applied to the spontaneous EEG in which the analysis is of some epoch in time. The use of assumptions about single, nonstationary (equivalent) dipoles can be significantly misleading about the location of sources. Most investigators utilize methods of estimating equivalent sources that take into account known neurophysiology. However, if there is no known neurophysiology the inverse solution could be entirely wrong. For example, an inverse solution that places the equivalent dipole for auditory stimulation in the center of the head, because the computed dipole is a vector sum of single symmetrical dipoles in each of the two hemispheres, would be misleading. It is for this reason that studies of visual, auditory, and somatic sensory systems have constituted a large body of MEG research. There are relatively well-known neuroanatomical and neurophysiological data associated with these sensory functions. Therefore, the localization sources associated with components of sensory evoked potentials can be validated with known neuroanatomical localization of sensory function. When MEG (or EEG) source estimates are validated against known neuroanatomy, no additional information is gained; however, confidence in the method is enhanced in respect to solutions that apply to anatomical systems that were not previously well defined, particularly functional systems that mediate higher cognitive processes.

If there is one fact that is almost certain, it is that widely distributed systems are involved in complex information processing. If ERPs or components of them are known to index complex memory processes, and particularly if they are modality nonspecific, it is very likely that a complex neurophysiological system underlies the generation of that ERP component.

25.3 Spontaneous MEG

Spontanoues alpha activity (8–12 Hz) was first measured by Cohen (1968), Reite et al. (1976), and others. Carelli et al. (1983) compared average spectral density of spontaneous EEG and MEG. They observed that strongest MEG amplitudes of alpha activity were measured over parieto-occipital areas and found no significant interhemispheric differences. Estimates of source location for spontaneous activity are difficult because signal averaging is required to increase the ratio of signal to noise. This is because single-channel SQUID systems require successive measurements. The background activity of MEG and instrument noise relative to the signal is sufficiently high to preclude the use of a single sample even if multichannel systems are available. Spectral analysis at each location does not give information about polarity. Chapman et al. (1984) used the basic idea that the covariance is positive when MEG and EEG are in phase, and it is negative when they are of opposite phase. Amplitude of the EEG and MEG is related to amplitude of their covariance. They computed relative covariance by dividing the covariance by the variance of the EEG. Isocontour maps of relative covariance were used to estimate the location of the sources of alpha activity. The sources identified were bilateral, near the midline, and in the vicinity of the calcarine fissure at a depth of 4–6 cm from the scalp. The same procedure was not applied to other spontaneous bands (beta, theta, or delta). A more recent paper by Vvedensky et al. (1985) reported the use of a four-channel magnetometer in a shielded room to study alpha activity. They observed easily identified trains of MEG alpha during which the fields came out of one side of the head and entered the other, inverting direction twice during every cycle. They estimated multiple sources, active one at a time, extending approximately 3 cm from the midline of the occipital cortex in either direction and at a depth of approximately 4 cm.

Many years ago Grey Walter (1953) expressed the idea that the spontaneous

EEG recorded from macroelectrodes inside or outside the brain resulted from the electrotonic and functional summation of "brain waves" different in frequency, amplitude, and phase. These frequency components of the spontaneous EEG came from different systems in the brain and reflected the function of systems. If one believes this hypothesis, then those frequencies can be represented by the components of a Fourier analysis of the spontaneous electrical activity. A legitimate question can be posed in respect to the source systems responsible for those Fourier components. For example, in the case of alpha activity, Fourier analysis of the spontaneous EEG can be used to identify an 8-Hz spectral "component" and an inverse solution applied to the scalp distribution of that component. In order to do this, the phase of the components computed from EEG at different scalp locations must be preserved. Following this logic one can then attempt an inverse solution to locate the sources of this component, in this case the alpha wave. This technique has been used by Brickett et al. (1986) to study the sources of alpha activity and also the sources of evoked steady state potentials. The sources of alpha activity are nicely located in bilateral striate cortex using this technique, a finding that fits well with other studies that conclude that alpha activity originates from striate cortex.

The same logic could be used to study other components of the MEG. One could, if one were bold enough, extend the logic further to ask whether there are sources associated with more abstracted mathematical derivations. For example, a multivariate analysis of MEG is possible to extract the factors that account for a predetermined percentage of the observed variance at each location for the purpose of a principal components analysis, and then the question can be posed as to whether there are different sources associated with those factors. We have tried this with data collected preceding speech production and have localized the principal components in the left temporal lobe.

25.4 Sensory Function

Retinotopic organization is mapped on the striate cortex. Brenner et al. (1981) estimated sources associated with full field and right and left hemifield stimulation. The stimulus was a vertical grating, the luminance of which varied sinusoidally across an oscilloscope with a spatial frequency of five cycles per degree. They estimated equivalent dipoles contralateral to hemifield stimulation that were oriented horizontally and symmetrically, in agreement with the cruciform model. The equivalent dipole for full-field stimulation was a linear sum of those estimated from the hemifield stimulation. Okada et al. (1982a), using essentially the same technique, showed that the phase lag between VEP and VEF was a function of spatial frequency of the stimulus and argued that the VEP was attributable in part to a different source than that measured by the VEF. Other studies of sinusoidally modulated gratings report sources in the Rolandic fissure near the sites from which efferent fibers influence eye movements (Lounasmaa et al,. 1985). Weinberg et al. (1985a, 1987a) recorded EF resulting from stereopsis of a random-dot binocular display. Their data suggested that the source of an early component of the EF associated with fusion was located in the striate area and the source of a later component associated with fusion in the right temporal lobe (Figure 25.6).

EEG auditory evoked potentials have been used extensively to distinguish between sources in brainstem and cortex. However, MEG studies have been, almost exclusively, concerned with cortical sources, with the exception of Cohen and Cuffin (1985) who reported that attempts to record 6–10 msec brainstem EF were unsuccessful. They explained the failure to observe brainstem MEG as attributable to the possible combination of excessive depth with respect to the location of the sensing coil and sources that were radial dipoles. It is, however, unlikely that the assumptions of a spherical model apply to sources as deep as cochlear nucleus. The question of whether MEG is capable of locating brainstem sources is still unresolved. Reite et al. (1978) first detected the auditory evoked field using brief clicks. They reported fields that were maximal in central and parietal regions, but did not sample the distribution of MEG fields sufficiently to establish reversals. Hari et al. (1980, 1984a) recorded auditory evoked fields elicited by a continuous tone of 800 msec duration. She distinguished between the sources of N100 (msec) and the slow field that immediately preceded offset of the stimulus. Both these sources were estimated to be in the superior surface of the temporal lobe in the primary auditory cortex. She postulated that the sources were sheets of dipoles consistent with the assumption that they were produced by apical depolarization of pyramidal cells with extracellular sinks in the superficial cortex (Figure 25.7).

Romani et al. (1982a) reported responses to single frequencies that were mod-

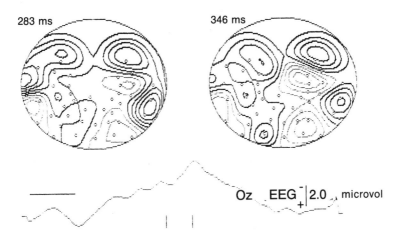

FIGURE 25.6A. Stereopsis study. MEG maps for two 21-msec time intervals centered at the time points indicated. The head is represented in equidistant projection with the vertex at center and the nose at the top. Recording positions are shown as small circles. Each contour line represents approximately 10 femtotesla; thicker lines indicate emergent flux. Grand average Oz EEG shown at bottom was recorded over a 766-msec epoch. Horizontal bar indicates stimulus duration. *Figure continued.*

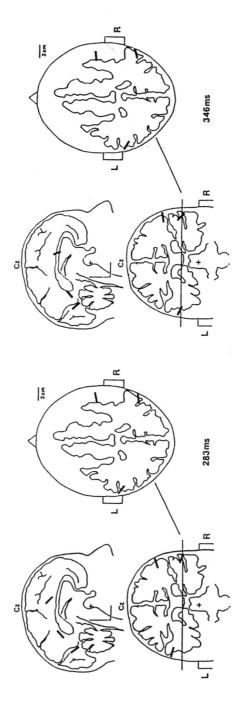

FIGURE 25.6B. Saguittal, horizontal, and coronal sections of the head showing the locations of the three equivalent dipoles for the two time intervals investigated.

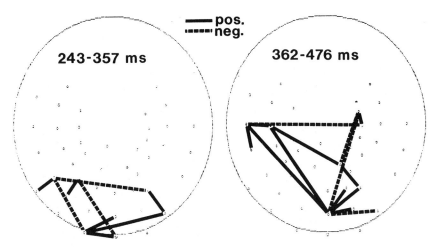

FIGURE 25.6C. Equidistant projections of the head with lines connecting those positions for which the recorded MEG showed the highest covariances over the time intervals indicated.

ulated at 32 Hz. They used steady-state averaging techniques in which the samples were time locked to the modulation and recorded averaged fields resulting from 200-. 600-, 2000-, and 5000-Hz stimulation. The sources estimated were located in the primary auditory cortex and increased in depth from 2.2 to 3.2 cm below the scalp as the frequency of the tone increased from 100 to 5000 Hz. Galambos, Makeig and Talamchoff (1981) described what they called the 40-Hz response. This steady-state response was a sinusoidal EEG following the response to re-petitive auditory stimulation that was of maximal amplitude at rates between 35 to 45 Hz. They suggested that the response was a superimposition of thalamic midlatency responses. Spydell et al. (1985) examined the relationship between the phase of the auditory brainstem component and the 40-Hz component of the averaged waveform, predicting that they should be different in normal controls and patients with brainstem lesions. They came to the conclusion that there were independent generators. They also observed that, in a group of patients with unilateral temporal lobe lesions, the 40-Hz response appeared unimpaired and concluded that temporal cortex was not involved in the response. However, Weinberg et al. (1987c) reported a study in which both EEG and MEG data were obtained from two healthy right-handed subjects with normal hearing during 40-Hz auditory stimulation. EEG recordings were taken from a vertex electrode; MEG recordings were obtained sequentially from a wide distribution. They sug-gested that a "source-system" was active during 40-Hz auditory stimulation that included bilateral temporal cortex. In accordance with Borda (1983), when the brain is in a steady-state as the result of being driven by repetitive auditory stimulation, it is likely that resonance within a limited portion of the auditory pathways could account for some of the enhancement at 40-Hz stimulus rates. We have proposed that a pathway consisting of reciprocal connections between audi-

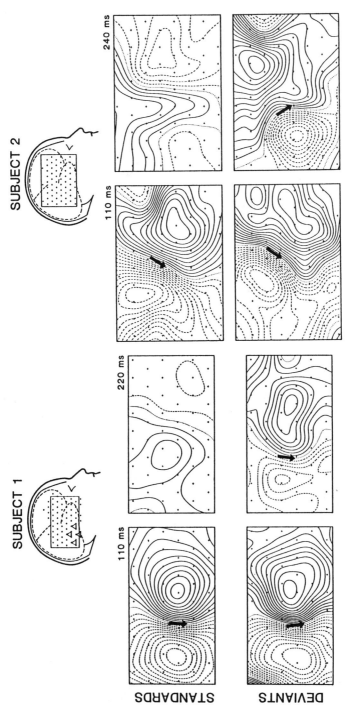

FIGURE 25.7. Auditory fields. Isocontour maps of responses to the standards and deviants at the peak latencies of N100m and MMNm. The maps were calculated from the integrated amplitudes (peak latency +/– 5 msec) using a weighted least squares approximation and bicubic spline interpolation. The isofield lines are separated by 35 fT. The continuous lines indicate flux out of the skull and the dotted lines flux into the skull. The interpoint distance in the measurement grid is 16 mm, and the crosses indicate the locations of the pickup coils (three of them recorded simultaneously; some positions of the probe are indicated by triangles). The arrows illustrate the approximate locations of the equivalent current dipoles.

tory cortex, medial geniculate nucleus, and inferior colliculus could be the basis for a resonance resulting in an enhancement of a response to 40-Hz auditory stimulation. This interpretation is supported by what is known about the organization of the thalamocortical auditory system in the cat (Anderson et al., 1980; Imig and Morel, 1983).

In a more recent study, steady-state stimulation has been used to study the processing of information delivered simultaneously and in phase to two modalities Weinberg et al. (1987c). Three conditions of steady-state stimulation were examined: tactile (T), auditory (A),and simultaneous (B). Fourier analysis was used to compute the amplitude and phase of the 40-Hz component from the MEG average at each recording position. These values were then plotted as vectors in polar coordinates. Because a phase difference of 180° is equivalent to a polarity reversal of the same signal, the average phase was calculated as that angle for which the root mean square amplitude of all vectors projected onto that angle was maximal. The resulting amplitude values were used to produce isocontour maps of the field over the surface of the head. The interpolated values for these maps were calculated as the weighted sum of all recording positions within a specified search radius, the weight for each position being proportional to the reciprocal of its distance. A least squares method similar to the procedure described above was used for estimating one or two dipole fits to the data as described by Harrop et al. (1986, 1987) for approximating the location of equivalent current dipole sources. The method we used also takes into account the number, size, spatial separation, and orientation of the gradiometer coils and utilizes multiple radii defined as the distance to each recording position from a known origin near the center of the head. The vertex EEG response to 30 and 40 Hz showed a clear following response, which was consistent with what would be expected from previous studies. The EEG response was consistent in all subjects for both unilateral and bilateral auditory stimulation. MEG isofield plots of the 40-Hz component of MEG responses for both binaural and monaural auditory stimulation bilateral resulted in dipolar fields observed over the temporoparietal regions. Two-dipole fits of the data estimate sources to be 2.0 cm posterior, 4.5 cm lateral, and 4.4 cm superior to the origin, placing bilateral sources roughly in the vicinity of either primary auditory cortex in superior temporal gyrus (Figures 25.8–25.11). The orientation of these dipoles is approximately vertical and directed toward midline, indicating that such sources may comprise equivalent sources oriented vertically in the superior plane of the temporal lobe. The vertex EEG from tactile stimuli for 40 Hz follows the stimulus, but its amplitude and phase are highly sensitive to variations in the area and position of the forefinger stimulated, and it is generally lower in amplitude than the auditory response. The MEG had a contralateral dipolar configuration. Using the contralateral data, dipole location estimates accounted for 80% of the variance of the observed fields. The position for this dipole was estimated as 3 cm posterior, 3 cm lateral (left hemisphere), and 8 cm above the origin, placing the equivalent dipole source in the area of the finger locations of somatosensory cortex (postcentral gyrus).

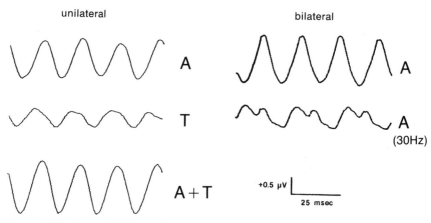

FIGURE 25.8. Average EEG responses to (A) bilateral sinusoidal tone bursts presented at rates of 30 and 40 Hz and unilateral tone bursts (right ear) at 40 Hz; (T) vibrotactile stimulation at 40 Hz (right index finger); and (A+T) combined (in phase) stimulation of right ear and right index finger at 40 Hz. Cz electrode referenced to linked mastoids.

Simultaneous, in phase, auditory, and tactile stimulation resulted in a EEG at the vertex that showed a following response that was similar in waveform and period to the tactile and auditory stimulation alone. The observed MEG data are shown (Figures 25.9, 25.10, 25.11). A complex distribution of fields with patterns of maxima and minima was observed in both hemispheres. The somewhat unexpected result was that the arithmetic sum of the tactile and auditory fields is very close to the empirically observed field resulting from stimulation.

The obvious interpretation of this finding is that the result of simultaneous stimulation is the sum of what would be expected from the stimulation of each of the modalities separately, as if the brain were simply doing a linear sum during simultaneous processing. Taken as a whole, the observed bilateral fields were what would be expected if the tactile and auditory stimulation were producing fields that summed on the left side, but not on the right. Because tactile stimulation does not result in ipsilateral fields, only the right-hemisphere auditory fields are available to sum in the right hemisphere. If it is the case that the combined fields for dual-modality stimulation are the sum of the fields for single-modality stimulation, it suggests that the 40-Hz response is confined to sensory systems related to the modality of stimulation; if there were an interaction of these modalities in the 40-Hz response, the combined fields would undoubtedly not be a simple linear sum of the two. From a more general perspective, the data suggested that the two modalities may be driven independently by modality-specific stimulation and support the interpretation that the 40-Hz response reflects a resonance in the sensory system stimulated. Plots of the topographical distribution of phase angles for an extracted spectral component may be used to support an assumption that more than one source is involved. Because the brain tissue, skin, and scalp are transparent to magnetic fields, there should be no phase difference over the head

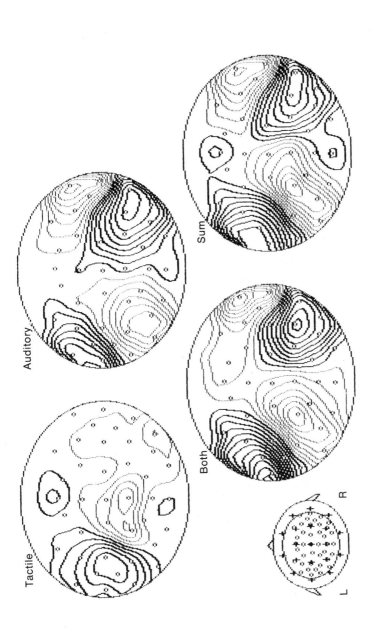

FIGURE 25.9. Isofield contour maps of the magnetic 40-Hz amplitude for three stimulus conditions and the sum of the tactile and auditory responses (subject U.R.). The maps are shown as equidistant projections of the head surface with vertex as the center and recording positions shown as small circles. The map border lies approximately at the level of T3/T4 indicated by the innermost circle on the inset diagram showing 10/20 system locations. Each contour level corresponds to 2.5 femtotesla, and light and dark lines indicate fields of opposite direction.

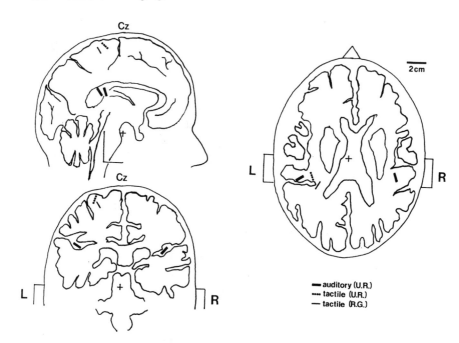

FIGURE 25.10. Three-dimensional representations of the location of equivalent dipole source estimates fitted to the observed frequency components using a least squares fitting routine. Two-dipole estimates were fitted for the auditory condition *(dark bars)* in homologous regions of superior temporal lobe, accounting for 86.2% of the variance in the observed values. Single dipoles were fitted to the tactile data for two subjects both in left postcentral gyrus and account for 70% (U.R.) and 81% (R.G.) of the variance. Drawings are taken from representative brain sections, and a small cross indicates the relative position of the origin of the coordinate system used in dipole localization.

if there is a single source; if only a single dipole were active, all magnetic data would be in phase, and there would be no topographical distribution. The systematic change in phase from posterior to anterior locations observed after auditory, tactile, and simultaneous stimulation is consistent with the interpretation that two or more dipoles summate to produce the sinusoidal 40-Hz field normally observed, even though the fields may appear dipolar. Therefore, although the estimated locations of equivalent dipole sources are consistent with the appropriate functional cortical sites, it does not mean that the source of the 40-Hz responses is entirely cortical. Because of known neuroanatomy of the auditory system a thalamocortical system must be involved; the reason we estimate sources in cortex is that they are closest to the gradiometer.

Other MEG studies have utilized transient and steady-state somatosensory stimulation because there is reasonable agreement about the anatomy of afferent somatosensory projections to the cortex. Brenner et al. (1978), Kaufman et al. (1981), and Okada et al. (1981) did the early experiments in which they showed that steady-state stimulation of the median nerve produced magnetic fields that suggested dipolar sources in the contralateral somatosensory cortex. Hari (1985)

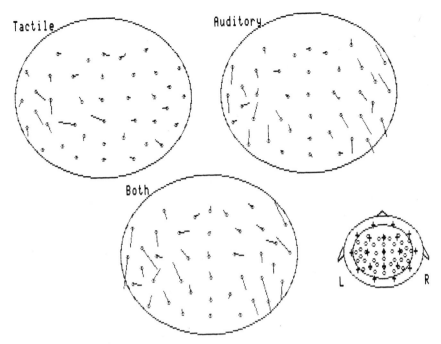

FIGURE 25.11. Phase plots of the 40-Hz frequency components used to produce the isocontour maps and dipole estimates for three conditions in subject U.R. Direction of lines indicates the relative phase of the 40-Hz component at each recording location. Length indicates the magnitude of the 40-Hz component, the largest magnitude (left-most position for condition Both) corresponding to approximately 36 fT.

studied both median nerve and peroneal nerve stimulation in the same experiment. Two different types of magnetic responses were reported, one type (SI) with sources near the primary projection areas of somatosensory cortex (30–80 msec and 150–180 msec latencies) and a second type (SII) near the Sylvian fissure (90–125 msec latencies). The SI response was lateralized with respect to contralateral stimulation. The SII response occurred after both ipsilateral and contralateral stimulation. Previous studies of these two cortical sites (somatosensory cortex and Sylvian fissure) suggest that they are not hierarchically organized anatomically or functionally, but that the Sylvian cortex functions to integrate somatosensory input with input from other sensory modalities and with memories. Magnetic field distributions differentiated between equivalent dipoles active in SI and SII, whereas electrical field distributions did not (Hari, 1985).

25.5. Motor Function

The search for sources of motor function using MEG has been primarily confined to study of the Bereitschaftspoiential (BP) of Kornhuber and Deecke (1965). The BP is a slow negative electrical cerebral potential that occurs preceding a voluntary movement and has been shown to be sensitive to both cortical and brainstem lesions of the motor system. Deecke et. al (1982) recorded magnetic and electric fields preceding finger movement and localized the source of the precentral gyrus in the contralateral area of the motor homunculus consistent with known anatomy (Figure 25.12). In electrical recordings the BP is not lateralized to contralateral hemisphere, but is widely distributed over both hemispheres. The MEG appears to allow much sharper localization to the contralateral hemisphere and is therefore more consistent with known anatomy and physiology. Okada et al. (1982b) also localized two active sources preceding finger movements to the precentral gyrus. These sources were anatomically separated for 75 msec and 110 msec components of the BP (preceding movement) and Hari et al. (1983) have also reported MEG sources in the motor cortex preceding foot movements. A good deal of evidence has been accumulating implicating regions of the mesial frontal lobes (SMA) in the process of preparation for complex, patterned movements. SMA are thought to participate with septohippocampal pathways in the programming of complex outputs, which consist partly of the development of intentional states that send instructions to primary motor cortex. Individuals with lesions of SMA often show a loss of spontaneity of movement and speech and in some cases loss of "voluntary" control of motor movements. Deecke et al. (1985) reported evidence for the location of systems associated with the control and programming of complex patterns of finger movements in the supplementary motor area of the mesial frontal cortex. Cheyne (1987) recorded the Bereitschaftsmagnetfeld preceding complex movements and showed some evidence for sources in the supplementary motor area of cortex (Figure 25.13).

25.6 Perception and Cognition

It is in the study of cognitive function that the concept of an "equivalent dipole" has met its most serious resistance. The issue is whether the strategy of defining a single (equivalent) dipole as a source of components of electrical potential that are known to index information processing is useful or misleading. Components of ERPs that are not modality specific and that are responsive to a variety of complex cognitive conditions are probably not localized within small regions of the brain. As a first approximation to a description of sources, however, estimates of equivalent single dipole sources have been applied to P300, N100, and CNV data. P300 is an electrical potential elicited by stimuli that require discrimination and judgment about the probability of their occurrence. One of the first reports of P300 observed by MEG was by Weinberg et al. (1985b) who reported a similar morphology of MEG and EEG responses; however, the MEG response, unlike the EEG response, showed reversed polarity over posterior hemispheres, suggesting deep midline sources, although no isocontour mapping was done. Okada et al. (1983) reported MEG data that were interpreted as evidence for the source of P300 being located in the hippocampal formation. There have, however, been reports that P300 can be observed in humans with bilateral temporal lobectomy (Knight,

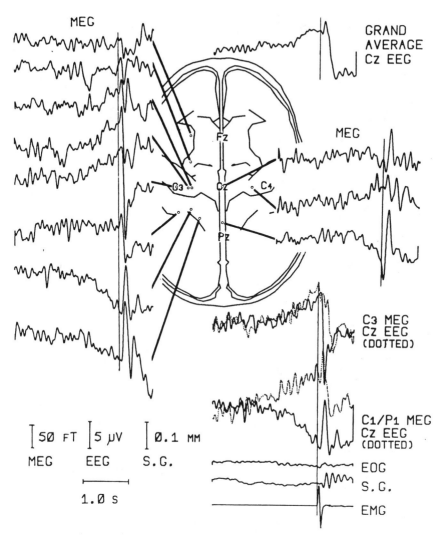

FIGURE 25.12. MEG BP. Averaged MEG recordings over different scalp locations as indicated by lines for 80 self-paced right finger flexions in one subject. EEG grand average from Cz is shown at upper right. EMG onset was used as trigger (shown on trace at bottom right). EOG and head movement (S.G.) were also monitored for artifact rejection of trials.

1986) and that it can be seen by intracranial electrodes widely distributed in the frontal and limbic brain of monkeys.

Weinberg et al. (1987e) conducted a study in which simultaneous EEG and MEG responses were recorded from subjects who were engaged in a task involving perceptual discriminations of moving visual stimuli. In this study in which the subjects detected and covertly counted rarely ($p=0.3$) occurring moving target shapes and ignored frequently ($p=0.7$) occurring moving nontarget shapes, a

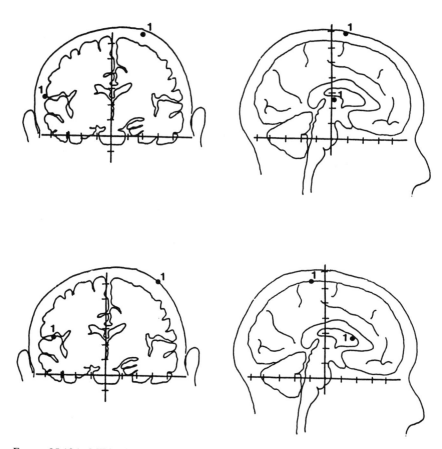

FIGURE 25.13A. MEG of SMA. *Top,* Projections of estimated dipole sources for readiness fields. Early component (pattern condition). *Bottom,* Projections of estimated dipole sources for readiness fields. "Motor" component (pattern condition).

600-msec ERP component that was symmetrically distributed (left-right) was identified as being related to the probability of appearance, detection, and counting of target stimuli. The simultaneously recorded MEG responses also showed an increase in amplitude at this same latency (600 msec) and revealed several reversals in magnetic sense over the scalp suggesting underlying dipolar sources (Figure 25.14). Unlike the EEG data, the scalp distribution of ERFs at 600 msec was quite left-right asymmetrical for target stimuli, whereas it was left-right symmetrical for nontarget stimuli. By applying a current dipole fitting procedure developed by Harrop et al. (1986, 1987), simultaneous current dipoles were estimated for the scalp distributions of magnetic flux 600 msec following the appearance of target and nontarget stimuli (Figure 25.15). The resulting source estimates were interpreted with a human neuroanatomy atlas deArmand et al. (1976) and were found to be located in brain regions that were consistent with

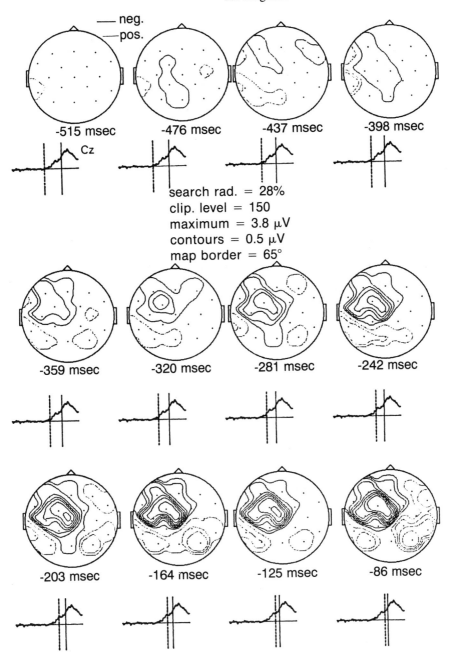

FIGURE 25.13B1. Isocontour maps of EEG (Laplacean). Pattern condition (Same subject as in Figure 25.13A). *Figure continued.*

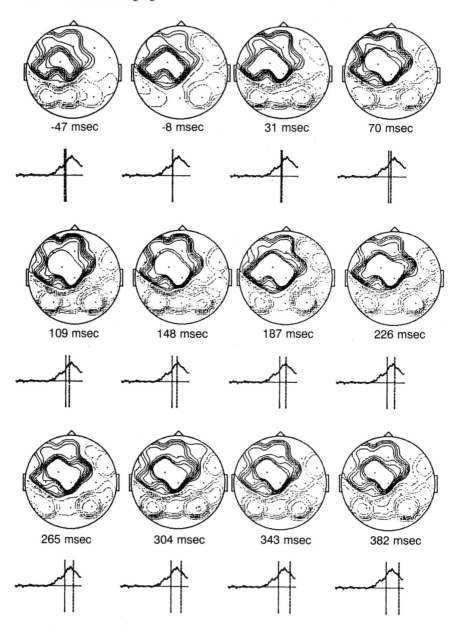

F<small>IGURE</small> 25.13B2. Legend same as for Figure 25.13B1.

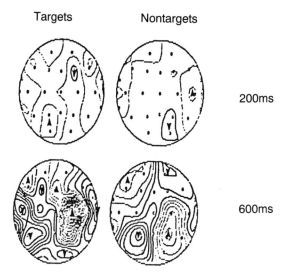

FIGURE 25.14. Isocontour maps. ERFs from subject TR at 200 and 600 msec following target and nontarget stimulus appearance. Lighter lines *(upward arrows)* indicate emerging flux, and darker lines *(darker arrows)* indicate ingoing flux. Each contour represents 15 fT. The subject's nose is located at the top of each map and Cz at its center. Recording positions are indicated by small circles on each map.

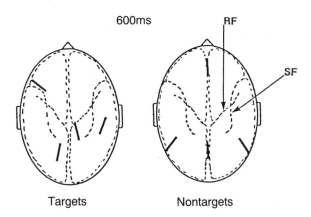

FIGURE 25.15. Diagram of the head. Subject TR (viewed from above) illustrating estimated positions of current dipole sources *(dark bars)* 600 msec after the appearance of target and nontarget stimuli. The locations of the Rolandic fissure (RF) and Sylvian fissure (SF) are indicated for reference. The 3-dimensional dipole coordinates were initially calculated with respect to an origin determined by four external landmarks on the head (inion, nasion, left, and right preauricular points). Using an atlas of 1:1 scale photos of brain slices (DeArmond et al. 1976), the locations and orientations of the dipoles were identified on appropriate brain sections in order to resolve their approximate locations with respect to cerebral anatomy.

current neuropsychological theory regarding cortical contributions to the performance of such a visual perceptual task.

These findings are of interest for two reasons. First, they illustrate an instance where the EEG and MEG results are apparently consistent in identifying the same component, i.e., a 600-msec amplitude peak related to task contingencies. Second, they illustrate an instance where additional information is gained from the MEG in that the symmetrical EEG scalp distribution suggests one central source associated with this component, whereas the MEG scalp distribution suggests that the 600-sec component is associated with multiple simultaneous sources that are asymmetrically distributed in the brain.

Another electrical potential that is known to index attention and information processing is the contingent negative variation, a slow increase in cortical negativity occurring between two stimuli, one of which is a warning stimulus and the second of which requires a decision or response. Some of the issues as yet unresolved include the nature of the distribution of sources underlying the shift and whether the shift is composed of overlapping components, each of which has different sources. It is clear from all MEG studies of the CNV that the maps of field distributions are not dipolar and designation of equivalent sources should be done with caution. Weinberg et al. (1983) reported studies in which the CNV was studied preceding the same movement that produced a BP. They reported two components of the CNV: the source of the early component was widely distributed and associated with attention, and a later component, the source of which was localized to motor cortex, was associated with response preparation. Fiumara et al. (1985), however, reported data suggesting that the source of the later component of the CNV was in frontal lobes. They comment on the complexity of the distributions of the fields and point out that the subject requires further study. Several laboratories are now in the process of investigating other components of ERPs that are associated with selective attention and information processing.

Weinberg et al. (1985a) recorded EFs resulting from stereopsis of a random-dot binocular display. The data suggested that the source of an early component of the EF associated with fusion was located in striate area, and the source of a later component associated with perception was in the right temporal lobe.

25.6.1 Pathological Function

The primary use of MEG in the study of brain pathology has been to investigate sources of focal epilepsy. Barth et. al. (1982, 1984) constitute a team who have studied the use of MEG as a tool for the localization of depth, orientation, and polarity of currents underlying paroxysmal discharges (Figure 25.16). Their data suggest that multiple sources having a defined temporal and spatial organization are responsible for different components of interictal (between seizure) electrical discharges. This group of researchers has also studied penicillin-induced focal epilepsy in rats and reported slow field shifts associated with the development of seizures Barth et al. (1984a). This finding has been corroborated in human epilepsy. Vieth and colleagues (1988) have reported slow shifts associated with

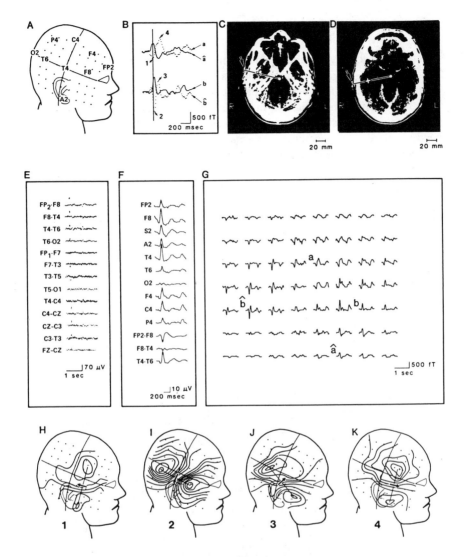

FIGURE 25.16. A, Rectangular MEG measurement. Matrix (2-cm spacing) oriented along the temporal axis of subject 1. All EEG electrodes are marked except the sphenoidal electrode (S2). B, Enlargements of averaged magnetic spikes from separate complementary regions (a and b) of the scalp show four components; the opposing polarities reflect the magnetic field simultaneously emerging from (upward) and re-entering (downward) the cranium. C and D, Computed tomography scan sections at the levels of sources a and b, respectively, show the depth of the source (*cross*) located along a line connecting the surface location of the source marked with a washer (*arrow*) to the center of the cranium. E, Spike in the raw EEG recorded from both hemispheres. F, Averaged EEG spike from three bipolar channels (*lower three traces*) and ten electrodes referenced to a noncephalic site (*upper ten traces*). G, Averaged magnetic spikes recorded from the MEG matrix exhibit two distinct complementary regions of differing morphology marked a and b (fT, femtotesla). H to K, Isocontour maps displaying the magnetic fields for each of the four temporal components of the magnetic spike complex. Reprinted with permission from Barth et al. (1984b).

ictal (seizure) activity in human patients and, perhaps even more interesting, slow shifts accompanying interictal discharges in patients with refractory epilepsy. Such interictal shifts are not seen in patients whose seizures are pharmacologically controlled, nor are they detected in normal resting subjects (Vieth et al., 1988). Slow electrical shifts during seizures, sometimes called DC shifts, have always been of interest, but technical problems related to electrode-scalp contact have limited the development of this line of inquiry. MEG recording, which does not require contact of the recording device with the scalp, provides a means for collecting these data without the difficulties inherent in electrical recordings.

Ricci et al. (1985) have also studied MEG activity associated with epileptiform discharges. They have reported that the equivalent source identified in one patient was verified by CT scan and surgery in that the localization predicted with MEG data was consistent with the location of a lesion in the structural image. Confirmation of the accuracy of MEG localization in epilepsy has come from our own laboratory (Crisp, 1986; Weinberg et al., 1987d). By plotting the MEG-estimated dipole position on the appropriate MRI "slice," we have shown that the location predicted by MEG corresponds to lesions shown on MR scans in two patients (Figure 25.17). Several other laboratories are now studying epilepsy in an attempt to identify the distribution of sources associated with both interictal and ictal states

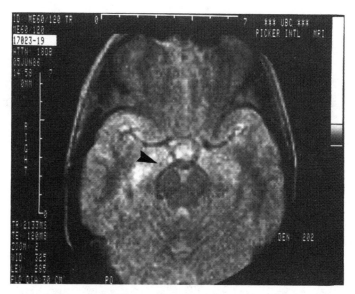

FIGURE 25.17. Single dipole fit of MEG data associated with averaged EEG interictal spike discharges. The predicted dipole was plotted on the patient's magnetic resonance image at a level consistent with the level estimated by the dipole localization procedure. The arrow indicates the predicted position and orientation of the dipole source, with source direction indicated by the arrow-head direction. The dipole fit is seen to be adjacent to the structural lesion detected by the MRI. Reprinted with permission from Weinberg et al., (1987): The combination of MEG and MRI in the estimation of sources associated with interictal discharges. In: *Functional localization: A challenge for biomagnetism*, Erne, S.N., Romani, G.L. (eds.). Singapore: World Scientific.

(Angianakis and Anninos, 1988; Cohen et al., 1988). If MEG data can be combined successfully with EEG and computed tomography (CT) or magnetic resonance imaging (MRI), as many laboratories are now attempting, the MEG technique could make a significant contribution to the noninvasive identification of pathology underlying epilepsy in individual patients.

A further contribution of MEG to the investigation of epilepsy is the observation, first reported in some of the earlier epilepsy studies, that it can detect abnormal activity not present in the scalp electrical recording (Modena et al., 1982; Ricci, 1983). Conversely, it has also been reported that, in one patient with bilateral epileptiform EEG discharges, organized MEG fields were only present over one hemisphere (Crisp, 1986). In this case, MEG fields associated with interictal discharges were clearly of a "dipole configuration" over the right hemisphere (the hemisphere in which a structural lesion was imaged on MRI), but when the left hemisphere discharges were investigated, no distinct pattern of magnetic activity emerged. This finding raises interesting questions about whether the left hemisphere electrical discharges were arising from a radial dipole that would not be computed with the MEG (using assumptions of sphericity of the head) or whether these signals were purely volume conducted from the other hemisphere.

MEG as an imaging device has the theoretical advantage of being noninvasive with both good temporal and spatial resolution. These advantages are not now combined in any other single imaging device. However, the combination of data from existing imaging devices into a single dynamic image would be an important breakthrough in the understanding of both normal and pathological brain function. The concept of a dynamic image is important. The future will see the development of methods of estimating distributed sources through a combination of mathematics and known anatomy and physiology. When that happens, a dynamic image of changing current patterns in the brain associated with memory and complex information processing will be possible. Then, science will have documented Sir Charles Sherrington's vision of the brain as "an enchanted loom where millions of flashing shuttles weave a dissolving pattern, always a meaningful pattern but never an abiding one."

References

Anderson, R.A., Snyder, R.L., Merzenich, M.M. (1980): The topographic organization of cortico-collicular projection from physiologically identified loci in the AI, AII, and anterior auditory cortical fields of the cat. *J. Comp. Neurol.* 191, 479–494

Angianakis, G., Anninos, P.A. (1988): Localization of epileptiform foci by means of MEG measurements. *Int. J. Neurosci.* 38, 141–149

Barth, D.S., Sutherling, W., Engel, J. Jr., Beatty, J. (1982): Neuromagnetic localization of epileptiform spike activity in the human brain. *Science* 218, 891–894

Barth, D.S., Sutherling, W., Beatty, J. (1984a): Fast and slow magnetic phenomena in focal epileptic seizures. *Science* 226, 855–857

Barth, D.S., Sutherling, W., Engel, J. Jr., Beatty, J. (1984b): Neuromagnetic evidence of spatially distributed sources underlying epileptiform spikes in the human brain. *Science* 223, 293–296

Borda, R.P. (1983): The 40/sec middle latency response in Alzheimer's disease, parkinson's disease, and age-matched controls. Unpublished doctoral dissertation, University of Texas

Brenner, D. Kaufman, L., Williamson, S.J. (1978): Somatically evoked fields of the human brain. *Science* 199, 81–83

Brenner, D., Okada, Y., Maclin E., Williamson, S.J., Kaufman, L. (1981): Evoked magnetic fields reveal different visual areas in the human cortex. In: *Biomagnetism.* Erne, S.N., Hahlbohm, H.-D., Lubbig, H. (eds.). Berlin: Walter de Gruyter, pp. 431–444

Brickett, P., Robertson, A., Crisp, D., Weinberg, H. (1986): Comparison of the magnetic fields related to alpha activity and visual evoked responses. EPIC VIII, Stanford, California

Carelli, P., Modena, I., Ricci, G.B., Romani, G.L. (1983): Magnetoencephalography. In: *Biomagnetism: An interdisciplinary approach.* Williamson, S.J., Romani, G.L., Kaufman, L., Modena, I. (eds.). New York: Plenum Press, pp. 469–482

Chapman, R.M., Ilmoniemi, R.J., Barbanera, S., Romani, G. L. (1984): Selective localization of alpha brain activity with neuromagnetic measurements. *Electroenceph. Clin. Neurophysiol.* 58, 569–572

Cheyne, D.O. (1987): Magnetic and electric field measurements of brain activity preceding voluntary movements: Implications for supplementary motor area function. Doctoral dissertation, Simon Fraser University

Cohen, D. (1968): Evidence of magnetic fields produced by alpha-rhythm currents. *Science* 161, 784–786

Cohen, D. (1972): Magnetoencephalography: Detection of the brain's electrical activity with a superconducting magnetometer. *Science* 175, 664–666

Cohen, D., Cuffin, B.N. (1985): Search for MEG signals due to auditory brainstem stimulation. In: *Biomagnetism: Applications and theory.* Weinberg, H., Stroink, G., Katila, T. (eds.). New York: Pergamon Press, pp. 316–320.

Cohen, D. Edelsack, E.A., Zimmerman, J. (1970): Magnetocardiograms taken inside a shielded room with a superconducting point-contact magnetometer. *Appl. Phys. Lett.* 16, 278–280

Cohen D., Cuffin, B.N., Kennedy, J.G., Lombroso, C.T., Gumnit, R.J., Schomer, D.L. (1988): Comparison of MEG versus EEG spike localization: Some results in a patient group with focal seizures. Poster presented at the Annual Meeting of the American Epilepsy Society, San Francisco

Cuffin, N.B. (1982): Effects of inhomogeneous regions on electric potentials and magnetic fields: Two special cases. *J. Appl. Phys.* 53, 9192–9197

Crisp, D. (1986): Neuromagnetic localization of current dipole sources in complex partial epilepsy. M.A. thesis, Simon Fraser University

DeArmond, S., Fusco, M., Dewey, M. (1976): *Structure of the human brain,* 2nd ed. New York: Oxford Press

Deecke, L., Boschert, J., Brickett, P., Weinberg, H. (1985): Magnetoencephalographic evidence for possible supplementary motor area participation in human voluntary movement. In: *Biomagnetism: Applications and theory.* Weinberg, H., Stroink, G., Katila, T. (eds.). New York: Pergamon Press, pp. 369–372

Deecke, L., Weinberg, H., Brickett, P. (1982): Magnetic fields of the human brain accompanying voluntary movement: Bereitschaftsmagnetfeld. *Exp. Brain Res.* 48, 144–148

Fiumara, R., Campitelli, F., Romani, G.L., Leoni, R. Caporali, M., Zanasi, M., Cappiello, A., Fioriti, G., Modena, I. (1985): Neuromagnetic study of endogenous fields related to the contingent negative variation. In: *Biomagnetism: Applications and theory.* Weinberg, H., Stroink, G., Katila, T. (eds.). New York: Pergamon Press, pp. 336–342

Galambos, R., Makeig, S., Talmachoff, P.J. (1981): A 40-Hz auditory potential recorded from the human scalp. *Proc. Nat. Acad. Sci.* 78, 2643–2647

Geselowitz, D.B. (1967): On bioelectric potentials in an inhomogeneous volume conductor. *J. Biophys.* 7, 1–11

Hari, R. (1985): Somatically evoked magnetic fields. *Med. Biol. Eng. Comput.* 22 (Supp. 1), 29–31

Hari, R. Aittoniemi, K., Jarvinen, M.L., Katila, T., Varpula, T. (1980): Auditory evoked transient and sustained magnetic fields of the human brain. Localization of neural generators. *Exp. Brain Res.* 40, 237–240

Hari, R. Antervo, A., Katila, T., Poutanen, T., Seppanen, M., Tuomisto, T., Varpula, T. (1983): Cerebral magnetic fields associated with voluntary limb movements in man. *Il Nuovo Cimento 2D* 1, 484–495

Hari, R. Hamalainen, M., Ilmoniemi, R. Kaukoranta, E., Reinikainen, K., Salminen, J., Alho, K., Naatanen, R., Sams, M. (1984a): Responses of the primary auditory cortex to pitch changes in a sequence of tone pips: Neuromagnetic recordings in man. *Neurosci. Lett.* 50, 127–132

Hari, R., Reinkainen, K. Kaukoranta, E., Hamalainen, M., Ilmoniemi, R., Penttinen, A., Salminen, J., Teszner, D. (1984b): Somatosensory evoked cerebral magnetic fields from SI and SII in man. *Electroenceph. Clin. Neurophysiol.* 57, 254–263

Harrop, R., Weinberg, H., Brickett, P., Dykstra, C., Robertson, A., Cheyne, D., Baff, M., Crisp, D. (1986): An inverse solution method for the simultaneous localization of two dipoles. Presented at the meeting of the Institute of Physics: Magnetism Subcommittee, Milton Keynes, England

Harrop, R., Weinberg, H., Brickett, P., Dykstra, C., Robertson, A., Cheyne, D.O., Baff, M., Crisp, D. (1987): The biomagnetic inverse problem: Some theoretical and practical considerations. *Phys. Med. Biol.* 32, 1545–1557

Imig, T.J., Morel, A. (1983): Organization of the thalamocortical auditory system in the cat. *Annu. Rev. Neurosci.* 6, 95–120

Kaufman, L., Okada, Y., Brenner, D., Williamson, S.J. (1981): On the relation between somatic evoked potentials and fields. *Int. J. Neurosci.* 15, 223–239

Knight, R.T. (1986): Neurophysiological mechanisms: Evidence from human lesion data. Presented at the Eighth International Conference on Event-Related Potentials of the Brain, California

Kornhuber, H.H., Deecke, L. (1965): Hirnpotenthalanderungen bei Willkürbewegungen und passiven Bewegungen des Menschen: Bereitschaftspotential und reafferente Potentiale. *Pflugers Arch.* 284, 1–17

Lounasmaa, O.V., Williamson, S.J., Kaufman, L., Tanenbaum, R. (1985): Visually evoked responses from non-occipital areas of the human cortex. In: *Biomagnetism: Applications and theory.* Weinberg, H., Stroink, G., Katila, T. (eds.). New York: Pergamon Press, pp. 348–353

Modena, I., Ricci, G.B., Barbanera, S., Leoni, R., Carelli, P. (1982): Biomagnetic measurements of spontaneous brain activity in epileptic patients. *Electroenceph. Clin. Neurophysiol.* 54, 622–628

Nunez, P.L. (1988): Methods to estimate spatial properties of dynamic cortical source activity. In: *Functional brain imaging.* Pfurtscheller, G., Lopes da Silva, F.H. (eds.). Toronto: Hans Huber Publishers, pp. 3–10

Okada, Y.C. (1985): Discrimination of localized and distributed current dipole sources and localized single and multiple sources. In: *Biomagnetism: Applications and theory.* Weinberg, H., Stroink, G., Katila, T. (eds.). New York: Pergamon Press, pp. 266–272

Okada, Y.C., Kaufman, L., Brenner, D., Williamson, S.J. (1981): Application of a SQUID to measurement of somatically evoked fields: Transient responses to electrical stimulation of the median nerve. In: *Biomagnetism.* Erne, S.N., Hahlmohm, H.-D., Lubbig, H. (eds.). Berlin: Walter de Gruyter, pp. 445–461

Okada, Y., Kaufman, L., Brenner, D., Williamson, S.J. (1982a): Modulation transfer functions of the human visual system revealed by magnetic field measurements. *Vision Res.* 22, 319–333

Okada, Y., Williamson, S.J., Kaufman, L. (1982b): Magnetic field of the human sensorimotor cortex. *Int. J. Neurosci.* 17, 33–38

Okada, Y.C., Kaufman, L., Williamson, S.J. (1983): The hippocampal formation as a source of the slow endogenous potentials. *Electroenceph. Clin. Neurophysiol.* 55, 417–426

Reite, M., Zimmerman, J.T., Edrich, J., Zimmerman, J.E. (1976): The human magnetoencephalograph: Some EEG and related correlations. *Electroenceph. Clin. Neurophysiol.* 40, 59–66

Reite, M., Edrich, J., Zimmerman, J.T., Zimmerman, J.E. (1978): Human magnetic auditory evoked fields. *Electroenceph. Clin. Neurophysiol.* 45, 114–117, 20

Ricci, G.B. (1983): Clinical magnetoencephalography. *Il Nuovo Cimento* 2, 517–537

Ricci, G.B., Buonomo, S., Peresson, M., Romani, G.L., Salustri, C., Modena, I. (1985): Multichannel neuromagnetic investigation of focal epilepsy. *Med. Biol. Comput.* 23 (suppl. 1), 42–44

Romani, G.L. Williamson, S.J. (eds.) (1983): Proceedings of the Fourth International Workshop on Biomagnetism. *Il Nuovo Cimento, 2D* 2, 123–664

Romani, G.L., Leoni, R. (1985): Localization of cerebral courses by neuromagnetic measurements. In: *Biomagnetism: Applications and theory.* Weinberg, H., Stroink, G., Katila, T. (eds.). New York: Pergamon Press, pp. 205–220

Romani, G.L., Williamson, S.J., Kaufman, L. (1982a): Tonotopic organization of the human auditory cortex. *Science* 216, 1339–1340

Romani, G.L., Williamson, S.J., Kaufman, L., Brenner, D. (1982b): Characterization of the human auditory cortex by the neuromagnetic method. *Exp. Brain Res.* 47, 381–393

Ryugo, D.K., Weinberger, N.M. (1976): Corticofugal modulation of the medial geniculate body. *Exp. Neurol.* 51, 377–391

Spydell, J.R., Pattee, G., Goldie, W.D. (1985): The 40 Hz auditory event-related potential: Normal values and effects of lesions. *Electroenceph. Clin. Neurophysiol.* 62, 193–202

Vieth, J. Schueler, P., Harsdorf, S.V., Fischer, H., Grimm, U. (1988): AC-MEG and AC-EEG at verified focal lesions and DC-MEG shifts during seizure and interictal periods. Presented at the Annual Meeting of the American Epilepsy Society, San Francisco.

Vrba, J., Fife, M., Burbank, M., Weinberg, H., Brickett, P. (1982): Spatial discrimination in SQUID gradiometers and 3rd order gradiometer performance. *Can. J. Phys.* 60, 1060–1073

Vvedensky, V.L., Ilmoniemi, R.J., Kajola, M.J. (1985): Study of the alpha rhythm with a 4 channel SQUID magnetometer. *Med. Biol. Eng. Comput.* 23 (Suppl. Part 1), 11–12

Walter, W.G. (1953): *The living brain.* London: Duckworth

Weinberg, H., Brickett, P.A., Deecke, L., Boschert, J. (1983): Slow magnetic fields preceding movement and speech. *Il Nuovo Cimento, 2D* 1, 495–504

Weinberg, H., Brickett, P.A., Vrba, J., Fife, A.A., Burbank, M.B. (1984): The use of a

SQUID third order spatial gradiometer to measure magnetic fields of the brain. *NY Acad. Sci.* 42, 743–752

Weinberg, H., Brickett, P., Neill, R.A., Fenelon, B., Baff, M. (1985a): Magnetic fields evoked by random-dot stereograms. In: *Biomagnetism: Applications and theory.* Weinberg, H., Stroink, G., Katila, T. (eds.). New York: Pergamon Press, pp. 354–359

Weinberg, H., Stroink, G., Katila, T. (eds.). (1985b): *Biomagnetism: Applications and theory.* New York: Pergamon Press.

Weinberg, H., Brickett, P., Coolsma, F., Baff, M. (1986): Magnetic localization of intracranial dipoles: Simulation with a physical model. *Electroenceph. Clin. Neurophysiol.* 64, 159–170

Weinberg, H., Brickett, P., Robertson, A., Crisp, D., Cheyne, D., Harrop, R. (1987a): A study of sources in the human brain associated with stereopsis. Presented at the Advanced Group for Aerospace Research and Development (NATO) Conference, Trondheim, Norway

Weinberg, H., Brickett, P., Robertson, A., Harrop, R., Cheyne, D.O., Crisp, D., Baff, M., Dykstra, C. (1987b): The magnetoencephalographic localization of source-systems in the brain: Early and late components of event related potentials. *J. Alcohol.* 4, 339–345

Weinberg, H., Cheyne, D., Brickett, P., Gordon, R., Harrop, R. (1987c): The interaction of thalamo-cortical systems in the 40 Hz following response. Presented at the Advanced Group for Aerospace Research and Development (NATO) Conference, Trondheim, Norway

Weinberg, H., Crisp, D., Brickett, P., Harrop, R., Purves, S.J., Li, D.K.B., Jones, M.W., Baff, M. (1987d) The combination of MEG and MRI in the estimation of sources associated with interictal discharges. In: *Functional localization: A challenge for biomagnetism.* Erne, S.N., Romani, G.L. (eds.). Singapore: World Scientific

Weinberg, H., Robertson, A., Brickett, P., Cheyne, D., Harrop, R., Dykstra, C., Baff, M. (1987c): Functional localization of current sources in the human brain associated with the discrimination of moving visual stimuli. In: *Current trends in event-related potential research.* Johnson, R.Jr., Parasuraman, R., Rohrbaugh, J.W. (eds.). New York: Pergamon Press

Weinberg, H., Cheyne, D.O., Brickett, P., Harrop, R., Gordon, R. (1988): An interaction of cortical sources associated with simultaneous auditory and somesthetic stimulation. In; *Functional brain imaging.* Pfurtscheller, G., Lopes da Silva, F.H. (eds.). Toronto: Hans Huber Publishers, pp. 83–88

Weinberg, H., Stroink, G., Katila, T. (1988). Biomagnetism. In: *Encyclopedia of Medical Devices and Instrumentation, vol 3.* Webster J. C. (ed.). New York: John Wiley and Sons

Williamson, S.J., Romani, G.L., Kaufman, L., Modena, I. (eds.) (1983): *Biomagnetism: An interdisciplinary approach.* New York: Plenum Press

26

Dynamic Functional Mapping of the Brain Cortex by Its Infrared Radiation

Igor A. Shevelev, Eugeny N. Tsicalov, Alexander M. Gorbach, Konstantin P. Budko, and George A. Sharaev

26.1 Introduction

About 30 years ago a new method of functional brain mapping—toposcopy or electroencephaloscopy—based on multichannel electroencephalogram (EEG) recording was introduced practically simultaneously in two laboratories: in Moscow (Livanov and Ananjev, 1960) and in Bristol (Walter, 1963). For that time, this pioneer study had the advantages of high flexibility and dynamics in spatial organization of the cortex activity, but it unfortunately did not influence profoundly the level and approaches of cortical neurophysiology. It must be noted that only recently have computer techniques of EEG mapping given new life to their approach.

Only 10 years later a new approach to the study of the distribution over the cortex of its functions was introduced. It was the +H– or multichannel isotope-clearance record of the changes in regional cerebral blood flow (rCBF) in the cortex during some of its functions (Ingvar and Lassen, 1962; Lassen et al., 1978).

A new era in functional neuroimaging began in the mid-1970s with the study by L. Sokoloff (1975, Sokoloff et al., 1977) on the distribution of brain metabolism, determined postmortem by 14C-deoxyglucose. Very soon, such new and powerful methods of neuroimaging as PET and SPECT appeared, with their possibilities to characterize, *in vivo* and separately, metabolic and rCBF distribution in the brain (for reviews see Greitz et al., 1985; Phelps and Mazziotta, 1985). Parallel to these developments, dynamic toposcopy of the brain activity was developed by optic methods with the use of potential-sensitive dyes (for review, see Grinvald, 1984). It must be noted that one more very powerful neuroimaging method—nuclear-magnetic-resonance tomography (NMRT—has not yet disclosed all its possibilities in functional mapping of the brain.

Evidently, each scientific method has its own peculiarities and limitations. Some are invasive, using isotopes, toxic dyes, or powerful magnetic fields, or have insufficient sensitivity and temporal or spatial resolution. All this does not mean that they are invalid for functional brain mapping; each of them has its own field of applications and such fields overlap only partially (for review, see Shevelev, 1987).

Nevertheless, it may be assumed that there is an obvious need for development of one more neuroimaging approach that is dynamic, noninvasive, distant, sensitive, and has sufficient spatial and temporal resolution. This chapter describes the principles and characteristics of a new neuroimaging method—thermoencephalo-scopy (TES)—that we developed in 1984 (Budko et al., 1984; Gulyaev et al., 1984; Tsicalov et al., 1984). It also discusses briefly some of the results obtained in rats with sensory and direct cortical stimulation, conditioning, fast waves spreading over the cortex, and light action in humans (Shevelev et al., 1985a–c, 1986a–g, 1987a,b, 1988, 1989a,b; Tsicalov et al., 1987, 1988; Gorbach et al., 1989).

26.2 Methods

As does any biological system, the brain produces and emits heat (about 10 mW/cm^2). By means of thermoconductivity and blood convection, this heat reaches an outer surface of the brain cortex and then the skull bones or scalp and is radiated into outer space. It can be detected there from various distances by the thermovision technique, in the waveranges of 3–5 or 8–14 μm (windows of transparency of the atmosphere's vapors for infrared radiation). A standard thermovisor camera (Figure 26.1C) has a pointed Sb/In infrared detector (F) attached to the container (G) with liquid nitrogen to lower the thermal noises of the detector. Through a mirror (B), special lens (D), and 2-D optical-mechanical scanning system (E), the infrared image of the object (A) was projected to the detector point-by-point sequentially. In our study we used a commercial thermovision camera (AGA-780M,AGEMA,Sweden) that has 40-msec scanning time and differential thermosensitivity of 0.2°C. As our experience showed, this sensitivity is not sufficient for recording of the very weak temperature changes in the working brain and must be increased 10–100 times. For this purpose, the camera was connected with a computer (Pericolor-2000E,Numelec, France) (Figure 26.1I) for digital image processing (Gulyaev et al., 1984) through a homemade fast interface (Figure 26.1H). The digitized thermoimage or thermomap, as we call it, in computer memory consists of 10,880 or 16,384 separate elements or pixels (128 × 85 or 128 × 128). Each map was measured during 40 msec, and temperature estimation of the part of an object that corresponds to one pixel lasts about 2.4 μsec. To increase the whole system's differential thermosensitivity, we used compute averaging of four successive maps and of four to nine sets of 72 maps each, received by repeated stimulation. Intervals between successive maps varied from 40 msec to 10 sec and between successive stimulations from 2 to 4 min. Synchronization of stimuli with one of the thermomaps was performed by computer through an interface (Figure 26,1G).

Spatial and temporal digital filtration was also used for improvement of the signal-to-noise ratio. Because we needed not absolute (this was also possible) but

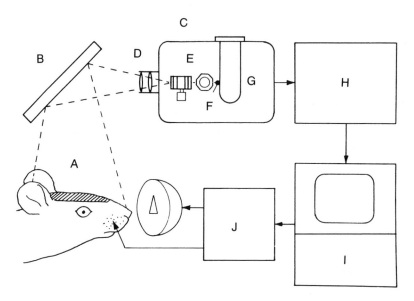

FIGURE 26.1. Scheme of the experiments. A, Anesthetized animal's head with intact skull; B, mirror; C, infrared camera; D, with lens; E, two-dimensional scanning system; F, detector; G, container with liquid nitrogen for cooling; H, electronic devices; I, computer; J, output interface and stimulators.

rather relative temperature measurements, two methods of comparative estimation were used: the reper method (background thermomap from each one) or the differential method (subtraction of the previous thermomap from each one). The second method was used for detection of spreading wave fronts.

During an experiment the thermovisor's camera was aimed through the mirror and focused on the dorsal surface of an animal's skull, previously scalped under local (Novocain Sol., 0.5%) or general (Nembutal Sol., 40mg/kg) anesthesia. An animal under local anesthesia was immobilized by d-tubocurarine (1 mg/kg) and was kept on artificial respiration. The body temperature was stabilized by heating. The head was softly fixed.

To date, TES study has been performed with white rats, rabbits, cats, monkeys, and humans. In the last case, the recording was performed from the skin of the scalp on the bald spot or on the head with shortly cut hair. During an experiment the subject, stimulators, and thermovision camera were located in a sound-proof chamber.

The parameters of TES now are: instrumental spatial resolution up to 70–100 μm/pixel, temporal resolution to 40 msec (up to 25maps/sec), and differential thermosensitivity up to 0.002°C, i.e., 100 times higher than in any standard thermovisor. Instrumental spatial resolution can be changed easily by putting different intermediate rings between the camera and lens, as in usual photographic techniques. It is evidant that in this case not only resolution but also map size is changed.

For each point on the map or for some zone on it, it was possible to obtain the curve of temperature change versus time (thermogram). Such calibrated curves serve as an exact quantitative estimation of thermoresponse and/or of the background thermoactivity.

26.3 Results

26.3.1 Sensory Stimulation

The main result of our TES study with sensory stimulation was the demonstration in the cortex of small and precisely localized zones of thermoactivation that were modally and regionally specific and depended on stimuli parameters. Shortly after stimulus onset, thermoresponse was generated in the specific primary cortical field and then in some other zones. Most often, it had a rather simple temporal pattern: the initial weak and short cooling was changed to fast and then slow heating. The size of heated zones in white rat cortex initially was small (200–300 μm), then widened during some seconds up to 1–2 mm, and finally (after tens of seconds or 1–2 min) disappeared or widened with loss of border sharpness.

Latency of temperature changes in the focus of thermoresponse varied from 160 msec to 3–5 sec, depending on stimulus modality and parameters. Often, after four to six rhythmic stimulus presentations (one every 2–4 min), the beginning of the response preceded the onset of the stimulus by 1–5 sec. In such cases it was possible to see the beginning of response generation prior to the moment of stimulation on averaged response maps and thermograms. Peak latency of the first (fast) phase of heating typically varied from 2 to 10 sec, whereas latency of the second (slow) one varied from 15 to 50 sec. The response duration changed significantly with stimulus parameters in the range of 35 sec to 2.5 min. The speed of the temperature changes was within the limits of 0.003–0.02°C/sec, being always slower for temperature decrease. Temperature contrast in the response focus did not exceed 0.2°C, but typically was within the limits of 0.02-0.1°C.

As an example, a typical set of thermomaps is shown on Figure 26.2 for monocular visual stimulation of the left retina of the white rat by a flash (0.16 sec, 9 angular degrees) localized centrally in the visual field. It may be seen that the responses in the 17th fields of both hemispheres (primary areas of the visual cortex) appeared simultaneously (map 2), but then the reaction in the contralateral cortex developed faster than in the ipsilateral and clearly predominated in amplitude and size. At the end of the demonstrated sequence of thermomaps (6), the spread of heating is seen mainly in the rostral and medial direction. For the latter direction, it is difficult to separate real spreading of the cortical activation from passive heat spread due to blood collection to the sagittal venous sinus.

Figure 26.3 demonstrates the thermograms of the same reactions as those shown in Figure 26.2. It is easy to see here the differences in the course and amplitude of these responses.

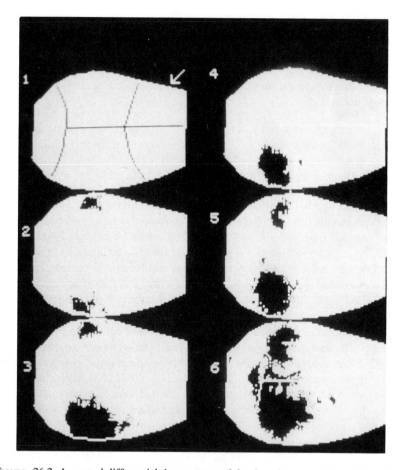

FIGURE 26.2. Averaged differential thermomaps of the dorsal aspect of the white rat brain cortex under light stimulation of the left eye. Frontal cortex appears at the right, cerebellum at the left. The background temperature distribution has been subtracted. The intervals in seconds between the light switching on and the maps 1-6 are equal to 1.3 (1), 2.6 (2), 3.8 (3), 6.4 (4), 9 (5), and 10.2 (6). The heated zones of the cortex are shown in dark; the cooled in white. The linear relative gray scale represents a total range of 0.1°C. The skull sutures are pictured over the map. The horizontal width of the entire map is equivalent to 20 mm.

Very often we saw thermoactivation not only in the specific visual cortex (areas 17, 18, 18a) but also in some other cortical fields—somatosensory (2), motor (10), associative (29), and in the cerebellum cortex. Thermoactivation occurred more frequently under somatosensory and auditory stimulation. The succession of switching on of all these zones was the same: more frontal fields and cerebellum were activated later than specific ones. The same is true for the relation between activation time in both hemispheres: the subdominant one for function showed later thermoactivation (2.7–4.5 sec), than the dominant one.

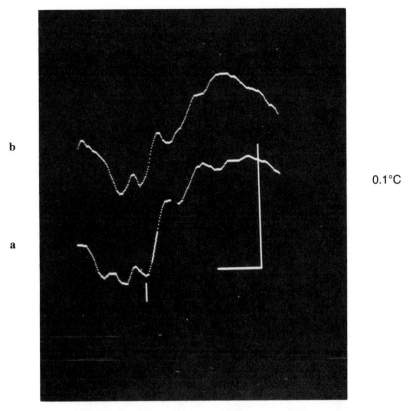

FIGURE 26.3. Thermograms for thermoresponses in Figure 26.2: (a) for a small zone (0.6 square mm) in the central part of heating in the 17th field of the cortex contralateral to the stimulated eye: (b) the same, as on (a) but for ipsilateral cortex.

The characteristics of thermoreactions (localization, number and size of the foci, temporal pattern and amplitude) evidently depended on the parameters of sensory stimulation: intensity, duration, frequency, localization on the receptor surface, size, and sharpness of the borders. It was possible to determine optimal stimulation parameters that evoked the biggest, most local, and sharpest response. For a visual stimuli they were intensity of 5–15 1× on the squint level, duration of 0.16–10 sec, intervals between successive flashes of 2– 4 min, size 9–20 angular degrees, localization in the central part of the retina, number of repetitions for summation, 6 –9. Such estimations were also made for other tested stimuli modalities.

26.3.2 Direct Cortical Stimulation

Bipolar silver electrodes with a tip separation of 2 mm were placed on the dura mater through small holes in the skull bone in the parietal part of the cortex. A single electric pulse of 10 volts and 0.3 msec duration was applied to the cortex of immobilized or anesthetized rat. Figure 26.4 shows a typical result of these experiments. It is seen that, at the first moment after shock, thermoresponse (small heating zone) appeared around the stimulating electrodes. Then, the symmetrical point in the contralateral hemisphere was activated through transcommissural connections. Only later, additional zones of heating were added in the paralyzed animal (Figure 26.4A). They were located in somatosensory, motor cortex, and in the parietal cortex (29th field). The number of activated zones on the dorsal surface of the rat cortex was up to five.

TES effects are greatly dependent on the state of an animal. Nembutal anesthesia diminished the number of thermofoci and the sizes and amplitudes of responses and increased the thresholds of response generation. Thus, under Nembutal (40 mg/kg) only three zones of heating were recorded: in the electrode site, in the symmetrical point of the other hemisphere, and in the ipsilateral motor cortex (Figure 26.4B). Under very deep anesthesia (140 mg/kg), heating existed only in the stimulation site (Figure 26.4C).

Direct electrical stimulation of the brain cortex gave the shortest time of thermoresponse appearance and peak latency: 0.16 sec for the first and 2 sec for the second. The speed of the temperature rise in the response focus was up to 0.03°C/sec.

26.3.3 Conditioning (Associative Learning)

In this part of the study, we elaborated in white rats conditional connections between a weak (subthreshold for pain) electric shock, applied to the skin of the animal's head from the right side (in the vibrissae region) or at the hind leg (unconditional stimuli), and weak light stimulation of the central part of the left Retina (conditional stimulus). The comparison of averaged responses to the shock (Figure 26.5A), to its combination with light (B) and to separate light presentation (C) after nine couplings of both stimuli shows that light alone, being the conditional signal, can produce the same temperature distribution over the cortex that the unconditional stimulus previously produced. Figure 26.5 shows that, in all three situations that are demonstrated the patterns of cortical thermoactivation were very similar. Thus, somatosensory (2nd field) and motor cortex (10th field), as well as cerebellum cortex, are activated earlier and more markedly in the hemisphere contralateral to the shock side. Only 2.7–4.5 sec later, the response of the symmetrical zones of the ipsilateral hemisphere appeared. In the case, the energy of the flash was lowered so it did not produce any separate heating of the cortex.

Differences between thermoresponses in these three situations were purely quantitative: the size of heated zones, amplitude of the response, and velocity of its growth are always lower to the conditional signal. It is interesting that in the coupling procedure and conditioning test (Figure 26.5B and C) the response in contralateral somatosensory cortex predominated, whereas in unconditional shock

A   B   C

-4s

0.16s

14s

22s

30s

38s

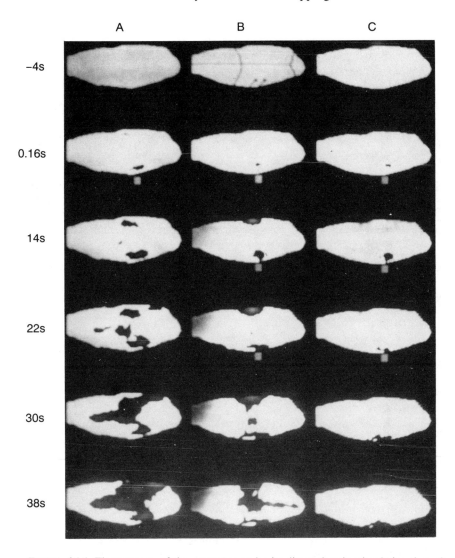

FIGURE 26.4. Thermomaps of the rat cortex under its direct electric stimulation through implanted electrodes in conditions of immobilization by d-tubocurarine and local anesthesia (A), Nembutal anesthesia 40 mg/kg (B), and 140 mg/kg (C). The first map in each column is a background one; the time of recording for other maps is shown from the left side in seconds. Averaging of nine responses by the reper method. Frontal cortex in this case appears at the left. The horizontal width of the map is equivalent to 15 mm. Other details as in Figure 26.2.

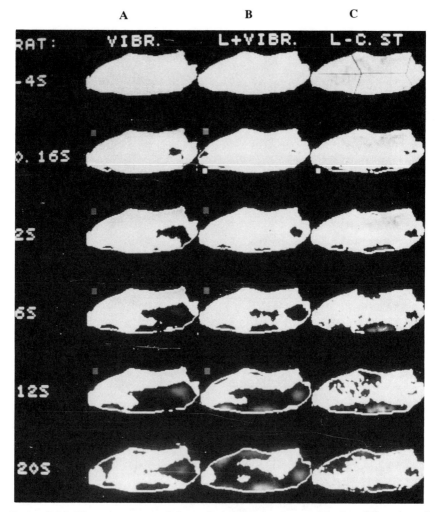

FIGURE 26.5. Thermoresponses of the rat brain cortex under conditioning: (A), action of the unconditional stimulus (weak electric shock at right vibrissae region); (B), coupling of conditional stimulus (light to the left eye) with shock; (C), action of the light alone after nine couplings. On maps A and B averaging of nine responses; on maps C, averaging of four responses. Other details as in Figure 26.4.

alone (A) the response is most pronounced and sharpened in the contralateral motor cortex. Conditional combination of the light with hind leg shock gave basically the same result, although the thermopattern evoked by shock appeared differently.

It is important to stress that we have never seen any significant thermoactivation between the described zones. This negative finding is in clear contradiction to the assumptions of classical conditioning theory about the transcortical spread of conditioning process that forms a specific path there.

26.3.4 Fast Thermowaves Spreading Over the Cortex

Through the dynamic monitoring of differential thermomaps (see Section 26.2) in real time or in a slowed manner, we have had the good luck of finding thermowaves spreading over the brain cortex. They had a set of typical trajectories (Figure 26.6) and relatively limited ranges of distance, duration, and speed (Figure 26.7). The probability of the wave occurrence increased significantly under light stimulation.

In the so-called background conditions, it was possible to see spreading thermowaves during some time after scalping of the head (performed under sufficient anesthesia), skull trepanation, or prolonged rhythmic sensory stimulation. The

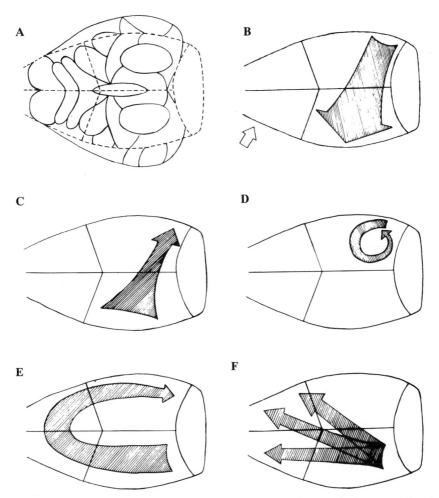

FIGURE 26.6. Citoarchitectonic fields of the dorsal surface of white rat brain cortex (A) and the main trajectories (types 1-5) of thermowaves spreading over the cortex (B-F).

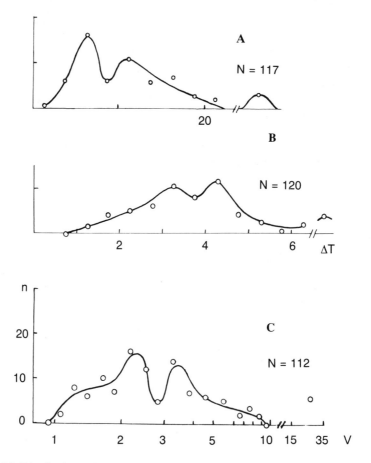

FIGURE 26.7. Distributions of 133 thermowaves spreading over the brain cortex under visual stimulation by the length of their trajectory (A), duration of movement (B), and speed (C). Abscissa in mm (A), sec in (B), and mm/sec in (C). Ordinate - number of waves.

probability of the occurrence of this phenomenon varied in all these conditions, but averaged 0.41 during the period of study.

During the rhythmic light stimulation, the probability of the phenomenon markedly increased to 0.92. Figure 26.8 shows that the number of spreading waves is connected with the stimulation moment: it increased mostly after the flash, but also during some time (10–12 sec) before it. It is seen that at the exact moment of stimulation (0 sec on abscissa), the probabilities of the wave generation and spreading are decreased.

Figures 26.9 and 26.11 show examples of two typical trajectories of the thermowave movement: big and local circles. In the first case (fourth type of trajectory on Figure 26.6), a spontaneous wave (zone of heating and successive cooling)

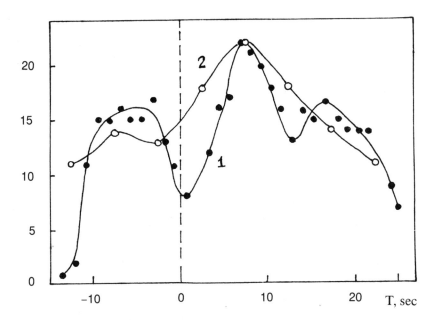

FIGURE 26.8. Distribution of thermowaves by the time of their appearance (1) and the time of their movement (2) in relation to the rhythmic visual stimulation (abscissa in sec, 0 sec - the flash moment). Ordinate - number of waves (N = 121).

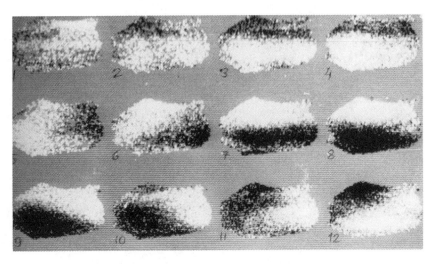

FIGURE 26.9. Circular thermowave movement over the cortex ("macrocircle") during 8.4 sec (intervals between maps are equal to 0.76 sec). The heated zone is darkened as it moves from the frontal part of the right hemisphere (upper half of each map) on the second map and finally returns to the same place (12th map).

traveled along one hemisphere, spread to the other, moved over it, and then returned to the initial point. The set of thermograms in Figure 26.10A (scheme on B) shows, for some points on the wave path, a successive time shift of the curve's maxima that indicated the spreading nature of the process.

The second example (Figure 26.11) corresponds to the third type of trajectory, shown on Figure 26.6. In this case the wave goes around the central part of the primary visual area of the cortex, contralateral to the stimulated eye. It is very interesting that this type of wave appeared only in contralateral cortex after light stimulation. The close connection observed between the type of wave trajectory and its hemisphere of origin (contralateral or ipsilateral) in relation to the side of stimulation (Figure 26.6) seems highly specific and intriguing.

26.3.5 Local Thermoresponses to Light on the Human Scalp

In this part of the study, the thermovisor camera was placed 75 cm from the head, and spatial resolution was lowered to 1.3 mm/pixel, while temporal resolution was equal to 0.76 sec. A short (5-sec) flash of round form and diameter of 6.5 angular

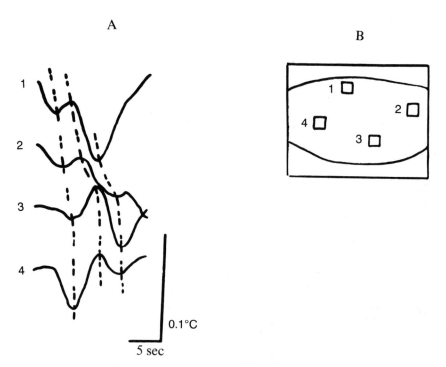

FIGURE 26.10. (A), Thermograms for the points on the path of the thermowave shown on Figure 26.9. (B), Scheme of the location of these points. Calibration: 0.1°C, 5 sec.

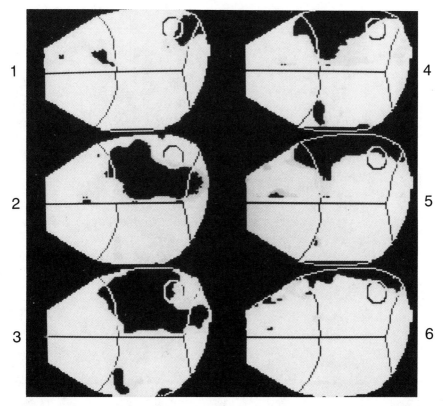

FIGURE 26.11. Circular thermowave movement over the cortex around the central part of primary contralateral visual projection (17th field) after light flash. Intervals between maps: 320 msec (1-2), 160 msec (between each two of the next).

degrees with an intensity of 18 1x at the squint level, presented to the central part of the right retina, evoked in the subject under study clear thermoresponses in the left occipitoparietal region (P3-C3 according to the EEG 10/20 system) and weaker heating in the right hemisphere (Figure 26.12). Latency of the heating in this response was 2.8 sec and peak latency, 9.4 sec; the size of the heated zone reached by this moment was 1×1.3 cm with temperatures up to +0.1°C.

26.4 Discussion

The data presented show that this new neuroimaging method—thermoencephalo-scopy (TES)—allows distant, dynamic, and noninvasive functional mapping of the brain cortex and thus the study of sensory events, associative learning (conditioning), and some spreading wave phenomena that may be in close relation with such higher cortical functions as selective sensory and common attention.

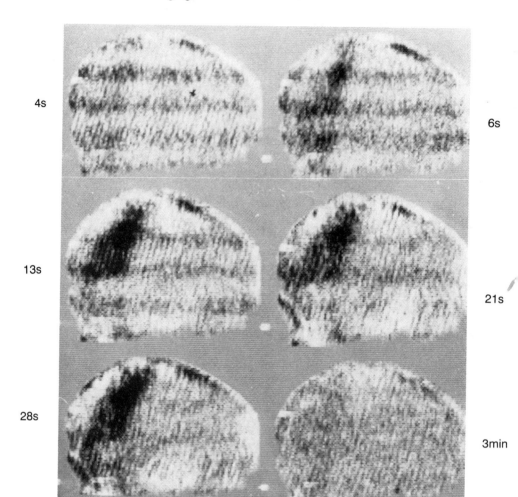

FIGURE 26.12. Averaged differential thermomaps of the occipitoparietal part of the human head (subject I.S.) from rear and slightly left side during light action on the right eye (first map) and after it (6 sec – 3 min). The recording time for each map is indicated in seconds after the beginning of light.

TES detects through the unopened bones of the skull the activation of small and precisely located cortical zones and measures their dynamics before, during, and after various brain functions. The multiple asynchronous activation of the primary and secondary projection fields of the cortex, as well as of other zones (motor, associative and frontal cortex, cerebellum cortex), was revealed under somatosensory, visual, and direct cortical stimulation. These effects are highly dependent on the state of an animal (anesthesia level). Thus, our experiments with TES help define the position, size, and sequence of operation of activated (heated) and deactivated (cooled) cortical zones during different types of function.

It is necessary to discuss here the mechanisms of the revealed TES phenomena and their relation to the traditional (mainly electrophysiological) and some new (PET) characteristics of the brain cortex activity. Mathematical and physical simulations have been used to quantitatively check supposed mechanisms of the TES-effects generation (Frolov et al., 1985; Shevelev et al., 1989a). They are changes in (1) the activity of cortical neurones and glia cells (transmembrane ion currents produce joule heat)(Howarth et al., 1975; von Muralt, 1981); (2) the local metabolism of units (Siesjo, 1978; Tasaki and Nakaye, 1985); (3) the local cerebral blood flow (lCBF) that is controlled by some metabolites (McCulloch, 1984); and (4) thermoconductivity in the activated cortical zones due to changes in blood flow (Betz and Hensel, 1962; Frolov et al., 1985).

There is no doubt that the factor in this chain that initiated events and is qualitatively the most important is a change in neuronal activity. Yet, quantitatively it cannot account for a significant amount and proportion of the heat recorded by TES. The same is true for the metabolic thermoproduction that can give no more than some percentage of the whole effect. Increase in the lCBF must lead at the first moment to a decrease in the cortex temperature because the arterial blood is 0.1–0.2°C cooler than the cortex. We have often seen initial short latency cooling after sensory stimulation. Yet, in the next period, carrying out of venous blood that is heated in the depth of the cortex must lead to an increase in its surface temperature (Demchenko, 1983; Hayward and Baker, 1969; McElligott and Melzack, 1967; Melzack and Casey, 1967). The modular organization of the radial blood vessels that are oriented normally to the cortex surface and have a regular (every 200–400 μm in different species) distribution over it (Baramidze et al., 1982) may result in the high efficiency and locality of the carrying out of heat from the cortex by the lCBF.

An additional but quantitatively very powerful mechanism of the TES effect is the change in thermoconductivity in the cortical cylinder, located around the radial vessel (Betz and Hensel, 1962; Frolov et al., 1985; Shevelev et al., 1989a). The constant thermogradient through the cortex (deep layers are hotter than outer ones) (Zeschke and Krasilnikov, 1976) gives a constant even outflow of the heat. In this condition, appearance of cylindrical zones of increased thermoconductivity around vessels leads to formation of heat channels through which more effective thermoconduction takes place. It is possible to name these channels "thermodipoles" with location of thermal sources in deep layers and thermal sinks in outer ones.

It is important to stress that TES is a two-dimensional method that reflects only intracortical events. Of course, contributions of the deep, subcortical sources of heat to the temperature distribution over the cortex may exist in principle but cannot give any local sharp-bordered thermoresponses as observed in this study. Some complications exist in the explanation of the human thermoresponse mechanisms. It is really difficult to determine exactly the role of the scalp thermoproduction in this response. Biophysical calculations show that the time of response growth must be longer for humans than for animals with thin skull bones. Yet, these characteristics are very similar, which contradicts the purely cortical origin of the human thermoresponse. How to explain in this case the localization and locality of this response is a problem.

Comparison of the temporal pattern of TES responses and electrical evoked potentials (EP) reveals pronounced differences. First, if EPs are events in the temporal range of hundreds of milliseconds, TES responses take up a range of tens of seconds and even minutes. Second, unlike EPs, they have a relatively simple monotonous pattern of prolonged heating. These peculiarities are possibly explained by two considerations. Long and slow development of the TES responses may be related either to the slow course of the metabolic and CBF events or to the participation in it of big groups of units, the activity of which has no reflection in the EPs. They may be, for instance, cortical interneurones and among them inhibitory cells that make no contribution to the EPs due to the nonoriented manner of their dipole packing. It is important that inhibitory cells generate IPSPs on excitatory neurons—signals of the opposite sign to EPSPs. Their algebraic summation gives the complex temporal pattern of EPs, where phases or waves of the opposite sign change one another successively. Meanwhile, for the TES responses, the activity of all these groups of units has the same energetic sign of the thermoproduction increase. This is a possible explanation for the relatively simple temporal pattern of the TES responses.

Comparison of the TES and PET characteristics of brain activity shows that the latter have the advantages of a three-dimensional nature, the possibility of separate estimation of the CBF and metabolism, a wide range of substances through which the brain distribution can be detected, and high sensitivity (Greitz et al., 1985; Phelps and Mazziotta, 1985). Yet, unlike TES, PET is invasive (isotopes), has lower spatial (2 mm/pixel versus 0.07 mm/pixel in TES) and much lower temporal resolution (tens of seconds to minutes versus 40 msec in TES). Last but not least, the cost of the TES device is only 3 to 5 % of that of the PET.

26.5 Conclusion

The widespread utilization of the neuroimaging methods (isotope-clearance method, 14C-DG, potential-depending dyes, EEG and MEG mapping, NMRT, SPECT, and PET) has demonstrated their great usefulness in the mapping of brain structures and functions. Meanwhile, there is a need for a new approach that is functional, dynamic, distant, and noninvasive, and has reasonable sensitivity, spatial, and temporal resolution. That is why we recently elaborated and tested in different experimental conditions a new method of neuroimaging, based on the recording of very weak changes in infrared radiation from the unopened skull that are connected with brain activity. The method—"thermoencephaloscopy" (TES)—is based on thermovision and digital image processing techniques. The parameters of the method are thermosensitivity of up to 0.002°C, instrumental spatial resolution of up to 70 μm/pixel, and temporal resolution of up to 40 msec (25 thermomaps/sec).

To date, very local, multiple, and modally and regionally specific cortical thermoresponses to various sensory stimuli, motor acts, associative learning (conditioning), and some wave processes spreading over the brain cortex and brain pathology have been studied in white rats, rabbits, cats, monkeys, and humans.

We conclude that the dynamic, noninvasive, and distant nature of TES in combination with its reasonable temporal and spatial resolution and sensitivity can make it a useful tool in neurophysiology, psychology, and medicine.

References

Baramidze, D.G., Reidler, R.M., Gadamski, R., Mchedlishvili, G.I. (1962): Pattern and innervation of pial microvascular effectors which control blood supply to cerebral cortex. *Blood Vessels* 19, 284-291

Betz, E., Hensel, H. (1962): Fortlaufende Registrierung der lokalen Durchblutung Im inneren des Gehirns bei wachen, frei beweglichen Tieren *Pflug. Arch.* 274, 608–614

Budko, K.P., Godik, E.E., Gorbach, A.M., et al. (1984): Thermoresponses of the brain to sensory stimulation. *Dokl. Acad. Sci. USSR.* 278, 486-488 (in Russian)

Demchenko, I.T. (1983): *Blood supply of the awakened brain.* Leningrad: Science (in Russian)

Frolov, A.A., Tsicalov, E.N., Kiseleva, et al. (1985): Estimation of the characteristics of the brain thermosources by temperature distribution on the brain skull. In: *Application of mathematical and computational methods in biology.* Puschino: Nauka, 132-133 (in Russian)

Gorbach, A.M., Tsicalov, E.N., Kuznetsova, G.D., et al. (1989): Infrared mapping of the cerebral cortex. *Thermology* 3, 108-111

Greitz, T., Ingvar, D.H., Widen, L., (eds.). (1985): *The metabolism of the human brain studied with positron emission tomography.* New York: Raven Press

Grinvald, A. (1984): Real-time optical imaging of neuronal activity. *Trends Neurosci.* 7, 143-150

Gulyaev, Y.V., Godik, E.E., Petrov, A.V., Taratorin, A.M. (1984): On a possibility of distant functional diagnostics of biological objects by their infra-red radiation. *Dokl. Acad. Sci. USSR.* 277, 1486-1491 (in Russian)

Hayward, J.N., Baker, M.A. (1969): A comparative study of the role of the cerebral arterial blood in the regulation of brain temperature in five mammals. *Brain Res.* 16, 417-440

Howarth, J.V., Keynes, R.D., Ritchie, J.M., vonMuralt, A. (1975): The heat production associated with the passage of a single impulse in pike alfactory nerve fibres. *J. Physiol.* 249, 349-368

Ingvar, D.H., Lassen, N.A. (1962): Regional blood flow in the cerebral cortex determined by Krypton 85. *Acta Physiol. Scand.* 54, 325

Lassen, N.A., Ingvar, D.H., Skinhoj, E. (1978): Brain function and blood flow. *Sci. Am.* 239, 50-59

Livanov, M.N., Ananjev, V.M. (1960): *Electroencephaloscopy.* Moscow: Medgiz (in Russian)

McElligott, J.G., Melzack, R. (1967): Localized thermal changes evoked in the brain by visual and auditory stimulation. *Exp. Neurol.* 17, 293-312

McCulloch, J. (1984): Perivascular nerve fibres and the cerebral circulation. *Trends Neurosci.* 7, 135-138

Melzack, R., Casey, K.L. (1967): Localized temperature changes evoked in the brain by somatic stimulation. *Exp. Neurol.* 17, 276-292

Phelps, M.E., Mazziotta, J.S. (1985): Positron emission tomography: Human brain function and biochemistry. *Science* 228, 799-809

Shevelev I.A. (1987): Functional mapping of the brain. *Uspekhy physiolog. nauk.* 18, 16-36 (in Russian)

Shevelev, I.A., Gorbach, A.M., Tsicalov, E.N., et al. (1985a): Thermovision study of reaction to light in the human brain. *Dokl. Acad. Sci. USSR.* 284, 1016-1019 (in Russian)

Shevelev, I.A., Tsicalov, E.N., Volovik, M.G., et al (1985b): Sensory mapping of the brain cortex. *Zhurn. Evol. Biochem. Physiol.* 21, 522-527 (in Russian)

Shevelev, I.A., Tsicalov, E.N., Gorbach, A.M., et. (1985c): Thermovision characteristics of human brain reaction to visual stimulation. *Physiol. Chelov.* 11, 538-543 (in Russian)

Shevelev, I.A., Kotlyar, B.I., Volovik, M.G., et al. (1986a): Finding of the motor cortex activation in cat during paw movement with the help of thermomapping of the brain. *Neurophysiologia* 18, 266-269 (in Russian)

Shevelev, I.A., Kuznetsova, G.D., Gulyaev, Y.V., et al. (1986b): Dynamic thermomapping of the rat brain under sensory stimulation and spreading depression. *Neurophysiologia* 18, 26-35 (in Russian)

Shevelev, I.A., Tsicalov, E.N., Budko, K.P., et al. (1986c): Thermowaves spreading over the white rat brain cortex. *Neurophysiologia* 18, 340-346 (in Russian)

Shevelev, I.A., Kuznetsova, G.D., Tsicalov, E.N., et al. (1989a): *Thermoencephaloscopy.* Moscow: Navka, 224 pp (in Russian)

Shevelev, I.A., Tsicalov, E.N., Budko, K.P., et al. (1986e): New method: Dynamic functional thermoimaging of the brain (thermoencephaloscopy). *Neurologija (Zagreb)* 35, 174-177

Shevelev, I.A., Volovik, M.G., Sharaev, G.A., et al. (1986f): Parameters of light stimulation and thermoresponse characteristics in the white rat brain cortex. *Neurophysiologia* 18, 332-340 (in Russian)

Shevelev, I.A., Volovik, M.G., Sharaev, G.A., et al. (1986g): Spatial characteristics and dynamics of activation loci in the cortex under conditioning (thermographic study). *Zhurn. Vyssh. Nervn. Deyat.* 36, 74-83 (in Russian)

Shevelev, I.A., Tsicalov, E.N., Gorbach, A.M., et al. (1987a): Dynamic thermoimaging of the brain functions. *Neuroscience* 22, (Suppl.), 1137

Shevelev, I.A., Tsicalov, E.N., Gorbach, A.M., et al. (1987b): Mapping of the primary visual cortex by thermoencephaloscopy. *Perception* 16, 273-274

Shevelev, I.A., Tsicalov, E.N., Gorbach, A.M., et al. (1988): Dynamic infra-red images of the brain. *IBRO News* 16, 8-9

Shevelev, I.A., Kuznetsova, G.D., Tsicalov, E.N., et al. (1989a): *Thermoencephaloscopy.* Moscow: Navka, 224 pp (in Russian)

Shevelev, I.A., Tsicalov, E.N., Gorbach, A.M., et al. (1989b): Fast thermowaves spreading over the brain cortex during visual stimulation. *Neurophysiologia* 21, 467-475 (in Russian

Siesjo, B.K. (1978): *Brain energy metabolism.* New York: John Wiley and Sons

Sokoloff, L. (1975): Measurement of local glucose utilization and its use in mapping local functional activity in the central nervous system. *Proceedings of the Fourth World Congress of Psychiatric Surgery*, Madrid

Sokoloff, L., Reivich, M., Kennedy, C., et al. (1977): The [14C] deoxyglucose method for the measurement of local cerebral glucose utilization: Theory, procedure and normal values in the conscious and anesthetized albino rat. *J. Neurochem.* 28, 897-916

Tachibana, S. (1966): Local temperature, blood flow and electrical activity correlations in the posterior hypothalamus of the cat. *Exp. Neurol.* 16, 148-161

Tasaki, I., Nakaye, T. (1985): Heat generated by the dark-adapted squid retina in response to light pulses. *Science* 227, 654-655

Tsicalov, E.N., Petrov, A.V., Taratorin, A.M., et al. (1984): Study of the own temperature fields related to excitation of the rat brain cortex. *Dokl. Acad. Sci. USSR* 278, 249-252 (in Russian)

Tsicalov, E.N., Kuznetsova, G.D., Shevelev, I.A., et al. (1987): Temperature distribution over the rat brain cortex under direct electric stimulation. *Neurophysiologia* 19, 216-223 (in Russian)

Von Muralt, A. (1981): Optical and thermal changes during electrogenesis. *Physiology of excitable membranes*. Salanki, J. (ed.). Budapest: Oxford University Press, pp. 183-187

Tsicalov, E.N., Shevelev, I.A., Gorbach, A.M., et al. (1988): Wave processes in the brain. In: Collective dynamics of excitations and structure formation in biological tissues. Gorkij: 26-34 (in Russian)

Walter, W.G. (1963): *The living brain.* London: Pelican Books

Zeschke, G., Krasilnikov, V.G. (1976): Decreases of local brain temperature due to convection (local brain blood flow) and increases of local brain temperature due to activity. *Acta Biol. Med. Germ.* 35, 935-941

27

Brain Imaging with Positron Emission Tomography (PET): Using a Method for Reformatting Three-Dimensional Physiological Data into a Standardized Anatomy

Torgny Greitz and Christian Bohm

27.1 Introduction

Positron emission tomography (PET) represents an enormous advance in the methodology of clinical brain research. In fact, PET has the capacity to estimate a variety of physical, chemical, and physiological parameters and their distribution in the living human brain. In this respect, no other method is able to compete with it.

PET is a type of computed tomography (CT) that examines slices of the body, but as opposed to roentgen CT, which measures the transmitted radiation, PET measures the radiation emitted from an injected radionuclide. The use of positron-emitting isotopes makes possible the analysis of the activity distribution by coincidence detection and image reconstruction, using the same type of algorithms as in roentgen CT.

The most commonly used isotopes are the bioisotopes: oxygen, nitrogen, and carbon, which are very short lived, having a half-life of 2 to 20 minutes. These isotopes may be incorporated into almost any biologically interesting substance to analyze and quantify almost any biochemical or physiological event. The most commonly studied parameters are blood-brain barrier permeability, cerebral blood volume distribution, cerebral blood flow, pH and cerebral metabolic rate of oxygen or glucose, and the cerebral rate of protein synthesis. PET also allows the study of the distribution and quantification of the density and the degree of occupancy of various receptors for neurotransmitters, hormones, and psychoactive drugs.

The description of physiological events in the brain is of course of limited interest without a pertinent anatomical allocation. To find out which areas are involved in different physiological processes, identification of anatomical structures is necessary. However, this is often difficult for several reasons. One is the fact that the appearance of a structure is often dependent on the parameter under study. Another problem is that of identification. For example, it may be very difficult to name an individual gyrus in a specific cut. Finally, due to individual anatomical differences, identical cuts can never be obtained from two individuals. All these difficulties can be overcome by using a computerized, individually adjustable brain atlas (Bohm, et al., 1983, 1985, 1986; Fox et al., 1985, 1988; Evans et al., 1988; Mazziotta, 1984).

27.2 Material and Methods

The atlas was based on information from digitized photos of cryosectioned cadaver brains that had been fixed in situ with formalin. CT examinations were made before and after fixation in order to rule out swelling or shrinkage. The brains were cut four slices to a millimeter, and all slices were photographed. So far, slices of three brains have been digitized, but the present data base is derived from one single brain. The intention is, however, to include material from the majority of the 13 brains available at present to account for individual variations, especially with regard to the configuration of the cortex.

Boundaries of different structures were drawn in on every fourth slice of the digitized photographs. Using a computer program, corresponding points in adjacent slices were connected or linked. The program then filled out the missing links to create a dense mesh that covered the entire structural boundaries and gave a complete three-dimensional description of the structure. The boundary surfaces were stored in the data base, from which the three-dimensional structures may be retrieved easily. (Figure 27.1).

The data base is under continuous development. To date, it contains the brain surface, the ventricular system, the cortical gyri and sulci, the Brodmann cytoarchitechtonic areas, and the major basal ganglia.

27.3 Use of the Atlas

27.3.1 As a Three-Dimensional Individualized Map

The atlas can be adapted to fit the anatomy of a given individual brain as it is presented in a set of CT or MR images (Figure 27.2). The adaptation is made by applying a set of rigid and nonrigid transformations to the atlas volume (Figure 27.3). This procedure has two steps: first automatically and then interactively. In the CT images, the program finds the longest sagittal and transverse diameters and the height perpendicular to the former.

Once the adaptation has been made, the modified atlas may be transferred to the PET images, and various structures, such as individual gyri or any of the major basal ganglia, may be called in and outlined (Figures 27.4 and 27.5). In this way,

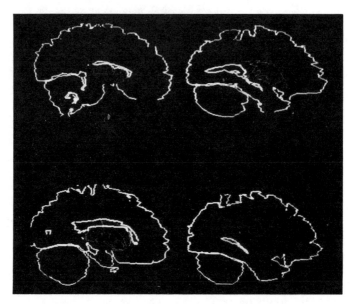

FIGURE 27.1. Four sagittal cuts through the three-dimensional atlas volume showing the brain surface and the ventricular system.

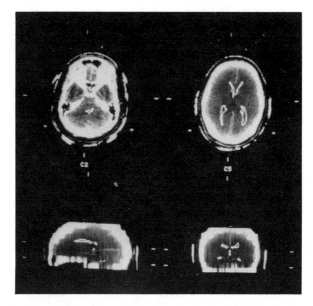

FIGURE 27.2. The atlas has been adapted to the anatomy of an individual brain as it appears on a set of CT images. Fitting the brain surface and the ventricular system is usually sufficient. In horizontal slices (two upper ones) front of head is up.

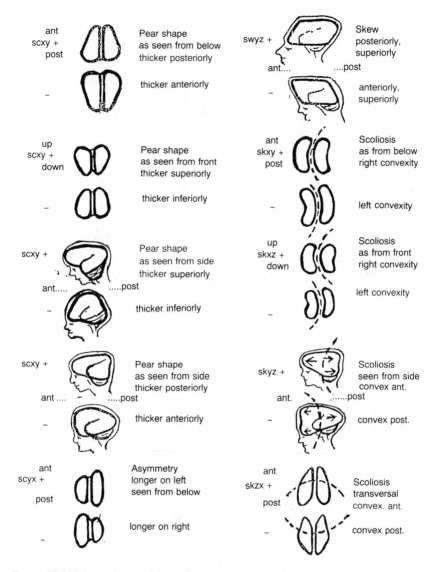

FIGURE 27.3 Schematic representation of some of the transformations used to adapt the atlas, showing their effect on the shape of the brain. sc = scaling; sk = scoliosis; sw = skew; y, z = Cartesian coordinates. For example, scxy = scaling of x in y-direction.

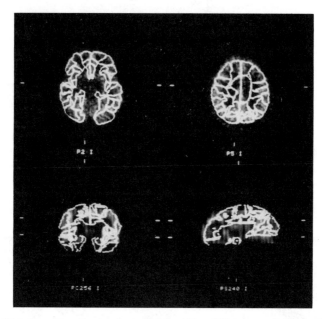

FIGURE 27.4. Once the atlas is adapted to the CT or MR images, it can be transferred to the PET images or any other three-dimensional representations of brain anatomy. Structures of interest—in this case the cortical gyri—may now be called in from the database and outlined.

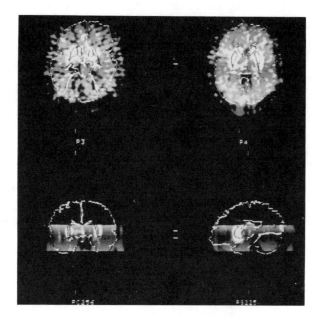

FIGURE 27.5. Following adaptation of the atlas, the basal ganglia (caudate nucleus and putamen) are outlined in a study of the binding of a dopamine-2 antagonist.

the atlas may be used for identification of structures and as a guide in selecting regions of interest.

27.3.2 As a Vehicle for Reformatting Three-Dimensional PET Data From Different Individuals Into a Standardized Anatomy

An interesting and promising aspect of using the atlas is to have it serve as a reference when comparing results from PET examinations of different individuals. By applying the inverse atlas transformations to the PET data volume, (i.e. running the transformations "backward") it is possible to relate the PET information to the reference atlas, instead of to the patient's anatomy (Bohm et al., 1986). In other words, the three-dimensional data of the PET study are transformed into a standardized anatomy (i.e., that of the atlas brain) (Figure 27.6). All brains will then have the same anatomical appearance in the PET studies, and all differences that we observe will be functional.

Reformatted PET data from different patients can thus be averaged, and averages from different categories of patients can be compared. This comparison is

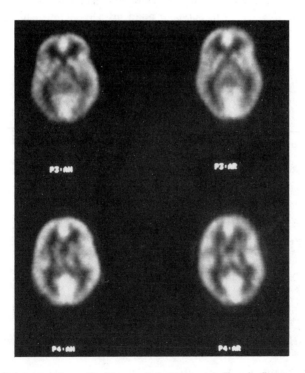

FIGURE 27.6. A comparison of averages of glucose metabolism in five normal volunteers (*left*) and ten schizophrenic patients (*right*). Note identical anatomical appearance of the two group averages.

facilitated by forming weighted differences between the two averages, each pixel value being divided by its standard deviation. The new image volume will then provide a measure of the significance of the differences point-by-point (Figure 27.7). In this way statistically significant differences between different groups of individuals may be identified. This type of image may be called a "significance image."

Similarly, two different functional states in a group of individuals may be compared. For example, the examination at rest may be compared with that of the same individual after an added stimulus. "Subtraction images" are then formed for each individual and transformed into the standardized anatomy, the reformatted data averaged, and the significance of differences measured.

One fascinating possibility made feasible by this method is the creation of a three-dimensional functional brain atlas. Before the advent of nuclear medicine and the introduction of techniques to measure regional cerebral blood flow (Lassen and Ingvar, 1963), no method was available to study brain function without interferring with the observed event. The application of image reconstruction methods allows for the first time a precise and accurate three-dimensional recording of the measured data. By reformatting images into a standardized anatomy, a variety of data from physiological experiments carried out at different centers with a standardized technique can now be transferred to a common data base. Merging information in this way will serve to build up a comprehensive and detailed three-dimensional functional brain atlas.

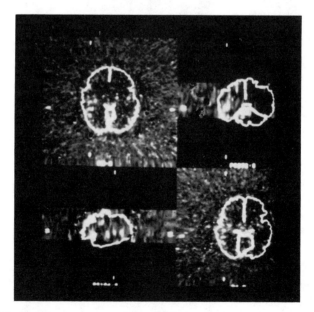

FIGURE 27.7. Significance images displaying differences in cerebral metabolic rate of glucose between the two groups shown in Figure 27.6. Marked significant differences are found in the right upper posterior temporal region and adjacent parieto-occipital area.

References

Bohm, C., Greitz, T., Kingsley, D., Berggren, B.M., Olsson, L. (1983): Adjustable computerized stereotaxic brain atlas for transmission and emission tomography. *Am. J. Neuroradiol.* 4, 731-733

Bohm, C., Greitz, T., Kingsley, D., Berggren, B.M., Olsson, L. (1985): A computerized individually variable stereotaxic brain atlas. In: *The metabolism of the human brain studied with positron emission tomography.* Greitz, T., et al. (eds.). New York: Raven Press

Bohm, C., Greitz, T., Blomqvist, G., Farde, L., Forsgren, P.O., Kingsley, D., Sjögren, I., Wiesel, F., Wik, G. (1986): Applications of a computerized adjustable brain atlas in positron emission tomography. *Acta Radiol.* 369 (Suppl.), 449-452

Evans, A.C., Beil, C., Marrett, S., Thompson, C.J., Hakim, A. (1988): Anatomical-functional correlation using an adjustable MRI-based region of interest atlas with positron emission tomography. *J. Cereb. Blood Flow Metab.* 8, 513-530

Fox, P.T., Perlmutter, J.S., Raichle, M.E. (1985): A stereotactic method of anatomical localization for positron emission tomography. *J. Comput. Ass. Tomogr.* 9, 141-153

Fox, P.T., Mintun, M.A., Reiman, E.M., Raichle, M.E. (1988): Enhanced detection of focal brain responses using intersubject averaging and change-distribution analysis of subtracted PET images. *J. Cereb. Blood Flow Metab.* 8, 642-653

Lassen, N., Ingvar, D.H. (1963): Regional cerebral blood flow measurements in man. *Arch. Neurol. Psychiatry* 9, 615-622

Mazziotta, J.C. (1984): Physiologic neuroanatomy: New brain imaging methods present a challenge to an old discipline. *J. Cereb. Blood Flow Metab.* 4, 481-483

Index of Cited Names

Subject Index